Disruptive Technologies and Digital Transformations for Society 5.0

Disruptive technologies and digital transformations for Society 5.0 aims to report innovations to enable a futuristic society in which new values and services are created continuously, making people's lives more conformable and sustainable. It aims to present how problems can be solved in different areas, including mobility, health, agriculture, food, manufacturing, disaster prevention, and energy to name a few. Society 5.0 framework is based on data captured by real-world sensors and sent to the virtual cloud world for Artificial intelligence (AI)-based analysis, which in turn will return to the real world in physical form through robots, machines, and motor vehicles. People, objects, and systems are all connected in Society 5.0 and converge in cyber and physical space to collect a large amount of data from a variety of sources using sensors and devices. In Society 5.0, new values created by social innovation eliminate regional, age, gender, and language disparities and enable the delivery of personalized products and services that meet many individuals and potential needs. Digital transformation marks a radical rethinking of how an organization uses technology, people, and processes to fundamentally change business performance. Disruptive technologies including AI, affective computing, Blockchain, biological computing, cloud computing, emotion theory, human-computer interaction, Internet of Things (IoT) predictive analysis, probabilistic methods, swarm intelligence, socio-cognitive neuroscience, quantum computing, web intelligence have monumental roles to play in digital reality and Society 5.0. These technologies are shifting the economic landscape and the time has come to imbibe these technologies and empower organizations to exploit them now and in the future. The Series accepts research monographs, introductory and advanced textbooks, professional books, and reference works.

Aim and Scope

- The series is focused to explore how disruptive technologies are helping in digital transformation and how organizations are changing the way they do business, concerning innovation processes and business model transformations.
- This series is focused on how various disruptive technologies are creating opportunities across the business landscape.
- This series provides a comprehensive guide to Industry 4.0 applications, not only introducing implementation aspects but also presenting conceptual frameworks to the design principles of Society 5.0. Besides, it discusses such effects in new business models and workforce transformation.
- Changing dynamics of global production, its complexities, high end automated processes, high-level competitiveness, and emerging technologies for new generation goods, products, and services.
- Special focus on AI, affective computing, Blockchain, biological computing, cloud computing, cognitive intelligence, digital business transformation, decision sciences, e-health services, emotion theory, Futuristic digital society, habitat Innovation, human-computer interaction, Internet of Things (IoT), Internet of Humans (IoH), IoT-oriented Infrastructure, mobile computing, neural computing, predictive analysis, probabilistic methods, resilience in cyber-physical systems, robotics and automation for futuristic applications, swarm intelligence, synergies, and tradeoffs of food, energy, and water (F-E-W) nexus, socio-cognitive neuroscience, smart homes, and smart buildings, smart mobility and transportation, smart factories, embedded devices, quantum computing, and web intelligence to name a few.

Sandeep Kautish · Prasenjit Chatterjee ·
Dragan Pamucar · N. Pradeep · Deepmala Singh
Editors

Computational Intelligence for Modern Business Systems

Emerging Applications and Strategies

Springer

Editors
Sandeep Kautish
LBEF Campus
Kathmandu, Nepal

Prasenjit Chatterjee
Department of Mechanical Engineering
MCKV Institute of Engineering
Howrah, West Bengal, India

Dragan Pamucar
Faculty of Organizational Sciences
Department of Operations Research
and Statistics
University of Belgrade
Belgrade, Serbia

N. Pradeep
Department of Computer Science
and Engineering
Bapuji Institute of Engineering
and Technology
Davangere, India

Affiliated to Visvesvaraya Technological
University
Belagavi, India

Deepmala Singh
Symbiosis Centre for Management Studies
Symbiosis International (Deemed
University)
Pune, India

ISSN 2730-9061 ISSN 2730-907X (electronic)
Disruptive Technologies and Digital Transformations for Society 5.0
ISBN 978-981-99-5353-0 ISBN 978-981-99-5354-7 (eBook)
https://doi.org/10.1007/978-981-99-5354-7

This Springer imprint is published by the registered company Springer Nature Singapore Pte Ltd.
The registered company address is: 152 Beach Road, #21-01/04 Gateway East, Singapore 189721, Singapore

Preface

Computational intelligence is a group of models and tools that encompass elements of learning, adaptation, and/or heuristic optimization. It is also referred to as techniques to derive more values from business operations, overcome real-world uncertainties and complexities, and make better decisions. Integration of modern key computational intelligence concepts is essential for the knowledge acquisition, and strategies for analysis, exploration, and knowledge generation using computational intelligent approaches. Computational intelligence plays a crucial role in modern business systems by providing powerful tools for data analysis, decision-making, and optimization. With the ever-increasing amount of data generated by businesses, computational intelligence techniques can help to extract meaningful insights and patterns from this data, which can then be used to inform strategic decision-making. By leveraging the power of computational intelligence techniques, businesses can make more informed decisions, optimize their operations, and ultimately achieve greater success.

This book is an exemplary document covering the applications of computational intelligent techniques in business systems and advocating how these techniques are useful in modern business operations. The book will redefine the Computational Intelligence Foundations: the three pillars, i.e. neural networks, evolutionary computation, and fuzzy systems. Also, emerging areas such as swarm intelligence, Artificial Immune Systems (AIS), support vector machines, rough sets, chaotic systems, and others have also been demystified in the book in order to strengthen the range of computational intelligence techniques, i.e. Expert Systems, Knowledge Based Systems, and Genetic Algorithms.

This book redefines the role of computational intelligence techniques in modern business system operations, i.e. marketing, finance, accounts, operations, personnel management, supply chain management, and logistics. In addition, this book guides the readers through using them to model, discover, and interpret new patterns that can't be found through statistical methods alone in various business system operations.

This book reveals how computational intelligence can inform the design and integration of services, architecture, brand identity, and product portfolio across the entire enterprise. The proposed book will also examine how to complement computational intelligence with visualization, explorative interfaces, and advanced reporting, thereby empowering business users and enterprise stakeholders to take full advantage of it. This book comprises well-structured chapters written by academic and industry researchers from across the world and all of them are experts in their respective fields of research. All the chapters have a common focus, i.e. how Computational Intelligence Techniques can help human society in combating COVID-19.

Organization of the Book

The book has been divided into three sections, namely Part I (**Computational Intelligence for Business Finance Applications**), Part II (**Computational Intelligence for Marketing, Business Process and Human Resource Applications**), and Part III (**Computational Intelligence for Operational Excellence, Supply Chain and Project Management**) based on the types of applications and case studies considered. The chapters cover a variety of topics, i.e. Industry 4.0, Internet of Things (IoT), Machine Learning, Deep Learning, Big Data Analytics for modeling COVID-19, Artificial Intelligence (AI) enabled social distancing for virtual assessments and diagnosis, and many more. The book is organized into 15 chapters. A brief summary of the chapters is presented below.

Chapter 1 provides a complete overview of the research conducted in the financial industry after implementing digital financial solutions over time, as well as a broad overview of this unique research topic. In Chap. 2, a comprehensive comparison of the stock market pre-COVID-19, during COVID-19, and post-COVID-19 for tech giants including Google, Apple, General Electric, IBM, and Microsoft has been implemented in order to analyze the impact of COVID-19 on the financial market. Chapter 3 aims to define the feelings-based hypotheses that are used to clarify financial exchange concerns and terminologies. In this study, it is recognized that emotions do not typically motivate investors, and that it is not critical for the property market to be adequate at its current state. A thorough analysis of the behavioral account hypothesis is required. The behavioral model is used in this study to better understand the investments. Behavioral research on the study of investment decisions in the commercial real estate market can be improved by utilizing digital and statical analysis with machine learning approaches. The role of AI in tackling difficulties related with money laundering in the banking sector is the subject of Chap. 4. The goal of Chap. 5 is to create a natural language processing model for identifying keywords in crowdfunded project descriptions that influence project success. Chapter 6 uses a variety of deep learning techniques for stock prediction, including Recurrent Neural Networks (RNN), Long Short-Term Memory (LSTM), Convolutional Neural Networks (CNN), and Bidirectional LSTM (BiLSTM). The goal of Chap. 7 is to employ several regression models of supervised machine learning to estimate the price of houses. Housing data from 5000 residences in the United States was analyzed for the proposed study. Random Forest Regression, XGBoost

Regression, Linear Regression, Artificial Neural Network, Gradient Boosting, and Ada Boosting are the six models employed. When the results are compared, Linear Regression and Artificial Neural Network surpass the other four algorithms with R2 Scores of 91.56 and 91.37%, respectively. In Chap. 8, a Software Defined Networking (SDN) architecture is considered, in which connections between users' devices and providers' nodes are defined based on resource needs and price. The findings imply that the ideal bidding strategy is determined by multiple factors, including (i) the competitor's bidding strategy; (ii) each participant's operational cost; and (iii) all players' available resources and the broker's requirements. The primary goal of Chap. 9 is to highlight the current status of AI applications in marketing, with a focus on digital marketing, using a desk research method. The chapter also seeks to pinpoint the digital marketing's future directions, with AI as a possible main driver. The World Book Fair, which was held in New Delhi by the National Book Trust (NBT), an Indian public-sector organization, is discussed in Chap. 10. The goal of this case study is to look at the public sector's large presence around the world, as well as its domination in particular areas. A three-stage approach for preventing attrition is presented in Chap. 11 (up to processing, processing, and post-processing). The case study's IBM HR dataset has been chosen. The selection strategy for the "maximum-out" feature is suggested for reducing the to-processing phase due to the large number of functions in the dataset. In Chap. 12, readers will learn about several data mining techniques and how to apply them to Customer Relationship Management (CRM). It also demonstrates how effective it is to convert a traditional CRM into a modern one. In Chap. 13, the benefits of using a machine learning algorithm to anticipate the behavior of current employees are discussed. Long Short-Term Memory (LSTM) algorithm and fuzzy rules are utilized in this study to assess worker attrition by projecting a scarcity of competent personnel by department and sending out a warning message about hiring new employees or redistributing workload among existing employees. Chapter 14 provides a new computational intelligence paradigm for psychological assessment based on q-Rung Orthopair Fuzzy (qROF). Chapter 15 develops a system for evaluating a candidate's personality using various strategies. With the use of machine learning and natural language processing, the suggested system asks CV-related and personality-based questions to anticipate and assess his/her personality, allowing the business to shortlist candidates based on the job profile and company requirements. Chapter 16 uses advanced image processing and deep learning algorithms to segment failure regions from disintegrated automobile assembly parts images. The impact of Industry 4.0 on making the arena digital is discussed in Chap. 17. In addition, this chapter covers the transition from Industry 4.0 to Society 5.0, as well as the future of Industry 4.0 and Industry 5.0. Chapter 18 discusses the roles of artificial intelligence and automation for Industry 4.0. The goal of Chap. 19 is to create a reliable decision-making tool in a long-term supply chain that incorporates the thinking process, the concept of six sigma, and lean thinking. Chapter 20 presents AI-based technological reforms of global projects. Chapter 21 adopted the Clark-Wright algorithm to optimize delivery vehicle routing in order to produce higher effects in the organization, boost efficiency, and quality of service, minimize transportation time, and reduce shipment delivery delays. The goal of

Chap. 22 is to determine how a country's diet influences its COVID-19 mortality rate. With so many different eating traditions around the world, determining which food groups can best predict a country's death rate would be fascinating. It estimated the proportion of fatalities caused by the corona-virus pandemic using linear regression and random forest/regression tree, taking into account statistical data on the population's eating habits. Chapter 23 presents a de-background multicolumn dynamic CNN for crowd counting. While crowd counting, the suggested model can manage crowd shape changes owing to viewpoint distortion and learn to minimize the background effect. In Chap. 24, a thorough examination of product bundling and its methods in various marketplaces throughout the world was conducted. Based on the presented literature review, it may be concluded that certain factors are much more essential than others. Observations are made in this study on how different factors compare to one another and which should be prioritized. The most vital aspect is information quality for the relevant bundling as per market criteria, and transportation, while important, is the least important. Chapter 25 focuses further on the prospect of building statistical techniques that deal with competitive science in order to construct a powerful decision-making tool suitable for long-term supply chain network research. The goal of Chap. 26 is to find the linked variable(s) with reverse logistics methods in bottle manufacturing businesses in the Indian state of Odisha and to rank those variables using the Technique for Order of Preference by Similarity to Ideal Solution (TOPSIS) method. Chapter 27 examines how numerous government agencies in developed countries throughout the world are incorporating AI into their procurement processes, as well as some of the capabilities that are being evaluated for future implementation.

Kathmandu, Nepal — Sandeep Kautish
West Bengal, India — Prasenjit Chatterjee
Belgrade, Serbia — Dragan Pamucar
Davangere, India — N. Pradeep
Pune, India — Deepmala Singh

Acknowledgements

The Editors of *Computational Intelligence for Modern Business Systems: Emerging Applications and Strategies* wish to acknowledge the efforts of the authors who have submitted their wonderful chapters to our edited textbook in the stipulated time.

This book would not have been possible without the valuable scholarly contributions of authors across the globe.

The editors avow the endless support and motivation from their family members and friends.

The editors take this opportunity to bestow their best regards to all reviewers for providing constructive comments for improving the quality of the book and to remain coherent in the content presentation of chapters.

Mere words cannot express the editors' deep gratitude to the entire Springer Nature Team, particularly Ms. Kamiya Khatter, Associate Editor, and Ms. Shalini Selvam, Project Coordinator, Springer Nature, for keeping faith on the editors, providing valuable support at every step, and showing the right path to accomplish this high-level research book.

The editors believe in "Cooperation, Coordination and Commitment can make any project to be success". For the successful completion of this edited volume, the editors sincerely acknowledge everyone who helped directly and indirectly.

Dr. Sandeep Kautish
Dr. Prasenjit Chatterjee
Dr. Dragan Pamucar
Dr. N. Pradeep
Dr. Deepmala Singh

Contents

Editors and Contributors

About the Editors

Sandeep Kautish is a Professor and Director-Academics with Lord Buddha Education Foundation, Kathmandu, Nepal, running in academic collaboration with Asia Pacific University of Technology & Innovation Malaysia. He has about 20 years of academic experience, including over six years in academic administration at various institutions in India and abroad. He earned his bachelor's, master's, and doctorate degrees in computer science on intelligent systems in social networks. He also holds a PG Diploma in management. His research areas are business analytics, machine learning, data mining, and information systems. He has over 40 journal publications and over authored/edited books to his credit.

Prasenjit Chatterjee is currently a Professor of Mechanical Engineering and Dean of Research and Consultancy at MCKV Institute of Engineering, West Bengal, India. He has over 6100 citations with an h-index of 40 and 125 research papers in various international journals and peer-reviewed conferences. He has been the Guest Editor of several special issues of SCI, SCIE, Scopus, and ESCI indexed journals. He has authored and edited several books on decision-making approaches, supply chain, and sustainability modeling. Dr. Chatterjee is one of the developers of two multiple-criteria decision-making methods called Measurement of Alternatives and Ranking according to Compromise Solution (MARCOS) and Ranking of Alternatives through Functional mapping of criterion sub-intervals into a Single Interval (RAFSI).

Dragan Pamucar is an Associate Professor at the University of Belgrade, Faculty of Organizational Sciences, Department of Operations Research and Statistics, Belgrade. Dr. Pamucar obtained his M.Sc. at the Faculty of Transport and Traffic Engineering in Belgrade in 2009, and his Ph.D. degree in Applied Mathematics with specialization in multi-criteria modeling and soft computing techniques at the University of Defence in Belgrade, Serbia in 2013. His research interests include the fields of computational intelligence, multi-criteria decision-making problems, neuro-fuzzy systems, fuzzy, rough, and intuitionistic fuzzy set theory, neutrosophic theory, with applications in a wide range of logistics problems. Dr. Pamucar has authored/co-authored over 150 papers published in international journals has been the guest editor of numerous special issues of Scopus and SCI-indexed journals. He has authored and edited books on decision-making approaches, optimization, and logistics.

N. Pradeep is an Associate Professor in Computer Science and Engineering, Bapuji Institute of Engineering and Technology, Karnataka, India. He has 18 years of teaching and research experience. His research areas are machine learning, pattern recognition, medical image analysis, knowledge discovery techniques, and data analytics. He has published over 20 research articles published in refereed journals, authored six book chapters, and edited several books. His one Indian patent application is published and one Australian patent is granted.

Deepmala Singh is serving as an Assistant Professor in Lord Buddha Education Foundation LBEF), Kathmandu, Nepal. She completed her Ph.D. from Banaras Hindu University in 2016. Her research focused on the digital initiatives of human resource development practices in BHEL. Before joining LBEF, she was associated with MNNIT Allahabad. Besides, she also served as a project fellow in a major research project funded by UGC in 2011. She has several publications in national and international journals.

Contributors

Himanshu Ahuja Department of Computer Applications, Manav Rachna International Institute of Research and Studies, Haryana, India

Kshitij Anand School of Mechanical Engineering, Kalinga Institute of Industrial Technology, Bhubaneswar, India

Gohad Atul Department of ETM, Robert Bosch Engineering and Business Solutions, Bengaluru, India

Md. Babul Islam School of Information Engineering, Huzhou University, Huzhou, China

Nirmala Baloorkar Department of Computer Engineering, K. J. Somaiya College of Engineering, University of Mumbai, Vidyavihar, Maharashtra, India

Bhimraj Basumatary Assistant Professor, Mathematical Sciences, Bodoland University, Kokrajhar, Assam, India

Mridula Batra Department of Computer Applications, Manav Rachna International Institute of Research and Studies, Haryana, India

Pradeep Bedi Galgotias University, Greater Noida, Uttar Pradesh, India

Suriya Begum An NIIT Venture, StackRoute, Bengaluru, India

Chiranjib Bhowmik Department of Mechanical Engineering, Techno India University, Kolkata, West Bengal, India

Sanjib Biswas Decision Sciences and Operations Management Area, Calcutta Business School, Kolkata, West Bengal, India

Prasun Chakrabarti ITM SLS Baroda University, Vadodara, Gujarat, India

Gomatam Mohana Charyulu Department of S&H, VFSTR Deemed to Be University, Valdamudi, Andhra Pradesh, India

Prasenjit Chatterjee Department of Mechanical Engineering, MCKV Institute of Engineering, Howrah, West Bengal, India

Shreya Chatterjee HR and OB Area, Calcutta Business School, Kolkata, West Bengal, India

Amrita Chaurasia Christ University, Gaziabad, India

Poushali Dey Indian Institute of Foreign Trade, Kolkata Campus, Kolkata, West Bengal, India

Gaurav Dhiman Department of Computer Science, Government Bikram College of Commerce, Patiala, India

David Edijala American University of Nigeria, Yola, Nigeria

Ashutosh Gaur Assistant Professor, Mangalamay Insitute of Management and Technology, Greater Noida, India

Mladen Gavranović Faculty of Transport and Traffic Engineering, University of East Sarajevo, Doboj, Bosnia and Herzegovina

Sudipta Ghosh Department of Mechanical Engineering, Durgapur Institute of Advanced Technology and Management, Durgapur, West Bengal, India

S. B. Goyal City University, Petaling Jaya, Selangor, Malaysia

Kamal Gulati Associate Professor, Amity University, Noida, Uttar Pradesh, India

Pragya Gupta SRM Institute of Science and Technology, Chennai, India

Swarna Hasibunnahar School of Life Science, Huzhou University, Huzhou, China

Bui Thanh Hung Data Science Laboratory, Faculty of Information Technology, Industrial University of Ho Chi Minh city, Ho Chi Minh City, Vietnam

Drishti Jain SRM Institute of Science and Technology, Chennai, India

K. Jayakrishna School of Mechanical Engineering, Vellore Institute of Technology, Vellore, India

Guneet Kaur Technology Editor, Cointelegraph, UK

Komalpreet Kaur Department of Electronics and Communication Engineering, Punjabi University, Patiala, India

Sukhpreet Kaur Department of Computer Science and Engineering, School of Technology, Pandit Deendayal Energy University, Gandhinagar, Gujarat, India; Department of Computer Science & Engineering, Chandigarh Engineering College, Landran, Mohali, India

Naman Kaushik Department of Mechanical Engineering, Indian Institute of Technology (BHU), Varanasi, Uttar Pradesh, India

Sandeep Kautish LBEF Campus, Kathmandu, Nepal

Kavita Kelkar Department of Computer Engineering, K. J. Somaiya College of Engineering, Vidyavihar, University of Mumbai, Mumbai, Maharashtra, India

M. Razaulla Khan School of Commerce & Business Management, Maulana Azad National Urdu University, Hyderabad, India

Jugnesh Kumar Echelon Institute of Technology, Faridabad, India

Yogesh Kumar Department of Computer Science and Engineering, School of Technology, Pandit Deendayal Energy University, Gandhinagar, Gujarat, India

Sumit Das Lala Department of Mechanical Engineering, Parul Institute of Engineering and Technology, Parul University, Vadodara, India

Rashima Mahajan Manav Rachna International Institute of Research and Studies, Faridabad, India

S. S. Mahapatra NIT Rourkela, Rourkela, India

Md. Mahiuddin Sabbir Department of Marketing, Faculty of Business Studies, University of Barishal, Barishal, Bangladesh

Shuvendu Majumder HR and OB Area, Calcutta Business School, Kolkata, West Bengal, India

Shilpa Malik Aravali College of Engineering & Management, Faridabad, India

Swagatika Mishra VSSUT, Burla, Sambalpur, Odisha, India

José André Moura Instituto Universitário de Lisboa (ISCTE-IUL), ISTAR, Lisboa, Portugal

Faisal Mungi Department of Computer Engineering, K. J. Somaiya College of Engineering, Vidyavihar, University of Mumbai, Mumbai, Maharashtra, India

T. Mydhili Department of Management Studies, Vignan's Nirula Institute of Technology and Science for Women, Guntur, Andhra Pradesh, India

Rohan Pal School of Mechanical Engineering, Kalinga Institute of Industrial Technology, Bhubaneswar, India

Dragan Pamucar Faculty of Organizational Sciences, Department of Operations Research and Statistics, University of Belgrade, Belgrade, Serbia

Bhakti Parashar Vellore Institute of Technology University, Bhopal, India

Jinit Punamiya Department of Computer Engineering, K. J. Somaiya College of Engineering, Vidyavihar, University of Mumbai, Mumbai, Maharashtra, India

Lakshmi C. Radhakrishnan Institute of Management and Technology (IMT Business School), Dubai, UAE

T. S. Ragesh Assistant Professor, Business Analytics, Prin. L. N. Welingkar Institute of Management Development and Research, Mumbai, India

Anand Singh Rajawat School of Computer Science and Engineering, Sandip University, Nashik, India

E. Fantin Irudaya Raj Department of Electrical and Electronics Engineering, Dr. Sivanthi Aditanar College of Engineering, Tiruchendur, Tamilnadu, India

K. Bhavana Raj Assistant Professor, Department of Management Studies, Institute of Public Enterprise, Hyderabad, India

G. Rajyalakshmi School of Mechanical Engineering, Vellore Institute of Technology, Vellore, India

Sandip Rakshit American University of Nigeria, Yola, Nigeria

S. Siva Venkata Ramana Department of Management Studies, Vignan's Nirula Institute of Technology and Science for Women, Guntur, Andhra Pradesh, India

Yash Rathod Department of Computer Engineering, K. J. Somaiya College of Engineering, University of Mumbai, Vidyavihar, Maharashtra, India

Paresh Rawat S. N. Technology, Bhopal, MP, India

Amitava Ray Department of Mechanical Engineering, Jalpaiguri Government Engineering College, Jalpaiguri, West Bengal, India

K. Saikumar Department of C.S.E., Malla Reddy University, Hyderabad (MRUH), Hyderabad, Telangana, India;
Researcher, Green Fields, Vaddeswaram, Guntur, Andhra Pradesh, India

E. Sandeep Kumar Department of ETM, Robert Bosch Engineering and Business Solutions, Bengaluru, India

Mihir Satra Department of Computer Engineering, K. J. Somaiya College of Engineering, Vidyavihar, University of Mumbai, Mumbai, Maharashtra, India

Umang Savla Department of Computer Engineering, K. J. Somaiya College of Engineering, University of Mumbai, Vidyavihar, Maharashtra, India

B. Ida Seraphim SRM Institute of Science and Technology, Chennai, India

Aashay Shah Department of Computer Engineering, K. J. Somaiya College of Engineering, University of Mumbai, Vidyavihar, Maharashtra, India

Prithvi Shah Department of Computer Engineering, K. J. Somaiya College of Engineering, University of Mumbai, Vidyavihar, Maharashtra, India

Siddharth Shahi Department of Computer Applications, Manav Rachna International Institute of Research and Studies, Haryana, India

Piyush Kumar Shukla Department of CSE, UIT RGPV, Bhopal, MP, India

Prashant Kumar Shukla Engineering School and Technology, Jagran Lakecity University, Bhopal, India

Ponduri Siddardha R.V.R. & J.C. College of Engineering, Chowdavaram, India

Anita Sikandar Shyamlal College, University of Delhi, Delhi, India

M. A. Sikandar School of Commerce & Business Management, Maulana Azad National Urdu University, Hyderabad, India

João Carlos Marques Silva Instituto Universitário de Lisboa (ISCTE-IUL), ISTAR, Lisboa, Portugal

Deepmala Singh Symbiosis Centre for Management Studies (SCMS), Symbiosis International (Deemed University) (SIU), Nagpur, Maharashtra, India

Uma S. Singh Associate Professor, Department of Commerce, ARSD College, University of Delhi, New Delhi, India

Deepak Singhal School of Mechanical Engineering, Kalinga Institute of Industrial Technology, Bhubaneswar, India

Nuno Manuel Branco Souto Instituto Universitário de Lisboa (ISCTE-IUL), ISTAR, Lisboa, Portugal

C. Sreerag School of Mechanical Engineering, Vellore Institute of Technology, Vellore, India

Rajeev Srivastava Computing and Vision Lab, Department of Computer Science and Engineering, Indian Institute of Technology (BHU), Varanasi, Uttar Pradesh, India

Subodh Srivastava Department of Electronics and Communication Engineering, National Institute of Technology, Patna, Bihar, India

Željko Stević Faculty of Transport and Traffic Engineering, University of East Sarajevo, Doboj, Bosnia and Herzegovina

Rashmi Ranjan Swain VSSUT, Burla, Sambalpur, Odisha, India

Khan Md. Raziuddin Taufique Oxford Brookes Business School, Oxford Brookes University, Oxford, UK

Santosh Kumar Tripathy Computing and Vision Lab, Department of Computer Science and Engineering, Indian Institute of Technology (BHU), Varanasi, Uttar Pradesh, India

Sushanta Tripathy School of Mechanical Engineering, Kalinga Institute of Industrial Technology, Bhubaneswar, India

Narasimha Rao Vajjhala University of New York Tirana, Tirana, Albania

Srinivas Viswanath Narayana College of Engineering, Gudur, Andhra Pradesh, India

Medhavi Yadav Department of Computer Applications, Manav Rachna International Institute of Research and Studies, Haryana, India

Part I
Computational Intelligence for Business Finance Applications

Chapter 1
Artificial Intelligence and Machine Learning in Financial Services to Improve the Business System

Komalpreet Kaur, Yogesh Kumar, and Sukhpreet Kaur

Abstract Machine learning is coming as a significant encroachment in the financial services industry. Finance has always been about data and is considered a complex field of study that includes knowledge from disciplines such as mathematics and statistics to human psychology and linguistics. Due to this, it is difficult to manage the various day-to-day challenges associated with finance, such as financial glitches attributed to human errors. The financial sector has employed machine learning for a myriad of purposes and its excellent applications. The work highlights the advancement of different learning techniques in financial services for data science. This chapter gives comprehensive prospects on a study accomplished in the financial industry after applying digital financial solutions over time and an extensive view of this distinctive research area. The organization comprises an introduction, motivation, and background that entails a block diagram of learning techniques, benefits, and various issues related to implementing machine learning techniques in the financial domain. Other things covered in this chapter are various datasets used by different researchers, and its focal point is to present a systematic survey of various applications of finance using artificial intelligence and finally expose a synthesis analysis based on the findings along with their benefits and issues. Overall, this chapter gives conscious and constructive assistance to researchers working towards the sustainable evolution of the finance industry.

K. Kaur
Department of Electronics and Communication Engineering, Punjabi University, Patiala, India

Y. Kumar (✉) · S. Kaur
Department of Computer Science and Engineering, School of Technology, Pandit Deendayal Energy University, Gandhinagar, Gujarat, India
e-mail: yogesh.kumar@sot.pdpu.ac.in; Yogesh.arora10744@gmail.com

S. Kaur
e-mail: er.sukhpreetkaur@gmail.com

S. Kaur
Department of Computer Science & Engineering, Chandigarh Engineering College, Landran, Mohali, India

S. Kautish et al. (eds.), *Computational Intelligence for Modern Business Systems*, Disruptive Technologies and Digital Transformations for Society 5.0,
https://doi.org/10.1007/978-981-99-5354-7_1

Keywords Data science · Financial services · Machine learning · Artificial intelligence · Deep learning · Datasets · Accuracy

1 Introduction

The data evaluation is based on test and error, which is an outlook that becomes unpractical in the case of large and diverse data sets. Machine learning (ML) comes as a besotted way to evaluate this massive amount of data. This rapid progress in ML increases its usage, demands, and importance in contemporary life. This has also replaced data mining in which explications are done using automating generic methods that have replaced the conventional statistical methods. So, precise results can be created using ML by evolving proficient and speedy algorithms and data-driven models for real-time processing [1]. The finance industry is broad and different segments have different use cases for ML. The use cases for ML in finance are both abundant and highly important. Fraud detection, credit quality assessment, algorithmic trading, and compliance regulation in real-time are examples of use cases [2]. ML in finance has refolded the financial services industry like never before. It helps businesses modify their customer experience and enables them to offer personalized products and services based on consumer actions [3]. ML techniques forecast and help diminish customer dissatisfaction, like an exact prediction of cash required in ATMs helps manage costs and improves the return on cash assets. Chabot offers 24*7 customer service with superhuman speed and constancy. ML underwriting agents assist human underwriters to make the best utilization of their time and effort. Fraud detection algorithms help find suspicious activity that the best human experts may fail to spot. Even business association is not wholly human activity any longer. Machine learning makes specific that these digital financial solutions persist in to carry out correctly, even as the wants of financial institutions grow significantly over time [1]. Due to its popularity in various fields, various researchers have also worked on machine learning in the finance sector. However, still, they were unable to give reviews on things related to it. Many researchers have presented the use of machine learning or artificial intelligence in the finance domain and presented using the technique for individual finance. However, there is no good survey or research presented in the area of artificial intelligence covering its different aspects and its applications in the field of finance, the role of data science in finance, and comparative analysis of previous research done in this field. There is a need for comprehensive study in the field of machine learning especially in the domain of finance. So, the objective and motivation behind this chapter are to outline the use of ML in the finance domain and the challenges faced while implementing it, and various advantages achieved by applying both together. Along with it, highlights a systematic review of various datasets or financial data used by machine learning algorithms to carry out the experiments on it.

1.1 Organization of the Chapter

The complete chapter is organized into various sections for terms related to the Finance sector. The sections are divided into numerous sections according to the main terms related to the use of AI in Finance. Started with Sect. 1 which contains the introduction of finance and the use of AI in it forwarded with the motivation behind writing this chapter given in Sect. 2 which also consists of challenges faced by the user while using AI in the finance sector. Further, in Sect. 3 Background is given about the main topic that contains data related to the role of data science in the finance sector, benefits and issues of AI in the finance sector, datasets used in financial applications along how AI is changing the financial services industry (Fig. 1).

Then, Sect. 4 contains the role of AI in the finance sector is included that further divided into various subheadings of applications of finance in which AI plays its role. The comparative analysis is also given in Sect. 5 in which a comparison of various applications of finance has been done in terms of the dataset used along with output achieved using different AI algorithms.

2 Motivation

Finance has always been about data. AI is used for digesting immense amount of data and learning from it by carrying out a particular task such as differentiating fraudulent legal documents from all authentic documents. Thus, ML learns from presented data to divide future and data-driven decision-making models from input data sets [1].

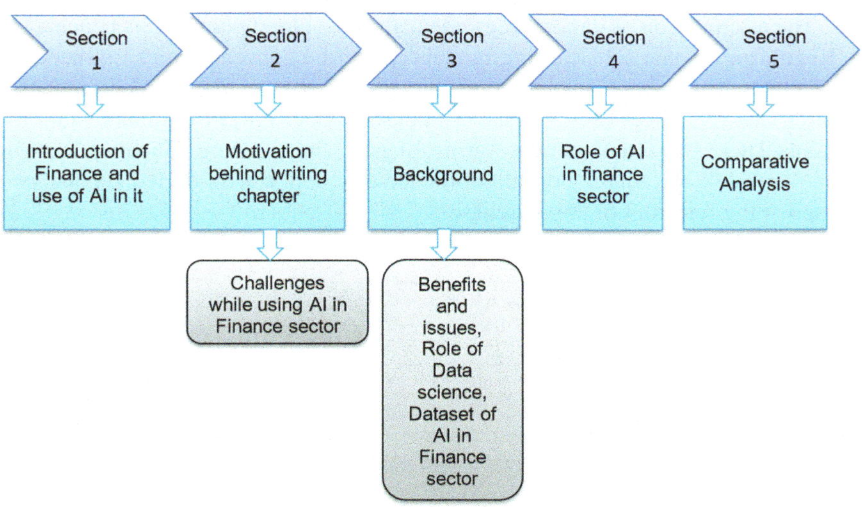

Fig. 1 Organization of the chapter

AI in finance gives numerous approaches for handling large and complex amounts of information in a better way. It also improves the ability of managing such large and complex data accessible by the finance industry [3]. Due to a large amount of historical financial data produced in the industry, machine learning has found many constructive applications like trade settlement, fraud detection, algorithmic trading, high-frequency trading, chatbots, loan underwriting, Robo-advisors, risk management, money laundering, and document analysis. AI using the latest technologies helps in different ways in the finance sector, from approving loans and executing credit scores to handling assets and estimating risk [4]. No doubt, machine learning has taken the finance industry to an all-new level. However, still, it still faces some challenges in its implementation in the industry like cost, security, updates, adopting challenges, etc. [2]. Along with these challenges, machine learning has various advantages like time-saving, less paperwork, improved workflow, etc., that attracted us to work on it. Moreover, its use in various areas to engross researchers for working on it improves the existing systems. It has been found that acceptable improvement is achieved in various areas of finance by using machine learning.

2.1 Challenges

As machine learning is enjoying a moment of resurgence, there are implementation challenges a finance sector should tackle to be successful. There are many challenges that are faced while implementing machine learning in the finance sector. Such challenges are shown in Fig. 2 [2, 5, 6].

Data-Banking or financial data is often poor quality and hard to find as it is stored in silos on various legacy systems. Algorithms flourish easily accessible large data sets. The assimilation of data sources, ideally onto a cloud platform, is hence a key.

Regulatory-Some self-learning models cannot be conventionally validated and consequently may be considered insufficient by the regulator. Comprehensive research into regulatory requirements is suggested in advance of implementation.

Tools-There exists an immense array of new and growing Machine Learning technologies. A systematic consultation course of action with digital specialists is advised at the forefront of any purchase.

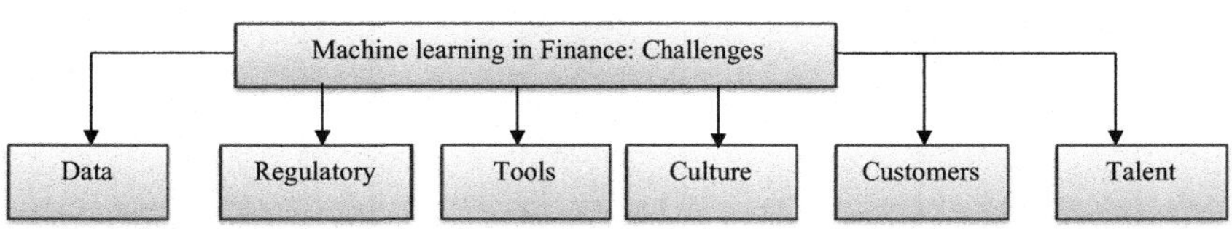

Fig. 2 Challenges of machine learning in finance

Culture-Judgment currently often trumps insights in firms- a cultural shift will be required. The democratization of analytics is required, and there should be a lure to sway data sharing between business divisions.

Customers-Older generations are not so much digital-savvy. Customers choose human interaction over communication with robots. A marketing /education session may be required to emphasize the benefits to the customer.

Talent-Initiating machine learning into business needs a change in skillset requirements from operational manager to analytics and data science.

3 Background Study

This section expounds on a brief overview of learning techniques and the role of data science in financial domain applications, along with its advantages, challenges, and objectives. Various application areas of machine learning in finance are also presented in this section.

Conventional computer science algorithms were application-specific, and time was necessary to develop a proficient algorithm for a specific application. Each application generally had different requirements, and as a consequence, there were not even single approaches that fit all methods. Artificial Intelligence (AI), ML, and Deep Learning assure a subset of algorithms that fit a range of sub-tasks under a specific application sphere [4].

Based on skills in using computers for surveying the data for structure, ML has evolved even in the cases of having no presumption about the structure. The experiments on it are a validation error on new data. On the other hand, a null hypothesis is proved by theoretical experiments. This happened due to frequent use of the iterative approach by ML for learning from data and automating the learning. Until a robust pattern is found, passes are run through the data [7] (Fig. 3).

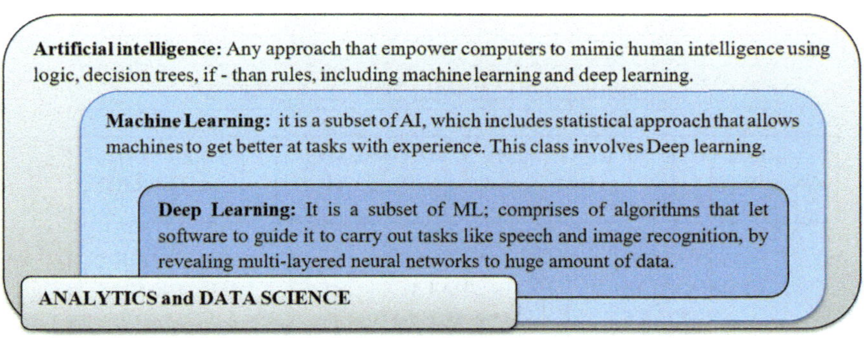

Fig. 3 Definition of various analytics and Data Science learning [4]

In the field of Deep Learning (DL), which is a subset of Machine Learning, the patterns of large amounts of data are studied. In DL, for dealing with data, neural networks are used which in turn also increases the computing power of the system. The current applications of DL include object recognition in images as well as the word recognition in various sounds. It is also being used in translating different languages automatically, medical diagnosis and other problems related to social issues by all the researchers [8]. That is why, the field of Data science is referred to as the area in which knowledge is extracted from huge data sets having data of unstructured and structured types. This knowledge is further being shared with their domains of business so that they can offer effective schemes and roadmaps in their business [9].

3.1 Role of Data Science in the Finance Sector

Industries observe data as a crucial commodity and energy. Raw data is used in it which is further converted into a meaningful product. This insight the enhancement in industry performance and another name of finance is the hub of data and its industries are considered as one of the preliminary users and explore data analytics. In fraud detection, customer management, algorithm trading and risk analytics data science is used [9, 10] (Fig. 4).

- Risk Analytics: It has an important place in different applications of finance related to data science. By using it, deliberate decisions can be taken by the company along with the rise in reliability as well as in-company security. In risk management, a core is data because it estimates the rate of loss recurrence and multiplies it with the magnitude of damage. There are different types of risks, which are evaluated by a company that arise from competitors, credits, market, etc. In managing risks, the primary step is its recognition along with

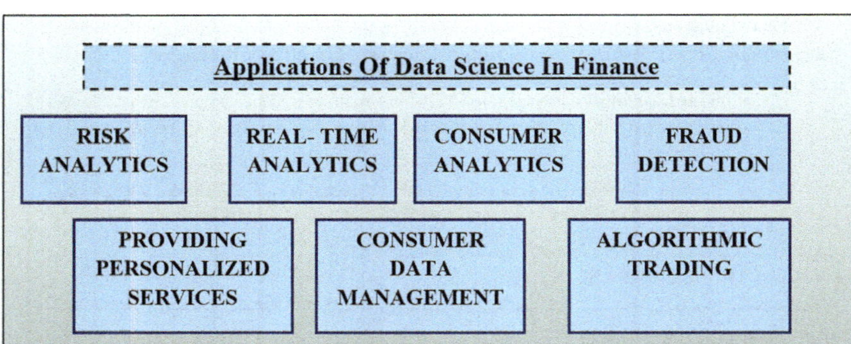

Fig. 4 Data Science for difference finance applications

its observation and prioritization. The data relating to customer information in the financial transaction is available in a massive manner. Due to it, institutions are trained on risk data, so that the risk scoring models can be implemented optimally with less expense. The creditworthiness of customer authentication is another important feature of risk management. For this purpose, data scientists are paid by companies that use machine learning algorithms to examine customer transactions [11, 12].

- Real-Time Analytics: Earlier the processing of data was performed in batches, which means it is chronological, not real-time. However, to get insights into present conditions, real-time data is required by most industries. That is why they faced problems. However, now with minimum latency, data can be accessed which happens with the evaluation of dynamic data pipelines and progress in technologies. In institutions, finances, this data science application can trace transactions; and generate credit scores along with other financial attributes without causing any latency problem [13].
- Consumer Analytics: Financial institution's primary function is consumer personalization. Data scientists are able to make suitable decisions and insights from consumer behaviour using the data in real-time analytics. The customer lifetime value is computed using consumer analytics by insurance companies like financial institutions. This will be added to their cross-sales that minimizes below zero customers for rationalizing the losses [14].
- Customer Data: Management financial Institutions require data. The working style of financial institutions has been transformed to much extent with the introduction of big data. Most of the data is provided through transactions and social media.

This data is presented in two forms as given below:

- Unstructured data
- Structured data.

Unstructured data causes many problems, whereas structured data is easier to handle.

The most significant part of Big Data is Business Intelligence (BI). BI helps industries in finding the important information about customers by applying machine learning. There are various other tools also which are used in AI for extracting meaningful information from the input data which includes data mining, text analytics, etc. By detailed analysis of customer data, the financial trends and market value changes are examined by machine learning [9].

- Providing Personalized Service: Personalized services to consumers are the liability of financial institutions that analyse consumer information using various techniques and further understand their interactions. Along with providing better communication to its users, financial institutions depend on software based on speech recognition and natural language processing. There is an increment in profits by taking actionable insights from financial institutions according to the

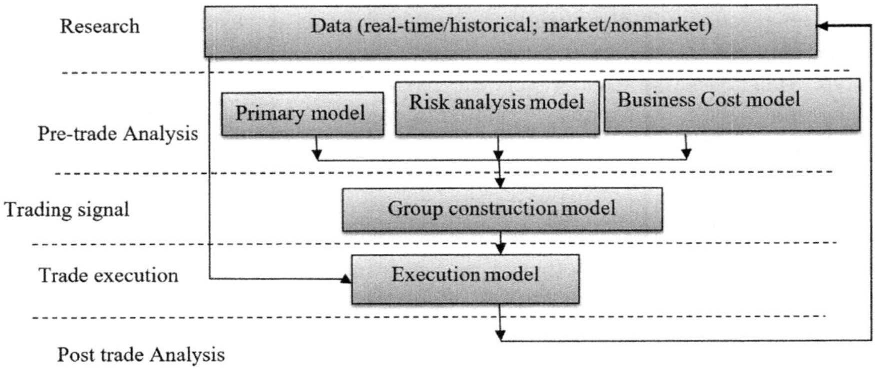

Fig. 5 Components of algorithmic trading [18]

customers' feedback data. This also optimizes the strategies of industries, which in turn helps them in improving their services [9].

- Fraud Detection: The industry of finance is majorly affected by frauds, and with an increase in several transactions, there is a rise in fraud risks. However, now the financial industries can keep track of fraud by introducing and increasing big data and analytical tools. In these industries, the most common fraud is credit card fraud. By advancing in algorithms, this type of fraud comes to know as it has increased the reliability of fraud detection that alerts companies about particularities in the financial purchase, ultimately encouraging them to block the account and reduce the losses [15, 16]. The unusual patterns of trading data can be spotted by various machine learning tools and the aware financial institutions to explore it more. Some insurance-related frauds are dealt with by banks [17].
- Algorithmic Trading: The components of algorithmic trading are shown in below Fig. 5. As can be seen there are three stages of algorithmic trading predate analysis, Trading signal generation, and Trade execution. Predate analysis takes in the mathematical model [18].
- To trade financial instruments, future activities are visualized by the risk alpha model.
- The risk model estimates the financial instruments-related risk levels.
- The transaction cost model computes the financial instruments' trading-related costs.

The portfolio construction model is included in trading signal generation, and its input is the results of alpha, risk, and transaction cost models. It further decides the amount of financial instruments portfolio allowed to go onward and in how many quantities.

The trades are executed at trade execution that makes various decisions by checking the transaction expenses and trading time. The trading strategy accompanied by venue and order type is the most general decision.

In financial institutions, the most significant part is played by algorithm trading which includes complex mathematical formulas and high-speed computations. This helps financial institutions develop new trading strategies, and it has a massive impact on high-speed computations. By this, data science is considered an essential feature of financial institutions.

There is a massive collection of multiple data operations in data science that involve statistics and machine learning reliant on data. Then further in the form of training and test set, these data are fed to a model that helped in implementing the module with algorithm parameters. This states that the future of Data Science is dependent on the advancement in Machine learning. [9]. Data science also includes [10]: Data Engineering, Automating Machine learning, automated data-driven decisions, Data Integration, Data visualization, Dashboards and BI and Automated data-driven decisions.

3.2 Benefits and Issues of Artificial Intelligence in Finance Sectors

This section covers various benefits and issues of using artificial intelligence in finance sectors. The logic to implement artificial intelligence in business seems evident. Nothing but machines can offer the following benefits:

- Process automation has reduced operational costs [4].
- Better productivity and enhanced user experiences have led to increased revenues [19].
- Better compliance and Strong security [12, 20].
- Cost-effective.
- Increasing Operational efficiency.
- Improve security.

Companies that are providing financial services can use a large number of AI algorithms that are open source. Also, these companies are also ready to spend any amount of money for purchasing state-of-the-art hardware which is required for increasing the computing power of the system. Various areas of the industries related to finance are improved with the application of AI that helped in dealing with large amount of existing data [21].

ML and deep learning algorithms in finance reduce labor costs by automating human labor, resulting in a significant saving of money in financial services. Operational efficiency is also improved by streamlining processes that increase the productivity and efficiency of financial operations. It also provides both network fraud prevention and security capabilities for financial institutions. The bank also benefits from the model's ability to select the financial indicators that are most relevant in the process of prediction and a high level of prediction accuracy. Human capabilities are improved, and AI's impact on business and the economy will be reflected in its direct contribution and ability to inspire complementary innovations.

The current development in AI mainly focuses on reducing the prediction cost and making the system faster and more accurate. These predicting action implications enable increased customer retention and prevent downtime through predictive maintenance on infrastructure or machinery. The benefits of AI are the ability of trades to execute at the best possible prices, increase accuracy, and reduce the mistakes likelihood with the ability to automatically and simultaneously check various market conditions. Human beings' emotional and psychological conditions cause some errors that also get reduced by replacing human work with AI systems. We have seen that various advantages are provided by AI technology, but still, some issues need to be addressed. Most of the financial services companies are not taking genuine value from AI technology for the reasons given below [2, 19].

- Research and Development in machine learning are expensive.
- The lack of machine learning engineers is another major issue.
- Selection of datasets for the experimentation.
- Financial incumbents are not smart enough when it comes to upgrading data infrastructure.

Sustainability and overfitting are two main issues in using AI models. Another possibility of an increase in the count of frauds using new approaches also increases the count of legal issues caused by mistakes of an algorithm. This also increases the count of privacy issues. Nowadays, AI models are used for preventing fraud, but there can be a possibility that the use of these tools will increase the number of cybercriminals defrauding users in the coming time. Also, the institute is legally responsible for whose person makes a mistake that also invokes legal questions on mistakes done by algorithms. The data is the primary source needed by these algorithms. However, the financial data is private information, so AI increases the concerns related to privacy.

3.3 Datasets Used in Financial Applications

In different studies or different applications of the financial domain, we use different datasets. A few of the datasets are available publicly that can be downloaded, such as Australian, Japanese, German, Korean, etc. These datasets are highly used in the prediction of bankruptcy and the scoring of credits. Some researchers have used publicly available data as well as some have collected data from specific countries also [22]. Table 1 shows some of the datasets and their description.

Table 1 Description of various datasets [22–25]

Name of dataset	Description
Quandl	For economic and financial data their dataset is considered as a good source that is mainly proving to be beneficial in model construction for stock prices or forecast economic indicators
World Bank Open Data	Their datasets enclose the analysis of population as well as a large number of economic and development indicators worldwide
IMF Data	This dataset has all data associated with international finances, rate of investment, rate of debts, and reserves dealing in foreign exchange
Financial Times Market Data	Provide updated details on financial markets worldwide, together with stock price indicators, property, and foreign exchange
Google Trends	They did a study and investigation of internet search activity data along with the different stories trending across the world
American Economic Association (AEA)	This data set contains the macroeconomic data of the United States of America
EU Open Data Portal	It was started in 2012 following European Commission Decision 2011/833/EU on commission document reuse. This dataset provides the required data without any cost and it contains the data of European Union Institutions. The data present in this dataset is collected from varying datasets and can be used anywhere in any sector
EOD Stock prices	EOD stands for the end of day and it consists of around 3000 US companies end of day stock prices, splits, and dividends curated by the Quandl community
Data.gov	It is a website of the US government that gives access to high-value, machine-readable datasets of various domains generated by the federal government's executive branch. The Vivek Kundra US federal CIO launched this website in 2009 that is powered by WordPress and CKAN open source applications
Global Financial Data (GDF)	The complete unabridged data series was generated by GDF that contains the combination of daily market data taken from historical and data feeds values. This can be accessed by anyone due to its free subscription and further can research on analysing the major global markets and economies
RBI Database	This dataset was launched by the Reserve Bank of India, RBI Data Warehouse which is a publisher's data platform for various Indian economy aspects. In this, the data is presented through a time series formatted report. It contains a dataset across financial markets, saving, money, employment, banking, national income, and others
PBGC Financial Summary Data	This consists of PBGC's key financial statements highlighted data of the period September 1992 to 2009. These statements can be accessed at PBGC.gov
Large Mutual Fund Facilities	This dataset contains a combined asset in US mutual funds that was $12.36 trillion in Oct 2007. There are around 8,000 US mutual funds that are significantly concentrated assets in a relatively small amount of mutual fund families which is 50% of all assets held by Top 10 mutual fund families

(continued)

Table 1 (continued)

Name of dataset	Description
US Stock Data	It contains US stock historical data since 2009 that is updated on a daily basis
CBOE Volatility Index (VIX)	It is a critical indicator of S&P's market forecasts for near-term volatility. This is a time series dataset that includes daily lows and highs, as well as closes and opens
Dow Jones Weekly Returns:	It contains a percentage of return that stock has every week for training their algorithm and measuring while producing the greatest rate of return by stock in the following week
Simin	The data provided by this dataset is available in one document so that it can be used in a timely manner. The data is collected from different financial statements. It is uploaded on the SEC website after its cleaning and organizing
CIA World Factbook	It includes the economic state of countries along with other military, communications, geography, and demographics stats

3.4 How Artificial Intelligence (AI) is Changing the Financial Services Industry?

In collecting and analysing data, AI is able to give much accurate future predictions and is very efficient in recognizing patterns in comparison to human ability. This helps in improving the effectiveness of banks in their routine work and helps in completing the tasks in less time. According to the remarks of the PWC's latest study, AI holds around 16 trillion dollars of the global economy in 2030. Moreover, its applications will cross 5 billion dollars of global investment, and by 2030 it is expected to save 1 trillion dollars in the banking industry [20]. There are various AI applications in the service industry. A few of them are given below.

Dedicated Services to HNI's (High Net Worth Individuals) Wealth and Portfolio Management: The main work of these services is to decide the trade-off between risk and return and in turn give warning to the users detailing various securities and assets with their possible returns. The advancement is given by AI to financial services companies which helps them give accurate and customized guidance to their well-off clients. BlackRock has helped the AI lab in its operations. It has been considered as the largest group of the world with higher investments having assets of 6 trillion dollars. Other global organizations are also using AI for improving predictions which is helping their clients.

Moreover, two AI systems have been brought by Swiss Bank UBS to renovate their trading floor. In this, after interpretation of market data, scope trading patterns can be found out by one. After that, it advises trading strategies to the bank's customers to help them achieve high returns. On the other hand, their customer's post-trade allotment preferences are conveyed by the second one [23].

- Virtual Financial Assistance and Automated Customer Support through Robo Advisors and Chatbots: The AI associate and other appropriate apps like Revolut are used by Banks that help in having quick services to the customers. For this purpose, they use Smart chat technologies that will help in transferring the queries of the customers to the supporting staff related to that query. It will require different processes of Natural Language Processing (NLP) for achieving this feature. After that, in 2016, an AI assistance Luvo was introduced by the Royal Bank of Scotland that answers customers' inquiries, and in some cases, it transfers them to human staff. This Robo advisor helps in reforming the experience and pleasure of customers. In India, the four top commercial banks are using Chatbots which are one of the major applications of AI. Along with it, they have advanced the customer experience by FinTech startups that reduce the cost and give better efficiency. On the other hand, to provide immediate feedback according to the captured facial expressions of clients, AI power-driven intelligent cameras are used by banks [24, 26, 27].
- Enhanced Insurance Experience: There are many applications for both instance of claim payment and underwriting policies in the data-driven insurance industry. The main requirements of insurance companies are to know more about the client's education, lifestyle, health, and character along with the filed claim occasion. However, these all things can be effectively captured using AI algorithms. In some US startups, there is a possibility of paying insurance claims within 3 s by doing multiple back-end procedures and checks using AI apps. They can check things while communicating with the customer at the front end [20, 28].
- Robotic Process Automation (RPA)—Repetitive Task Automation: There is a need for repetitive front and middle office processing in repetitive activities like deposit and withdrawal processing, billing, statement generating, cheque clearing, etc. However, with RPA and AI software, this can be accomplished better, resulting in better efficiency, improved time management, and expense savings. Human intelligence and skills are progressively imitating robotics technology from industrial robots to self-driving cars, which may become a game-changer in the financial services industry. There is speedy growth for investment in the robotic sector, and there is almost double the number of funding deals in robotics. According to CB Insights, in 2014 it was 273 million dollars, then in 2015, it became 587 million dollars. On the other hand, the expansion of investment was 115% in 2015 compared to 55% in 2014 [4, 20].
- Use of alternate data for analysis of credit scoring and predictive one: There is large count of SMEs and financially barred individuals that are unable to use bank credit in cases when there is very little or no credit account history. So, giving a loan to such customers becomes very challenging for banks. However, now loans can be sanctioned with the usage of AI by fintech startups. They can gather and process data such as educational background, social media, police records, employment history, age, location, spending habits, and other things. With predictive analysis (using AI), one can compute the credit score, avoid bad loans, and give insight into the current demand of a client's credit and the next.

Now, several FinTech companies are using AI power-driven algorithms that are disturbing the loan industry with their AI solutions that increase markets [15, 25].

Regulatory Compliance, Fraud Prevention, and Detection and Prevention of Money Laundering: After the financial calamity in 2009, there was stress of compliance and risk management on the banks and financial services firms. Basel Accords provided risk management and unwieldy capital sufficiency compliance I, II, and II together with AML and KYC processes that are required for managing all types of risks in the system. It is also required when banks are using the credit system and various practices that can lead to fraud. Although this process requires time in the units of person hours due to huge paperwork, it opens the way for the usage of AI. If AI is implemented in this work, then this work can be completed within seconds by recognizing the patterns and reading the data in very less time. Using JP Morgan's COIN, millions of hours of work can be completed in minutes. The doubt is indicated and prediction of human activities done by anti-fraud-based AI products that mark variance. Further, for image recognition at ATMs, deep learning like AI techniques can be used along with real-time camera images that help expose and avoid crimes and fraud [17, 20].

4 Role of AI in the Finance Sector

This section presents the contribution of various researchers in finance that have used AI algorithms to boost performance in the respective domain. Various Artificial Intelligence, deep learning applications in finance are presented in given sub-sections that have used Support vector machine (SVM), Recurrent Long Short Term Memory (LSTM), Backpropagation neural network (BPNN), Particle swarm optimization (PSO), DBN, K-nearest neighbour (KNN), Naïve Bayes, decision tree, Convolution neural network (CNN) and hybrid of one or two algorithms in financial distress, predicting credit card risk, Sentiment analysis, algorithm trading and stock price description that all comes under finance sector.

4.1 Financial Distress in Finance Sector Using Artificial Intelligence

For both practitioners and scholars, a great interest has been found by financial distress forecasting. The probability of financial distress can be estimated using a number of AI and statistical approaches. In this case, the prediction of financial distress for a system having a probability greater than the cutoff value is considered. This was improved by Bae [29] by collecting the financial data of MNCs located in the Korea Credit Guarantee Fund (KODIT) annually for the prediction of accuracy of financial distress problems [29].

Further, they have developed a radial basis function SVM (RSVM) based financial distress prediction model. For justification, they have compared their proposed RSVM with AI techniques and suggested a better financial distress predicting model that helps a chief finance officer make better decisions incorporating financial distress. Then, Hsieh et al. [30], proposed the SVM method and examined its predictive ability. Their proposed method uses the characteristics of a penalty function for generating predictions in a better way [30] Further, presented an evolutionary artificial bee colony (EABC) algorithm for including the properties of Particle Swarm Optimization (PSO) in which a velocity and flying direction is given by each bee that optimize their proposed penalty guided SVM (PGSVM). For public industrial firms in Taiwan, an EABC-PGSVM was used for constructing a reliable prediction model and compared the proposed EABC-PGSVM with backpropagation neural network (BPNN), PGSVM optimized by the ABC algorithm (BPGSVM), and classic SVM optimized by the ABC algorithm (BSVM). Indifference to existing methods, Lin et al. [31], have proposed an approach for feature selection of FDP problems that combines expert knowledge with the wrapper approach [31]. Based on experts' domain knowledge, the financial features are categorized into seven classes. Fengyi has applied the wrapper method to search subsets of good features containing top candidates from each feature class. They have compared various scholarly models for concept verification that lead to feature selection methods. Their experiment indicates that the proposed method has selected feature set-based prediction model that gives a better outcome in terms of prediction accuracy than standard feature selection-based models. Yu-Pei Huang et al. [32], have also reviewed work done in predicting financial distress using ML algorithms [32]. Among all four supervised algorithms, the XGBoost gives a more accurate outcome in terms of FD prediction. The hybrid of the DBN-SVM model gives more accurate forecasts than using SVM or DBN classifiers individually.

Furthermore, a novel meta FDP framework was proposed by Wang et al. [33], it consists of feature regularising modules for the identification of discriminatory predictive power of number of features and enhances the aggregation over base classifiers using a probabilistic fusion module [33]. The results obtained from it show that the proposed RS2_ER method can give an effective prediction on FDP. Then Sun et al. [34], have focused on effectively constructing class-imbalanced data streams based on dynamic FDP models [34]. The combination of SMOTE and AdaBoost SVM ensemble integrated with time weighting (ADASVM-TW) in which SMOTE stands for synthetic minority oversampling technique was utilised by them. In case of SMOTE, the class of every data batch is balanced before applying another approach for prediction modelling of dynamic financial distress. In the second one, the SMOTE is embedded with ADASVM-TW and designs a new sample weighting mechanism. For testing purposes, financial data of 2628 Chinese listed companies' dataset has been used that show both simple and embedding integration model that are able to improve the recognition ability for minority financial distress samples.

Numerous models are proposed for the detection of an occurrence of significant events in financial systems, but there is a need to automate significant events. In this concern, Rönnqvist et al. [35], have presented a method for recognizing relevant text and extracting natural language descriptions of events using deep learning [35]. Their model is leveraged by semantic vector representations unsupervised learning and supervised by entity names and dates a small set of event information on extensive test data. They have demonstrated the applicability of their news-based financial risk mainly related to bank distress and government interventions. Along with this, various researchers have proposed models for predicting contractor financial crises. So, Choi et al. [36], have proposed ensemble models based on voting that predict the financial distress of contractors for 2 or 3 years ahead of the prediction point using a financial distress definition based on finance [36]. The South Korean financial contractors' statements for the period 2007 to 2012 were used to evaluate the proposed model's performance. The results obtained show that the 0.940 and 0.910 receivers operating characteristic curve (AUC) values predict financial distress for each prediction year. In 2020, Said MARSO et al. [37] did an analysis on the performance of advanced cuckoo algorithm in terms of getting optimum weight of feedforward neural network and further named it CSFNN [37].

Further, to investigate the efficiency, they have compared the CSFNN with back-propagation feedforward neural network (BPNN) and Logistic Regression (LR) by applying it to two different periods of manufacturing sector collected data. The outcome in case of one year before bankruptcy gives an accuracy of 90.30% in the case of the CSFNN model and 88.33% and 82.15% in the case of the BPNN and LR model. It was 82.79% for CSFNN and 81.05% and 73.27% for BPNN and LR, respectively, in the case of three years before the bankruptcy.

4.2 *Prediction of Credit Card Risk in the Finance Sector Using Artificial Intelligence*

In the analysis of credit card, a predictive performance of a broad class of binary classifiers was examined by Jones et al. [38], in which they used large sample of global credit ratings for the period of 1983–2013 [38]. The study discovered that the new classifiers outperform existing ones in cross-sectional and longitudinal test samples and are robust to a variety of data types and assumptions. They have concluded that simple classifiers can be used in more sophisticated approaches, mainly in having the main objective of interpretability of modelling exercise. The study has shown that financial credit scoring is crucial in the finance industry sector for assessing individuals' creditworthiness and enterprises. For performing this task, various statistics-based ML techniques have been employed, but in ML techniques, one of the significant challenges is the curse of dimensionality. So, to improve classification Jadhav et al. [39], have investigated feature selection in credit scoring problems and proposed a novel approach. Information Gain Directed Feature Selection algorithm (IGDFS) for it [39]. The proposed approach performs the feature ranking based on the information gain and GA wrapper (GAW) algorithm for propagating the top m features and then classifies it using SVM, KNN, and Naïve Bayes ML algorithms. The outcome achieved inaccuracy shows that SVM gives the best results from the other two ML algorithms. This show that SVM models give good outcome in credit scoring performance so, Tian, et al. [40], have also proposed state-of-the-art kernel-free fuzzy quadratic surface SVM model approach [40]. The results show that their proposed method performs well in classification and handling searching proper kernel function and complex model issues related to classical SVM models. Further, Masmoudi et al. [41] have modelled the payment default of loan subscribers using a discrete Bayesian network that includes a built-in clustering feature with latent variables [41]. The model was calibrated on loan contracts describing the actual dataset, and results obtained from it highlight a regime-switching of a default probability distribution. To deduce various researchers have employed the possible repayment behaviour of rejected credit applicants ML and statistical methods. Shen et al. [42], have used unsupervised transfer learning and 3-way decision theory to propose a novel 3-stage reject inference learning framework [42]. The usability of the proposed framework shows its applicability in rejecting inference and handling adverse transfer learning problems. For validation of the proposed framework, Chinese credit data was considered, and the outcome obtained from it shows the superiority in credit risk management applications. Further, Wang et al. [43], have focused on a comparative assessment of five popular classifiers' performance in credit scoring [43]. The classifiers considered were LR, Random Forest, KNN, Naïve Bayes and Decision Tree. The study found that all classifiers have their pros and cons, so saying which one is best is emphatic. However, in terms of AUC, accuracy, recall, and precision, a better outcome is achieved using Random Forest.

Furthermore, for the credit scoring problem, the suitability of dynamic selection techniques was evaluated by Leopoldo Melo [44]. They have also presented Reduced Minority k-Nearest Neighbors (RMkNN) to enhance the state-of-the-art dynamic selection techniques in local regions for imbalanced credit scoring datasets. As compared to state-of-the-art, better prediction performance is achieved using the proposed technique. The other main benefit of RMkNN is that there is no need for any sampling or pre-processing method for generating a dynamic selection dataset (called DSEL). For predicting whether the loan will be repaid in the P2P platform or not, a benchmarking study of various credit risk scoring models was proposed by Vincenzo [45]. For analysis of the experiment, an 877,956 samples real social lending platform (Lending Club) dataset was used and evaluated results in terms of specificity, AUC, and Sensitivity. In the end, the best three approaches have been evaluated using various eXplainable Artificial Intelligence (XAI) tools.

For the prediction of bankruptcy and finance activity, credit risk assessment is a critical task that has been explored using ML and statistical methods. To further enhance the performance of credit modelling use of ensemble strategies has been suggested in recent works. So, Florez-Lopez et al. [28], have explored various complementary sources of diversity for optimizing the model's structure that leads to a manageable number of decision rules without affecting the performance [28]. The empirical results suggest that CADF is a good solution compared to individual classifiers and RF, gradient boosting-like ensemble strategies for credit risk problems. By seeing the improvement using AI/DL approaches, Huang et al. [46], have also used a probabilistic neural network (PNN) that gives a minimum error rate and the second type of error and the highest AUC value [46].

Further, the fraud detection problem was phrased as LSTM and sequence classification task to incorporate transaction sequences. Jurgovsky et al. [47], have also integrated the traditional feature aggregation approach and reported results in traditional retrieval metrics [47]. Comparing the proposed algorithm with the baseline RF classifier shows an improvement in detection accuracy on offline transactions. From manual features, aggregation approaches benefits are achieved by both sequential and non-sequential learning approaches.

To construct a credit risk assessment model Zhang et al. [48], have proposed a new approach for peer to peer lending market [48]. They first used a Transformer encoder to extract the textual features from the loan description then combined them with load application-derived challenging features and final loan features. Then send the combined features into a two-layer Feedforward NN for predicting the probability of default loans. The proposed approach was tested on Renrendai loan data from the Chinese market and LendingClub loan data from the American market datasets. The results show that a better outcome is achieved by the model in which the textual loan description is considered compared to loan default prediction, and the best outcome is achieved under AUC and G-mean metrics.

Further, Golbayani et al. [49], did a survey and gave a comparative outcome of results obtained by various ML and AI techniques in predicting credit rating [49]. Then they applied RF, SVM, Multilayer perceptron, and Bagged decision tree techniques on the same datasets and evaluated the results using 10 tenfold cross-validation techniques. The results show the best performance is achieved using a Decision tree-based model.

4.3 Sentiment Analysis in the Finance Sector Using Artificial Intelligence

The importance of analysing the massive volume of text from social networks and websites has been raised by developing online virtual communities. To develop a public mood dynamic prediction model has been developed by Chen et al. [50], by analysing online news articles and financial blogs [50]. This has been done concerning behavioural finance perspectives and characteristics of online financial communities. To Taiwan sentiment analysis investors opinion mining and big data approaches are applied in their work and verified their proposed model using China-Time.com, Google stock market news, Yahoo stock market news, and cnYES.com experimental datasets. The results obtained from it show that big data analysis techniques for assessing the emotional content of commentary on financial and current stock issues can be forecasted effectively. Further, Twitter data 1 2 has been considered by company stock prices and served the need for scoring the impression carried out for a particular firm. Das et al. [51], have made a classifying model from historical data that can improve outcomes [51]. For humongous data processing, spark streaming has been considered, along with Apache Flume and Twitter API-like data ingestion tools used for further implementation and analysis. Xu et al. [52], have also presented a continuous naïve Bayes learning framework to review multi-domain and large-scale e-commerce platform product sentiment classification [52]. They have also extended the naïve Bayes parameter estimation mechanism to a continuous learning style and then proposed various ways of fine-tuning the learned distribution based on three types of assumptions for adapting better to different domains. The experiment was conducted on the movie reviews and the Amazon product sentiment dataset.

Various news articles are engaged in prediction processes, but combining the technical indicators from news and stock price sentiments and making prediction models learn sequential information within time series better is still of concern. So, Li et al. [53], have proposed a new stock predicting system that is able to represent numerical price data using technical indicators and analysis and further represent the textual news articles using sentiment vectors of sentiment analysis [53]. Further, they have set up the deep learning model for learning the sequential information within the series of market snapshots that are constructed by news sentiments and technical indicators.

Rich source of information is represented by textual materials that improve the decision making of organizations, businesses, and people. Pröllochs et al. [54], have proposed an approach in which they have taken document level labels as input and then learn a document level labels-based negation policy [54]. There are various limitations in existing models. To address these problems, Mohammad Ehsan Basiri et al. [55], have proposed an attention-based Bidirectional CNN-RNN Deep Model (ABCDM) [55]. In this, they have used two independent bidirectional GRU, and ABCDM extracts LSTM layers and both past and future contexts by considering the flow of temporal information in both directions. Further, to put more or less emphasis on various words an ABCDM bi-direction layer was used that reduces the dimension of feature and extract the position invariant local features. On sentiment polarity detection, ABCDM effectiveness detection is considered the most common and necessary sentiment analysis task. For experiment purposes, three Twitter and five review datasets are used, and ABCDM results are compared with six proposed DNNs for sentiment analysis.

4.4 Algorithm Trading for Finance Traders Using Artificial Intelligence

The motive of Rys et al. [56], analysed and formulated the machine learning approaches with fixed strategy optimization specificity parameters [56]. Sensitivity performance is the most critical problem for little change in parameter and number of local extrema distribution over the solution space in a distinctive way. The approaches were designed for significant shortening of computation time without affecting the substantial strategy quality of loss. Their method was operated on 20 years of daily price sample data and presented three sets of two asset portfolios. The strategy was traded on DAX and SPX index futures in the case of first case, and in the second case, it was done on MSFT and AAPL stocks, and then the final case was done on CBF and HGF commodities futures.

For financial markets trading like forex and stock, AI has been increased, and out of all, reinforcement learning has become prevalent for financial traders. Meng et al. [57], have reviewed all current forex/stock predictions in which reinforcement learning has been used as a direct ML approach [57]. All the articles reviewed in this work have various unrealistic assumptions like no bid or ask spread issues and no liquidity and transaction costs. On reinforcement learning algorithms profitability, a significant impact has been seen in transaction costs compared to baseline-tested algorithms. They have also given a performance comparison between reinforcement learning and other DL or ML models and assessed the impact of bid/ask spread on the profitability of transaction costs. From overall work, it has been found that reinforcement learning in forex or stock trading helps in early development and also stated that there is a need for a reliable approach in the same domain. Then, Fengqian et al. [58], have used real-time financial data and processed it using K-line theory and candlesticks as a generalization price movement for a period that helps in de-noising [58].

Further, a decomposition of candlesticks is done into various subparts by using a specified Spatio-temporal relationship based on which subparts cluster analysis was obtained for getting the features of learning. Along with this, K-lines are used to clustered learning features that are added into the model, and unknown environment adaptive control parameters are realized using a deep reinforcement learning approach for realizing the high-frequency transaction strategy. To verify model performance, they have used various financial derivatives transactions like financial features, commodity features, and stocks. They also compared the proposed approach with fuzzified price, K-lines, and price-based methods. Fuzzy neural networks and recurrent neural network-like prediction-based approaches are used to verify the proposed method's accuracy which shows a higher prediction accuracy and robustness of the proposed method.

4.5 Prediction of Stock Price Indexing

There are various problems associated with the prediction of direction of movement of the stock and stock price index for Indian stock markets. Jigar Patel et al. [59], have compared SVM, NB, RF, and ANN four prediction models with two approaches for input to these models [59]. Computation of ten technical parameters computation is involved with the first approach for input data in which they have used open, low, close, and high prices stock trading data. On the other hand, these technical parameters are represented as trend deterministic data in the second approach. Then for these two input approaches, prediction model accuracy was evaluated for which ten years of historical data was used that is taken from 2003 to 2012 of Infosys and Reliance industries stocks. The results obtained from it show that out of the other three prediction models, the random forest can give better performance for the first approach of input data. Further, Bisoi et al. [60], have focused on two objectives, namely daily trend prediction and day ahead stock price prediction using a Robust Kernel-based Extreme Learning Machine (RKELM) integrated with VMD in which Differential evolution algorithm was used for the optimization of kernel function parameters DE-VMD-RKELM [60]. In the end, trend prediction was compared with SVM, ANN, and Naïve Bayes classifier that shows the superiority of the proposed model over other predictive methods.

With the use of AI and an increase in computational capabilities programmed prediction approaches are proven to be more efficient when used in the prediction of stock prices. For the prediction of next day closing of 5 different sectors of companies, ANN and RF techniques were used by Vijh et al. [61]. New variables are created by close, low, high, and open stock prices as financial data used as the input to the model. MAPE and RMSE standard strategic indicators are used to evaluate the models.

Solutions to various challenging problems have been provided by the current flow in the research of DL. New methods for these problems have been adopted in quantitative analysis. However, due to non-stationary financial data-like problems, significant challenges must be overcome before using DL. So, Tsantekidis et al. [62], have proposed a new approach for constructing stationary features that allow DL models to be applied efficiently [62]. The tasks of mid-price limited order book movements task are used for thoroughly testing these features. They have evaluated Convolutional Neural networks (CNN) and Recurrent Long Short Term Memory (LSTM) networks like DL models. The author has evaluated the novel model in which LSTM and CNN useful features' ability are extracted for analysing the time series. The outcome achieved from it shows that the combined model gives better results than tested individual CNN and LSTM models in the prediction horizons. Then, Chalvatzis et al. [63], have used tree-based models to test their proposed approach along with deep LSTM neural networks [63]. The results obtained on testing methods for the period of 2010–2019 show a 350, 403, 497 and 333% of the overall model achieved cumulative returns using S&P 500, Dow Jones Industrial Average (DJIA), NASDAQ, and Russell 2000 stock indices, respectively.

Furthermore, Hulu et al. [64], have proposed a novel stock closing price forecasting framework with higher prediction than traditional models (Hulu et al. 2020). This deep hybrid framework contains the predictor optimization method, deep learning predictor part, and data processing part components. In this pre-processing is done using empirical wavelet transform and in data processing a post processing is done using an outlier robust extreme learning machine. The primary part of this composite frame is an LSTM network-based deep learning predictor network that is jointly optimized by the PSO algorithm and dropout strategy. In their hybrid framework, every algorithm plays its functions for getting better prediction accuracy [64]. For forecasting experiments, three challenging datasets are used to verify the performance of their proposed model, and used various comparative models to prove the proposed framework's effectiveness.

5 Comparative Analysis

This section shows the comparative analysis of learning technologies applied in finance applications and classifiers employed by researchers for a particular application and their results. The data presented below in Table 2 will help understand which machine learning technology is better to implement in which finance application. This table can help in making Hybrid Models for improvising results in various aspects of finance applications.

Finance is all about data, and we have seen handling such big data and processing it. Both are challenging tasks [81–83]. Machine learning or AI is a process of learning that is training from the data and making predictions on testing based on the training data, which means training is a crucial step in the machine learning process. As data is big data, it makes the processing speed low. AI is an expensive technology [84, 85]. As we do not have a universal AI algorithm that can be applied to most finance applications, every time, a new application requires a new version of algorithms. It makes it very costly. As implementation cost is high, sometimes this technique fails to give accurate results and gives results with a high false rate. Acceleration in the access of databases or financial data is vital because accuracy somehow depends on it. So it can be suggested that this survey chapter can help researchers to have a brief overview of all available machine learning techniques applied in finance, and this can help them to build a hybrid model which can, in return, help in improvisation of accuracy in terms of processing time, cost and as well as performance.

Table 2 Comparative analysis of applications and technologies implemented

Application	Author	Learning technique	Dataset used	Classifiers	Parameters
Predicting stock price index	Kyoung-jae and Han [65]	Supervised learning	Korea	Neural networks, GA	Accuracy = 61.70%
	Kara et al. [66]	Supervised learning	Turkey	Neural networks, SVM	Accuracy = 75.74%
	Chiang et al. [67]	Supervised learning	Multiple	Neural networks	Accuracy = 87.80%
	Patel et al. [68]	Supervised learning	India	Neural networks, SVM, RF,NB	ANN= 86.69%, SVM = 89.33%, RF=89.98% NB = 90.19%
	Wang [69]	Supervised learning	Taiwan	Fuzzy logic	Accuracy= 93%
Financial distress prediction model	Chen et al. [70]	Supervised learning	Taiwan stock exchange corporation	ANN and DM	Accuracy = 82.14%
Algoritmic trading rule	Rodrıguez et al. [71]	Supervised learning	Spain	Neural networks	Accuracy = 54–58%
	Giacomel et al. [72]	Supervised learning	North American and Brazilian stock markets	Ensemble of neural networks	Capital hit rate = 60% And capital Gain = 56% in 166 days
	Taylor et al. [73]	Supervised learning	S&P 500	Artificial neural network	Accuracy = 74.45%
Fraud detection	Maes et al. [74]	Supervised learning	Europay International (EPI)	ANN & BBN	Accuracy of ANN = 60% BNN = 68% (±10% False POS)
	Raj et al. [75]	Supervised learning	FDS	Fuzzy logic	Accuracy = 100%
	Aleskerov et al. [76]	Supervised learning	Tested on synthetically generated data	Neural network based data mining	Accuracy = 85%
Credit scoring	Sayjadah et al. [77]	Supervised learning	Real-time data	Random forest	Accuracy = 81.81%

(continued)

Table 2 (continued)

Application	Author	Learning technique	Dataset used	Classifiers	Parameters
	Bing Zhu et al. [78]	Supervised learning	Real-world dataset from a Chinese consumer finance company	Relief algorithm + CNN	91.64%
Chatbots	Muangkammuen et al. [79]	Supervised learning	Thai text	RNN	Question recognition accuracy = 86.36% Answering accuracy = 93.20%
Portfolio management	Jiang et al. [23]	Supervised learning	Poloniex	Deep reinforcement learning method	tenfold returns in 1.8 months
High-frequency trading	Yang et al. [80]	Supervised learning	E-Mini S&P 500 future market	Inverse Reinforcement learning	Accuracy = 90%

6 Conclusion

In the presented finance chapter, we have seen that machine learning or AI is a good subset of data science, rapidly undergoing development. The financial market is exceptionally well suited for it, and its potential applications in finance are constantly growing. With the introduction of ML and AI in financial systems, number of data can be analysed, store, interpreted and calculated without explicit programming. This chapter addressed machine learning, artificial intelligence techniques and briefly commented on popular models, such as ANN, SVM, CNN, BFNN and RF, and presented a systematic review of various terms related to machine learning and artificial intelligence techniques that concern the finance sector. This provides an abstract view to the reader of artificial intelligence and its usage in various domains of the financial domain—the benefits and issues related to implementing digital solutions in the financial system. From the chapter, we also tried to present the earlier work done in this research field and have done comparative analysis that helps other researchers to use it efficiently and appropriately in future developments.

References

1. Aziz S, Dowling M (2018) Machine learning and AI for risk management. Springer International Publishing, pp 33–50
2. Prado MLD (2019) Ten applications of financial machine learning. SSRN Electron J, 1–16
3. Chen K, Yadav A, Khan A, Zhu K (2020) Credit fraud detection based on hybrid credit scoring model. International conference on computational intelligence and data science (ICCIDS 2019), 2–8
4. Buchanan BG (2019) Artificial intelligence in finance. Work was supported by The Alan Turing Institute
5. Klein KJ, Knight AP (2015) Innovation Implementation: overcoming the challenge. Curr Dir Psychol Sci 14:243–246
6. Mukherjee S, Shaw R (2016) Big data—concepts, applications, challenges and future scope. Int J Adv Res Comput Commun Eng 5:66–74
7. Henrique BM, Sobreiro VA, Kimura H (2019) Literature review: machine learning techniques applied to financial market prediction R. Expert Syst Appl 124:226–251
8. Heaton JB, Polson NG (2016) Deep learning in finance. Appl Stochastic Models Bus Industry, 1–20
9. Provost F, Fawcett T (2013) Data science and its relationship data-driven decision making. Foster provost and tom fawcett. Big Data 1:51–59
10. Marrara S, Azzini S (2018) Opportunities and risks for data science in organizations: banking, finance, policy, special session overview, 612–613
11. Yang B (2019) Construction of logistics fi nancial security risk ontology model based on risk association and machine learning. Saf Sci 123:1–11
12. Zhu Y, Zhou L, Xie C, Wang G, Nguyen TV (2019) Forecasting SMEs credit risk in supply chain finance with an enhanced hybrid ensemble machine learning approach. Int J Prod Econ 211:22–33
13. Fan Y (2015) Domain ontology construction method and empirical research. Wuhan University Press, Wuhan, pp 39–40
14. Florez-lopez R (2007) Modelling of insurers' rating determinants. An application of machine learning techniques and statistical models. Europ J Oper Res 183:1488–1512
15. Dorronsoro R, Ginel F, Carmen S, Cruz CS (1997) Neural fraud detection in credit card operations. IEEE Trans Neural Network 8:827–834
16. Wang C, Han D, Luo S (2019) A deep learning approach for credit scoring of peer-to-peer lending using attention mechanism LSTM. IEEE Access 7:2161–2168
17. Duhart B, Hernández-gress N (2016) Review of the principal indicators and data science techniques used for the detection of financial fraud and money laundering. International conference on computational science and computational intelligence (CSCI), 1397–1398
18. Harvey C, Liu Y (2015) Backtesting. J Portf Manag 42:13–28
19. Agarwal P (2019) Redefining banking and financial industry through the application of computational intelligence. 2019 Advance science engineering technology international conference, pp 1–5
20. Mehrotra A (2019) Artificial intelligence in financial services—need to blend automation with human touch. 2019 International conference automobile computer technology management, pp 342–347
21. Zhang P, Yu K, Yu JJ, Khan SU, Member S (2017) QuantCloud: big data infrastructure for quantitative finance on the cloud. IEEE Trans Big Data 4:368–380
22. Lin W, Hu Y, Tsai C (2012) Machine learning in financial crisis prediction: a survey. IEEE Trans Syst Man Cybern. Part C (Appl Rev) 42:421–436
23. Jiang Z (2017) Cryptocurrency portfolio management with deep reinforcement learning, IntelliSys
24. Cheng T (2018) AI Robo-advisor with big data analytics for financial services. 2018 IEEE/ ACM international conference advance social networks analysis mining, pp 1027–1031

25. Guo S, He H, Huang X (2019) A multi-stage self-adaptive classifier ensemble model with application in credit scoring. IEEE Access 7:78549–78559
26. Ben-David, Arie, Frank E (2009) Accuracy of machine learning models versus "hand crafted" expert systems–a credit scoring case study. Expert Syst Appl, 5264–5271
27. Snihovyi, Oleksandr, Ivanov O, Kobets V (2018) Implementation of Robo-advisors using neural networks for different risk attitude investment decisions. 2018 International conference on intelligent systems (IS). IEEE
28. Florez-Lopez R, Ramon-Jeronimo JM (2015) Enhancing accuracy and interpretability of ensemble strategies in credit risk assessment, A correlated-adjusted decision forest proposal. Expert Syst Appl, 5737–5753
29. Bae JK (2012) Predicting financial distress of the South Korean manufacturing industries. Expert Syst Appl, 9159–9165
30. Hsieh TJ, Hsiao HF, Yeh WC (2012) Mining financial distress trend data using penalty guided support vector machines based on hybrid of particle swarm optimization and artificial bee colony algorithm. Neurocomputing, 196–206
31. Lin F, Liang D, Yeh CC, Huang JC (2014) Novel feature selection methods to financial distress prediction. Expert Syst Appl, 2472–2483
32. Huang YP, Yen MF (2019) A new perspective of performance comparison among machine learning algorithms for financial distress prediction. Appl Soft Comput J, 1–14
33. Wang G, Ma J, Chen G, Yang Y (2020) Financial distress prediction: regularized sparse-based Random Subspace with ER aggregation rule incorporating textual disclosures. Applied Soft Comput J, 1–17
34. Sun J, Li H, Fujita H, Fu B, Ai W (2020) Class-imbalanced dynamic financial distress prediction based on Adaboost-SVM ensemble combined with SMOTE and time weighting. Inf Fusion, 128–144
35. Rönnqvist S, Sarlin P (2016) Bank distress in the news: describing events through deep learning. Neurocomputing, 1–32
36. Choi H, Son H, Kim C (2018) Predicting financial distress of contractors in the construction industry using ensemble learning. Expert Syst Appl, 1–44
37. Marso S, Merouani ME (2020) Predicting financial distress using hybrid feedforward neural network with cuckoo search algorithm. International workshop on statistical methods and artificial intelligence (IWSMAI 2020), pp 1134–1140
38. Jones S, Johnstone D, Wilson R (2015) An empirical evaluation of the performance of binary classifiers in the prediction of credit ratings changes. J Banking Finance, 1–43
39. Jadhav S, He H, Jenkins K (2018) Information gain directed genetic algorithm wrapper feature selection for credit rating. Appl Soft Comput, 1–50
40. Tian Y, Yong Z, Luo J (2018) A new approach for reject inference in credit scoring using kernel-free fuzzy quadratic surface support vector machines. Appl Soft Comput J, 1–26
41. Masmoudi K, Abid L, Masmoudi A (2019) Credit risk modeling using Bayesian network with a latent variable. Expert Syst Appl, 157–166
42. Xu F, Pan Z, Xia R (2020) E-commerce product review sentiment classification based on a naive Bayes continuous learning framework. Information processing and management, pp 1–7
43. Wang Y, Zhang Y, Lu Y, Yu X (2020) A Comparative Assessment of Credit Risk Model based on Machine Learnin: a case study of bank loan data. International conference on identification, information and knowledge in the internet of things, pp 141–149
44. Junior LM, Nardini FM, Renso C, Trani R, Macedo JA (2020) A novel approach to define the local region of dynamic selection techniques in imbalanced credit scoring problems. Expert Syst Appl, 1–16
45. Moscato V, Picariello A, Sperlí G (2021) A benchmark of machine learning approaches for credit score prediction. Expert Syst Appl, 1–8
46. Huang X, Liu X, Ren Y (2018) Enterprise credit risk evaluation based on neural network algorithm. Cognit Syst Res, 1–11
47. Jurgovsky J, Granitzer M, Ziegler K, Calabretto S, Portier PE, He-Guelton L, Caelen O (2018) Sequence classification for credit-card fraud detection. Expert Syst Appl, 1–30

48. Zhang W, Wang C, Zhang Y, Wang J (2020) Credit risk evaluation model with textual features from loan descriptions for P2P lending. Electronic commerce research and applications, pp 1–31
49. Golbayani P, Florescu I, Chatterjee R (2020) Comparative study of forecasting corporate credit ratings using neural networks, support vector machines, and decision trees. North Amer J Econ Finance, 1–16
50. Chen MY, Chen TH (2017) Modeling public mood and emotion: Blog and news sentiment and socio-economic phenomena. Future Gen Comput Syst, 1–27
51. Das S, Behera RK, kumar M, Rath SK (2018) Real-time sentiment analysis of twitter streaming data for stock prediction. Procedia Comput Sci, 956–964
52. Shen F, Zhao X, Kou G (2020) Three-stage reject inference learning framework for credit scoring using unsupervised transfer learning and three-way decision theory. Decision Support Syst, 1–10
53. Li X, Wu P, Wang W (2019) Incorporating stock prices and news sentiments for stock market prediction: a case of Hong Kong. Inf Process Manag, 1–19
54. Pröllochs N, Feuerriegel S, Lutz B, Neumann D (2020) Negation scope detection for sentiment analysis: A reinforcement learning framework for replicating human interpretations. Inf Sci, 205–221
55. Basiri ME, Nemati S, Abdar M, Cambria E, Acharrya UR (2021) ABCDM: an attention-based bidirectional CNN-RNN deep model for sentiment analysis. Future Gen Comput Syst, 279–294
56. Rys P, Slepaczuk R (2019) Machine learning in algorithmic trading strategy optimization—implementation and efficiency. Central Europ Econ J, 206–229
57. Meng TL, Khushi M (2019) Reinforcement learning in financial markets. MDPI, 1–17
58. Fengqian D, Chao L (2020) An adaptive financial trading system using deep reinforcement learning with candlestick decomposing features. IEEE Access, 63666–63678
59. Patel J, Shah S, Thakkar P, Kotecha K (2014) Predicting stock and stock price index movement using trend deterministic data preparation and machine learning techniques. Expert Syst Appl, 1–14
60. Bisoi R, Dash PK, Parida AK (2018) Hybrid variational mode decomposition and evolutionary robust kernel extreme learning machine for stock price and movement prediction on daily basis. Appl Soft Comput J, 1–66
61. Vijh M, Chandola D, Tikkiwal VA, Kumar A (2020) Stock closing price prediction using machine learning techniques. International conference on computational intelligence and data science (ICCIDS), pp 599–606
62. Tsantekidis A, Passalis N, Tefas A, Kanniainen J, Gabbouj M, Iosifidis A (2020) Using deep learning for price prediction by exploiting stationary limit order book features. Appl Soft Comput J, 1–10
63. Chalvatzis C, Hristu-Varsakelis D (2020) High-performance stock index trading via neural networks and trees. Appl Soft Comput J, 1–22
64. Carta S, Ferreira A, Podda AS, Recupero DR, Sanna A (2020) Multi-DQN: an ensemble of deep Q-learning agents for stock market forecasting. Expert Syst Appl, 1–46
65. Kyoung-jae K, Han I (2000) Genetic algorithms approach to feature discretization in artificial neural networks for the prediction of stock price index. Expert Syst Appl 19:125–132
66. Yakup K, Boyacioglu MA, Baykan OK (2011) Predicting direction of stock price index movement using artificial neural networks and support vector machines: the sample of the Istanbul Stock Exchange. Expert Syst Appl 38:5311–5319
67. Chiang WC, Enke D, Wu T, Wang R (2016) An adaptive stock index trading decision support system. Expert Syst Appl 59:195–207
68. Patel J, Shah S, Thakkar P, Kotecha lK (2015) Predicting stock and stock price index movement using trend deterministic data preparation and machine learning techniques. Expert Syst Appl 42:259–268
69. Wang YF (2003) Mining stock price using fuzzy rough set system. Expert Syst Appl 24:13–23
70. Chen W-S (2009) Du YK. Using neural networks and data mining techniques for the financial distress prediction model. Expert Syst Appl 36:4075–4086

71. Fernandez-Rodrıguez F, Gonzalez-Martel C, Sosvilla-Rivero S (2000) On the profitability of technical trading rules based on artificial neural networks: Evidence from the Madrid stock market. Econ Lett 69:89–94
72. Giacomel F, Galante R, Pereira A (2015) An algorithmic trading agent based on a neural network ensemble: A case of study in North American and Brazilian stock markets. IEEE/WIC/ACM international conference on web intelligence and intelligent agent technology (WI-IAT), 2
73. Taylor B, Kim M, Choi A (2014) Automated stock trading algorithm using neural networks. Proceedings of the 2nd international conference on intelligent technologies and engineering systems (ICITES2013). Springer, Cham, pp 849–857
74. Maes S, Tuyls K (2002) Credit card fraud detection using Bayesian and neural networks. Proceedings of the 1st international naiso congress on neuro fuzzy technologies, 1–8
75. Raj BES, Portia AA (2011) Analysis on credit card fraud detection methods. International conference on computer, communication and electrical technology (ICCCET), IEEE, pp 152–156
76. Aleskerov E, Freisleben B, Rao B (2002) Cardwatch: A neural network based database mining system for credit card fraud detection. Proceedings of the IEEE/IAFE computational intelligence for financial engineering (CIFEr). IEEE, pp 220–226
77. Yashna S, Hashem IAT, Alotaibi F, Kasmiran KA (2018) Credit card default prediction using machine learning techniques. 2018 Fourth international conference on advances in computing, communication & automation (ICACCA), IEEE, pp 1–4
78. Zhu B, Yang, W, Wang H, Yuan Y (2018) A hybrid deep learning model for consumer credit scoring. International conference on artificial intelligence and big data (ICAIBD), IEEE, pp 205–208
79. Muangkammuen P, Intiruk N, Saikaew KR (2018) Automated Thai-FAQ Chatbot using RNN-LSTM. 22nd International computer science and engineering conference (ICSEC), IEEE, pp 1–4
80. Steve Y, Paddrik M, Hayes R, Todd A, Kirilenko A, Beling P, Scherer W (2012) Behavior based learning in identifying high frequency trading strategies. IEEE Conference on computational intelligence for financial engineering & economics (CIFEr), pp 1–8
81. Unadkat V (2018) Deep learning for financial prediction. 2018 International conference circuits system digital enterprise technology, pp 1–6.
82. Gong XL, Liu XH, Xiong X, Zhang W (2019) Financial systemic risk measurement based on causal network connectedness analysis. Int Rev Econ Financ 64:290–307
83. Luo J, Xin Y, Tian Y (2020) Unsupervised quadratic surface support vector machine with application to credit risk assessment. Europ J Oper Res, 1008–1017
84. Randhawa K, Loo CK, Seera M, Lim CP, Nandi AK (2018) Credit card fraud detection using AdaBoost and majority voting. IEEE Access 6:14277–14284
85. Shenyong X, Yu H, Wu Y, Peng Z, Zhang Y (2017) Self-evolving trading strategy integrating internet of things and big data. IEEE Internet Things J 5:2518–2525

Chapter 2
Covid-19 Related Ramifications on Financial Market: A Qualitative Study of the Pandemic's Effects on the Stock Exchange of Big Technology Companies

Pragya Gupta, Drishti Jain, B. Ida Seraphim, and Rashima Mahajan

Abstract Due to the sudden outbreak of novel Coronavirus (Covid-19), the world economy came to an abrupt standstill. The financial market almost collapsed during quarantine, workers were left jobless and many companies ran bankrupt. One of the methods for investigating the impact of Covid-19 on financial market is by analysing the stock market. The daily fluctuations of stock prices help the investors to get an insight about the overall stock market. Hence, in order to study the impact of Covid-19 on financial market, a comprehensive comparison of the stock market pre-Covid-19, during Covid-19 and post-Covid-19 for tech giants like Google, Apple, General Electric, IBM and Microsoft has been implemented. The dataset has been imported using the Yahoo Finance API (Application Programming Interface).

Keywords Covid-19 · Stock market · Yahoo finance API · Forecasting · Linear regression · Time series

---

P. Gupta (✉) · D. Jain · B. I. Seraphim
SRM Institute of Science and Technology, Kattankulathur, Chennai, India
e-mail: pg8978@srmist.edu.in

D. Jain
e-mail: dj7278@srmist.edu.in

B. I. Seraphim
e-mail: idaserab@srmist.edu.in

R. Mahajan
Manav Rachna International Institute of Research and Studies, Faridabad, India
e-mail: rashima.fet@mriu.edu.in

S. Kautish et al. (eds.), *Computational Intelligence for Modern Business Systems*, Disruptive Technologies and Digital Transformations for Society 5.0,
https://doi.org/10.1007/978-981-99-5354-7_2

1 Introduction

The sudden outbreak of novel Coronavirus (Covid-19) led the world to come to an abrupt standstill. The virus was initially reported by WHO (World Health Organization) on December 30, 2019, and thereafter, WHO declared it a global health emergency followed by a global pandemic outbreak [1]. Governments and world organizations ordered the civilians to maintain social distance, as well as to stay in quarantine for months. The strict quarantine rules and regulations exhibited an adverse impact on the global financial market as many people and industries were unable to function in a "work-from-home" world riddled with travel bans. The world economy was affected severely. In order to get a bigger picture, the stocks of five leading brands of the world, namely, Google, Apple, General Electric, IBM and Microsoft have been thoroughly studied. Despite being huge multinational corporations, these tech giants experienced a sudden decline in stocks. So, it is safe to say that even smaller and medium-sized companies would've incurred major risks and losses. The daily fluctuations in the stock prices help to analyse the overall stock market scenario, which can further be used to forecast future stock prices. Looking at the past trends, it is evident that the world hasn't seen a sudden drop in the stock market in past 10 years. Inspecting the U.S. stock market through the lens of Dow, it is evident that the market was already in decline around Covid-19. Thereafter, by mid-March 2020, the Dow reached its minimum point, registering a 2,999 points' drop in a single day on March 16th, which summed up to a drop of 36% within the course of 2 months and took 8 months to recover [2]. Accurate prediction of stock market returns is a very challenging task due to the non-linear and volatile nature of the financial stock markets. With the introduction of artificial intelligence and increased computational capabilities, programmed methods of prediction have proved to be more efficient in predicting stock prices [3]. In order to forecast the stock prices, Linear Regression model can be used. The financial data, namely, High, Low, Open and Close prices of stock can be used for creating new variables (high-low percentage, percentage change) which are further used as inputs to the model.

The primary aim of the findings is to forecast the financial growth of big tech giants based on existing trends. If we consider that a massive part of the world economy is dependent on these companies, it will enable us to get a brief insight as to what will be the overall scenario of the market. To study the patterns, we plotted and studied the graphs under three timelines: pre-Covid, during Covid and post-Covid. These helped us to examine the trends in detail and predict the expected returns and find the correlations among these tech giants.

2 Related Work

Dinesh and Girish [4] cited in their paper that irregular stock price variations make it difficult for investors to estimate the risk factor and the return values of the stocks, hence, in order to help stock brokers for investing money in stock market, the daily behaviour of stock market can be forecasted based on certain parameters to encourage investors to invest with more confidence by taking risk factor and fluctuations in consideration. Upon comparing various machine learning models to forecast the stock price, they came to the conclusion that the linear regression model provides the best result compared to polynomial and RBF regression. One major shortcoming of their work was that it couldn't help to detect the variations of the stocks based on two of the major parameters, i.e., risk factor and expected return values.

Xu [5] provided great insight on the necessity of both fundamental and technical analysis to predict the economic pattern of a company by forming an algorithm that outperforms conventional time series analysis in stock market forecasting. He combined time series analysis technique with information from the Google trend website and the Yahoo Finance API to predict weekly changes in stock price. The experimental setup was implemented by applying ARMA (Autoregressive–moving-average) time series analysis to propose an algorithm in order to evaluate different parameters, then regress changes and study relationships to improve the performance of forecasting process. One major shortcoming was that the Google trend index record of just one company was extracted for the prediction pattern of the entire stock market, which couldn't give a proper insight into the entire stock market. On contrary, more refined data could be used to increase accuracy. In general, the weekly stock price changes within different intervals of new values were predicted correctly with some exceptions.

3 Methodology

The historical data for the five companies (Google, Apple, General Electric, IBM and Microsoft) has been collected from Yahoo Finance [6]. We used dataset for three different timeframes, i.e., pre-Covid-19, during Covid-19 and post-Covid-19. The pre Covid-19 data is dated from 1 January 2015 to 1 April 2020. The during Covid-19 data is dated from 1 January 2020 to 1 May 2020. The post-Covid-19 data is dated from 1 November 2020 to 19 November 2021. The data contains information about the stock such as High, Low, Open, Close, Adjusted close and Volume. Only the day-wise Adjusted Close price of the stock has been extracted because it incorporates actions, such as stock splits, dividends, and rights offerings which give a better overview of the stock trends. The methodology used here mainly revolves around Predictive Modelling which is a method of predicting future outcomes by using data modelling.

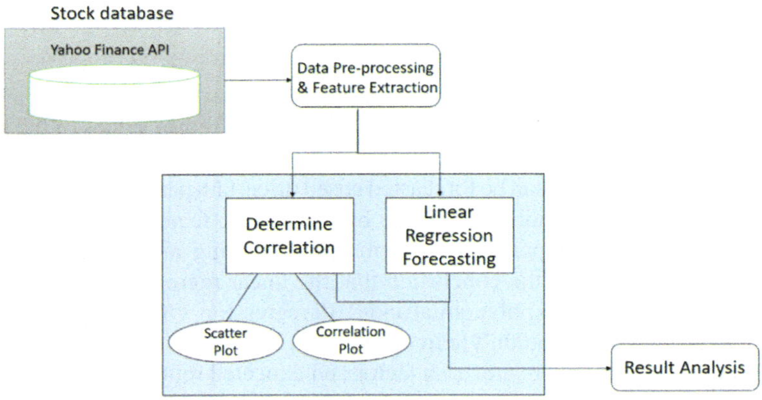

Fig. 1 Stock price forecasting and correlation model

3.1 Architecture

The model follows a simple architecture, which involves data cleaning and pre-processing using various data mining techniques. The dataset has been imported using the Yahoo Finance API. The relevant features were extracted, which involved High, Low, Open and Close prices of stock, along with adjusted close value. The main parameter used is Adjusted Close Price feature for the accuracy of the stock prediction. The pre-Covid-19 dataset is further used to train and plot the linear regression forecast for stock prices. Further, various correlations are plotted to study the impact of Covid-19 on the five tech giants. Scatter Plot and Correlation plots are used to depict the relation between the stock prices of various companies, which is further used to plot the risk factor and expected return value for the three different timelines (Fig. 1).

The Prediction Model using Linear Regression Method has been built using Python Programming. The aim is to predict a stock's Adjusted Close using historical data.

3.2 Data Representation

The dataset has been imported using the Yahoo Finance API. Yahoo Finance is a media platform that provides financial news, data about stock quotes, press releases, and financial reports. Various features like High, Low, Open and Close prices of stock, along with the adjusted close value were used. High and low refer to the maximum and minimum prices in a given time period. Opening and closing prices are those at which a stock began and ended trading in the same period. Closing price is the raw price of the cash value of the last transacted price before the market closes.

Adjusted closing price factors in corporate actions, such as stock splits, dividends and rights offerings.

The historical data for the five companies (Google, Apple, General Electric, IBM and Microsoft) has been collected from Yahoo Finance. The Adjusted Closing Price dataset for three different timeframes, i.e., pre-Covid-19, during Covid-19 and post-Covid-19 has been used. The pre-Covid-19 data is dated from 1 January 2015 to 1 April 2020. The during Covid-19 data is dated from 1 January 2020 to 1 May 2020. The post-Covid-19 data is dated from 1 November 2020 to 19 November 2021. Yahoo Finance has a set of specific symbols denoted for each company, as follows: AAPL for Apple, GE for General Electric, GOOG for Google and MSFT for Microsoft.

(For ease of understanding, the following tables show the tail elements of dataset for the given timeframes).

Pre-Covid-19 Adjusted Closing Price. (1 Jan 2015 to 1 April 2020) (Table 1).

During Covid-19 Adjusted Closing Price. (1 Jan 2020 to 1 May 2020) (Table 2).

Post-Covid-19 Adjusted Closing Price. (1 Nov 2020 to 19 Nov 2021) (Table 3).

Table 1 Pre-Covid-19 adjusted closing price

Date	Symbols				
	AAPL	GE	GOOG	IBM	MSFT
2020–03-26	63.8154	64.5491	1161.75	98.6546	153.607
2020–03-27	61.1733	60.5744	1110.70	94.4074	147.300
2020–03-30	62.9190	62.7208	1146.81	98.6895	157.661
2020–03-31	62.7906	63.1182	1162.81	96.9417	155.182
2020–04-01	59.5468	55.9638	1105.61	91.8818	149.671

Table 2 During Covid-19 adjusted closing price

Date	Symbols				
	AAPL	GE	GOOG	IBM	MSFT
2020–04-27	69.9218	51.1146	1275.88	110.041	171.260
2020–04-28	68.7884	54.0559	1233.67	110.347	167.088
2020–04-29	71.0478	52.307	1341.47	112.462	174.586
2020–04-30	72.5467	54.0559	1348.66	109.726	176.337
2020–05-01	71.3787	51.6711	1320.60	106.502	171.772

Table 3 Post-Covid-19 adjusted closing price

Date	Symbols				
	AAPL	GE	GOOG	IBM	MSFT
2021–11-15	150.000	106.669	2987.76	118.870	335.456
2021–11-16	151.000	103.349	2981.52	118.459	338.890
2021–11-17	153.490	101.989	2981.23	118.059	339.119
2021–11-18	62.7906	100.669	3014.17	116.660	341.269
2021–11-19	157.869	99.959	2999.05	116.050	343.109

3.3 Additional Variables

In order to forecast the predicted stock price values, some additional variables are supposed to be added. One of them is High-Low Percentage (HL_PCT), i.e., the breadth indicator which measures the percentage of Net New Highs for a particular group of stocks, such as ETF or an index. The other required variable is Percentage Change (PCT_change), i.e., the percentage difference between closing and open stock price.

Formulae:

(i) High-Low Percentage (HL_PCT) = ((High-Low) / Close) * 100
(ii) Percent Change (PCT_change) = ((Close-Open) / Open) * 100

The linear regression forecasting has been implemented for the pre-Covid-19 dataset, so a new table is created with the additional variables as follows (Table 4).

By taking Adjusted Close, Volume, High-Low Percentage, Percentage Change and Label, the stock prices for the next 10 months based on the historical data during the peak of Covid-19 can be forecasted.

Table 4 Forecasting dataset

Date	Symbols				
	Adjusted close	Volume	HL_PCT	PCT_change	Label
2020–03-26	1161.75	3,571,700	6.5797	4.49271	1142.31
2020–03-27	1110.70	3,208,500	4.0298	-1.3289	1175.76
2020–03-30	1146.81	2,574,100	4.8089	1.93592	1193.19
2020–03-31	1162.81	100.669	3.1965	1.35187	1193.31
2020–04-01	1105.61	2,344,200	2.9160	-1.4598	1185.55

3.4 Linear Regression Forecasting

Linear regression forecasting has been used for predicting the exchange return of the stocks. It is utilized in science, business, and just any other field where forecasting and predictions are relevant. Linear regression model helps to identify the relationship between dependent variable and one or more independent variables. Simple linear regression model is defined by using a feature to predict an outcome. The features used in this case are Adjusted Close, Volume, High-Low Percentage, Percentage Change. The pre-Covid-19 dataset was used, out of which 20% of the data had been segregated in order to train it and forecast the prediction for the next 10 months, i.e., From April 2020 to January 2021. The adjusted close value graph is plotted from May 2019 to April 2020. It shows steady growth till February 2020, but as soon as March 2020 hits, a sudden decrease in the trend from 1550 to 1050 price value is evident. This sudden decrease in adjusted close price was experienced because of the sudden outbreak of Covid-19 during its first wave. On the basis of this trend, the adjusted close price for upcoming months can be predicted. Now based on the historical data of the adjusted close price till 1 April 2020, the forecast for the next 10 months is plotted as follows (Fig. 2).

The forecasting model gives an accuracy of 75%. It is evident that the sudden drop in the stock price during February–March 2020 leads to a decreased prediction for the follow-up months.

Fig. 2 Stock price forecasting and correlation model

3.5 Correlation Between Google, Apple, General Electric, IBM & Microsoft

For comparing the scenario of the stock market pre-Covid-19, during Covid-19 and post-Covid-19, the correlation between the adjusted closing stock prices of the five tech giants throughout three different timelines has been studied. The correlation has been calculated using the Pandas library in Python. This gives the correlation matrix for each company with one another, which helps to study the similarities or disparities in the trends of the Adjusted Close Price values of all the companies. Furthermore, seaborn and Matplotlib packages of Python have also been used in order to get a visual representation of the correlation matrix, hence a heat map for all three timeframes has been plotted. The correlation of each company is plotted against one another for pre-Covid-19, during covid-19 and post-Covid-19 scenarios. Furthermore, these correlations have been thoroughly studied to understand the impact of Covid-19 on the stock market of these tech giants.

Pre-Covid-19 correlation. (1 Jan 2015 to 1 April 2020) (Table 5).

The correlation between the five tech giants can be visualized using correlation plot, where, lower the correlation between two entities, darker the color of the plot. From the first scenario, it is evident that the overall correlation between the Adjusted Close Price of all the companies is very low, with an average of 0.4. Google and Microsoft exhibit the highest correlation of 0.7214 which shows that they have almost similar increasing or decreasing trends in the adjusted close values of their stock prices. While, General Electric and Google have the least correlation, i.e., 0.3877 which means that the change in their adjusted close price values is not similar (Fig. 3).

During Covid-19 correlation. (1 Jan 2020 to 1 May 2020) (Table 6).

The correlation can be visualized using a correlation plot, where higher the correlation between two entities, lighter the color of the plot. It is evident from the correlation plot that the trend of all the companies is quite similar, i.e., there is some external factor that has led to the sudden increase or decrease in their adjusted close price values at the same time. The overall correlation of all the companies is quite high, with an average of 0.7 (higher than pre-Covid-19 scenario). This implies that because of Covid-19, all the companies faced a similar trend in their adjusted close

Table 5 Pre-Covid-19 correlation

Symbols	Symbols				
	AAPL	GE	GOOG	IBM	MSFT
AAPL	1.0000	0.4264	0.6122	0.5206	0.6751
GE	0.4264	1.0000	0.3877	0.4867	0.4294
GOOG	0.6122	0.3877	1.0000	0.5118	0.7214
IBM	0.5206	0.4867	0.5118	1.0000	0.5889
MSFT	0.6751	0.4294	0.7214	0.5889	1.0000

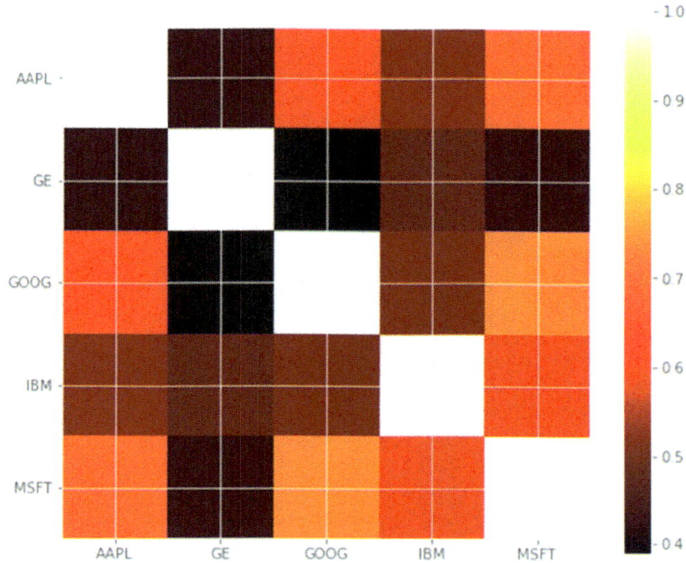

Fig. 3 Correlation plot for pre-Covid-19

Table 6 During Covid-19 correlation

Symbols	Symbols				
	AAPL	GE	GOOG	IBM	MSFT
AAPL	1.0000	0.7303	0.8982	0.8050	0.9335
GE	0.7303	1.0000	0.6775	0.7270	0.7169
GOOG	0.8982	0.6775	1.0000	0.7970	0.9122
IBM	0.8050	0.7270	0.7970	1.0000	0.7980
MSFT	0.9335	0.7169	0.9122	0.7980	1.0000

price values of the stocks, which concludes the fact that Covid-19 indeed had a huge impact on the financial market and global economy (Fig. 4).

Post-Covid-19 correlation. (1 Nov 2020 to 19 Nov 2021) (Table 7).

From the post-Covid-19 scenario, it is evident that the overall correlation of the adjusted close price values for all the companies has greatly decreased, with an average of 0.3. Some companies have correlation in negative (Microsoft and General Electric). This shows that after the first wave of Covid-19, some companies recovered from their losses at a faster rate than others (Fig. 5).

After comparing the three timelines, it is observed that in the pre-Covid-19 scenario, there isn't much correlation in the adjusted close values of the five companies. This is because every company worked on its own terms and under different circumstances. But as we come across the scenario during Covid-19, it can be

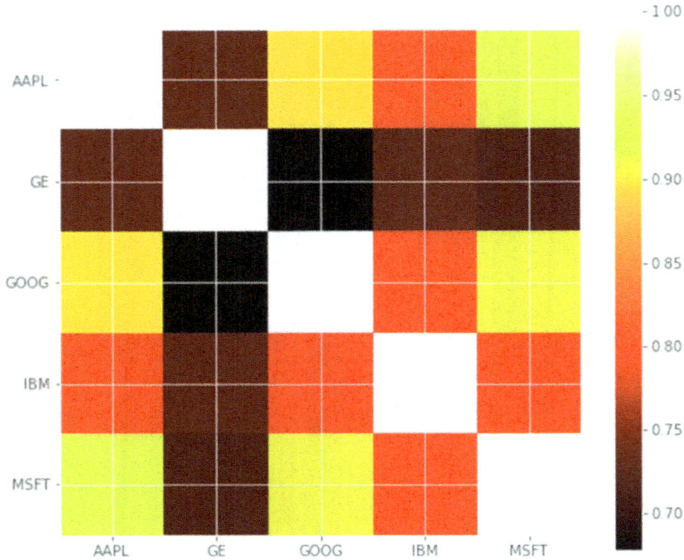

Fig. 4 Correlation plot for during Covid-19

Table 7 Post-Covid-19 correlation

Symbols	Symbols				
	AAPL	GE	GOOG	IBM	MSFT
AAPL	1.0000	0.0542	0.5372	0.0323	0.6578
GE	0.0542	1.0000	0.0185	0.4123	-0.1120
GOOG	0.5372	0.0185	1.0000	0.0525	0.6977
IBM	0.0323	0.4123	0.0525	1.0000	-0.0161
MSFT	0.6578	-0.1120	0.6977	-0.0161	1.0000

observed that suddenly there is an increase in the correlation among the five companies as it was during this time when all the companies faced the same circumstances and external factors (Covid-19). So, since Covid-19 was the common external factor that influenced the stock prices of all the companies, it is evident that Covid-19 had an adverse effect on the financial market.

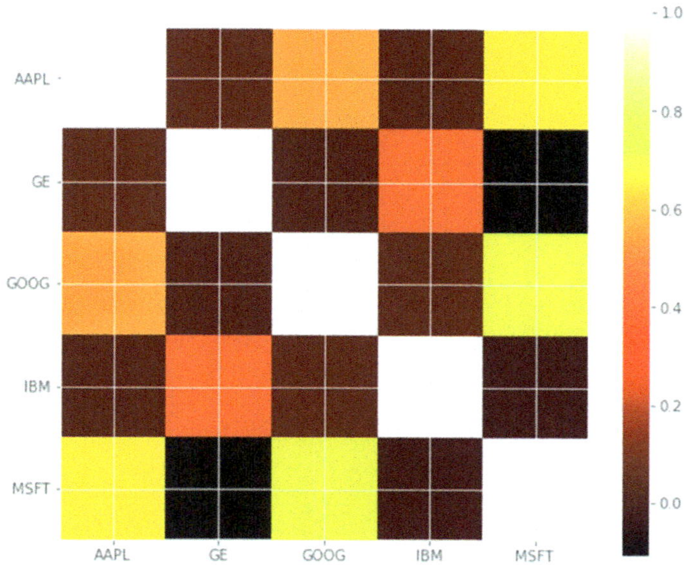

Fig. 5 Correlation plot for during Covid-19

4 Findings and Discussions

Based on the correlation table and correlation plot, a correlation scatter plot across two parameters, i.e., Risk Factor and Expected Return Value [7] has been plotted. These two parameters help in determining whether a business is running at loss or profit. Higher risk factor accompanied with lower expected return value exhibits that a company is running in loss or a decreasing market value, whereas, lower risk factor and higher expected return values determine that the company is running in profit or an increasing market value. The correlation scatter plot between the five companies during three different timelines gave a clear picture about the impact of Covid-19 on financial market. All the companies faced the same manner of decline during March 2020-May 2020, with their increase in risk factor up to an average of 40% and decrease in expected return values which were almost negative. This shows that some external factor has led to this sudden change in the market values of these companies. In order to study these variations in depth, correlation scatter plots for all three timelines have been plotted.

4.1 Pre-Covid-19 (1 Jan 2015 to 1 April 2020)

The pre-Covid-19 scenario shows a less risk factor for all five companies, with an average of 18%, while the expected return is overall positive. This scenario is right

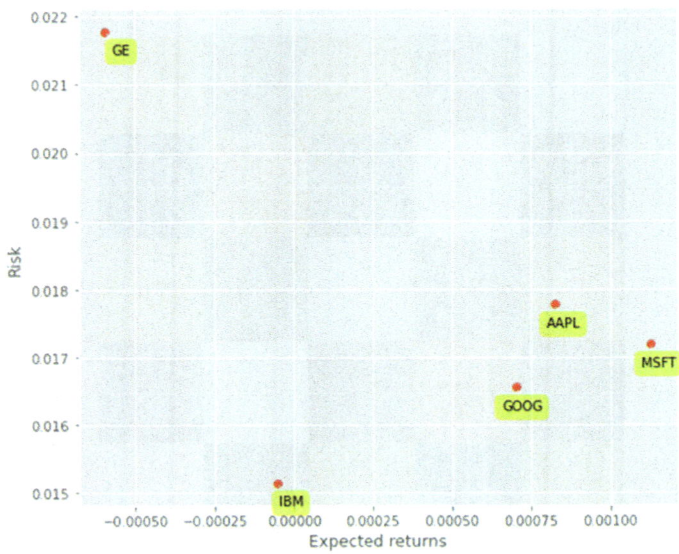

Fig. 6 Correlation scatter plot for pre-Covid-19

before when Covid-19 pandemic spread globally bringing the whole world to a standstill (Fig. 6).

4.2 During Covid-19 (1 Jan 2020 to 1 May 2020)

It is evident from the during Covid-19 scenario, that the companies incurred a sudden loss in their overall official functioning. This was during the time when Covid-19 outbreak led to the decision of shutting down offices and workplaces in order to follow strict social distancing rules and regulations, which further resulted in an adverse effect on the industries and businesses. Companies like Google, Apple, General Electric, IBM and Microsoft incurred a decline in their expected return (negative return value) and sudden increase in the risk factor (up to 50%), so it is safe to say that even the small and medium level business would've incurred the loss in similar pattern, where chances are that many of those companies might've run into bankruptcy.

4.3 Post-Covid-19 Dataset (1 Nov 2020 to 19 Nov 2021)

The post-Covid-19 correlation scatter plot is plotted to give an insight about the aftermath of the global pandemic. The plot depicts that the companies restored whatever losses they incurred during Covid-19, and are again functioning with a decreased risk factor, which dropped to an average of 17% and a positive increase in expected return value.

5 Conclusion

From the final findings and results, it is concluded that Covid-19 pandemic had a hard-hitting effect on the world financial market during the course of five months (Jan 2020 to May 2020) when the pandemic was at its peak and the civilians were in panic without availability of any immediate cure to the virus. The sudden outbreak forced the companies to reduce their workforce on-site and practise strict work-from-home policy. Every industry faced sudden loss in their overall functionalities, and it is evident from the inspection of the increased risk factor and reduced expected return of the five tech giants (Fig. 7). But after the course of 7–8 months, the industries were back on their forefeet as it can be concluded from Fig. 8. The industries learned and evolved to keep working with the "new normal" scenario of social distancing and work-from-home by coming up with various new policies and revolutionary technologies [8]. Moreover, soon after countries like USA, UK, India, Israel, etc., started working towards the cure of coronavirus. With the availability of vaccinations and rapid testing drives held by the governments, the strict social-distancing rules were relieved [9]. In order to conclude, it can be said that Covid-19 had an adverse effect on the global financial market for the span of almost 7–8 months, but the losses were soon recovered after the constant efforts of countries, governments and organizations to tackle the pandemic.

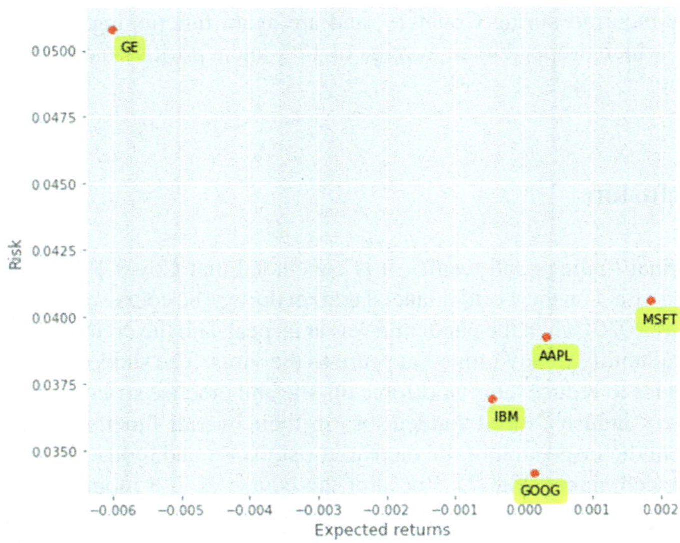

Fig. 7 Correlation scatter plot during Covid-19

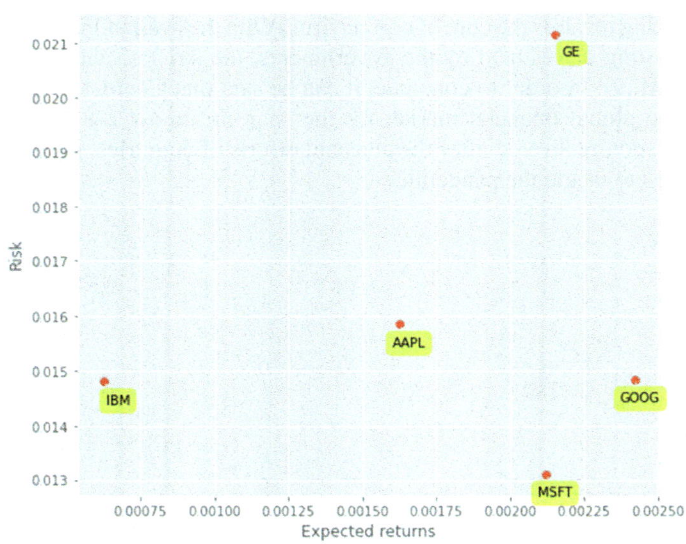

Fig. 8 Correlation scatter plot for post-Covid-19

References

1. Fauci AS, Lane HC, Redfield RR (2020) Covid-19—navigating the uncharted. In: The New England Journal of medicine
2. UNDP Data Futures Platform. https://data.undp.org/content/covid-19-stock-market-impact/
3. Vijh M, Chandola D, Tikkiwal VA, Kumar A (2019) Stock closing price prediction using machine learning techniques. In: International conference on computational intelligence and data science (ICCIDS 2019)
4. Bhuriya D, Kaushal G, Sharma A, Singh U (2017) Stock market prediction using a linear regression. In: 2017 International conference of electronics, communication and aerospace technology (ICECA)
5. Xu SY, Stock price forecasting using information from yahoo finance and google trend. https://www.econ.berkeley.edu/
6. Yahoo Finance, https://au.finance.yahoo.com/
7. Fischer Black, Estimating Expected Return. Financial Analyst J, October 1993
8. Khurana S, Haleem A, Luthra S, Huisingh D, Mannan B (2020) Now is the time to press the reset button. In: Cleaner production
9. Haynes BF, Corey L, Fernandes P,. Gilbert PB, Hotez PJ, Rao S, Santos MR, Schuitemaker H, Watson M, Arvin A (2020) Prospects for a safe COVID-19 vaccine. In: Science translational medicine

Chapter 3
Computational Intelligence Techniques for Behavioral Research on the Analysis of Investment Decisions in the Commercial Realty Market

S. Siva Venkata Ramana, T. Mydhili, Ponduri Siddardha, Gomatam Mohana Charyulu, and K. Saikumar

Abstract The real estate market shows huge behavioural dispositions recorded in the customary financial markets. The principal goal of this examination is to recognize the Behavioral Factors that impact the assessment of investment of investors in the realty market. The primary target of this examination is to characterize the feelings-based hypotheses utilized to clarify the financial exchange issues and terms. In this paper, it is realized that feelings can't generally spur investors, and it isn't vital that the property market effectively be adequate at the feeble structure. There is a need for a profound examination of the hypothesis of behavioural account. This investigation is helpful to comprehend the investments by utilizing the behavioural model. Using digital and statical analysis using ML techniques get a chance to improve Behavioral Research on The Analysis of Investment Decisions in The Commercial Realty Market and further analyse the stock. Investors consistently need to put resources into those tasks with more prominent benefits and the capital's base odds of risk or loss.

S. S. V. Ramana (✉) · T. Mydhili
Department of Management Studies, Vignan's Nirula Institute of Technology and Science for Women, Guntur, Andhra Pradesh, India
e-mail: ritikabateja@gmail.com

P. Siddardha
R.V.R. & J.C. College of Engineering, Chowdavaram, India

G. M. Charyulu
Department of S&H, VFSTR Deemed to Be University, Valdamudi, Andhra Pradesh, India

K. Saikumar
Department of C.S.E., Malla Reddy University, Hyderabad (MRUH), Hyderabad, Telangana, India

Researcher, Green Fields, Vaddeswaram, Guntur 522302, Andhra Pradesh, India

K. Saikumar
e-mail: saikumarkayam4@ieee.org

S. Kautish et al. (eds.), *Computational Intelligence for Modern Business Systems*, Disruptive Technologies and Digital Transformations for Society 5.0,
https://doi.org/10.1007/978-981-99-5354-7_3

Keywords Realty market · Financial experts · Behavioural account · Financial matters · Human behaviour · Financial exchange issues · Financial markets · Computational intelligence techniques

1 Introduction

Behavioural Finance, which differs from a conventional account and its uncertainties of financial specialist reasonability and market effectiveness, is based upon intellectual psychology science of how individuals think and the cutoff points to circumstances when markets are uneconomical. Instead of utilizing all accessible data to assess investments, financial specialists sift through some data. Various observational investigations have been completed on behavioural account factors affecting investment choices [1]; in an examination of the well-known perceptual mistakes among financial specialists and its association with their behaviour utilizing 200 investors in securities exchange, found that there is an immediate positive connection between's behavioural factors and investment choices. In another investigation, [2] examined the impact of behavioural budgetary information on the practices of the investors in Stock exchangers found that behavioural factors, for example, similarity, commonplace idea and instance direction, emphatically and essentially impact basic venture leadership [3], investigated how overconfidence may influence the basic investment leadership of the individual investment that found no connection between overconfidence and investment choices, as shown in Fig. 1.

The term real estate generally represents a business that is not considered a profession. A profession applies science or figuring out how to be useful to other people, the assistance to the educator or individual pertaining it as individual dealings. In contrast, a business is fundamentally connected to gain profits from the one involved

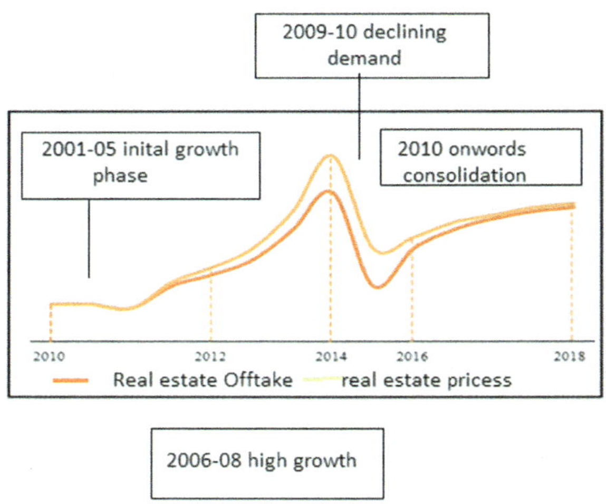

Fig. 1 Real estate Price and off-take levels

in the business. A profession suggests claimed achievement in exceptional information. An individual may take part in business with or without extraordinary information. Nobody is worried about the inquiry whether he has any information on the business because the outcome influences nobody else. In any case, the Real estate profession does not correlate with an assortment of morals administering it. Every business can maintain morals for the growth of their business for gaining benefits in the long term. Persons who have achieved success in the Real estate business have proficient achievement since they applied to their business maintaining standards in managing human resources and the public.

The real estate market in India is growing continuously as development is in progress. The persons involved in the business have to provide assets to the growing needs without any disputes. The level of the real estate market has a huge increment and has risen from 20–40% in the years between 2010–2020. The retail property advertises a wide scope of investment objects, running from littler retail to enormous retail edifices. The property showcase also comprises various retail focuses, from small neighbourhood malls to enormous strip malls. Whatever retail focus the speculator is keen on, the base for the result of all business Real estate ventures is its occupants and, in this way, the obtaining power these can draw in. The resources available in the real estate market need complete information in today's economy. The constraints and factors are to be specified in the asset factors also to be addressed.

The real estate market needs to follow several factors to exhibit the available assets to sell or buy assets. The benefits for the investors are to be specified compared to other real estate specialists. The real estate investor needs to monitor the asset factors keenly, and the benefits gained when investing in a particular asset. The important factors need to be clarified for gaining the benefits with proper investment. The investor can support the real estate market financially only when the market level is high and if investors are gaining reasonable benefits from the assets they financed, as shown in Fig. 2.

Corrado and Jordan [4] contend that a mixture exists between the company's Real estate technique and its corporate system, which brings about more vulnerable

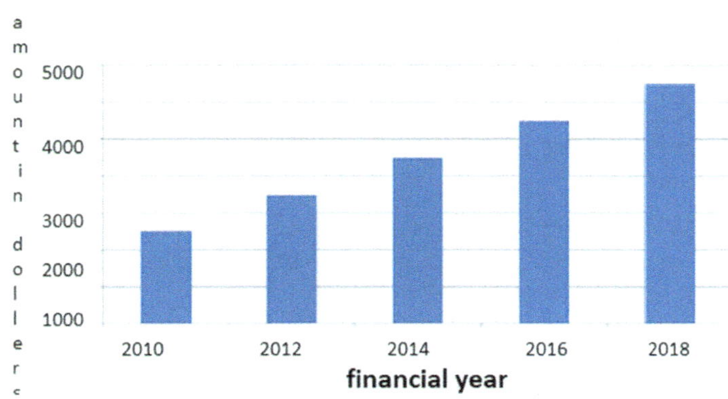

Fig. 2 India market size in Real estate

Real estate choices. The authors require a progressively key arranging corporate Real estate market [5]. Further, compete that accepting retail occupants in Real estate is the most intense consequence, and assets are being spent reviewing probable ventures. Kahuthu [6] indicates that the retail inhabitants may contrast in size, the executives, deals, their introduction to customer prototype, and their situations in their market and locations towards their competitors.

2 Need of Investing in the Realty Market

As of late, real estate has increased more interest from investors looking for venture openings that offer a chance and effect between the more dangerous and the less dangerous assumptions. Real estate has gone from being simply a space provider to being viewed as an important asset because of expanded worldwide competition and I.T. industry and made up around 10–30% of Indians all-out resources between 2014–2020.

The retail property platform offers a wide scope of speculation objects, going from littler retail centers to big retail markets. Likewise, the retail property market comprises various kinds of retail centers, from little malls to large malls. Whatever retail place the financial specialists are keen on, the base for the result of all business speculations is its inhabitants, and in this manner, the investment can draw in.

Hence, putting resources into retail property requires a wide base of information that includes information on the economy and all the locational factors that impact the property and the explicit retail attributes. Along these lines, there are numerous viewpoints to consider when putting resources into retail properties, particularly how the retail property platform is unique from other market fragments.

Nonetheless, assumption issues have emerged because of the expanding distribution of real estate in venture portfolios as there is by all accounts an expanding interest for progressively key and definite examination of a genuine legacy when settling on investments.

The explanation behind analysing market investigation is that there is space for and requirement for such an investigation in the scholarly condition. The market examination performed by consultancy firms is commonly classified and not accessible to general society.

Having represented the point and reason with the examination just as the thought with some issues like:

- What kinds of assumption issues emerge when putting resources into business properties and how do these influence the market examination before the investment.
- How is the market investigation in principle and the most significant interest factors in the massive measure of real estate information to consider when investing in it.

- To what degree do these ongoing request factors hypothetically viewed as significant in examining real estate market performance levels.

3 Investment in Real Estate

In India, having properties had become a fantasy for each potential financial specialist anticipating large benefits. All are peering toward the real estate market in India for a wide assortment of reasons:

- It's a consistently developing economy on a persistent ascent, and the flash in economy increments buying intensity and makes interest for the Real estate segment.
- Many alumni from various reputed organizations will be expected to be 2.7 million specialists who can make office space that can reach 250 million square feet.
- Fortune 500 and other presumed companies can expand their business levels in India to make more real estate.
- The financial specialists with a good reputation can become the market leaders, and such kinds of professionals gain 60% of the market profits. The remaining percentage can simply balance the initial investment with the gained profits.
- Business Process Outsourcing (B.P.O.) in India has indicated its ability in divisions like auto-segments, synthetic inventions, pharmaceuticals, and decorations to coordinate the best in the world.

These positive characteristics of India will certainly pull in increasingly remote financial specialists sooner rather than later.

It is essentially an objective to maintain since borrowing money for future property is deemed realty investments. Since investors earn your payoff by borrowing dollars in custom interest earned, this is comparable to a bond. There is no fast way to earn money or become wealthy in property investment, but you may create wealth over time by investing properly. Everyone is probably aware that there are many methods to get wealthy, but property investment is one of the most effective.

4 Literature Survey

The key divisions of the Real estate business are venture, activity and organization. To handle both, capital is required. Investment is the work of capital in obtaining Real estate or interests for the changeless proprietorship of the individual securing it. Activity is the work of capital in the procurement or improvement of the real estate or interests for business tasks.

The behavioural explanation is the expansion of behavioural financial aspects, which manages feelings and psychology research of investors. Through this investigation, we can comprehend the reasonable and impractical behaviour of financial specialists just as psychological and behavioural psychology science to clarify the behaviour of investors towards real estate. Regularly the impractical behaviour of financial specialists impacts the basic leadership procedure of property investors. Since this behavioural doesn't think about the foundation and accessible data to decide. The hypothesis of behavioural financial discussions about how investors make choices and their feelings and intellectual psychology encourage them to settle on wise choices.

Kisii County [7] contend that there is a requirement for an increasingly advanced and proficient site determination process for firms in which Real estate has a key capacity. To perform better site determination choices, the greater network explicit factors likewise must be thought about in the income examination so firms can more readily maintain a strategic distance from momentary choices and end up with the most invaluable Real estate allotments [8] further contend that characterizing the retailers' business volumes, purchaser socioeconomics, and target clients are extremely significant when assessing a Real estate venture opportunity and playing out a site selection. Mydhili [9] mentioned that the association between reality and financial markets has long been documented and acknowledged in developed and emerging economies.

Nairobi Muthama [10] clarifies that the Real Estate Investment Trust (REIT) is a section that changes like a stock and puts resources into Real estate straightforwardly or in a roundabout way. Subsequently, there are various REITs; the value REITs put resources into and possess properties, and their income is connected to the lease. Then again, contract REITs go through venture and responsibility for contracts, and their income is connected to the premium they gain on the home loan advances.

Real estate is centred on the decisions and choices regarding property markets based on human behaviour. This examination essentially studies how humans respond to predispositions and how human psychology experience these inclinations or when to make decisions. After the experience of overall investigations, we have concluded that behavioural investments are best in clarifying stock value peculiarities identified with the idea of under and eruption and so forth. These are called inconsistencies because these disregard the exchanging rules. Nyariboet [11] has given a concise portrayal of the oddities. Nobody can clarify aside. The author portrayed that stock estimation is undervalued while the development is overrated. This is because of the impractical behaviour of the financial specialist. This financial specialist can clarify the different evaluating peculiarities of property because investors don't think about the accessible data and foundations to decide.

Onsomu [12] investigated the significant chances and difficulties of Indian firms alongside firms from numerous different nations. Likewise, it is discovered that Real estate is still fundamentally affected by nearby factors. Nearby information, neighbourhood economies, neighbourhood entertainers and neighbourhood establishments will keep on assuming the noteworthy job, although to some degree influenced now by firms, shoppers and financial impacts from other, far-off regions of the world.

Dichev et al. [13] deciphers that Real estate as a benefit class is inconceivably not the same as a capital market resource. It is a characteristic fence against expansion, encounters low instability and consequently produces positive long haul that comes back with various valuation strategies and systems followed, distinctive revelation and standards, bookkeeping guidelines; Following the accepted procedures proposed could prompt standardization of exposure of data which could, in the end, lead to better choices.

Edmanset [14] dissects that outside investment's significance in real estate improvement can't be contested. It is likewise discovered that the consistently expanding interest for lodging, business space, townships and foundation in India can be conveniently taken into account just if outside venture into this part is permitted. Isolated support will likewise get quality and polished methodology in real estate improvement in the nation.

Mydhili [15] found that the Rationality in decision making of property buyers in the residential realty market is a fact. It could also know that the buyers prefer their analysis rather than completely depending on others in decision making.

French [16] represents that the Indian Real estate segment is determined to be a rewarding goal for remote investors. The development of the Indian re-appropriating industry gives great chances to Real estate investors [17]. The Indian realty part, whenever channelized appropriately, could launch the development of a few different areas in India through its regressive and forward linkages [18]. However, there are potential requirements for local just as outside interests in India [19]. The Nonappearance of a private controller to screen strategic policies winning in the Indian Real estate platform is a hazard factor by financial specialists [20].

5 Methodology

Considering the multifaceted nature of human psychology research, it is difficult to recognize what drives a specific activity with every person in some cases. Along these lines, psychology research frequently utilizes open-finished meetings or investigations by which an individual's responses could be brought into a full set. Then again, if ends are to be summed up, more individuals ought to be analysed to make an exploration agent for a specific population. Even though behavioural financial aspects have a bigger number of components of psychology science than standard financial aspects, its proposes were grounded through basic behavioural tests and overviews that have been demonstrated to produce precise outcomes. In addition, by leading overviews and straightforward tests, behavioural financial experts can reach determinations that have prompted another hypothesis of monetary science [21].

Since the point of this method is to see the outcomes across the realty market, the most helpful approach to lead explore was through an online overview by which the greater number of examinees could become. Besides, considering the point was to test distinctive psychology levels in people, online selection permits individuals to react with more genuineness guaranteeing classification for the members [22–26].

The advancement of the study was organized through a couple of stages which contained:

- Design of overview that relies upon the theory that required testing.
- Sharing of the overview online to a different number of individuals who were approached to incorporate some other invested individual.
- Gathering of the reactions and receiving information for the investigation
- Data investigations exceed expectations results.

Two reviews were shaped with just one inquiry of distinction, which was to test the loss aversion [27–34]. The two overviews contained 32 inquiries and were conveyed to the working population to partake effectively in the Real estate platform exchanges. Moreover, the reason is that the two people are similarly ignorant of the behavioural inclinations and are inclined to psychological mistakes; in this manner, the study included the two orientations of various instructive foundations [35]. The technique utilized for investigation is the near strategy to build up whether contrasts among various gatherings of respondents exist. On the off chance that the distinctions demonstrate to exist, further investigation for similitudes and contrasts will be directed [23].

Monetary science has been created depending on exceptionally levelheaded individuals, otherwise called "homo economicus". The term is broadly spread in the books on financial ideas [36–41]. It recognizes a person who acts splendidly objectively and is driven solely by his advantages to boost his prosperity. Why homo economics is constantly balanced for two reasons: it has all the important data expected to settle on an ideal choice [42]. Other than testing inclinations, the point is likewise to see the degree of comprehension of the financial markets since individuals will generally overestimate their degree of information which can be hurtful when settling on a budgetary choice [24].

Furthermore, it can recognize every one of the choices together with their expenses and advantages [43–46]. This implies that every individual is intelligently reliable, knows the standards of likelihood, and utilizes those guidelines for basic leadership. Besides, as do all others, this individual forms data exclusively through explanation while feelings and psychology levels have no impact.

The trust factor is a nonconventional instrument utilized by leaders who may participate in impractical or error-inclined and one-sided behaviour. In the same way, like other such factors, passionate and spontaneous factors influence the trust factor. Trust has been part of the human decision-making point for several years. Without legitimate assurances for implementation, the trust factors substitute for such lawful certifications. It likewise brings down the exchange expenses of participating in authoritative courses of action and purchasing products and ventures, in any event, when lawful assurances and reviews are set up. Trust-based exchanges are frequently upheld through the negative reputational impact. At the same time, go back on an exchange or a relationship and the beneficial outcome when maintaining an agreement or a relationship.

Being dependable turns into a behavioural standard and advances into instinctive behaviour. Legitimate and casual authorizations and monetary prizes energize trust. Financial specialists as often as possible utilize the trust factor as the best and most proficient approach to execute exchanges. While utilizing the trust factor, leaders find intermediaries for the reliability for basics, complex, and dependable data that can't be considered at all or just at generous financial and time costs. Among these intermediaries are the rating offices of budgetary resources, government authorization of financial resources, and ethnic, neighbourhood, strict, and national groupings one relates to or trusts. For instance, if Standard and Poor's doles out a triple-A rating to a specific monetary resource, people and financial associations are probably going to believe that such a rating is solid. People will contribute with family, companions, and individuals from their surroundings or strict gatherings since they accept that they can be trusted. This trust can be upheld on the off chance that one accepts that if the obligations of trust are broken, the gathering being referred to endure reputational and financial expenses.

Financial specialists frequently expect portfolio supervisors to be better educated in a universe of complex and regularly deceptive data. In addition, people frequently follow the pioneers they trust and participate in grouping behaviour when they are uncertain about what item or budgetary resource to purchase or sell. Crowding can bring about market wasteful aspects called value falls or resource value gaps that can be serious, particularly when dependent on misleading the data.

6 The Connection Between Behavioral Factors and Realty Market Performance

A relationship investigation was directed to set up whether there was a connection between the factors. The connection investigation shows the route, quality, and criticalness of the connections among the examination factors. A positive connection shows that different factors will likewise increment as one variable expands. Then again, a negative connection shows that as one variable build, the other variable reductions. The examination model that was utilized in the investigation was:

$$\text{REIP} = \lambda 0 + \lambda 1\,\text{OVRCONF} + \lambda 2\text{FRAD} + \lambda 3\text{GRP} + \lambda 4\text{REPRE} + \varepsilon.$$

where: REIP is Real estate investment execution,

OVRCONF is overconfidence,

FRAD is Frame Dependence,

G.R.P. is Grouping and

REPRR is Representativeness.

λ0 is the consistent term λ1, λ2 λ3 and λ4 are the coefficient for overconfidence, outline dependence, grouping and representativeness separately.

ε is the error term that was thought to be regularly conveyed.

7 Data Analysis

In light of the fact and the clear significance of psychological science for the choice-making process, an attempt was made to check the effect of the inclinations on the private financial specialists in the realty market. The parameters analysed were loss incompatibility, overoptimism, inflation, location, and grouping. Opinions considered are represented as.

1. Segment factors don't influence the impact of inclinations on the private financial specialists in the realty market, and gender doesn't impact the behavioural factors with finance investors. Also, age doesn't impact the exhibition of conduct of the investors.
2. Information on behavioural financial matters represents the impacts of social predispositions on private financial specialists.
3. Financial specialists and investors in the realty market show Loss repugnance predisposition
4. Private speculators in the realty market exhibit Overoptimism predisposition
5. Private speculators in the realty market are overconfident in assuming rising market values.

Regarding the real estate inclinations with investors, a 4-point scale was utilized to appraise the sites they lean toward the most. The point was to perceive how judicious their decisions are concerning the present market circumstance and how high or low land speculations extended in their rundown of favoured ventures. The respondents were approached to extend the four most basic kinds of speculations–bank stores, shares, land, and bonds dependent on their degree of inclination in the scope of 1 (generally wanted) to 4 (the least liked).

The limit of respondents, which is 26.75% of the respondent, expresses that they first make pre-research of the task without anyone else and afterwards meet the salesman in the wake of taking all the data, site visit and documentation checking, they make instalment to the designer and next gathering of 39.25% of the respondent express that, they meet the representative at engineer's office and in the wake of talking about with venture chief and site visit, they make instalment to the engineer, as shown in Fig. 3.

After an assortment of information, the examination is one of the primary issues that emerge before each investor. It is occasionally very hard to survey and mastermind the colossal sum of information. The information has been gathered for four regions relating to Delhi N.C.R., such as Gurgaon, Noida, Hyderabad and Mumbai.

The information has been gathered for two factors: Consumer behaviour and Marketing Strategy in Real Estate Sector, and two separate respondents are gathering. There are further four separate gatherings of respondents in each gathering according to the location. In this way, generally speaking, there are eight separate gatherings of respondents.

The second significant investigation variable is the Marketing technique in the Real domain segment. The inquire about informational index would be Real Estate

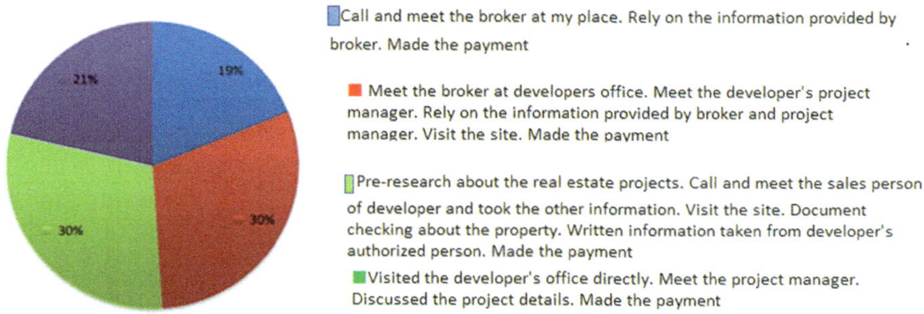

Fig. 3 Process of buying the property

Fig. 4 Behaviour of Investor during Property Finalization

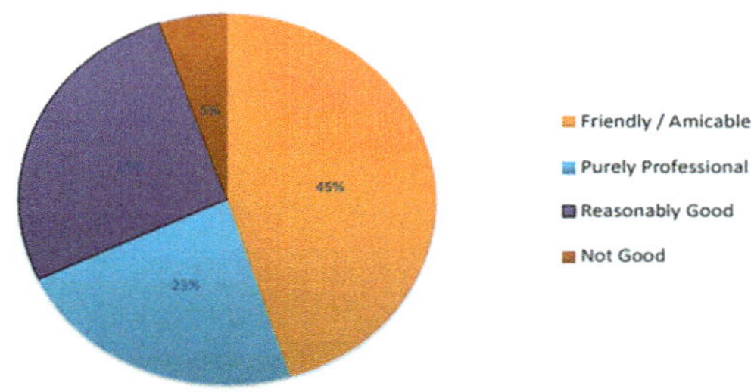

designer's showcasing group and Independent Showcasing group managing asset items, as shown in Fig. 4.

- There is no past investigation identified with Marketing methodology in the Real Estate part, so there is no characterized rule of studying Marketing techniques and procedures related to the Real Estate Sector.
- Current Real Estate promoting strategies are not fully evidenced and hypothetically characterized; in any case, each land association utilizes those strategies. These techniques are filling in as 100% accomplishment for some associations and a few associations; these techniques are 100% disappointments.
- Target Data set for Real Estate advertising variable is little when contrasted with Informational target index for consumer behaviour. In this way, it is simpler to gather the information by the individual meeting.

The concept, implementation, and evolution of physiologically and culturally driven computing models is known as computational intelligence (CI). Machine learning is now the most widely used artificially intelligent technique, and in reality, CI is used for some of the most effective A.I. technologies.

In this comparative work, $\lambda 0$ is consistent, but $\lambda 1$, $\lambda 2$ $\lambda 3$ and $\lambda 4$ are terms that are separately differentiated through importance and non-importance measures. Moreover, Gurgaon and Noida have more suitability of property site; therefore, using this importance, and non-importance model's user is differentiated.

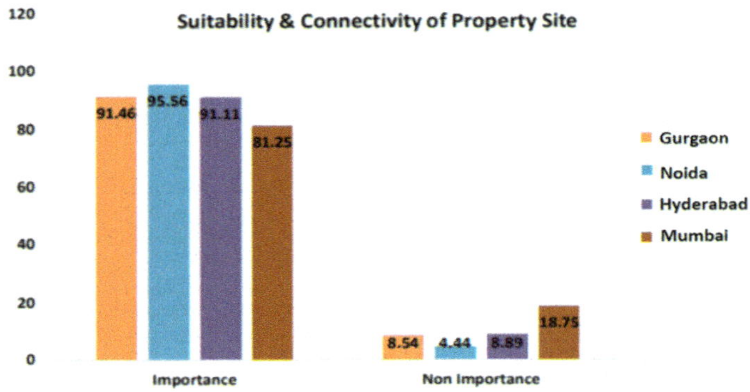

Fig. 5 Comparative Study of Suitability & Connectivity of Property Site

The principle and fundamental contribution the finance have been analysed (Fig. 5). Be that as it may, the strategies in human science are overviews, interviews, member perceptions. Center gathering isn't impacted as a similar estimation of different techniques. So this paper utilized optional information techniques to gather the information. Information concerning this subject is gathered from the optional source. This subject is another and very not quite the same as other research points. Financial support is headway in the account of this time.

Behavioural Factors account is a significant development idea nowadays. Presumably, past examines have clarified the speculation conduct of the individuals; however, a specific name is given to these behaviours and practices in financial investments.

To, one way or another, this is another idea of its application in property advertising. Practices are principal predecessors to change the financial specialist choices in any venture. So, practices ought to be consistently undermined at whatever point we watch the speculation change.

The descriptive Statistics on the ranking of different attributes related to the real estate market are depicted in Table 1.

The real estate developer's behaviour is analysed when finalizing the property with the investor, and the parameters are represented and depicted in Table 2.

The problems faced by the investors during the valuation of the property based on the features indicated by the owner is depicted in Fig. 6.

Table 1 Different attributes statistics

Attribute	N samples	Mean	Std. deviation
Location of the property	100	5.74	0.624
Infrastructure	100	4.96	0.847
Quality of construction	100	6.32	0.235
Comfortability of living	100	5.95	0.874
Demand and supply	3.54	4.23	0.258

Table 2 Behavior of real estate developer during finalizing of property

Parameter	Frequency	Percentage
Friendly/Amicable	124	49.22
Purely professional	51	25.52
Reasonably good	68	31.52
Not good	19	6.25

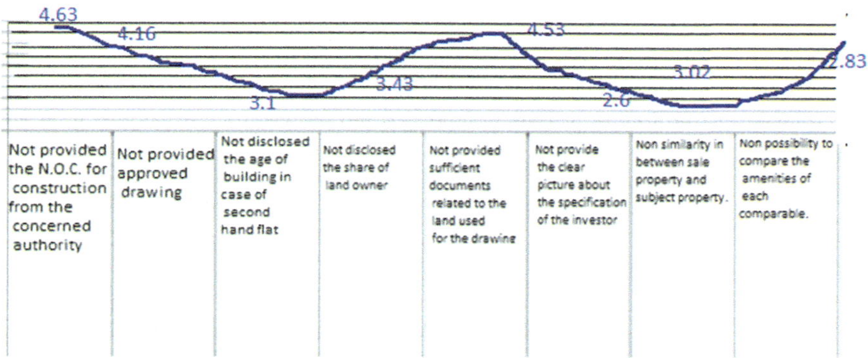

Fig. 6 Problems Faced by the Investors

The above Fig. 6 clearly explains about Problems Faced by Investors in various scenarios. Various discussions are analysed and differentiated through the scale of values from 5 to 0.

8 Conclusion

The behavioural account is a significant development idea nowadays. Almost certainly, past inquiries about the behaviour of the individuals are considered; however, a specific name is given to these dispositions and practices as behavioural accounts. This is another idea of its application in property advertising in some way or another. Practices are the primary precursor to changing the financial investment choices in any real estate investment. So, practices ought to be consistently undermined at whatever point we watch the investment phenomena. The motivation behind this paper is likewise to communicate and pull up the significance of behavioural funds in nearly every field of life with budgetary choices. The investment test uncovers no connection between the explanation behind picking a Real estate business and their age. There is no connection between age and class of business utilized, and individuals differ in the classification of Real estate dealings they enjoy independent of age. It further indicated no connection between instructive capability and contribution to business. Thus, their instructive capability does not impact the real estate professionals in doing the business full-time or low maintenance.

9 Future Scope

This research work is most suitable for Investment Decisions in the Commercial Realty Market for future decisions. This work is helping out customers as well as marketing organizations. This investigation is helpful in comprehending the investments by utilizing the behavioural model. Investors consistently need to put resources into those tasks with more prominent benefits and the capital's base odds of risk or loss.

References

1. Andonov A, Eichholtz P, Kok N (2012) Value added from money managers in private markets? An examination of pension fund investments in real estate. Working Paper, Maastricht University. Mejudice, pp 1–45
2. Chaffai M, Medhioub I (2014) Behavioural finance: an empirical study of the Tunisian stock market. Int J Econ Financ Issues 4(3):527–538
3. Graham JR, Harvey CR (2009) Investor competence, trading frequency, and home bias. INFORMS 55(7):1094–1106
4. Corrado CJ, Jordan BD (2020) Fundamentals of investments: valuation and management. http://elibrary.gci.edu.np/bitstream/123456789/685/1/BM-783%20The%20Essential%20Guide%20to%20Business%20Etiquette%20by%20Lillian%20H.%20Chaney%2C%20Jeanette%20S.%20Martin.pdf. Accessed 25 May 2021
5. Kahneman D, Tversky A (2013) Prospect theory: an analysis of decision under risk. In: Handbook of the fundamentals of financial decision making: Part I, pp 99–127
6. Kahuthu DG (2011) Effects of herd behaviour on trading volume and prices of securities at NSE. Unpublished M.B.A. Research Project. The University of Nairobi
7. Kisii County (2014) Kisii County Annual Report, 2014. Ministry of Housing. Sessional Paper No. 3 of 2004 on National Housing Policy for Kenya
8. Kothari CR (2004) Research methods, quantitative and qualitative approaches
9. Mydhili V, Sundaridadhabai A (2017) Glimpse on realities of realty sector in India: a special focus on A.P. J Adv Res Dyn Control Syst, 9:18
10. Nairobi Muthama AK (2012) Effect of Investor psychology on real estate market prices in Nairobi. An Unpublished M.B.A. report. University of Nairobi
11. Nyamute W, Lishenga J, Oloko M (2015) The relationship between investor behaviour and portfolio performance at the Nairobi securities exchange. Int J Multidisciplinary Res Dev 2(5):248–551
12. Onsomu ZN (2014) The impact of behavioral biases on investor decisions in Kenya: male vs female. Int J Res Humanities Arts Literat 2(6):87–92
13. Dichev, Ilia, Piotroski (2021) The long-run stock returns following bond ratings changes. J Finance 56:173–203
14. Edmans et al. (2007) Sports sentiment and stock returns. J Finance 62:1967–1998
15. Mydhili V, Sundaridadhabai (2019) Rationality in decision making of residential property buyers: a Myth or Fact. Int J Innov Technol Exploring Eng 8(5)

16. French S (1986) Decision theory: an introduction to the mathematics of rationality. Ellis Horwood, Chichester
17. Glaser M, No¨th M, Weber M (2004) Behavioral finance. In Koehler DJ, Harvey N (Eds), Blackwell handbook of judgment and decision making, Blackwell, Oxford, pp 527–46
18. Gallimore P (2004) Behavioural real estate research, Retrieved January 17, 2004, from http://construction.ntu.ac.uk/graduate_school/Research/Property
19. Johnsson M, Lindblom H, Platan P (2002) Behavioral finance and change of investor behavior during and after the speculative at the end of the 1990s. School Econ Manag (2002)
20. Kim KA, Nofsinger JR (2008) Behavioral finance in Asia. Pacific-Basin Finance J 16:1–7
21. Guiso L et al (2008) Trusting the stock market. J Finance' 63(6):2557–2600
22. Lakonishok J, Shleifer A, Vishny R (1994) Contrarian investment, extrapolation, and risk. J Finance 49:1541–1578
23. Millington AF (1988) An introduction to property valuation, 3rd edn. Estates Gazette, London
24. Montier J (2009) Behavioural investing: a practitioner's guide to applying behavioural finance. Wiley
25. Żywiołek J, Rosak-Szyrocka J, Jereb B (2021) Barriers to knowledge sharing in the field of information security. Manag Syst Prod Eng 29:114–119
26. Żywiołek J, Rosak-Szyrocka J, Mrowiec M (2021)K nowledge management in households about energy saving as part of the awareness of sustainable development. Energies 14(24):8207
27. Żywiołek J, Schiavone F (2021) Perception of the quality of smart city solutions as a sense of residents' safety. Energies 14:5511
28. Almahirah MS, Jahan VNM, Sharma S, Kumar S (2021) Role of market microstructure in maintaining economic development. Empirical Econ Lett 20(2)
29. Nayak NR, Kumar S, Gupta D, Suri A, Naved M, Soni M (2022) Network mining techniques to analyze the risk of the occupational accident via bayesian network. Int J Syst Assur Eng Manag
30. Kumar S (2021) Relevance of Buddhist philosophy in modern management theory. Psychol Educ 58(3):2104–2111
31. Kumar S, Baag PK, SKV (2021) Impact of E.S.G. Integration on equity performance between developed and developing economy: evidence from S and P 500 and NIFTY 50. Empirical Econ Lett 20(4)
32. Kumar S, Baag PK, SKV (2021) Financial engineering and quantitative risk analytics. SYBGEN Learning 1(1):1–360
33. Kakti A, Kumar S, John NK, Ratna VV, Afzal S, Gupta AD (2021) Impact of patients approach towards healthcare costs on their perception towards health: an empirical study. Tobacco Regulatory Sci 7(6–1):7380–7390
34. Roland G, Kumaraperumal S, Kumar S, Gupta AD, Afzal S, Suryakumar M (2021) P.C.A. (Principal Component Analysis) approach towards identifying the factors determining the medication behavior of indian patients: an empirical study. Tobacco Regulatory Sci 7(6-1):7391–7401
35. .Kumar S, Baag PK (2021) Ethics erosion in capital market: Lehman Brothers' case study of Repo 105. In: AIMS-18, March 4, India, AIMS Texas U.S.A., A1872-Final.pdf (aims-international.org)
36. Kumar S, Baag PK (2021) Erosion of Ethics in credit derivatives: a case study. In: AIMS-18, March 4, India, AIMS India & Texas U.S.A., A1873-Final.pdf (aims-international.org)
37. Parnaik VK, Chaturvedi P (2015) Fluorescence recovery after photobleaching studies reveal complexity of nuclear architecture. Int J Chem 4(4):297–302
38. Singh B, Jalil NA, Sharma DK, SR, Kumar K, Jebakumar Immanuel D (2021) Computational systems overview and random process with theoretical analysis. 2021 7th international conference on advanced computing and communication systems, pp 1999–2005
39. Setiawan R, Leonardo Cavaliere LP, Koti K, Ogunmola GA, Jalil NA, Kalyan Chakravarthi M, Suman Rajest S, Regin R, Singh S (2021) The artificial intelligence and inventory effect on banking industrial performance. Turkish Online J Qual Inquiry 12(6):8100–8125

40. Roespinoedji D, Juniati S, Hasan H, Jalil NA, Shamsudin MF (2019) Experimenting the long-haul association between components of consuming renewable energy: ARDL method with special reference to Malaysia. Int J Energy Econ Policy 9:453–460
41. Sharma DK, Jalil NA, Nassa VK, Vadyala SR, Senthamil LS, TN (2021) Deep learning applications to classify cross-topic natural language texts based on their argumentative form. 2021 2nd international conference on smart electronics and communication, pp 1580–1586
42. Sharma DK, Jalil NA, Regin R, Rajest SS, Tummala RK, TN (2021) Predicting network congestion with machine learning. 2021 2nd international conference on smart electronics and communication, pp 1574–1579
43. Zhao B, Chaturvedi P, Zimmerman DL, Belmont AS (2020) Efficient and reproducible multi-gene expression after single-step transfection using improved bac transgenesis and engineering toolkit. A.C.S. Synth Biol 9(5):1100–1116
44. Kubiczek J, Hadasik B (2021) Challenges in reporting the COVID-19 spread and its presentation to the society. J Data Inf Qual 13(4):1–7
45. Bieleń M, Kubiczek J (2020) Response of the labor market to the needs and expectations of Generation Z. e-mentor 86(4):87–94
46. Sehgal P, Chaturvedi P (2022) Survival strategies in cold-adapted microorganisms, 1st ed. Singapore: Springer Singapore

Chapter 4
Trust the Machine and Embrace Artificial Intelligence (AI) to Combat Money Laundering Activities

Guneet Kaur

Abstract The research work is focused on examining the role of artificial intelligence (AI) in addressing challenges associated with money laundering in the banking sector. Money laundering is a global issue that threatens financial stability and international security, making anti-money laundering research crucial. Furthermore, just 0.2% of money laundered through the financial system is estimated to be seized. The crime is growing increasingly sophisticated and intricate, and the amount of the crime increases banks' vulnerability. Researchers have begun to investigate the possibility of artificial intelligence approaches in this setting. However, a thorough assessment has identified a systematic knowledge deficit that systematically examines and synthesizes artificial intelligence techniques for anti-money laundering efforts in the banking industry. Therefore, this chapter is focused on a systematic review of key technologies categorized into artificial intelligence or machine learning (AI or ML), natural language processing (NLP), robotic process automation (RPA), and cloud-based solutions. However, various challenges concerned with these techniques, such as data quality, the nature of money laundering and data volume, and data heterogeneity, are also discussed. As a result, the findings add to the total knowledge base in anti-money laundering from the banking sector's perspective. Additionally, future study directions were narrowed even further based on the limitations discovered.

Keywords Money laundering · Terrorist financing · Distributed ledger technology · Anti-money laundering · Artificial intelligence · Machine learning · Natural language processing · Robotic process automation · Cloud-based · SaaS

G. Kaur (✉)
Technology Editor, Cointelegraph, UK
e-mail: askguneet@gmail.com

S. Kautish et al. (eds.), *Computational Intelligence for Modern Business Systems*, Disruptive Technologies and Digital Transformations for Society 5.0,
https://doi.org/10.1007/978-981-99-5354-7_4

1 Introduction

The current coronavirus (COVID-19) pandemic has already proven to be one of the worst crises in recent history, wreaking havoc on several sectors of the global economy. The banking industry is one of the hardest hit by the economic downturn brought on by the COVID-19 epidemic [1]. For instance, financial crime is one of the most noticeable repercussions in the financial technology (FinTech) sector.

The COVID-19 epidemic has been used by some criminals to perpetrate fraud and exploitation schemes using a variety of ways, including cyber-enabled fraud. National and international authorities warn citizens and businesses against imposters, investment, product scams, and insider trading concerning the coronavirus. Those looking to launder criminal proceeds or fund terrorism are likely to take advantage of any flaws in a company's system. Furthermore, the United Nations believes that up to $2 trillion is unlawfully transferred each year [2]. Big banks are used by criminals to hide money that is often tied to organized crime, with funds being used to pay for assets in order to conceal the money's origin. According to the National Crime Agency (NCA), money laundering costs the UK economy more than £100 billion every year [3].

To address these issues, governments and financial institutions focus on responding to the coronavirus pandemic, affecting their ability to carry out anti-money laundering and counter-terrorist financing (AML/CFT) obligations in areas such as suspicious transaction reporting supervision and regulation, policy reforms, and international cooperation. This could lead to criminals discovering new ways to get around customers due to diligence measures; exploiting economic stimulus measures and insolvency schemes as a way for natural and legal persons to conceal and launder illicit proceeds; increasing the misuse of online financial services and virtual assets to move and conceal illicit funds; and increasing the use of the unregulated financial sector, opening up new avenues for criminals to launder illicit funds. In undeveloped countries, criminals take advantage of the coronavirus and the resulting economic crisis to enter new cash-intensive and high-liquidity business sectors.

The fight against money laundering (ML) among the various financial sectors has prompted financial organizations to develop technologically advanced weapons. As a result, these organizations want to successfully deploy sophisticated analytical technologies like machine learning (ML) and artificial intelligence (AI) to tackle various financial crimes. In addition, financial institutions are dealing with criminals and authorities who are enforcing increasingly harsh penalties. Fear of supervisory consequences and reputational damage, branch offices are more committed than ever to combating money laundering and terrorist financing. The key motivations for using AI and machine learning in AML operations are to increase the quality of investigations and regulatory filings and reduce false positives and associated operating expenses. However, the pandemic's dramatic shift in consumer behavior has led many financial institutions to recognize that static, rules-based monitoring tactics aren't as accurate or adaptable as behavioral decision-making systems.

In recent years, regulators have imposed massive fines on banks that have fallen short on anti-money laundering tactics. According to research published in February 2021 by business-to-business (B2B) information services firm Kyckr, 28 financial institutions throughout the world were penalized £2.6 billion in 2020 for AML offenses [4]. Swedbank was fined €347 million by Swedish and Estonian regulators in March of that year for breaking money laundering legislation [5].

In 2018, the Dutch regulator fined ING €775 million after the bank failed to prevent the laundering of hundreds of millions of euros between 2010 and 2016 [6]. In the same year, British and US authorities fined Deutsche Bank $650 million for letting affluent clients transfer $10 billion out of Russia [7].

As AI and machine learning technologies are dynamic by nature and are able to respond intelligently to market changes and developing hazards, they can be swiftly and painlessly integrated into current compliance programs. As a result, early adopters are gaining significant efficiencies while also assisting their institutions in meeting expanding regulatory requirements. In this context, this chapter is focused on the following exploratory questions:

(a) What are the issues in the money laundering sector that require analytical attention?
(b) What artificial intelligence technologies can be used to handle the money laundering issues in the banking sector?
(c) What are some research gaps and areas for development that should be looked at further?

The remaining sections of this chapter are organized as follows. Section 2 delves into the background of money laundering. Section 3 depicts the major challenges that banking products and services face, as well as various artificial intelligence (AI) techniques that can help solve these problems. Section 4 covers the challenges associated with AI techniques followed by the need to use distributed ledger technology in Sect. 5. Section 6 reviews significant research contributions and Sect. 7 examines potential study discrepancies and opportunities for improvement followed by a conclusion in Sect. 8.

2 Background

This section throws some light on the history of money laundering.

2.1 History of Money Laundering

Money laundering is a term coined in the United States to describe the Mafia's attempt to "launder" illegal funds through cash-intensive washing salons controlled by corporate acquisitions or business creations. According to estimates, illicit sources account for two to five percent of global gross domestic product (GDP). Drug selling generates a large portion of the economy, with 810 billion dollars in revenue in 2003 [8]. In 2005, Austrian police seized drugs worth 49,266,800 Euro (drug seizures in terms of street prices) and accused 25.892 people of violating the Austrian Narcotics Act [8]. Although there is a clear desire to misuse the internet in order to conduct unlawful transactions in the form of online banking, Cyber money, and Electronic Purse, the majority of illegal transactions are handled by cash because there is the least risk of leaving one's mark.

Moreover, money laundering economic study necessitates a proper characterization of the phenomena [9]'s approach is based on two essential aspects of money laundering:

(a) Illegality (general feature): money laundering refers to the use of any revenue derived from criminal or unlawful activities;
(b) Concealment (particular feature): money laundering's principal purpose is to conceal the illegal source of such profits.

The first of these two features emphasizes money laundering's criminal autonomy, as it does not examine the specific illegal or unlawful activity that generated the earnings that were later laundered (kidnapping, drugs markets, corruption, fraud, etc.). The second feature indicates money laundering's critical economic function of converting unlawful capital to legal capital. The unique economic role of this activity is to convert prospective purchasing power into adequate purchasing power, as it cannot be employed directly for consumption, investment, or saving.

2.1.1 Stages of Money Laundering

The three stages of money laundering are as follows: The first step is to place illegal gains (the placement), which entails physically infiltrating cash (derived from crime) into the banking system. After that, the money is turned into book money (primary and secondary deposits), which is then layered (stacking of illegal funds). These sophisticated acts are used to hide the source of money by arranging complex financing transactions between states and stacking multiple levels of transactions. The third stage involves reintegration and parking this unlawful money, which has no ties to organized crime and is converted into a publicly visible asset through investments in businesses, tourism initiatives, and industrial operations.

Placement

In the first stage, referred to as placement, ill-gotten gains from criminal acts are infiltrated into a legal banking/economic system; at this point, there is a higher danger of being discovered.

This refers to the immediate transfer of criminal funds into a legal, financial system without drawing the attention of regulatory authorities. Limiting quantities are undercut with structuring and smurfing to circumvent identification, reporting duties, and paperwork requirements. Furthermore, money is divided into partial quantities in an organized manner to allow payment into many bank accounts below the corresponding identification and declaration limits.

Another way of placement entails exerting influence over financial sector institutions to purchase existing banks or establish new banks in offshore jurisdictions. Furthermore, co-worker corruption is a regularly utilized criminal instrument to deposit incriminated money: attempts are made to socialize with bank employees in order to allow direct infiltration of money while avoiding the attention of supervisory authorities. In addition, depositing money into foreign bank accounts allows criminals to participate in the financial or economic cycle.

Secondary deposit, unlike primary deposit, is an indirect penetration of money supply into the bank system and, as a result, a conversion into book money through the connectivity of a natural or legal entity. This can be accomplished by switching the medium. At the same time, the incriminated cash is converted into other assets or by using front men who, while acting on their own behalf, trade for the account of a third party or grant the use of their name to carry out (realize) an account opening, the formation of a company, or the conclusion of an insurance policy without being visible on the surface. Forward displacement of the money laundering location onto life insurances, financial service providers, and exchange offices can also be used for indirect placement.

Layering

Criminals use the so-called layering stage to try to hide the source of unlawful income by making a large number of transactions and transferring black money around. Withal (multiple transfer and transaction), transaction intensity and speed rise; electronic payment technology, as well as diverging jurisdiction and inadequate criminal prosecution cooperation, typically simplify/facilitate the layering procedures. Over- and under-invoicing in international commercial transactions, charging fictitious goods or services in the form of winding up fake transactions, or back-to-back loaning are all ways to legitimize capital transfer.

In this situation, money launderers deposit a specified amount of money into a bank account (or convincingly verify securities acceptable at value with another bank) in order to reclaim the same amount as a bank loan. If this is the case, a money launderer raises his own funds and has proof of origin for the bank loan. Rather than banks, connected companies can act as lenders.

Misuse of financial derivatives and swaps is also frequent among money launderers to obscure the source of damning funds. Two offshore firms, which are de jure independent but de facto controlled by the same individual, enter into an option contract by each taking a "long" or "short" position. The profit of the other compensates for the loss of the first. The financial volume of offshore centers is estimated to be between 10 and 12 billion dollars, with an annual growth rate of 15% [8].

Integration

In this third stage, converted and transferred capital is predominantly infiltrated into the official economy through financial investments (particularly deposits, stocks) or property (direct investment in firms and real estate) in countries that provide extraordinary short odds.

While the above three-stage model of money laundering is a simple way of explaining the activity, it is uncommon for a single bank or its money laundering reporting officer (MLRO) to notice or be able to detect all three steps in a single case. Criminals do not even need to use all three stages of the paradigm to legitimize their criminal property. Furthermore, criminals would frequently try to hide their activities by using a variety of institutions, often in separate jurisdictions. Each institution will only see bits of a broader puzzle, which is why the suspicious activity reporting process is crucial to the successful seizure of assets resulting from illicit activity and the prosecution of money launderers and other criminals.

It is necessary to study what money laundering is not in order to fully comprehend the notion. Money laundering is a misnomer in and of itself. To begin with, it isn't just money that is involved; it can be any tangible or intangible property that has resulted directly or indirectly, in whole or in part, from illegal action. Second, the term 'laundering' supports the idea that it is a process by which criminals attempt to 'wash' or 'clean' their criminal property to seem different from its original state at the end of the procedure. The consensus is that criminals will attempt to accomplish this by utilizing various services, products, currencies, and jurisdictions. This isn't always the case, however. Consider the following illustration.

The earnings of a drug trade are deposited into a bank account by a criminal. He takes some of the money the next day and spends it. Is it possible that the criminal used the bank to launder the funds? Yes, even though the bank does not appear to have assisted him in "washing" the money or even changing its appearance, the proceeds of his crime were turned into credit in his bank account. The classic model of money laundering is useful in some ways, but it fails to explain a variety of scenarios where the facts do not fit the usual paradigm.

This places a clear need to utilize technology in order to combat money laundering. Furthermore, the fight against money laundering among the various financial sectors has prompted financial organizations to develop technologically advanced weapons. As a result, these organizations want to deploy sophisticated analytical technologies like machine learning and artificial intelligence to successfully tackle various financial crimes. In addition, financial institutions are dealing with criminals and authorities who are enforcing increasingly harsh penalties. Fear of supervisory consequences and reputational damage, branch offices are more committed than ever to combating money laundering and terrorist financing (AML / CFT).

2.2 *Anti-money Laundering Policies*

To combat money laundering and terrorist financing, the Financial Action Task Force (FATF) was established in 1989 [10]. It is a multi-jurisdictional organization with 35 member jurisdictions and two regional organizations. According to the FATF, all anti-money laundering and counter-terrorist financing programs must contain specific data analysis and reports. They also demand that an institution be able to identify and authenticate its customers, a requirement known as "know your customer." This bans anonymous accounts and fictitious account holder names and requires financial institutions to take precautions while dealing with correspondent and shell banks. Banks are also required to preserve records of all transactions for a minimum of five years. The records must include:

- The names of clients and/or beneficiaries and their addresses.
- The nature of the transactions.
- The dates of transactions.
- The types of currency.
- The amounts of currency.
- The types of accounts.
- The identification numbers of any account utilized.

The FATF requires two forms of reporting: suspicious transaction reports (STRs), which are filed with the national financial intelligence unit, and currency transaction reports (CTRs), which are filed with the national financial intelligence unit and indicate transactions above a specific value.

2.2.1 Rule-Based and Risk-Based Policies

In 1990, the "Forty Recommendations on Money Laundering" were published as an essential foundation for preventing, identifying, and suppressing unlawful financial transactions [11]. The "Nine Special Recommendations on Terrorist Financing" were released after September 11, 2001, terrorist attacks in New York City to focus efforts on CTF [12]. In addition, financial organizations may quickly implement and demonstrate compliance with such rule-based procedures since they are transparent and straightforward. These rules are used in the automatic layer of current AML systems to determine whether a transaction is suspicious. However, because transactions can be engineered to circumvent the constraints, the clarity of these regulations makes it more difficult to identify fraud [13]. The guidelines also result in an over-reporting of questionable activity, which is costly and time-consuming to investigate.

A risk-based approach offers private actors more discretion over what to disclose by having more ambiguous reporting criteria. This places the decision in the hands of private companies, making them accountable for the reporting's success. However, in certain nations, this has exacerbated over-reporting. It's a higher-risk strategy that can be arbitrary or haphazard, resulting in banks losing customers or being fined for underreporting. Despite this, several financial institutions have implemented risk-based procedures [14]. By measuring client and transaction risks and recognizing outlier behavior, they transcend the constraints of rule-based systems.

2.2.2 General Data Protection Regulation (GDPR)

The European Union (EU) recently established specific crucial regulations (officially known as GDPR) concerning the collection, storage, and use of personal data; these regulations took effect in May 2018 and superseded the EU's 1995 Data Protection Directive (DPD). The new legislation applies uniformly to industry, EU organizations, and entities that deal with EU data in terms of scope. The data-driven policies focus on a few specific issues, such as data ownership, openness, explainability, and the trustworthiness of algorithms trained or developed with such data. For example, GDPR requires data-driven automated systems, including AML systems, to comply with the following requirements during implementation:

- Data and algorithm explanation frameworks.
- Legal data processing and data ownership.
- Ethical compliance.

2.3 AML Frameworks

Currently, AML frameworks can be broken down into four layers [15]. The *data layer* is where all critical data is collected, managed, and stored. This comprises data from the financial institution and data from outside sources such as regulatory bodies, authorities, and watch lists. From a technology standpoint, traditional systems have architectural flaws, such as data quality, data management, and data governance concerns. In the AML community, an enterprise data hub that incorporates the technology necessary for efficient processing often maintains the Data Layer. Hadoop is used for parallel processing and data acquisition, for example [16].

The *screening and monitoring layer* then looks for suspicious activity in transactions and clients. Financial organizations have mainly automated this layer into a multistage approach based on rules or risk assessments. The monitoring procedure is carried out continually, and it uses analytical models to assess transactions and client profiles. This layer's components work together in a collaborative framework that includes various tools. For data retrieval and post-operational storage, they maintain bidirectional contact with the Data Layer.

If suspicious activity is discovered, it is forwarded to the *alert and event layer* to be investigated further. The significant number of transactions to be examined—along with the lack of supporting data—increases the time necessary to analyze each transaction. Financial institutions are incorporating cutting-edge statistical and data-related technology into their AML strategies to reduce the risks and expenses of manual inspection. This layer makes decisions based on prior decisions' historical data and comparisons to similar transactions and decisions.

Finally, a human analyst determines whether to block or authorize a transaction on the *operational layer*. It is a legal necessity that a human agent makes the final decision in a transaction. The previous layers are used to keep track of all transactions and flag any that may be fraudulent. On the other hand, the final choice is made by a person based on the knowledge provided by the previous steps in the process.

3 Various Artificial Intelligence (AI) Technologies for AML/CFT Compliance

Money laundering's scope is difficult to determine, but it is thought to be large. For example, the United Nations Office on Drugs and Crime (UNODC) estimates that 2 to 5% of global GDP is laundered each year [17]. This amounts to between EUR 715 billion and EUR 1.87 trillion every year [17].

The financial incentive is a common element in most organized crime. Organized crime groups use various money laundering strategies to increase their assets and inject them into the legitimate economy. Tracing these assets entails tracing the networks they are connected to.

Various AI technologies that assist in addressing these challenges are discussed in the section below.

3.1 AI and Machine Learning (ML)

AI and machine learning techniques, which are at the heart of AI, are revolutionizing and will enhance how we manage financial risk, including credit, operational, market, and compliance risks [18]. Everything from assessing how much a bank should lend to a customer to delivering warning signals to financial market traders about position risk, detecting consumer and insider fraud, and increasing compliance and decreasing model risk is now possible due to the advent of AI-driven solutions. This chapter will discuss the application of AI and ML techniques for managing compliance risk.

Financial firms must comply with risk management requirements, especially in the aftermath of the financial crisis. While risk management professionals frequently try to establish a distinction between what they do and the often-bureaucratic condition of regulatory compliance, the two are inexorably intertwined since they both pertain to risk management systems as a whole. To that extent, compliance is most closely associated with enterprise risk management, albeit it touches on each of the risk functions of credit, market, and operational risk separately.

RegTech, a compliance-focused industry akin to FinTech, has thus emerged to aid firms in managing the rising needs of compliance [19]. The sheer volume of data that needs to be examined and the non-conventional nature of this data are driving AI and machine learning to play a significant role in this sector. Furthermore, the capacity to continuously monitor a firm's activity is a fundamental advantage of machine learning from a pure RegTech perspective. This ability for real-time insights, according to [20], allows for the prevention of compliance breaches rather than having to deal with the consequences of violations after they have occurred.

Other benefits mentioned include the opportunity to free up regulatory capital due to improved monitoring and automation, decreasing some of the estimated $70 billion spent annually on compliance by giant financial institutions [18]. IBM is a prominent player in this field, having acquired Promontory (a 600-person RegTech firm) and now offering a range of AI-driven solutions for lowering RegTech expenses, illustrating the broad industry interest in the subject, which is not limited to startups [21]. For example, a combination of IBM's Watson AI skills and Promontory's domain-specific experience is being utilized to assure compliance through real-time speech dialogue analysis. This entails converting voice interactions to text and then using natural language processing to classify the text into categories that identify potential non-compliance. Other applications of machine learning include the intuitive understanding and interpretation of regulatory consequences, which a London-based company called Waymark now provides to financial institutions using natural language processing [18].

According to the compliance teams, between 1 and 2% of AML alerts result in a Declaration of Suspicion (DS) [22]. Machine learning and AI will be the most transformational, assisting in identifying and deactivating 98% of false positives [22]. This will free up more resources for the 2% of cases that are more likely to be suspicious. Using customer information files to identify high-risk entities likely to represent an accurate positive result, statistical analysis can lower the likelihood of false positives. Furthermore, the outcomes of analysts' decisions can be reintroduced into the system to test machine learning-based prioritization algorithms in order to reduce possible false positives during the subsequent transaction monitoring. Furthermore, using big, annotated bank datasets, standard machine learning approaches such as support vector machines (SVMs) and random forest (RF) can be utilized to classify fraud transactions [23]. Because the bank monitors the transaction data, these data-driven systems are typically employed during the placement and stacking phases. Because funds have already passed through fraud-detection procedures, the final integration phase is difficult to detect.

3.2 Natural Language Processing (NLP) and Soft Computing Techniques

Financial institutions are increasingly seeking to understand the customer's professional, institutional, political, and social context by analyzing large amounts of external data, such as information and media, public archives, social networks, and other open-source data sources, in order to implement a risk-based approach to customer knowledge (KYC). External data can be found with a typical name search. Despite this, it cannot provide context for the name, determine linkages with politically exposed persons (PEP) or high-risk entities, or analyze other risk indicators based on these sources. To evaluate unstructured data and establish these linkages, natural language processing, and AI approaches can be used. Advanced analysis of unstructured data is critical because it improves efficiency by automating increased vigilance operations and uncovers linkages and threats that could otherwise go unreported.

Natural language processing is an area of artificial intelligence that allows computers to comprehend, interpret, and manipulate human speech. Fuzzy logic is an analytical technique that uses several values to analyze inaccurate or approximate data, resulting in a usable (but imperfect) output [24]. Nonbinary logics use a range of values rather than just 0 or 1. Fuzzy Logic systems can generate sound output in response to incomplete, ambiguous, distorted, or inaccurate (fuzzy) input, emulating human decision-making more closely than classical logic and extracting more helpful information from data that is too imprecise to yield definite results using classical logic. Hardware, software, or a combination of both can be used to implement fuzzy logic. Natural language processing and fuzzy matching tools also help reduce false positives and negatives (for example, in sanction screening processes) [25]. Still, they primarily solve data quality issues by improving the programs' ability to link data elements, such as connecting search engine results with PEP lists, detecting fraud attempts, monitoring sanctions lists, etc. For instance, the Financial Intelligence Unit (UIF) of Italy, in collaboration with the Bank of Italy's Directorate-General for Financial Supervision and Regulation, developed a fuzzy logic application to generate anti-money laundering indicators for non-banking financial intermediaries [26]. The suggested fuzzy system—which is still in the experimental stage—enables the elaboration of quantitative data (i.e., cross-border payments from/to higher risk countries) to help the periodic AML/CFT risk assessment of such intermediaries.

Other applications include:

- Customers' identification and verification: AI, including biometrics, machine learning, and liveness detection techniques, can be used to perform microexpression analysis, anti-spoofing checks, fake image detection, and human face attribute analysis in the context of remote onboarding and authentication.
- Business relationship monitoring and behavioral and transactional analysis: This can be performed using-
  - Supervised machine learning algorithms: Enable faster and real-time data analysis following the applicable AML/CFT rules.
  - Unsupervised machine learning algorithms: These algorithms are used to group customers into cohesive groupings based on their behavior, and then create controls that can be set more appropriately based on a risk-based approach (e.g.: transaction threshold settings), allowing for a tailored and efficient monitoring of the business relationship.
  - Alert Scoring: Alert scoring assists in focusing on patterns of behavior and issuing notifications, as well as the requirement for increased due diligence.
- Automated data reporting (ADR): It is the process of employing standardized reporting templates and automated digital applications (data pooling tools) to

make the underlying granular data of regulated organizations available in bulk to supervisors.

- Identification and implementation of regulatory updates: Machine Learning techniques with natural language processing, cognitive computing capabilities, and robotic process automation (RPA) can scan and interpret large volumes of unstructured regulatory data sources on an ongoing basis to automatically identify, analyze, and then shortlist applicable requirements for the institution; or implement (to a degree) the new or revised regulatory requirements (via codification and generation of impeachment documents).

3.3 Robotic Process Automation

RPA, or Robotic Process Automation, is software that replicates a human's actions when doing repetitive, rule-based tasks. The robot works quickly and precisely, using the same applications that your staff does every day. All activities in traditional automation depend on programming/scripting or other integration techniques to backend systems or internal applications. RPA, on the other hand, automates software that can migrate work from humans to computers, allowing companies to stop paying people to do the job that could be automated, speed upfront and back-office transaction processing, achieve near "Instant On" integration at the lowest cost, and reduce errors, all while increasing productivity by making workers smarter [27].

Financial Crimes Officers may find that various types of intelligent automation can help decrease costs, boost efficiency and effectiveness in three areas within a financial crimes compliance program, as discussed in the sections below.

3.3.1 Transaction Monitoring

It is an excellent illustration of how improved technology may help with financial crime compliance. Typical AML transaction monitoring solutions are built to consider rules-based typologies and situations requiring continual tuning and updates. Because they are often more rudimentary and rule-based, they might miss a slew of risk variables. This frequently results in many false positives that must be resolved by humans. Institutions can use bots to scan the internet and specific public due diligence sites for essential data and collect data from internal sources and approved sources (as identified by the Institution). They can also put the due diligence findings into an electronic case file for examination by an analyst. Finally, the analyst saves time by deploying the bot to do these research and record-keeping duties.

3.3.2 Know Your Client

Performing KYC during onboarding and regular review periods takes financial institutions a lot of effort and resources. Institutions may devote hundreds of hours each month to KYC tasks, typically reinforced by contractors and consultants, depending on their size and an annual turnover of new customers. KYC processes are typically made up of very repetitive tasks that Intelligent Automation can supplement and speed up. As a result, many financial institutions have already identified areas in their KYC process where RPA can help. This would involve duties such as document extraction from public news sources, negative news screening requests and retention of the findings based on pre-defined search parameters, and data importation from the documentation into a KYC system.

RPA, when correctly set up, can save a lot of time, allowing KYC analysts to focus on areas of onboarding that require more in-depth study, such as clearing a smaller list of lousy news outcomes or reviewing any remaining gaps in information or documents. For example, RPA could minimize or eliminate the need to contact clients repeatedly, resulting in a better customer experience because bots may eventually attain more accuracy in collecting due diligence information.

Financial crimes compliance personnel would continue to perform targeted testing of results obtained using automation to understand the precision with which the machines perform or refine the parameters being used to achieve greater accuracy when using Intelligent Automation to supplement KYC work. However, as the accuracy and consistency improve to an acceptable level, reducing the testing or quality assurance (QA) reviews may be possible over time.

3.3.3 Compliance Testing

Financial institutions also devote a significant amount of time and resources to AML and sanctions compliance testing. Testing data feeds, system validations, third-party vendor or outsourced processes, and other internal methods are examples of this. Intelligent Automation can also be helpful here because some of the duties related to testing are repetitive. Furthermore, sampling is no longer necessary because a machine performs the tests. Therefore, it is possible to test the entire population, removing sampling errors.

RPA can assist people in swiftly identifying flaws in early data sets (including documentation) as part of their testing scope job. RPA could also be used to execute simple testing methods to determine data completeness, depending on how structured the data is at a given Institution. For example, RPA could quickly analyze if KYC files include required data points like address, date of birth, source of wealth, citizenship, and following the Institution's standards on 100% of the files while evaluating KYC compliance, identifying identification outliers for further root cause analysis.

3.4 Cloud-Based Solutions

The supply of computing services such as servers, storage, databases, networks, and software over the internet to promote speedier innovation and economies of scale is known as cloud computing technology. Anti-money laundering firms use cloud computing, and financial institutions can employ private cloud computing technologies to access cloud computing services. Private cloud computing technology services and infrastructure are retained on a private network in a private cloud. The risk of cybercrime is decreased due to this private cloud computing technology, which uses data kept on a private network to prevent money laundering and terrorist financing. Financial institutions can employ cloud computing technology for money laundering solutions through these systems.

Cloud computing technology consists of three core services: software as a service (SaaS), platform as a service (PaaS), and infrastructure as a service (IaaS). Financial institutions can boost their productivity and reduce the expenses of creating an IT infrastructure due to the advent of cloud computing technology Anti-Money Laundering (AML) solutions. For instance, Financial Crime and Compliance Management Cloud Service (FCCM Cloud Service), an end-to-end anti-money laundering SaaS for midsized financial institutions, was launched by Oracle [28]. Oracle is the first company to offer "AML as a Service" on the cloud.

Similarly, statistical analytics software's (SAS') comprehensive, end-to-end anti-money laundering solution provides unprecedented prediction and detection capabilities, the lowest false positives, and decreased investigation times—all while lowering total compliance costs [29]. The features of SAS's cloud-based solution are listed below.

- Comprehensive solution: AI, machine learning, robotic process automation, and advanced network analytics support transaction monitoring, customer due diligence, onboarding, watchlist screening, case management, and regulatory reporting on a single, integrated platform.
- Nimble platform: All analytical tasks—from data access, preparation, interactive exploration, optimization, machine learning, AI, and other advanced model building through deployment operations—are brought together in a unified, open, and high-performance environment.
- Data Orchestration: To make data analytics ready, it centralizes the enrichment and transformation of a range of data kinds, including transaction data, nonmonetary event data, geographic data, risk lists, third-party data, and a variety of additional customer data.
- Entity-centric investigations: With entity analytics, AML analysts and investigators may visualize whole networks using the network viewer and dive deep into relationships via link expansion, giving them a comprehensive perspective of risk.

4 Challenges of Using AI to Detect Money Laundering

Various challenges concerned with the techniques, as mentioned earlier, are data quality, the nature of money laundering and data volume, and data heterogeneity [30].

At the instance level, datasets in banking and finance have a different set of quality issues. Missing, dummy, and null values are only a few examples. This would occur in most data fields in all databases, except the customer type and fund name, misspellings, typos, and phonetic errors. Furthermore, banking datasets are typically managed in a distributed manner for the sake of flexibility and security. When an integrating activity is required, the independence and heterogeneity of each data source can be data quality challenges, as all conflicts must be resolved. Finally, the data preparation stage is mostly used to address data quality issues.

Whether it's cash, check, credit card, or electronic transactions, most sectors deal with money in some form. All mediums are employed in a banking and finance setting, so developing an AML solution is difficult because ML instances are not self-revealing. ML reporting will likely be uncommon. As a result, ML activities are becoming increasingly advanced. Because it mimics normal behavior, ML crimes are readily disguised within a normal distribution. As a result, they are present in most legal transactions. As a result, data quantities and the nature of machine learning present obstacles to the first generation of AML solutions, which are rule-based mechanisms based on preset sets of fixed criteria, such as employing mean and standard deviation rules for volume and quantity of transactions over time.

Analyzing big datasets is a time-consuming process for AML professionals; the large and growing volume of datasets in financial institutions concerning the relatively limited number of questionable ML instances in them becomes a challenge. Furthermore, the huge amounts of data required for analysis are rarely available in a single location. The dissemination of datasets necessitates a data preparation integration process, resulting in data quality difficulties, as discussed in the previous section. Furthermore, financial datasets for ML research are typically large and heterogeneous.

5 Distributed Ledger Technology (DLT) to Prevent Money Laundering

The financial sector is paying attention to distributed ledger technology, both because of its use in crypto-asset transactions and because of the proliferation of initiatives that have the potential to improve the efficiency, speed, transparency, and resilience of processes underlying financial transactions. Distributed ledger technology refers to a database with several identical copies dispersed among various participants and is updated in a synchronized manner by the parties' agreement [31]. DLT can increase transaction traceability on a cross-border and global scale, making identity verification easier. Consumers can authenticate themselves and be automatically accepted or rejected through smart contracts that validate the data, so a responsible and controlled usage of DLT for data and process management could speed up the CDD process. DLT technology may also provide benefits for handling CDD needs, such as reducing user worries about the process, increasing private sector cost-effectiveness, and creating a more accurate and quality-based data pool. In China, for example, financial institutions are using DLT to share watch lists or red flags based on the level of confidentiality allowed by the system [32].

Regardless of its virtues, DLT appears to continue to offer issues and cause major worry regarding AML/CFT, as seen by virtual assets' regulation and/or supervision. Transactions in virtual assets (VA) based on DLT are decentralized in nature, allowing un-intermediated peer-to-peer transactions to take place without any examination, unlike transactions through traditional intermediaries such as banks [33]. If no entity or apparent location is responsible for the action, they lead to jurisdictional issues. In addition, traditional FATF rules that have focused on regulating/supervising intermediaries may face issues. As a result, the usage of this technology should be closely monitored and scrutinized by FATF members. Finally, authorities may wish to think about the carbon footprint of DLT versus traditional tools.

6 Key Research Contributions by the Researcher

Although combating money laundering is a difficult endeavor, the findings of this chapter point to a number of ways in which the banking sector's fight against it can be aided by the deployment of numerous strategies across multiple application domains. To begin, the evaluated papers all focused on methodologies categorized into five major technologies, including AI/ML, NLP, RPA, and cloud-based solutions that can aid in transaction monitoring, KYC, and compliance testing domains. As a result, incorporating AI approaches into these highlighted application domains can help banks boost their anti-money laundering operations by giving crucial assistance in dealing with money launderers. The research also highlighted the issues associated with AI-based techniques and the need to adopt DLT to tackle money laundering-related issues. However, AI-based technologies have the following advantages:

- Understanding of potential money laundering exposure, allowing for the implementation of appropriate mitigation control measures and, thus, prevention,
- Efficient detection of hidden patterns and case prioritization allows for timely and accurate identification of suspicious transactions, prevents further integration of illegal money, and alerts filtering.
- The overall global identification of suspicious activity relationships and enlightened investigative decision-making, enable investigators to obtain a clear overview of correlations and trends in money movement, as well as to make more efficient and accurate decisions, and thus to conduct investigations.

7 Discussion and Future Work

Although the research effectively captured the challenges associated with financial crime in addressing them effectively, the number of AI technologies along with their disadvantages have been discussed in this chapter. Further recommended future work is discussed below.

Despite the fact that money laundering studies are on the rise, it is still a relatively new area, with a study on the subject being limited in comparison to other crimes, necessitating more concentrated research. As a result, the following recommendations for future research are made. First, while most studies have focused on detection, continued efforts should extend to other application domains such as risk evaluation, case prioritization, pattern/anomaly discovery, visual analysis, and decision support, allowing researchers to explore and experiment with new techniques in those areas, resulting in more literature.

More research should be done, for example, on the focus area investigation, where different strategies for assisting AML investigators/analysts in their everyday job through visual analysis and decision support should be investigated. In terms of visual analysis, it is suggested that future research focuses on integrating AI and visualization in AML, where input from investigators or analysts might be gathered directly to evaluate the solution. Furthermore, given the diversity of regulation across nations, future research could do a comparative study to see how it affects the performance of a certain AI technique's performance in the AML area.

8 Conclusion

This chapter summarizes and navigates the voluminous quantity of research that employs AI approaches to combat money laundering. The use of artificial intelligence tools and how they can aid in the fight against money laundering in the banking sector were studied. As a result, the chapter's key research question is: how do AI techniques may aid in the fight against money laundering in the banking sector? Based on the investigation, it can be concluded that applying AI in the field of AML can help the banking sector battle money laundering in various ways. For example, banks may use AI in various application domains, and it can be used to supplement several AML approaches. As a result, banks will be able to limit risk, detect suspicious activity, and examine suspicious situations comprehensively in their growing role in countering money laundering.

Finally, the findings of this chapter provide a systematic review for other researchers, which may be useful in directing future research on highlighted gaps in the literature in this area. In contrast, the findings of this chapter provide practitioners with a knowledge base in terms of an evidence-based overview of relevant techniques in the realm of AML and recommendations for appropriate solutions for various application domains. As a result, the findings of this chapter provide theoretical and practical contributions. In this regard, the findings of this chapter contribute to the total knowledge base in anti-money laundering from the perspective of the banking industry, thus widening the inquiry in areas where prior research has focused less.

References

1. Wronka C (2021) Impact of COVID-19 on financial institutions: navigating the global emerging patterns of financial crime. J. Finance Crime 2(1):1–3
2. United Nations (2021). Overview. [Online] United Nations: Office on Drugs and Crime. Available at: https://www.unodc.org/unodc/en/money-laundering/overview.html. Accessed 13 December 2021
3. NCA (2019) National Economic Crime Centre leads push to identify money laundering activity. [Online] Nationalcrimeagency.gov.uk. Available at: https://www.nationalcrimeagency.gov.uk/news/national-economic-crime-centre-leads-push-to-identify-money-laundering-activity. Accessed 13 December 2021
4. Pai S (2021) Anti-money laundering: current state of play. Governance Directions 73(2):56–60
5. Gottschalk P (2020) Sweden: Swedbank money laundering. In: Corporate responses to financial crime, pp 125–130. Springer, Cham, pp 2–19
6. Ravenstijn B (2000) The Netherlands: anti money-laundering programmes—the case of ING. J Money Laundering Control 3(2):6–16
7. Rose KJ (2020) De-risking or recontracting–the risk dilemma of EU money laundering regulation. J Risk Finance 3(1):3–6
8. Schneider F, Windischbauer U (2008) Money laundering: some facts. Eur J Law Econ 26(3):387–404
9. Masciandaro D (1999) Money laundering: the economics of regulation. Eur J Law Econ 7(3):225–240
10. Alexander K (2001) The international anti-money-laundering regime: the role of the financial action task force. J Money Laundering Control 1(2):1–12
11. Tyre C (2010) Anti-money laundering legislation: implementation of the FATF forty recommendations in the European Union. J Prof Law, 69
12. Force FAT (2001) Nine special recommendations on terrorist financing. Paris: FATF/GAFI, October, www.fatf-gafi.org/dataoecd/55/16/34266142.Pdf
13. Han J, Huang Y, Liu S, Towey K (2020) Artificial intelligence for anti-money laundering: a review and extension. Digital Finance 2(3):211–239
14. Helmy THE, zaki Abd-ElMegied M, Sobh TS, Badran KMS (2014) Design of a monitor for detecting money laundering and terrorist financing. Int J Comput Netw Appl 1(1):15–25
15. Le Khac NA, Markos S, Kechadi MT (2010) A data mining-based solution for detecting suspicious money laundering cases in an investment bank. In: 2010 Second International conference on advances in databases, knowledge, and data applications, pp 235–240. IEEE
16. White T (2009) Hadoop: the definitive guide. 1st edn, OReilly Media
17. Çeku SO, Kutllovci SS, Emini SA, Petrit S (2019) Money laundering as a form of economic criminality–the case of the Republic of Kosovo. Transactions 12(3)
18. Aziz S, Dowling M (2019) Machine learning and AI for risk management. In: Disrupting finance Palgrave Pivot, Cham, pp 33–50
19. Butler, T. and O'Brien, L., 2019. Understanding RegTech for digital regulatory compliance. In Disrupting Finance. Palgrave Pivot, Cham, pp. 85–102.
20. Arner DW, Barberis J, Buckey RP (2016) FinTech, RegTech, and the reconceptualization of financial regulation. Nw J Int'l L Bus 37:371
21. Bektenova GS (2018) Are Regtech, Fintech, Blockchain the Future?. KnE Soc Sci, 61–67

22. Grasshoff G, Gehra B, Villafranca V, Gittfried N, Grataloup V, Hefter K, Leiendecker J, Pauly O (2017) Transforming bank compliance with smart technologies
23. Gyamfi NK, Abdulai JD (2018) November. Bank fraud detection using support vector machine. In: 2018 IEEE 9th annual information technology, electronics and mobile communication conference (IEMCON), pp 37–41. IEEE
24. Heidarinia N, Harounabadi A, Sadeghzadeh M (2014) An intelligent anti-money laundering method for detecting risky users in the banking systems. Int J Comput Appl 97(22):1–14
25. Zhang Z, Salerno JJ, Yu PS (2003) Applying data mining in investigating money laundering crimes. In: Proceedings of the ninth ACM SIGKDD international conference on Knowledge discovery and data mining, pp 747–752
26. Gara M, Manaresi F, Marchetti DJ, Marinucci M (2019) The impact of anti-money laundering oversight on banks' suspicious transaction reporting: evidence from Italy. Bank of Italy Occasional Paper, (491)
27. Geetha T, Malini A, Indhumathi M (2020) Robotic process automation. Int J Comput Techniques 7(5):1–12
28. Bostel A (2021) Building better transaction monitoring for anti-money laundering. J Financial Compliance 4(4):313–335
29. Palshikar GK, Apte M, Baskaran S (2014) Analytics for detection of money laundering. In: TCS technical architects conference (TACTiCS 2014) Affilication: tata consultancy services limited
30. Le Khac NA, Markos S, O'neill M, Brabazon A, Kechadi MT (2009) An investigation into data mining approaches for anti money laundering. In: Proceedings of international conference on computer engineering and applications (ICCEA 2009)
31. Romero Ugarte JL (2018) Distributed ledger technology (DLT): introduction. Banco de Espana Article 19:18
32. Zhu H, Zhou ZZ (2016) Analysis and outlook of applications of blockchain technology to equity crowdfunding in China. Financial Innov 2(1):1–11
33. Kerimov A, Koibichuk V, Mynenko S (2020) Blockchain technology in bank's anti-money laundering. Economic and social development: book of proceedings, pp 874–883

Chapter 5
Predictive Analysis of Crowdfunding Projects

Aashay Shah, Prithvi Shah, Umang Savla, Yash Rathod, and Nirmala Baloorkar

Abstract Crowdfunding has become the social media version of fundraising campaigns whose underlying principle is to raise funds for a project from multiple people and to collectively accrue the required resources to make the project successful. The principal aim of crowdfunding platforms is to introduce budding entrepreneurs to an expanded pool of investors rather than traditional financial investors. Kickstarter is the largest reward-based crowdfunding platform which has successfully funded more than 2,00,000 projects and raised more than $6 billion. However, scarcely one-third of the projects are successful in reaching the funding goal before the deadline. Hence, reckoning the probability of success of a project is an interesting challenge. The proposed system helps classify a project as a success or failure. Supervised Machine Learning Models are implemented from which Random Forest provides the highest accuracy score of 90%. Regression Algorithms are implemented to estimate the funding a project is capable of achieving. Furthermore, BERT, spaCy, and TF-IDF are implemented to find keywords that affect the success of the project.

Keywords Crowdfunding · Kickstarter · Supervised machine learning models · Random forest · Regression algorithms · BERT · spaCy · TF-IDF

A. Shah · P. Shah (✉) · U. Savla · Y. Rathod · N. Baloorkar
Department of Computer Engineering, K. J. Somaiya College of Engineering, University of Mumbai, Vidyavihar, Maharashtra, India
e-mail: prithvi.shah@somaiya.edu

A. Shah
e-mail: aashay.shah@somaiya.edu

U. Savla
e-mail: u.savla@somaiya.edu

Y. Rathod
e-mail: yash.gr@somaiya.edu

N. Baloorkar
e-mail: nirmalashinde@somaiya.edu

S. Kautish et al. (eds.), *Computational Intelligence for Modern Business Systems*, Disruptive Technologies and Digital Transformations for Society 5.0,
https://doi.org/10.1007/978-981-99-5354-7_5

1 Introduction

Crowdfunding provides a fundraising channel that helps entrepreneurs raise funds from multiple people to finance a new business venture. There are different types of crowdfunding models: Equity-Based (for financial returns), Reward-Based (for non-monetary rewards), and Donation-Based (for philanthropy or sponsorship), out of which Reward-Based Models are predominantly used [1, 2]. Kickstarter is the leading reward-based crowdfunding platform for aspiring entrepreneurs anticipating to raise capital and reach a larger target audience. Despite being such a large platform, two-thirds of the projects fail to reach the required goal amount before the deadline. Hence, reckoning the probability of success of a project can help both investors to choose a valuable and potentially successful project and entrepreneurs by providing them guidelines to improve their chances of success.

The system proposed below provides a machine learning solution to classify a project as a success or failure as well as a natural language processing approach to analyze the project description. To train the machine learning models, four datasets are employed namely, Description Dataset having 2,00,000+ records, General Dataset having 2,00,000+ records, Rewards Dataset having 45,000+ records, and Social Media Dataset having 18,000+ records. To train natural language processing models, Description Dataset having 2,00,000+ records is employed. Machine learning models enable the system to classify the projects based on processed data and provide an output that can be extrapolated to various cases and Stratified K-Folds Cross Validation is performed on all algorithms. With respect to natural language processing, keyword extraction is performed using BERT, spaCy, and TF-IDF to find keywords that affect the success of the project for each project category.

2 Literature Survey

A research in 2017 proposed using the random forests learning model by assigning optimal weights to the individual classifiers and then performing weighted majority voting on them to classify the projects [3]. Another research proposed performing prediction modeling for crowdfunding projects using a deep learning approach named Multi-Layer Perceptron and performing exploratory data analysis to get decisive conclusions about relationships between factors that affect the success of projects [4]. A research aimed at using Support Vector Machines Recursive Feature Elimination (SVM-RFE) to find keywords that may affect the success of the project [5]. A separate research focused on the impact of rewards on the success of the project. Their analysis employed various statistical methods, including Pearson correlation tests, Kolmogorov–Smirnov test, and Kaplan- Meier estimation, to study the impact [6]. Additional research tried to analyze relationships between products on crowdfunding platforms and their subsequent launch on e-commerce websites considering parameters such as their project goal amount [7].

3 Proposed Methodology

A comprehensive literature survey revealed that not much work has been done on the descriptions of the projects, which plays a significant role in understanding the project at a single glance. Thus our scope of research focused more on keyword extraction from project descriptions using Natural Language Processing, thereby trying to find patterns from project description that contributes to the probability of the success of the project. We further applied statistical analysis thereby calculating a score for each project description and using the score in machine learning models for predicting the success of the project which has never been done before. Furthermore, the statistical theory on the importance of rewards was given a more practical approach by including them in machine learning models which predicts the success of the project, thereby bolstering the earlier findings. We also focused on the digital presence of the project, which plays a key role in the success of the project to a great extent and has never been talked about before. Even established companies shell out a significant amount of their revenue on social media presence to always be among the talks, so for these projects to be successful, social media presence plays a remarkable role (Fig. 1).

The four datasets that were mentioned above are pre-processed, transformed, and cleaned using various data mining techniques and the result of this phase gives us a clean dataset that is free of anomalies capable of being utilized for natural language processing and machine learning models. The next phase entails natural language processing implementation where BERT, spaCy, and TFIDF have been used to perform keyword extraction on project descriptions to procure optimistic and pessimistic words for each project category that affects the success of a project. The next phase involves employing various supervised machine learning algorithms to classify the project into two categories namely Successful and Failure. Three classification models were employed using General Dataset, Rewards Dataset, and Social Media Dataset respectively to classify projects. In the next phase, the same datasets are used to predict the estimated funding a project is likely to receive, using regression models.

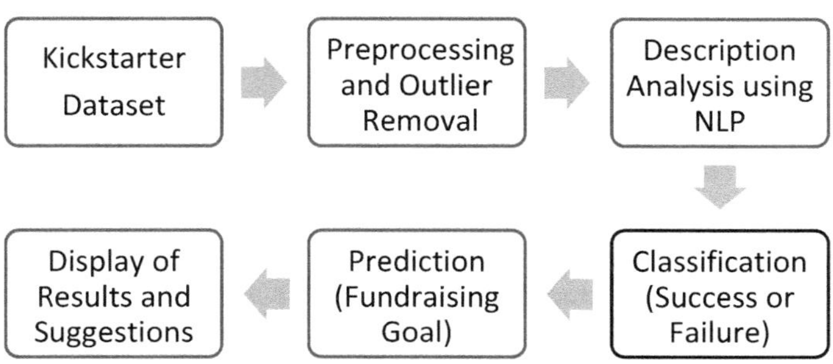

Fig. 1 Proposed method design

4 Datasets and Preprocessing

4.1 Understanding the Datasets

The General Dataset and Description Dataset considered for implementation is from Web Robots, which provides web crawling and scraping services. The General Dataset has around 2,00,000 tuples and around 35 attributes whereas the Description Dataset has around 2,00,000 tuples and 3 attributes. The General Dataset focuses more on the generic aspects of a project like the project category, the funding required and the time period in which funding is desired; these aspects constitute the most essential facet of the projects. It also talks about the location of the project, which plays a key role as more industrial locations will attract more funding. The Rewards Dataset and Social Media Dataset considered for implementation is from Kaggle. The Rewards Dataset has around 45,000 tuples and around 15 attributes whereas the Social Media Dataset has around 18,000 tuples and around 35 attributes. The Rewards Dataset gives an insight into different aspects of rewards that the investors will be entitled to on the success of the project, these rewards determine the profitability of the projects, hence play a vital role. The Social Media Dataset gives a perception into the digital presence of these projects, their digital presence helps us estimate the popularity of the projects which in turn directly impacts the revenue of the projects.

4.2 Data Preprocessing

We start data preprocessing by removing null values from the datasets. The tuples that are null throughout are dropped, whereas the other null values are replaced by the mean values of their column. In order to avoid overfitting, duplicate tuples were eliminated and ineffectual attributes were also eliminated from the dataset. The categorical attributes were converted into numerical attributes using Label Encoding to make them compatible with machine learning models.

4.3 Project Attributes

Each dataset taken into consideration consists of numerous features such as "Project Description", "Country," "Currency," "Goal Amount," "Funding Period," "Number of Backers," "Number of Reward Levels," "Projects Created," "Projects Backed," "Images," "Videos," "Websites" and many more. The target attribute for classification models is "Project State" with values successful or failure whereas the target attribute for regression models is "Pledged Amount" which determines the funding a project is likely to receive [8]. Table 1 provides a description of some important attributes.

Table 1 Project attributes and their descriptions

Attribute	Description	Attribute	Description
Goal amount	Expected funding desired by the creator	Funding period	Time period specified to receive the funding
Number of backers	Number of people who fund the project	Pledged amount	Funding received when the funding period is over
Images	Number of images on project description page	Videos	Number of videos on project description page
Website	Number of web pages related to the project	Rewards	Number of rewards provided by the creator
Minimum reward	Minimum funding to be eligible for reward	Maximum reward	Maximum funding which gives the supreme reward
Projects created	Number of projects previously created by the creator	Projects backed	Number of previous projects that were successful in receiving the funding
Facebook friends	Number of Facebook friends of the creator	FAQ's	Number of FAQs related to the project

5 Keyword Extraction Using Natural Language Processing

Keyword Extraction is a method to perceive significant words that can be used to represent the text and it is a very effective method to get insights from unstructured data [9]. In this step, we identified the most optimistic and pessimistic words for each project category using the project descriptions as input. We represented them in the form of word clouds and provided them to the creator so that he can use those to increase the chances of success. We used Description Dataset for the same, which consists of 3 attributes namely Project Description, Project Category, and Project State.

5.1 Methodology

As per the below system displayed in Fig. 2, we made use of three natural language processing models namely spaCy, BERT, and TF-IDF. These three models cover three different aspects of project description and extract keywords for each project category.

The 3 aspects are as follows:

1. Presence of successful keywords: The spaCy algorithm works on finding the keywords present in the description and on the basis of word frequency across the dataset we calculated whether the keyword is common among successful projects [10].

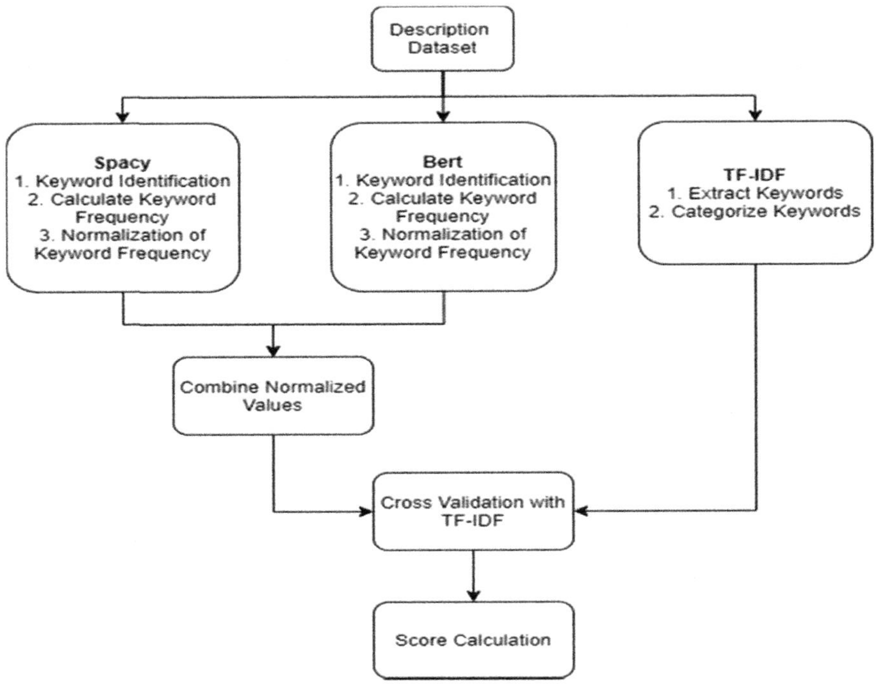

Fig. 2 Keyword extraction methodology

2. Keywords being significant for a specific project: The BERT algorithm is used for its bidirectional understanding with respect to the language to identify keywords that hold the most significance in the description based on the context [11].
3. Comparing against other descriptions to identify uniqueness: TF-IDF identifies keywords unique to the description by comparing them against pre-existing descriptions [12]. We further calculated the frequencies of the keywords extracted by spaCy and BERT for each project category and we normalized the frequency to avoid skewing of data. These normalized frequencies are combined and cross-validated with the keywords extracted by TF-IDF. Using this, we calculated scores for each project description, a higher score signifying higher chances of success and a lower score signifying fewer chances of success.

5.2 Implementation

Identify Keywords Using spaCy, BERT, and TF-IDF

We used the "en_core_web_lg" model of spaCy to identify keywords from project description by considering only the proper nouns, adjectives, and common nouns and avoiding prepositions and articles as they are not important parts of the project description. We used the "distilbert-base-nlimean- tokens" model of BERT which is the base model for Natural Language Inference (NLI) and Natural Language Processing using the MEAN pooling strategy for CLS tokens [13]. The pooling strategy is simply the way the model combines the different inputs (and thus information) from the different layers. We used the TF-IDF Vectorizer to identify how unique a keyword from a description is when compared to other descriptions in the corpus. TF-IDF stands for Term frequency-inverse document frequency. This is a statistical measure used to assess the significance of the word in the document with respect to the entire corpus. The result of all three models will be respective lists of keywords extracted.

Calculate Keyword Frequency for Keywords Extracted Using spaCy and BERT

We splited the data on the basis of project categories and using the list of keywords extracted by spaCy and BERT, we found the frequencies of optimistic and pessimistic words based on whether the project was successful or unsuccessful for both of them. We stored the keywords along with their frequencies in a dictionary. The final output of this step will be a collection of optimistic and pessimistic keywords along with their frequency of occurrence for each project category for spaCy and BERT respectively.

Once the data was split on the basis of project categories, the keywords obtained from TF-IDF are also categorized into optimistic and pessimistic words based on the successful or unsuccessful state of the project. We didn't calculate frequencies of these words, as TF-IDF takes into account the entire dataset while extracting keywords. Hence, we simply used it for cross-validation at a later stage.

Normalize Keyword Frequency

We used the frequency of keywords obtained from the previous step and normalize them in order to avoid skewed data. In this step, we simply divided the frequencies of keywords by the number of projects in that category having that particular state. For example, suppose the category is "Technology," the frequencies of optimistic keywords of this category are divided by the number of successful projects of this category to get a normalized frequency, and similarly, the frequencies of pessimistic keywords of this category are divided by the number of unsuccessful projects of the same category.

Merge Normalized Frequencies

In this step, we combine normalized frequencies of spaCy and BERT obtained above, as follows:

i. If the keyword is present in both the results, we simply take the average and store the value.
ii. If the keyword is present in only one result, the other frequency is considered to be 0 and we took the average and stored the value.

The output of this step will be a merged solution list of optimistic and pessimistic words for each project category respectively.

Cross-Validation with TF-IDF

The result of the previous step is cross-validated with the keywords obtained using TF-IDF, if the keyword obtained from spaCy and BERT is a part of the TF-IDF solution, the word along with its frequency is retained, else it is discarded. This is the final step for keyword extraction whose output will provide a final collection of optimistic and pessimistic keywords along with their normalized frequencies for each project category which can be used to calculate a score for each project description.

Calculation of Description Score

In this step, we calculated a score for the description using the previously found normalized values to identify the overall quality of the description. It works by assigning positive weight to keywords of successful projects and assigning negative weight to keywords of unsuccessful projects. Hence for each word in the description, the function goes through all the keywords extracted for that particular category and uses their normalized value to compare whether the word is more common in successful or unsuccessful projects. The total weight of the description is calculated by adding all the calculated values and the overall positive or negative weight is used to represent the success or failure of the project. The entire description can be evaluated using this method to detect flaws and shortcomings in the project description.

Results

In Fig. 3, we can see the result of the above methodology for the parent category, "Technology." We analyzed that Hardware Projects have higher chances of success as compared to Software Projects. Similarly, we performed the same procedure for all other categories.

Fig. 3 Optimistic and pessimistic keywords for the category "Technology"

6 Classification of Projects Using Machine Learning

We used supervised machine learning algorithms namely Logistic Regression, Decision Tree, Random Forest, Support Vector Machine, and K-Nearest Neighbor to identify whether a project is likely to succeed or fail. We used the "score" method from the sci-kit learn library to measure the accuracy of all the above-mentioned models. After employing Stratified K-Folds Cross-Validation to bolster our findings, we will select the model with the best accuracy. We created three classification models using General Dataset, Rewards Dataset, and Social Media Dataset. Stratified K-Folds Cross Validation is employed to bolster our findings. It selects folds based on the mean response value such that it is the same in all the folds. We have used 10 folds for all the models. Each fold represents all strata of the datasets, hence irrespective of the training and testing sample the results are unbiased. We observed that the accuracy after applying Stratified K-Folds remains almost the same as the accuracy obtained by running the individual algorithms on the dataset. This signifies that the algorithm provides persuasive results irrespective of the training and testing samples, hence proving that our research is reliable [14].

1. Classification Models using General Dataset: We used the following attributes as inputs for the classification models trained on General Dataset: Country, Currency, Parent Category, Sub Category, Funding Period, Goal Amount, and Description Score (calculated above while performing keyword extraction). After performing training on 80% of the data, i.e., 1,23,101 tuples, we test the remaining 20% of the data, i.e., 30,776 tuples, we can conclude that the random forest algorithm provides the best accuracy of 90% after performing Stratified K-Folds Cross-Validation.

We observed that Random Forest provides the best result out of all the chosen algorithms. A research showed that Naïve Bayes provided an accuracy of 60%, Logistic Regression gave an accuracy of 69% on a similar dataset [15]. As per another research, XG Boost provided an accuracy of 75%, Random Forest with Ada Boost provided an accuracy of 80% and Cat Boost with a learning rate of 0.21 resulted in an accuracy of 83% [16]. All these previous research works didn't take into account the description of the project, whereas our research considered the attribute, Description Score, which was calculated above using Natural Language Processing, thereby increasing the accuracy of the results.

2. Classification Models using Rewards Dataset: We used the following attributes as inputs for the classification models trained on Rewards Dataset: Country, Parent Category, Sub Category, Goal Amount, Project Duration, Number of Reward Levels, Minimum Reward, Maximum Reward, Mean Reward, Median Rewards, Updates in Reward, Project Comments. We performed training on 80% of the data, i.e., 32,383 tuples, and test for accuracy on the remaining 20% of the data, i.e., 8096 tuples. After training the data using all the supervised learning algorithms mentioned above, we can conclude that random forest provided the maximum accuracy of 86% after performing Stratified K-Folds Cross-Validation.

As mentioned in earlier research works that late added rewards have more impact on the success of the project, we included the attribute of the number of times the rewards have been updated in our machine learning model thereby getting an accuracy much higher than the previous research which showed an accuracy of just 68% excluding that attribute [20].

3. Classification Models using Social Media Dataset: We used the following attributes as inputs for the classification models trained on General Dataset: Parent Category, Sub Category, Currency, Goal Amount, Project Duration, Facebook Connected, Facebook Friends, Facebook Shares, Has Video, Has Website, Projects Created, Projects Backed, Number of Videos, Number of Images, Number of Words in Description, Number of FAQ [17]. We performed training on 80% of the data, i.e., 14,452 tuples, and test for accuracy on the remaining 20% of the data, i.e., 3,613 tuples. After training the data using all the supervised learning algorithms mentioned above, we conclude that random forest provided the maximum accuracy of 84% after performing Stratified K-Folds Cross-Validation.

As mentioned in a research work Logistic Regression provided an accuracy of 75% on a similar type of dataset [17]. Similarly, another research showed that Logistic Regression gave an accuracy of 77%. Our classification model using the random forest algorithm provides the best result on Social Media Dataset to date [18].

6.1 Results and Discussion

Table 2 summarizes the findings of the various Classification Models on all three datasets. It encapsulates the accuracies of various Classification Models. It helps us understand that Random Forest proves to be the best algorithm for all the datasets and these models also surpass the accuracies obtained by previous research.

The intellect behind the boosting of accuracy using General Dataset can be attributed to the inclusion of the concept of Project Description, thereby proving that the description of the project plays a very significant role in the success of the project. Similarly, the increment in accuracy using the Rewards Dataset can be attributed to how rewards affect the chances of success. It has already proved statistically that rewards play a significant role toward the success of projects. We simply made use of that statistical concept to practical use and thereby bolstered the earlier finding. The classification model created using Social Media Dataset proves that

Table 2 Classification models and their accuracies

Algorithm	General dataset (%)	Reward dataset (%)	Social media dataset (%)
Random forest	90	86	84
Decision tree	84	78	74
Logistic regression	87	83	77
SVM	87	80	73
KNN	87	77	67

Table 3 Regression models and their R2 scores

Algorithm	General dataset	Reward dataset
Random forest regressor	0.67	0.73
Decision tree regressor	0.63	0.67
Linear regression	0.35	0.65

the social media engagement of the project also helps in increasing the chances of success of the project.

7 Prediction of Estimated Project Funding

We used regression methods, namely Linear Regression, Decision Tree Regression, and Random Forest Regression, to predict the project's funding. We used the "score" method from the sci-kit learn library to measure the R2 score of all the above-mentioned models. The percentage of variance in the dependent variable that is predictable from the independent variables is resolved by the R2 score. After employing Stratified K-Folds Cross-Validation to bolster our findings, we selected the model with the best R2 score.

We created two regression models using General Dataset and Rewards Dataset. We did not use the Social Media Dataset here since a large number of attributes in the dataset will increase the variance thereby not providing accurate results.

Using the General Dataset, we used the same attributes as mentioned in the classification models above. Here we trained 70% of the data and tested the remaining 30% of data, thereby concluding that random forest provides the best results with an R2 score of 0.71, whereas Stratified K-Folds Cross-Validation provides an R2 score of 0.67.

Using the Rewards Dataset, we used the same attributes as mentioned in the classification models above. Here we trained 70% of the data and test the remaining 30% data, thereby concluding that random forest provides the best results with an R2 score of 0.73 after Cross-Validation.

Table 3 clearly shows that Random Forest Regressor is the most suitable model for both datasets. There is not much substantial research in this domain to compare our results, but some research states that R2 scores of 0.75 in scholarly research are considered substantial whereas an R2 score of 0.5 is considered moderate [19].

8 Conclusion and Future Work

The aim of this research chapter is to develop a natural language processing model for discovering keywords from the project descriptions that affect the success of the projects. We used BERT, spaCy, and TF-IDF to include all three aspects of keyword extraction to suggest to the entrepreneurs a list of optimistic words that when used increase the chances of success and a list of pessimistic words that should be avoided in the project descriptions as they tend to have a negative impact. The aim of this chapter is to develop classification models to predict the success of crowdfunding projects using various machine-learning techniques. To design the model, a variety of supervised algorithms have been tested out of which Random Forest Classifier provides the finest results. Since two-thirds of the projects fail to reach their goal amount in the desired period of time, this tool can help prospective entrepreneurs to predict whether their project will reach the desired goal amount. If not, they can make plans to increase the promotion activity. This research can also help investors to choose a valuable and potentially successful project.

Also, the hypothesis made earlier that the descriptions have a significant impact on the success of the project was proved with the help of the results of classification models. The efforts to calculate the score of each project description using natural language processing and use them in classification models were definitely not in vain. It increased the accuracy of the models to a significant extent as compared to those models which didn't take into account this very aspect. Similarly, the prominence of rewards for the success of the project was bolstered by a practical approach.

In future work, we would suggest the above research be extrapolated for other crowdfunding platforms so that the results are more specific pertaining to the respective platform.

References

1. Böckel A, Hörisch J, Tenner I (2021) A systematic literature review of crowdfunding and sustainability highlighting what really matters. Manage Rev Q 71:433–453.https://doi.org/10.1007/s11301-020-00189-3
2. Zhao L, Ryu S (2020) Reward-based crowdfunding research and practice. In: Shneor R, Zhao L, Flåten BT (eds) Advances in crowdfunding. Palgrave Macmillan, Cham. https://doi.org/10.1007/978-3-030-46309-0_6
3. Ahmad FS, Tyagi D, Kaur S (2017) Predicting crowdfunding success with optimally weighted random forests. In: 2017 international conference on infocom technologies and unmanned systems (Trends and Future Directions) (ICTUS), pp 770–775. https://doi.org/10.1109/ICTUS.2017.8286110.10
4. Yu P, Huang F, Yang C, Liu Y, Li Z, Tsai C (2018) Prediction of crowdfunding project success with deep learning. In: 2018 IEEE 15th international conference one-business engineering (ICEBE), pp 1–8. https://doi.org/10.1109/ICEBE.2018.00012
5. Chen L, Shen E (2019) Finding the keywords affecting the success of crowdfunding projects. In: 2019 IEEE 6th international conference on industrial engineering and applications (ICIEA), pp 567–571. https://doi.org/10.1109/IEA.2019.8714815
6. Lin Y, Lee W, Chang CH (2016) Analysis of rewards on reward-based crowdfunding platforms. In: 2016 IEEE/ACM international conference on advances in social networks analysis and mining (ASONAM), pp 501–504.https://doi.org/10.1109/ASONAM.2016.7752281
7. Sharma V, Lee K (2018) Predicting highly rated crowdfunded products. In: 2018 IEEE/ACM international conference on advances in social networks analysis and mining (ASONAM), pp 357–362.https://doi.org/10.1109/ASONAM.2018.8508797

8. Tian Z, Guan L, Shi M (2018) The key factors of successful internet crowdfunding projects—an empirical study based on different platforms. In: 2018 15th international conference on service systems and service management (ICSSSM), pp 1–6. https://doi.org/10.1109/ICSSSM.2018.8465009
9. Thushara MG, Mownika T, Mangamuru R (2019) A comparative study on different keyword extraction algorithms. In: 2019 3rd international conference on computing methodologies and communication (ICCMC), pp 969–973. https://doi.org/10.1109/ICCMC.2019.8819630
10. Schmitt X, Kubler S, Robert J, Papadakis M, LeTraon Y (2019) A replicable comparison study of NER software: StanfordNLP, NLTK, OpenNLP, SpaCy, Gate. In: 2019 sixth international conference on social networks analysis, management and security (SNAMS), pp 338–343. https://doi.org/10.1109/SNAMS.2019.8931850
11. arXiv:1810.04805v2 [cs.CL]
12. Guo A, Yang T (2016) Research and improvement of feature words weight based on TFIDF algorithm. In: 2016 IEEE information technology, networking, electronic and automation control conference, pp 415–419.https://doi.org/10.1109/ITNEC.2016.7560393
13. arXiv:1910.01108v4 [cs.CL]
14. Refaeilzadeh P, Tang L, Liu H (2009) Cross-validation. In: Liu L, Özsu MT (eds) Encyclopedia of database systems. Springer, Boston, MA. https://doi.org/10.1007/978-0-387-39940-9_565
15. Mukherjee P, Badr Y, Karvekar SN (2020) Prediction of success in crowdfunding platforms. In: 2020 international conference on decision aid sciences and application (DASA), pp 233–237.https://doi.org/10.1109/DASA51403.2020.9317273
16. Jhaveri S, Khedkar I, Kantharia Y, Jaswal S (2019) Success prediction using random forest, CatBoost, XGBoost and AdaBoost for Kickstarter campaigns. In: 2019 3rd international conference on computing methodologies and communication (ICCMC), pp 1170–1173. https://doi.org/10.1109/ICCMC.2019.8819828
17. Lu C, Xie S, Kong X, Yu PS (2014) Inferring the impacts of social media on crowdfunding. In: Proceedings of the 7th ACM international conference on Web search and data mining
18. Kaur H, Gera J (2017) Effect of social media connectivity on success of crowdfunding campaigns. Procedia Comput Sci 122, Elsevier B.V, pp 767–74. https://doi.org/10.1016/j.procs.2017.11.435
19. Henseler J, Ringle C, Sinkovics R (2009) The use of partial least squares path modeling in international marketing. Adv Int Market (AIM) 20:277–320
20. Patil S, Mehta JM, Salunkhe HS, Shah HV (2021) Kickstarter project success prediction and classification using multi-layer perceptron. In: Prateek M, Singh TP, Choudhury T, Pandey HM, Gia Nhu N (eds) Proceedings of international conference on machine intelligence and data science applications. Algorithms for intelligent systems. Springer, Singapore. https://doi.org/10.1007/978-981-33-4087-9_60

Chapter 6
Stock Prediction Using Multi Deep Learning Algorithms

Bui Thanh Hung, Prasun Chakrabarti, and Prasenjit Chatterjee

Abstract The stock market has an important role in the development of modern society. They allow the deployment of economic resources. Changes in stock prices reflect changes in the market. With powerful data processing capabilities in many fields, deep learning is also widely used in the financial field such as: stock market prediction, optimal investment, financial information processing, and execute financial trading strategies. Therefore, stock market prediction is considered one of the most popular and valuable areas in the financial sector. In this study, we propose using multi deep learning algorithms for stock prediction: RNN, LSTM, CNN, and BiLSTM. We do experiments on a stock that has a wide range of trading days and use them to predict daily closing prices. The experimental results show that the multi deep learning models can achieve good results in predicting stock prices compared to many traditional prediction models.

Keywords Stock prediction · Deep learning · RNN · LSTM · CNN · BiLSTM

1 Introduction

The stock market is a network that provides the foundation for most of the world's major economic transactions with volatility known as stock value based on market equilibrium. Capturing a stock's value even a fraction of a second in advance can be highly profitable thanks to potentially huge price spreads. The appeal of finding

B. T. Hung (✉)
Data Science Laboratory, Faculty of Information Technology, Industrial University of Ho Chi Minh city, Ho Chi Minh City, Vietnam
e-mail: buithanhhung@iuh.edu.vn

P. Chakrabarti
ITM SLS Baroda University, Vadodara, Gujarat 391510, India

P. Chatterjee
Department of Mechanical Engineering, MCKV Institute of Engineering, Howrah, West Bengal 711204, India

S. Kautish et al. (eds.), *Computational Intelligence for Modern Business Systems*, Disruptive Technologies and Digital Transformations for Society 5.0,
https://doi.org/10.1007/978-981-99-5354-7_6

predictive solutions has prompted researchers, in both industry and academia, to find ways to overcome issues such as volatility, seasonality, time dependence, economy, and the rest of the market.

Predicting the stock market is a challenging problem since the issues confronting stock prices are numerous. Many analysts and researchers in the financial field believe that the market could be predictable. Investment theory analysis techniques refer to the methods to directly predict stock prices through the study of stock market data in the past.

We present in this paper multi deep learning methods that are: RNN, LSTM, CNN, BiLSTM to predict stock values. Validating this approach is done by testing on stocks of many stocks that have a wide range of trading days. Also, we compare the performance of these methods to some other methods of machine learning. The predicted results by multi deep learning algorithms are well-agreed with the actual data.

This paper includes five sessions. Session 1 introduces the problem. Session 2 presents the related works. Section 3 presents the operating principle and implementation of the multi deep learning algorithms. The results achieved when forecasting stock prices will be outlined in Sect. 4. Finally, Sect. 5 is some conclusions and future direction.

2 Related Works

There are many approaches for stock prediction problem. The main approach is based on machine learning. In the past, traditional machine learning such as artificial neural networks [1], Fuzzy logic [2], and support vector machine [3] was used to solve these problems. Among these, the artificial neural network is the most successful, but it is still limited and unreliable [4].

Several different time series forecasting methods have been proposed, such as Autoregressive [5], Autoregressive integrated moving average [6], and neural network-based methods [7, 8]. However, because time series data, specifically stock price data, is often non-linear and non-stationary, and these existing approaches cannot efficiently extract sufficient characteristics of series data to achieve accurate time series forecasting results.

In recent years, deep learning model has been a new approach that can give multi-level representation of the original input by combining simple but non-linear modules. Each module of a deep neural network converts a representation level into a more abstract level. With enough of these transformations, the deep learning model can learn all the features from the input data. Deep learning methods have been used in various areas such as computer vision, speech recognition, and natural language processing.

In the era of deep learning, recurrent neural network (RNN) and Convolutional Neural Network (CNN) are two widely used methods of forecasting. Faustryjak et al. [9] have shown that RNN models are one of the best models for extracting features and they have used it to transform a long sequence. They can model problems seamlessly with multiple input variables, which is of great benefit in time series forecasting, where classical regression methods can be difficult to adapt to multiple inputs and multivariate forecasting problems.

One of the approaches of the RNN model is the Long Short-Term Memory (LSTM) model, which has a lot of flexibility compared to other methods. LSTMs are capable of storing information for a long time with the help of cells within them, which can carry information unchanged. The LSTM network is one of the most advanced deep learning architectures for sequence learning tasks, such as handwriting recognition, speech recognition, or time series prediction. Nguyen et al. [10] applied this method to predict the development of mechanical cracks. Xiong et al. [11] used the LSTM to forecast movements in the S&P 500 index and exchange rates. BiLSTM is composed of two LSTMs: forward and backward. This model has some advantages compared with LSTM. Some authors used CNN to predict stock price [12–14]. Another studies proposed GAN, Reinforcement learning for stock prediction [15–18].

Our research focuses on using multi deep learning algorithms. We use all famous deep learning approaches for stock prediction such as: RNN, LSTM, CNN, BiLSTM. We apply them in various stocks and compare the results with traditional machine learning approaches as well as analysis the advantages and disadvantages of each deep learning approach for this problem.

3 Methodology

3.1 The Proposed Model

Our methodology consists of three-step processes: Data collection, Preprocessing, and Training. The functionality and responsibility of each process are described in detail in Fig. 1.

3.2 Data Collection

The data is collected from historical stock prices of Vietcombank company, the largest bank in Vietnam with the code VCB. The dataset includes 10000 records collected in 10 years from 2009 to 2019.

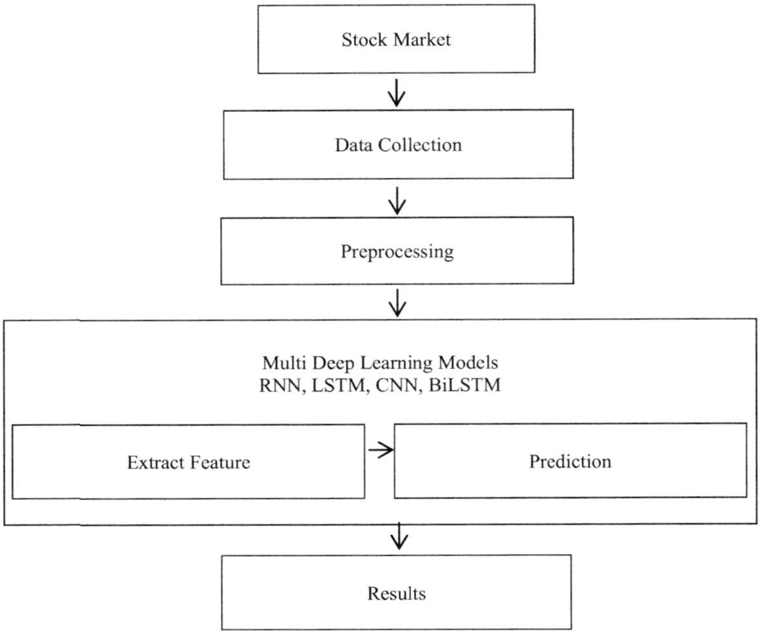

Fig. 1 The proposed model

3.3 Preprocessing

To make the prediction, we do preprocessing to reduce the noise of initial data. After preprocessing, we normalized the data with the data missing at random (MAR) assumption as follows:

$$\mathbf{X_i} = (\mathbf{X_i} - \mu^t)/\tau^t \tag{1}$$

In this formula, μ^t and τ^t describe the standard deviation of X. Because of predicting data for the next day with data a week before (transactions are limited weekend), we empirically choose $t = 5$. For example, computing the 5-day standard deviation description of the data is used to predict the data of 6th day.

3.4 Training

We build a deep learning model that processes input data with 5 financial factors to predict future closing prices. The five factors of stock data for a day are High Price, Low Price, Open Price, Close Price, and Volume. These 5 factors are important features and are not included in the price prediction of Mean Reversion, or MAR analysis methods. These factors can therefore be used as five data features for price prediction.

Suppose we enter $\mathbf{X} = \{\mathbf{X}_1, .., \mathbf{X}_t\}$, which includes daily stock data of t days. Each $\mathbf{X}_k$ in X is a vector generated from 5 features as follows:

$$[X_{k,i}]_i^k = [X_{k,high}, X_{k,low}, X_{k,open}, X_{k,close}, X_{k,volume}]$$

We extract the h_t of the deep learning algorithm and put it in 5 fully connected layers with 5 neurons to generate $\widehat{\mathbf{X}}_{t+1}$ to get an approximation of $\mathbf{X}_{t+1}$ and we take $\widehat{\mathbf{X}}_{t+1}$ to predict the closing price in t + 1 days. Next, we will present all deep learning models in our proposed method: RNN, LSTM, CNN, and BiLSTM. Next, we will present all deep learning models in our proposed method: RNN, LSTM, CNN, and BiLSTM.

RNN

Recurrent Neural Network (RNN) was introduced by John Hopfield (1982). RNNs are called regression because they perform the same task for all elements of each sequence, with the output dependent on all previous calculations.

RNN is a memory model, capable of remembering previously calculated information. Unlike previous traditional neural network models, the input information is completely independent of the output information.

Most RNNs are designed as a series of iterative modules. These modules usually have a simple structure with only one layer of mesh. RNN training is similar to traditional ANN training. The value of each output depends not only on the calculation result of the current step but also on the calculation result of the previous steps. The architecture of RNN is shown in Fig. 2.

LSTM

One notable method of recurrent neural networks is the long-term short-term memory method, which is capable of learning long-term dependencies. Compared to a forward-propagation neural network, LSTM has both forward and backward loops connecting the nodes of the network. This model is suitable for solving problems

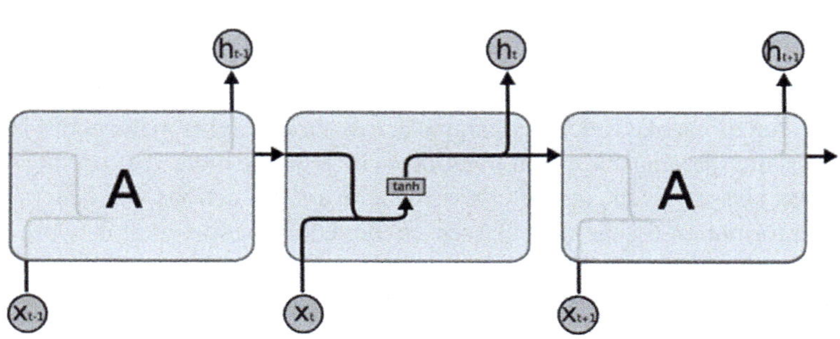

Fig. 2 RNN architecture

Fig. 3 LSTM cell

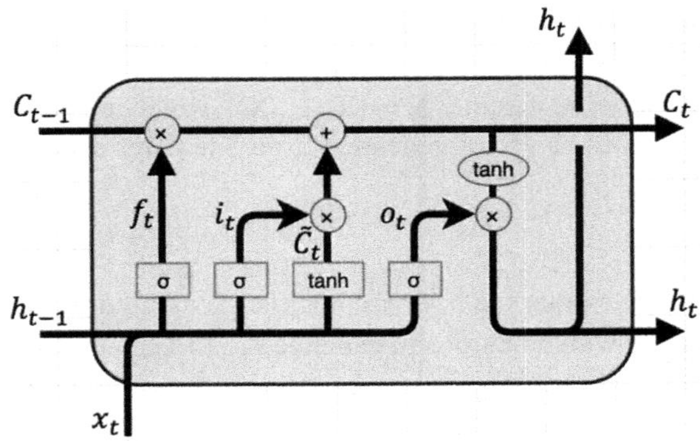

of classification, storage, and prediction of time series information, because some time intervals during separation of important data in the series may be delayed. The LSTM has been created to deal with the bursting and loss of gradients, which can be encountered when training a conventional RNN. LSTM is introduced by Hochreiter and Schmidhuber in 1997 [19] (Fig. 3).

In the given sequence of vectors (x_1, x_2, ..., x_n), state h_t of the LSTM at time t is calculated as follows:

$$h_t = o_t * \tan h(c_t) \tag{2}$$

$$c_t = f_t * c_{t-1} + i_t * \tan h(Wx_c x_t + Wj = h_c h_{t-1} + b_c) \tag{3}$$

$$o_t = \tan h(Wx_O x_t + Wh_O h_{t-1} + Wc_O c_t + b_o) \tag{4}$$

$$i_t = \sigma(Wx_i x_t + Wh_i h_{t-1} + Wc_i c_{t-1} + b_i) \tag{5}$$

$$f_t = \sigma\left(Wx_f x_t + Wh_f h_{t-1} + Wc_f c_{t-1} + b_f\right) \tag{6}$$

CNN

A CNN network is a collection of Convolution layers that overlap and use nonlinear activation functions like ReLU and tanh to activate the weights in the nodes [12–14]. Each class after passing activation functions will generate more abstract information for the next classes. Each class after passing activation functions will generate more abstract information for the next classes. In the feedforward neural network model, each input neuron gives each output neuron in subsequent layers. Fig. 4 shows the architecture of CNN.

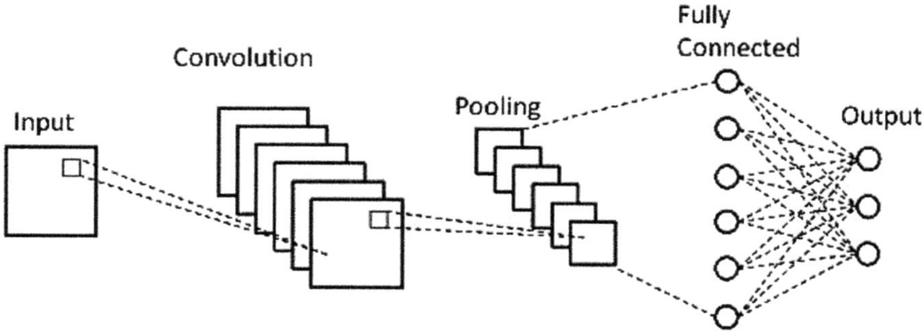

Fig. 4 CNN model

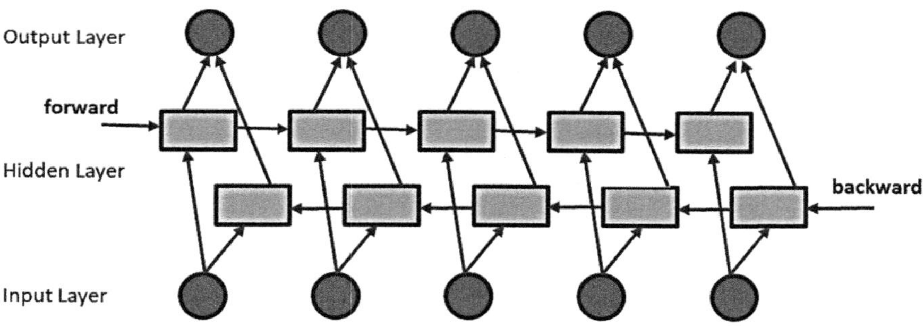

Fig. 5 Bidirectional LSTM model

Bi-LSTM

Bi-LSTM is a distinguishing model. In particular, the hidden layer consists of two single LSTM network layers: forward LSTM and backward LSTM. The activation function is the Leaky ReLU and in the output layer, the sigmoid function is used. To optimize Bi-LSTM, we chose cross entropy as the loss function. The BiLSTM model is shown in Fig. 5 [20, 21].

4 Experiments

4.1 Dataset

We applied the multi deep learning models to predict the closing stock price of the VCB Stock. The data is collected from historical dataset stock prices of Vietcombank company. Some data of VCB stocks are shown in Table 1. Figure6 visualizes the data by four features: Open, High, Low, and Close. Volume feature is visualized in Fig. 7.

The dataset is divided into three parts: training, testing, and validation, presented in detail in Table 2. 80% of the data is used for training, 10% for validation, and 10% for testing the models. We used activation by Sigmoid, optimization with Adam, other parameters used in our proposed model are presented as follows:

RNN:

Hidden node: 64
epoch: 64

Table 1 Statistics of dataset

Stock	Time	Open	High	Low	Close	Volume
VCB	10/30/2019	81.1	82.5	80.5	80.6	809,820
	10/31/2019	81	81.7	80.5	81.5	527,230
	11/1/2019	81.5	82.8	81	81	2,794,080
	11/4/2019	81.6	81.9	79	79	1,187,100
	11/5/2019	79.5	80.8	78.9	80.6	791,470
	11/6/2019	80	80.9	79.4	80.7	729,340
	11/7/2019	80.7	81.3	80.3	81	1,040,190
	11/8/2019	81.3	83.1	80.8	82.9	1,388,490
	11/11/2019	82.9	83.3	81.9	82.1	775,770
	11/12/2019	82.1	83.6	81.8	83.4	795,100

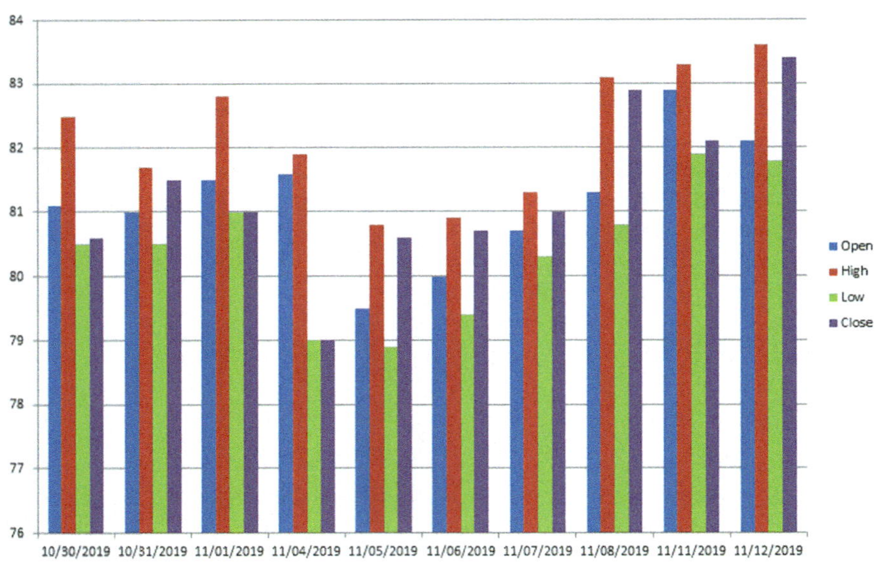

Fig. 6 Data visualization by 4 features: open, high, low and close

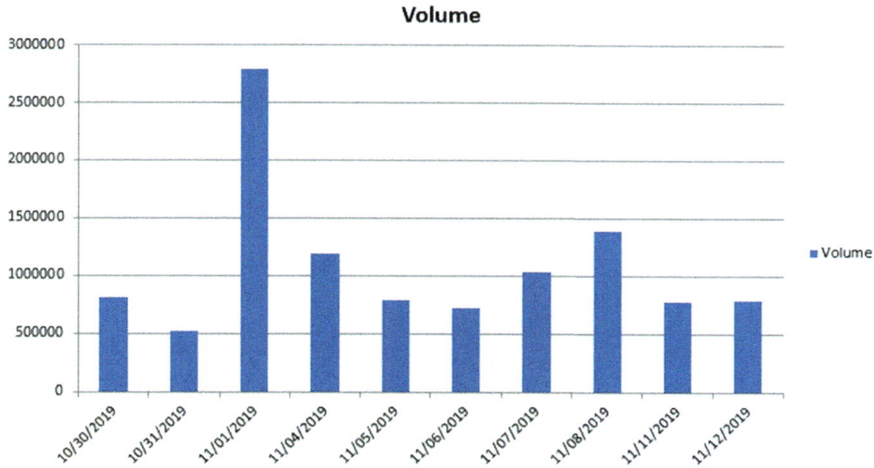

Fig. 7 Data visualization by volume feature

Table 2 Statistics of dataset

Dataset	Percentage
Training	80
Testing	10
Validation	10

dropout: 0.2

LSTM:

Hidden node: 64
epoch: 128
dropout layer: 0.2

CNN

Convolution layer: 1D
kernel size: 3
Max Pooling: pool size of 3.
dropout layer: 0.4

BiLSTM

number of LSTM: 2
hidden node: 64
dropout layer: 0.2
epoch: 128.

Table 3 The results of RNN, LSTM, CNN and BiLSTM models with MAE metric

Model	MAE
RNN	4.0268
LSTM	3.4201
CNN	5.2134
BiLSTM	6.2035

We used all deep learning models of Keras [22] and Tensorflow [23] frameworks. We used RMSE (Root Mean Square Error), MAPE (Mean Absolute Percentage Error), MAE (Mean Absolute Error), and AR (Average Revenue) metrics to evaluate our proposed models. Assuming that the predicted closing price and the actual and per k days are $\hat{y}_k$ and y_k, we can calculate the metrics as follows:

$$\mathrm{MAE} = \frac{1}{N}\sum_{k=1}^{N}\left|\hat{y}_k - y_k\right| \tag{7}$$

$$\mathrm{MAPE} = \frac{1}{N}\sum_{k=1}^{N}\frac{\left|\hat{\mathbf{y}}_k - \mathbf{y}_k\right|}{\mathbf{y}_k} \tag{8}$$

$$\mathrm{RMSE} = \sqrt{\frac{1}{N}\sum_{k=1}^{N}(\hat{y}_k - y_k)} \tag{9}$$

$$\mathrm{AR} = \frac{1}{N-1}\sum_{k=1}^{N-1}(y_{k+1} - y_k), \text{ if } \hat{y}_{k+1} > \hat{y}_k \tag{10}$$

MAE, MAPE and RMSE will be small numbers if the real data is approximate with close price prediction. Based on 4 methods, AR is scored by the average value of the stocks.

Table 3 and Fig. 8 show the results of multi deep learning models evaluated by MAE metric. Table 4 and Fig. 9 show the results of multi deep learning models evaluated by MAPE metric. Table 5 and Fig. 10 show the results of multi deep learning models evaluated by RMSE metric. Table 6 and Fig. 11 show the results of multi deep learning models evaluated by AR metric. Table 7 shows the results of multi deep learning models evaluated by all metrics.

From the results, we saw that MAE, MAPE, RMSE and AR scores of RNN are 4.0268, 0.0168, 5.3526, 0.6725; LSTM are 3.4201, 0.0145, 4.0346, 0.7478; CNN are 5.2134, 0.0571, 6.4632, 0.6326, and score of BiLSTM are 6.2035, 0.0608, 7.0264, 0.5068. The illustrations of the prediction dataset compared to the real dataset in RNN, LSTM, CNN, and Bi-LSTM are shown in Figs. 12, 13, 14, and 15. So, the best method is LSTM, follows is RNN, CNN, and BiLSTM is the lowest score model.

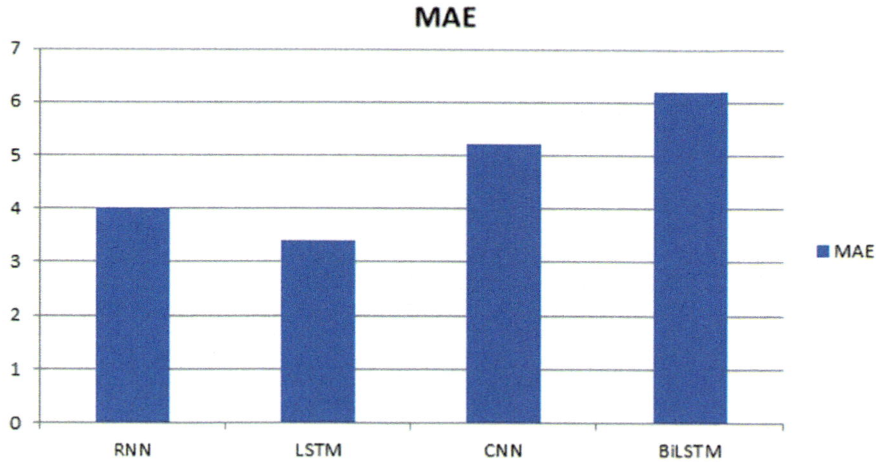

Fig. 8 MAE scores of multi deep learning models: RNN, LSTM, CNN, and BiLSTM

Table 4 The results of RNN, LSTM, CNN, and BiLSTM models with MAPE metric

Model	MAPE
RNN	0.0168
LSTM	0.0145
CNN	0.0571
BiLSTM	0.0608

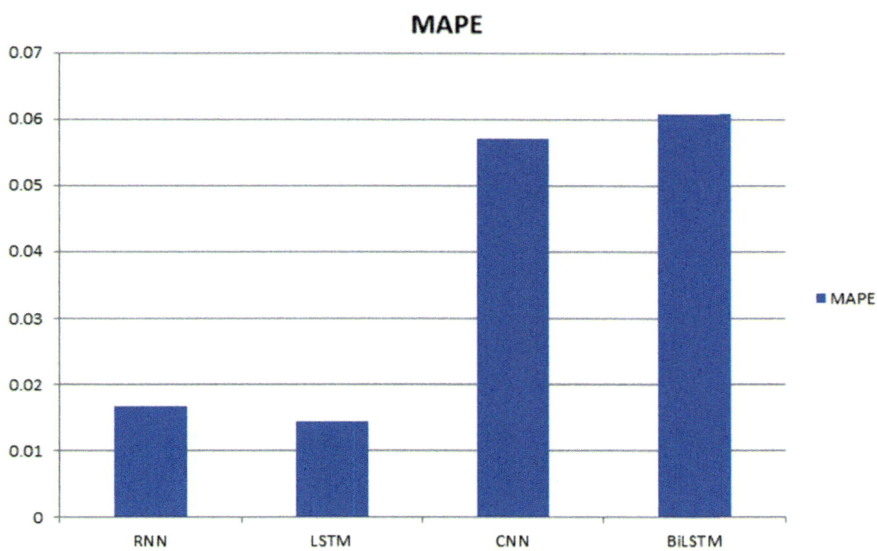

Fig. 9 MAPE scores of multi deep learning models: RNN, LSTM, CNN, and BiLSTM

Table 5 The results of RNN, LSTM, CNN, and BiLSTM models with RMSE metric

Model	RMSE
RNN	5.3526
LSTM	4.0346
CNN	6.4632
BiLSTM	7.0264

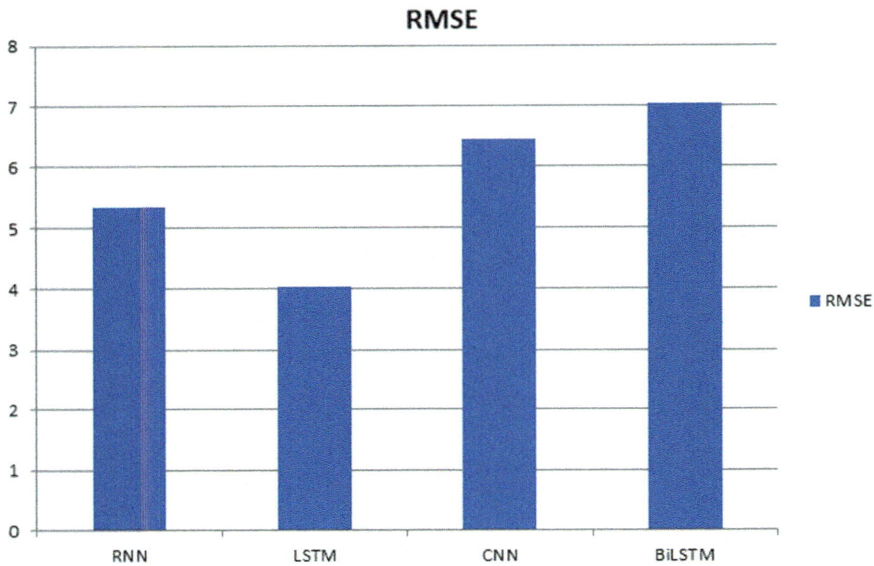

Fig. 10 RMSE scores of multi deep learning models: RNN, LSTM, CNN, and BiLSTM

Table 6 The results of RNN, LSTM, CNN, and BiLSTM models with AR metric

Model	AR
RNN	0.6725
LSTM	0.7478
CNN	0.6326
BiLSTM	0.5068

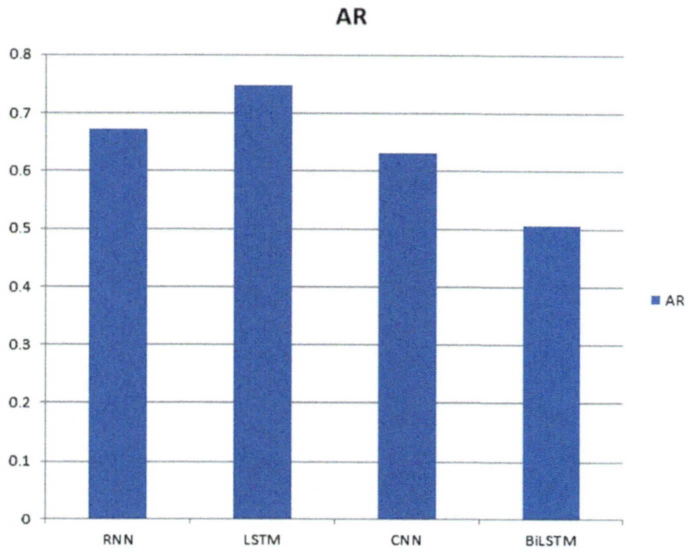

Fig. 11 AR scores of multi deep learning models: RNN, LSTM, CNN, and BiLSTM

Table 7 The results of RNN, LSTM, CNN, and BiLSTM models with all metrics

Model	MAE	MAPE	RMSE	AR
RNN	4.0268	0.0168	5.3526	0.6725
LSTM	3.4201	0.0145	4.0346	0.7478
CNN	5.2134	0.0571	6.4632	0.6326
BiLSTM	6.2035	0.0608	7.0264	0.5068

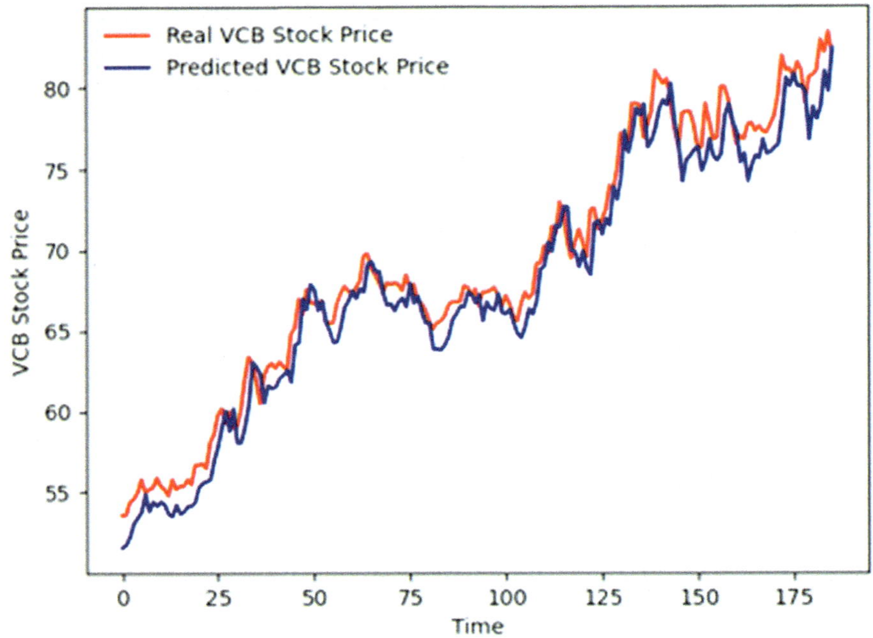

Fig. 12 The results of RNN model

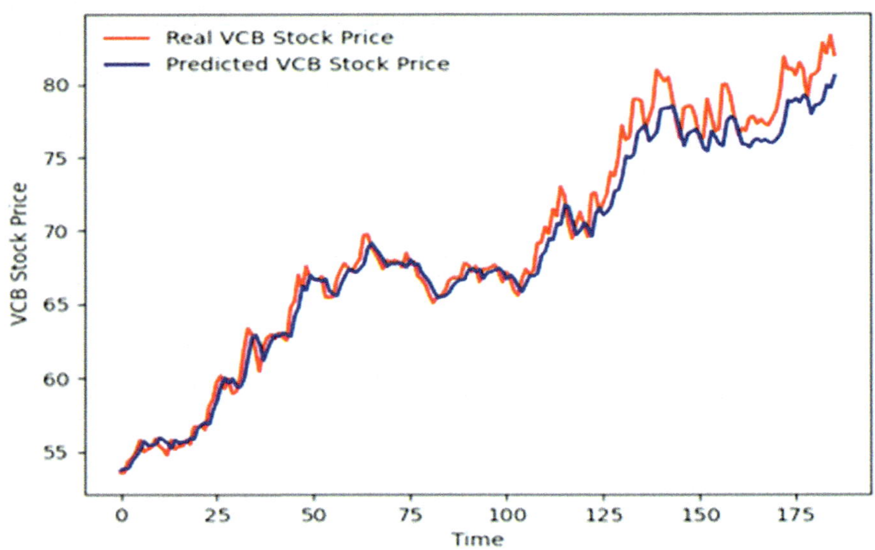

Fig. 13 The results of LSTM model

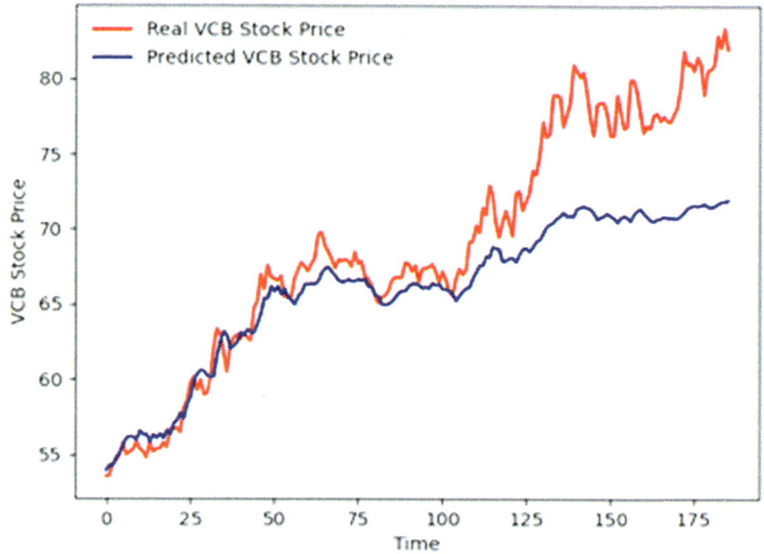

Fig. 14 The results of CNN model

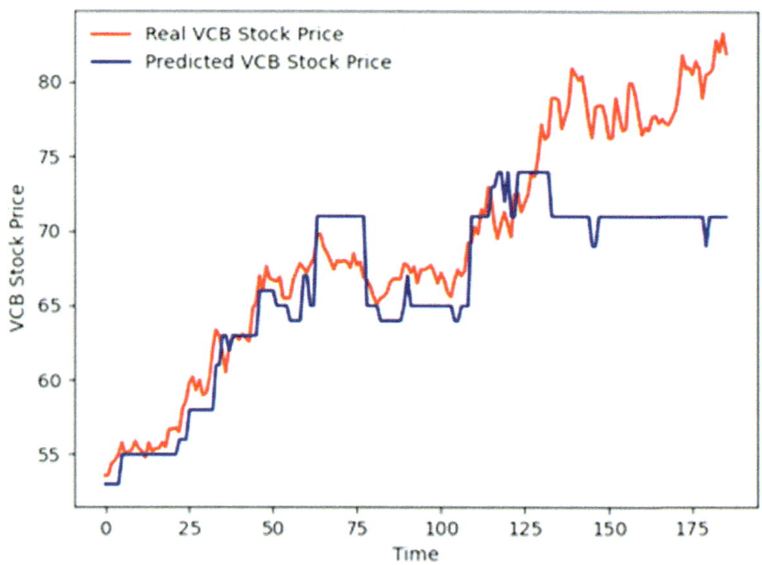

Fig. 15 The results of BiLSTM model

5 Conclusion

We have applied multi deep learning methods: RNN, LSTM, CNN, BiLSTM to predict the price of VCB stock in Vietnam stock market. The model is assumed to be independent of the other. The results show that LSTM is the best deep learning model for stock prediction.

Under the current approach, we assume that the pattern for a particular stock is independent of other stocks in the market, but in practice these stocks are strongly correlated and to some extent, with stocks in other markets. In the future, it may be intuitive to try and build a model that considers these correlations. In addition, currently, the data used is end-of-day observations. Predictive performance can be improved by taking full stock values by the minute or hour.

References

1. Zhang GP, Berardi V (2001) Time series forecasting with neural network ensembles: an application for exchange rate prediction. J Oper Res Soc 5(6):652–664
2. Maguire LP, Rocher B, McGinnity TM, McDaid L (1998) Predicting a chaotic time series using a fuzzy neural network. Inf Sci 112(1–4):125–136
3. Kim K (2003) Financial time series forecasting using support vector machines. Neurocomputing 55(1):307–319
4. Hassan MR (2009) A combination of hidden markov model and fuzzy model for stock market forecasting. J Neurocomput 3439–3446
5. Jain A, Kumar AM (2007) Hybrid neural network models for hydrologic time series forecasting. Appl Soft Comput 7(2):585–592
6. Aladag CH, Egrioglu E, Kadilar C (2009) Forecasting nonlinear time series with a hybrid methodology. Appl Math Lett 22(9):1467–1470
7. Pacurar M (2008) Autoregressive conditional duration models in finance: a survey of the theoretical and empirical literature. J Econ Surv 22(4):711–751
8. Al-Shiab M (2006) The predictability of the amman stock exchange using the univariate autoregressive integrated moving average (ARIMA) model. J Econ Administrative Sci 22(2):17–35
9. Faustryjak D, Jackowska-Strumiłło L, Majchrowicz M (2018) Forward forecast of stock prices using LSTM neural networks with statistical analysis of published messages. Interdisciplinary PhD workshop (IIPhDW) 2018:288–292
10. Nguyen DLH, Do DTT, Lee J, Rabczuk T, Nguyen-Xuan H (2019) Forecasting damage mechanics by deep learning. Comput, Mater Continua 61(3):51–77
11. Xiong R, Nichols EP, Shen Y (2015) Deep learning stock volatility with Google domestic trends. arXiv e-prints arXiv: 1512.04916
12. Tsantekidis A, Passalis N, Tefas A, Kanniainen J, Gabbouj M, Iosifidis A (2017) Forecasting stock prices from the limit order book using convolutional neural networks. In: 19th IEEE conference on business informatics, CBI 2017, Thessaloniki, Greece, July 24–27, 2017, vol 1: Conference Papers, pp 7–12
13. Hung BT, Chakrabarti P (2022) Parking lot occupancy detection using hybrid deep learning CNN-LSTM approach. algorithms for intelligent systems book series (AIS)
14. Hung BT, Tien LM (2021) Facial expression recognition with CNN-LSTM. Research in intelligent and computing in engineering. Springer Series in Advances in Intelligent Systems and Computing

15. Rather AM, Agarwal A, Sastry VN (2015) Recurrent neural network and a hybrid model for prediction of stock returns. Expert Syst Appl 42:3234–3241
16. Hung BT (2022) Using deep unsupervised method for stock prediction. Lecture notes in networks and systems book (LNNS, vol 288)
17. Hung BT, Semwal VB, Gaud N, Bijalwan V (2021) Violent video detection by pre-trained model and CNN-LSTM approach. In: Proceedings of integrated intelligence enable networks and computing. Springer series in algorithms for intelligent systems
18. Ding X, Zhang Y, Liu T, Duan J (2015) Deep learning for event-driven stock prediction. In: Proceedings of the twenty-fourth international joint conference on artificial intelligence, IJCAI 2015. Buenos Aires, Argentina, July 25–31, 2015, pp 2327–2333
19. Hochreiter S, Schmidhuber J (1997) Long short-term memory. Neural Comput Appl 9(8):1735–1780
20. Hung BT (2020) Combining syntax features and word embeddings in bidirectional LSTM for vietnamese named entity recognition. Further advances in internet of things in biomedical and cyber physical systems
21. Hung BT (2019) Domain-specific versus general-purpose word representations in sentiment analysis for deep learning models. Front Intell Comput: Theory Appl 252–264. Springer
22. Chollet F, Keras (2015). https://github.com/fchollet/keras
23. Abadi M, Barham P, Chen J, Chen Z, Davis A, Dean J, Devin M, Ghemawat S, Irving G, Isard M, Kudlur M, Levenberg J, Monga R, Moore S, Murray DG, Steiner B, Tucker P, Vasudevan V, Warden P, Wicke M, Yu Y, Zheng X (2016) Tensorflow: a system for large-scale machine learning. In: Proceedings of the 12th USENIX conference on operating systems design and implementation, OSDI'16, pp 265–283. https://doi.org/10.1007/s10107-012-0572-5

Chapter 7
House Price Prediction by Machine Learning Technique—An Empirical Study

Suriya Begum

Abstract Depicting the price of a home is becoming crucial day by day, as the cost of land and houses rises year after year. House prices are significantly associated with features such as locality, region, house, age, and people; forecasting price of an individual house needs a lot of information. The House Price Index (HPI) is one of the standard tools for assessing house price variations. The objective of the study is to predict the price of houses with the use of various regression approaches of supervised machine learning. In the proposed study, housing data of 5000 homes in USA have been analyzed. Six models are used, namely, Random Forest Regression, XGBoost Regression, Linear Regression, Artificial Neural Network, Gradient Boosting and Ada Boosting. The results are compared, of which Linear Regression and Artificial Neural Network outperformed with R2_Score of 91.56 and 91.37% compared to the other four algorithms.

Keywords House price · Random forest regressor · Gradient boosting · MLP regressor · Linear regression · Ada boosting

1 Introduction

Investment can be incorporated into various business components, like fixed deposits, gold coins, share market, land, etc. Investing in houses is also one of the market trends. Before investing in houses or lands a complete survey about the prices in various locations is necessary for better profit gains. Estimating the house prices will help people who are planning to purchase a house [1]. This estimation helps them to know the cost range in the future so that they can plan well in advance about their investment. It is also useful for property investors to identify the trend of housing prices in various locations of their choice. Identifying the cost of a property is crucial

S. Begum (✉)
An NIIT Venture, StackRoute, Bengaluru, India
e-mail: suriyabegumstore@gmail.com

S. Kautish et al. (eds.), *Computational Intelligence for Modern Business Systems*, Disruptive Technologies and Digital Transformations for Society 5.0,
https://doi.org/10.1007/978-981-99-5354-7_7

these days since the cost of land and houses rises year after year [2]. Price of houses is a key pointer from a business point of view, and cost ranges have become a hot topic among purchasers and vendors [3]. Figuring out a better technique to calculate property prices that truly represent market prices has become a major topic [4]. Real estate industry is one of the most cost-sensitive and least transparent in the world, and it is always changing [5, 6]. Predicting house prices is an important feature of real estate. The literature tries to take out relevant information from historical real estate marketplace data [7]. The amount of daily information in the real estate industry is enormous and growing rapidly [8]. The phenomenon of dropping or growing housing values has piqued the researcher's curiosity, as well as that of many other relevant parties. The demand for renting and buying homes has increased as a result of increased urbanization [9].

House prices have a big impact on the economy and price fluctuations are a big problem for clients and real estate agents. Every year, housing prices rise, reinforcing the necessity for a methodology or technique that can forecast future house values [10]. People who are purchasing a new home are careful about their investments and market policies. The existing system integrates cost-of-house calculations without the required forecasting of future cost engagements and cost increases [11].

The House Price Index is a standard method for appraising increases in home prices. Day by day, the cost of homes is increasing and is connected with significant features such as locality, region, and people [12]. Home purchases are calculated by using OFHEO. These tendencies are reflected in the home market in the United States. Aside from these home values indices, the construction of a home values forecasting model can substantially aid in the estimate of future home values and the formulation of real estate policy [13].

Although a huge number of articles have used typical machine learning algorithms to properly predict house prices; they rarely analyze the assessment of different models and ignore the less renowned yet sophisticated models [12]. To effectively raise the accuracy, a variety of algorithms are applied [14]. In real estate and appraising analysis, machine learning provides a better, alternative technique, particularly with respect to property market prediction [15]. Experiments show that a competitive technique is Multiple Regression Analysis focused on mean squared error measures [16]. Related to geographical criteria; the approach adopted assists in determining a beginning price for a unit. Customers can utilize the various frameworks to help them find a home that meets their needs [17]. Future costs can be predicted by analyzing past monetary exercises and value ranges, as well as anticipated developments [5].

The rest of this paper is divided into seven sections. A concise related work is carried out on house price in Sect. 2. The objectives are proposed in Sect. 3. Methodology is explained in Sect. 4. Proposed model is discussed in Sect. 5. In Sect. 6, analysis and findings of the proposed model are discussed. In Sect. 7 future works are described. Finally, Sect. 8 concludes the paper.

2 Background

The classification performance of machine learning techniques such as C4.5, RIPPER, Nave Bayesian, and AdaBoost in developing a home price prediction system is examined. The trials showed that the RIPPER method constantly surpasses the other algorithms related to accuracy when it comes to house price prediction [18]. New idea to forecast the price of a certain unit based on market values using various Machine Learning algorithms is described in the project [19]. AdaBoost and J48 tree modeling techniques and efficient data mining methods such as Weka and Rapid Miner were used to get a greater impact on price forecasting [14]. Multiple Regression techniques are compared based on various factors in their study [9].

Applying a diversity of regression methods, the objectives are achieved on a weighted average of the numerous methods to obtain the most precise findings. The outcome showed that this technique produces minimal error and the maximum level of accuracy in comparison with other individual methods. They also employed Google maps to even get exact real-world prices by employing real-time neighborhood facts [6]. Both traditional machine learning techniques and new methods are employed for predicting sales prices of houses [8]. The author has participated in a Kaggle competition, and placed ranked 35th out of 2221 outcomes, and the RMSE of model test data is 0.12019, indicating strong performance and little over-fitting [20]. The large discrepancy between housing costs in Melbourne and also best affordable districts has been revealed during their study. Furthermore, investigations show that combining stepwise and SVM both of which are dependent on mean squared error assessment is a competitive method [7]. Regression analysis such as simple linear regression, multiple linear regression, and neural networks have been applied to accurately determine the home value. The algorithm with the lowest MSE is picked as the best for estimating the price of a house [21]. Home value prediction model which is built on the Tensor Flow is applied to address the challenges. The investigational findings suggest that the suggested strategy forecasts individual property prices significantly than the SVR approach [22]. The market value of a home in the Boston area is estimated by preparing the best model [23].

The differences between many advanced models using both classic and modern machine learning methodologies were investigated. Also, presented an accurate solution for housing price forecast by thoroughly validating numerous strategies in experimental analysis on regression [12]. Relevant models for home purchasers are discovered and linear regression is used to analyze past property sales in Australia [11]. Models on house availability based on required house attributes and house value projection are proposed and the model is built for a tiny city in Andhra Pradesh's West Godavari region. Decision tree classifier, decision tree, and multiple regressions are all part of the project [17]. Property values are appraised and employed three machine learning techniques: SVM, RF, and GBM. SVM is still a valuable method in data fitting, according to our research, because it can give reasonably precise forecasts within a short time frame for Hong Kong homes [15]. Support vector machines, Feed-forward Neural Networks, and extended regression techniques are used to generate three different modules in this work. The trials show also that Feedforward Neural Network algorithm regularly outperforms all other algorithms in terms of accuracy when it comes to home price prediction for Turkish homes [3]. Multiple regression, Lasso and Ridge regression algorithms, SVM, and Extreme Gradient Boost Regression are used in modeling explorations for Bengaluru homes [24]. XGBoost is used to forecast housing values in this study. The proposed home value prediction model has a prediction accuracy of 98 percent for Karachi homes [10]. An overview is given about how to anticipate house costs with use of python language and several regression algorithms [25]. The models that manage with regression and classification problems are studied and presented the results in the study [26]. Spatial prediction of home values in Beijing is developed by introducing a collection of machine learning approaches, including XGBoost, linear regression, Random Forest Regression, Ridge and Lasso Model, bagging, and boosting. XGBoost outperforms when compared to other models [13].

Multiple supervised regression algorithms are implemented using Python. The explanatory variables are used to forecast future values, and regression is utilized to do so. Calculating the prediction error is used to evaluate the model. When a tiny error happens, the regression analysis would gain a lot of accuracy [2]. Multiple Linear Regression techniques are compared in the project [16, 27]. Decision Tree Regression model is implemented to forecast housing costs in Mumbai. It will assist clients in putting money into a direct request without having to go through a broker. The decision tree regression model has an accuracy of 89 percent, according to the findings of this study [5, 28]. Multiple prediction approaches are compared to arrive at the best outcomes. Because of its linearity, linear regression was selected as the approach [4].

3 Objectives

- To predict the sales price of a house.
- To reduce the difference between actual and predicted rating.
- To visualize the data using various plots to get better insights.

4 Methodology

Various steps of the methodology are:

- Collected of data in the form of.csv file.
- Analysis, cleaning, and manipulation of data.
- Data Analysis is accomplished to observe the result of each independent feature on the dependent feature.
- Selection of Models.
- Evaluation of models to discover the most excellent one.

4.1 Preprocessing of Data

Various steps of preprocessing the data are:

- Features are converted to categorical type by using Label Encoding technique and data was then transformed to numerical variables so that it could be used in model construction.
- Cleaning of data will be performed and missing values are deleted in preparation for analysis and model creation. Scaling will be performed on some features for the model building.
- K-Fold with value 10 is used for training and testing the data.

Various steps of analysis of data are as shown in Figure 1.

4.2 Analysis of Data

In this stage, the relationship between the independent and the dependent variables is investigated for 5000 houses. Figure 2 shows the Pair plot of "Price" and "Avg. Area Income", "Avg. Area House Age", "Avg. Area Number of Rooms", "Avg. Area Number of Bedrooms", "AreaPopulation", "Address".

From the above Fig. 3, it is observed that the Avg. Area Income increases and prices also increase. From Fig. 4, it is observed that the Avg. Area House Age increases, prices slightly increase.

From Fig. 5, it is observed that the Avg. Area Number of Rooms increases, prices also increase. From Fig. 6, it is observed that the Avg. Area Number of Bedrooms increases, prices slightly increases.

From the above Fig. 7, it is observed that as the Area Population increases, prices also increase.

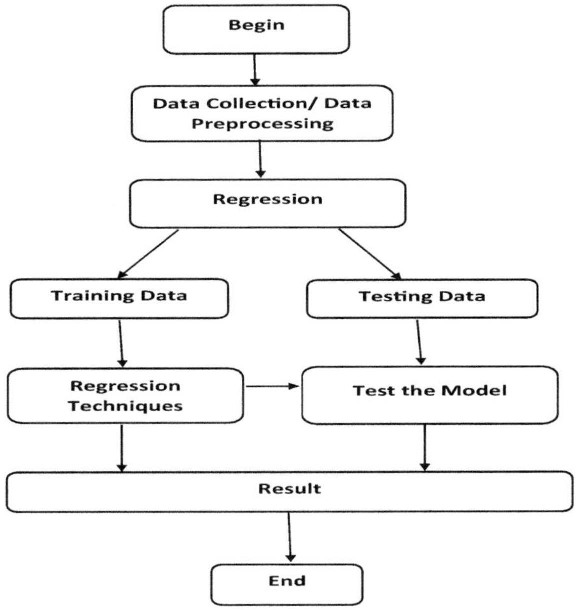

Fig. 1 Data analysis steps

4.3 *Feature Engineering*

Feature Abstraction: Label Encoding method is used to encode the categorical features. **Visualization**: Histogram of the independent variables and target variable is shown in Fig. 8.

4.4 *Predictive Modeling*

It makes predictions about unknown data based on historical data.

Steps of Predictive Modeling:

- Collection of data.
- Cleaning, preparing and manipulating of data.
- Train the model
- Test the model
- Validation of model.

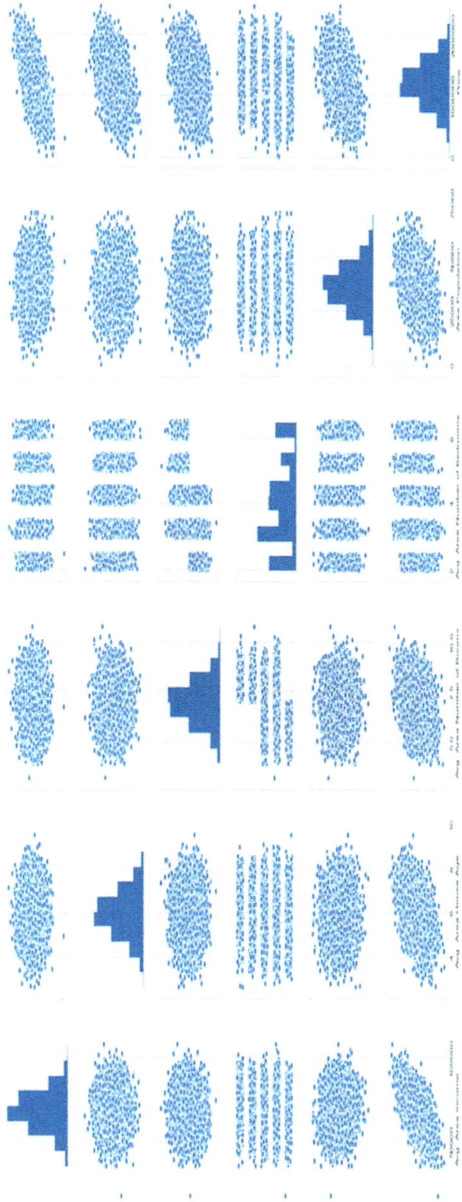

Fig. 2 Pair plot of "Price" and "Avg. Area Income", "Avg. Area House Age", "Avg. Area Number of Rooms", "Avg. Area Number of Bedrooms", "Area Population", "Address"

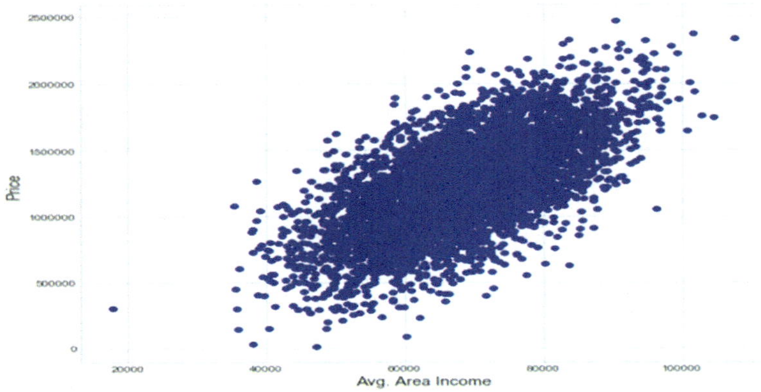

Fig. 3 Scatter plot of avg. area income versus price

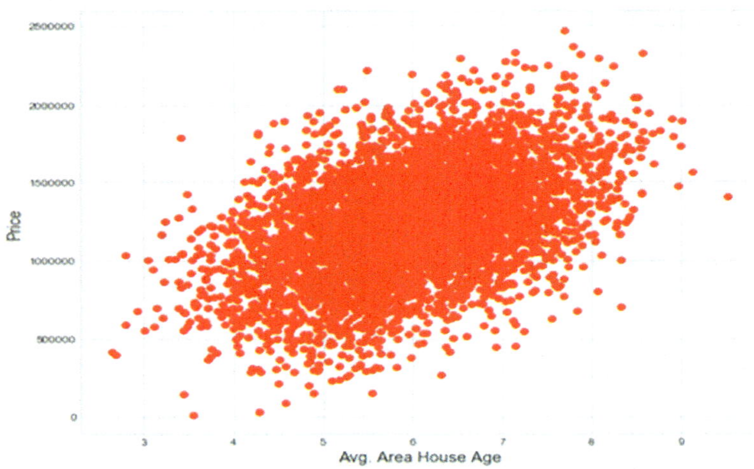

Fig. 4 Scatter plot of avg. area house age versus price

4.5 Methodology

Multiple Regression Models are used in the study. The hyper-parameters of the model are tuned to obtain maximum accuracy.

Regression

Regression models Random Forest Regressor (RF), Gradient Boosting, XGB Regressor, Ada Boosting, Linear Regression, and MLP Regressor are used for predictive modeling.

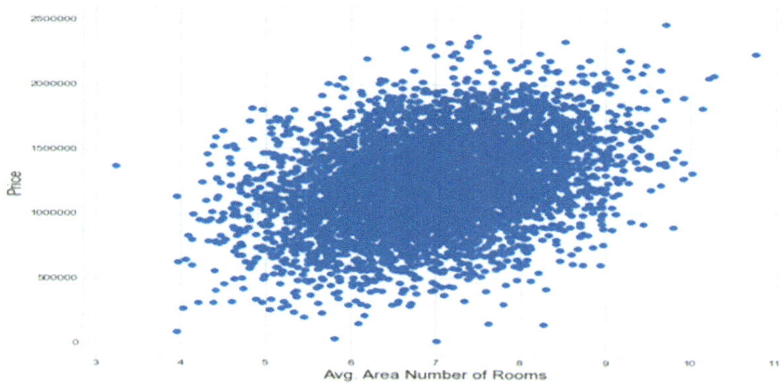

Fig. 5 Scatter plot of avg. number of rooms versus price

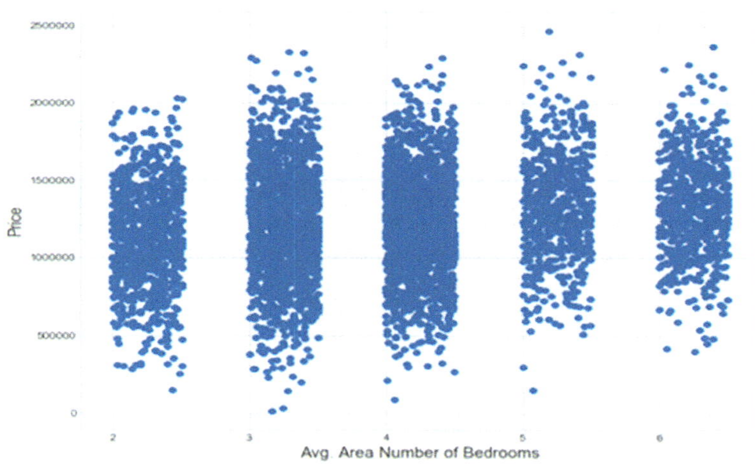

Fig. 6 Scatter plot of avg. number of bedrooms versus price

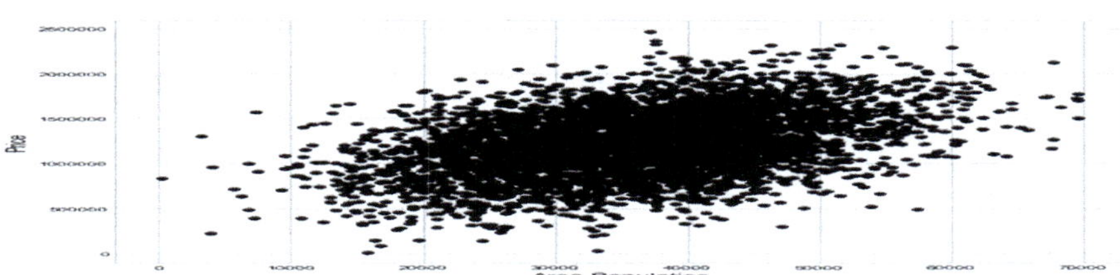

Fig. 7 Scatter plot of area population versus price

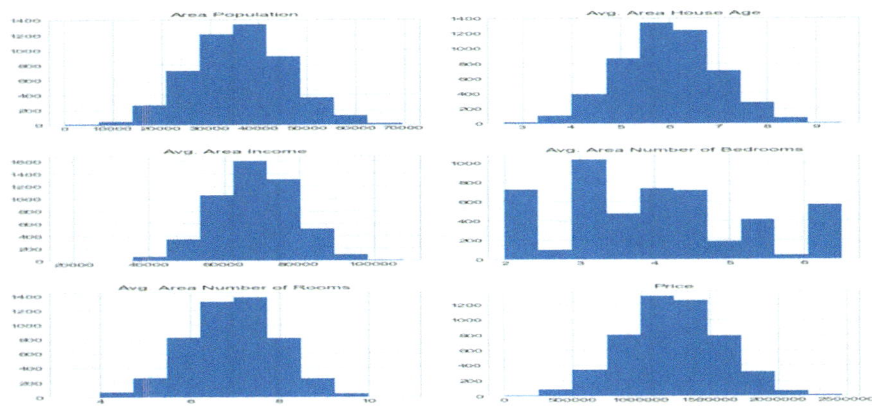

Fig. 8 Histogram of the independent variables and target variable

5 Proposed Model

Characteristics of the Dataset

The dataset is taken from kaggle for study purpose. The dataset contains 5000 records and 7 columns. Table 1 shows features, descriptions, and its values. Correlation coefficient and feature importance are employed to drop the features. Correlation matrix of the six variables is shown in Fig. 9.

6 Analysis and Findings

Finally, we have taken 5000 records and 4 features for the experiment. Based on Correlation between variables and feature_importances the final features are taken for modeling model 0.70:30 ratio is taken for splitting the training and evaluating dataset. Performance is calculated and evaluated for all the algorithms based on different metrics. Metrics evaluation of the six algorithms is shown in Table 2.

Coefficient of determination: This metric is denoted by R^2 and also called are "R squared". It is a statistical measure that shows the closeness of the data on the regression line. This metric computes the amount of the difference in the target feature that is anticipated from the input variable(s). 1.0 is considered as the best possible score and it can be negative also. A model that all the time predicts the anticipated value of y, disregarding the independent variables, would get an R^2 score of 0.0 (1).

Table 1 Features, description and values of the dataset

Sl	Feature	Description	Values
01	'Avg. Area Income'	It refers to the average income of people of the town where the house is situated	Min = 17,796.63 Max = 107,701.74
02	'Avg. Area House Age'	It refers to the average age of homes in a given town	Min = 2.64 Max = 9.51
03	'Avg. Area Number of Rooms'	It is the average number of rooms per house in same town	Min = 3.23 Max = 10.75
04	'Avg. Area Number of Bedrooms'	It is the average number of bedrooms of houses people living in the same town	Min = 2.0 Max = 6.5
05	'Area Population'	It is the population of town where the house is located	Min = 172.61 Max = 69,621.71
06	'Address'	It is the address for the house	It's a categorical Data
07	'Price'	It is the price at which the house is sold	Min = 15,938.65 Max = 2,469,065.59

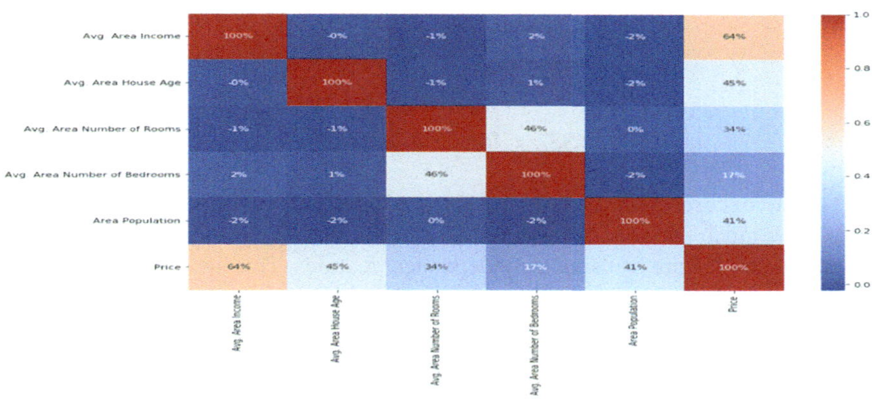

Fig. 9 Correlation matrix of the six features

Table 2 Evaluation of metrics of the six algorithms

Sl. No	Algorithm	R2_ Score (%)	Mean absolute error	Mean squared error	Root mean square error	Training accuracy (%)	Testing accuracy (%)
01	Linear regression	91.56	80,145.09	10,030,535,589.43	100,152.56	91.41	91.56
02	XGBoost regression	89.97	87,296.84	11,925,794,287.81	109,205.28	93.33	89.97
03	Random forest regression	88.00	94,679.84	14,263,930,716.31	119,431.69	96.45	88.00
04	Artificial neural network	91.37	80,896.75	10,256,356,585.70	101,273.67	91.45	91.37
05	Gradient boosting	89.97	87,189.83	11,919,574,400.22	109,176.80	92.98	87.97
06	Ada boosting	83.71	110,457.52	19,364,755,411.57	139,157.30	84.96	83.71

$$R^2 = 1 - (RSS/\ TSS) \tag{1}$$

R^2 = Coefficient of determination

RSS = Residual Sum of Squares

TSS = Total Sum of Squares

Variance score: While determining the dependent variable, first we need to understand how much variance is observed in it. This metric explains the fluctuation of the dependent feature by using the independent features. This measures how far a set of numbers is spread out from their mean value. Finest possible score is 1.0, lesser values are poorer.

Root mean square error (RMSE): It takes the difference of each observed and predicted value (2).

$$\text{Residual} = \text{observed value} - \text{predicted value} \tag{2}$$

The mean squared error (MSE) is calculated by first squaring and summing the residuals and then dividing by the number of samples using residuals (3).

$$\mathbf{MSE} = \sum (residuals)2/n \tag{3}$$

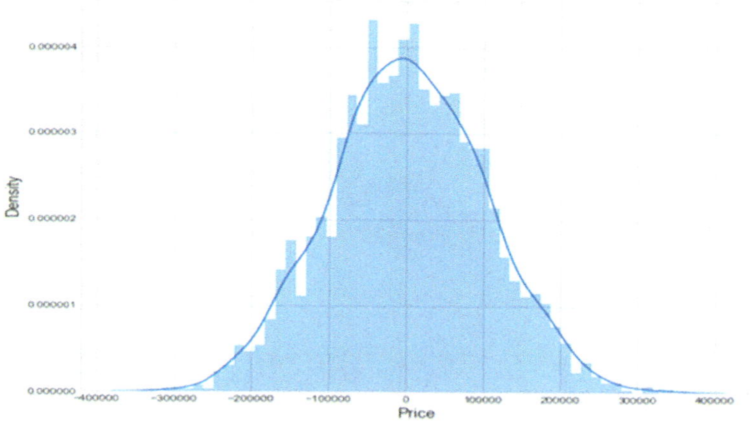

Fig. 10 Linear regression distplot

where 'n' is the number of samples.

RMSE is then calculated by taking a square root of the MSE so that it is in the same unit as original data (4).

$$\mathbf{RMSE} = \sqrt{\sum (residuals)2/n} \tag{4}$$

The value of RMSE indicates relatively far the forecasts are to the genuine values (on an average).

MAE = sum of all absolute errors/number of errors (n) (5)

$$\mathbf{MAE} = \sum (\mathrm{X_i} - \mathrm{X})/\mathrm{n}, \ \text{where i} = 1 \text{ to n} \tag{5}$$

The distribution as well as scatter plots of Linear, XGBoost, Artificial Neural Network, Gradient Boosting, Ada Boosting, and Random Forest models are as shown in the below Figs. 10, 11, 12, 13, 14, 15, 16, 17, 18, 19, 20 and 21.

7 Future Work

In the future study, modeling on real-time Indian dataset will be carried out.

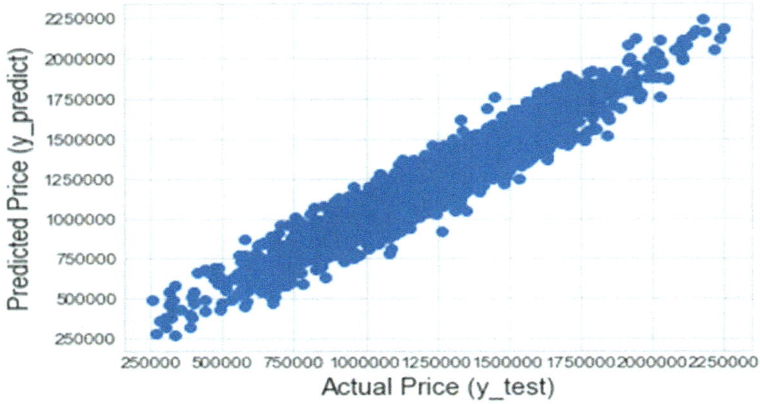

Fig. 11 Linear regression scatter plot

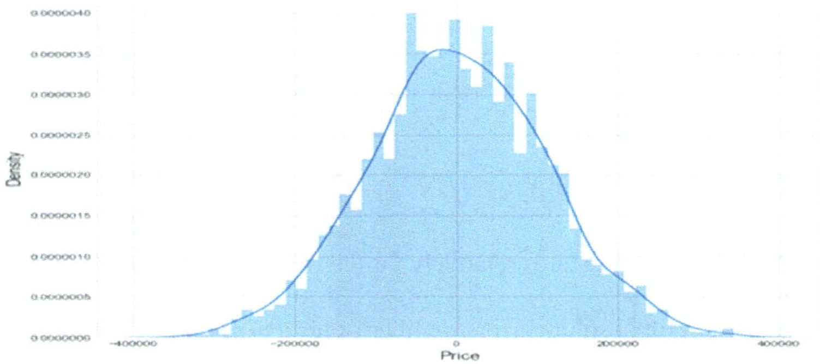

Fig. 12 XGBoost regression distplot

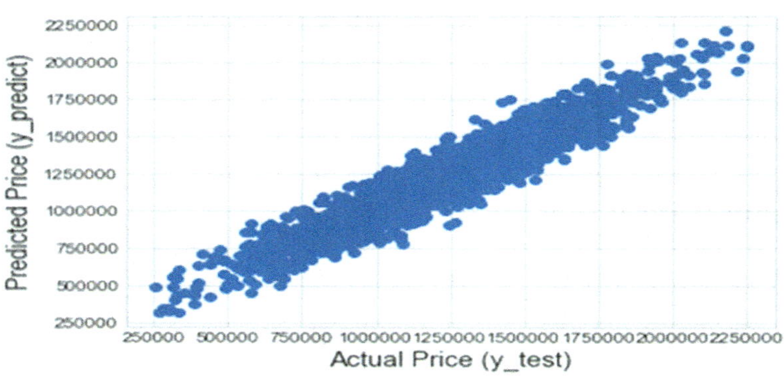

Fig. 13 XGB regression scatter plot

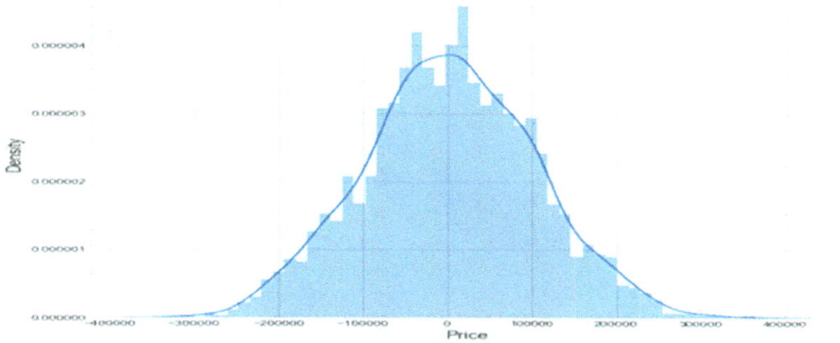

Fig. 14 ANNdistplot

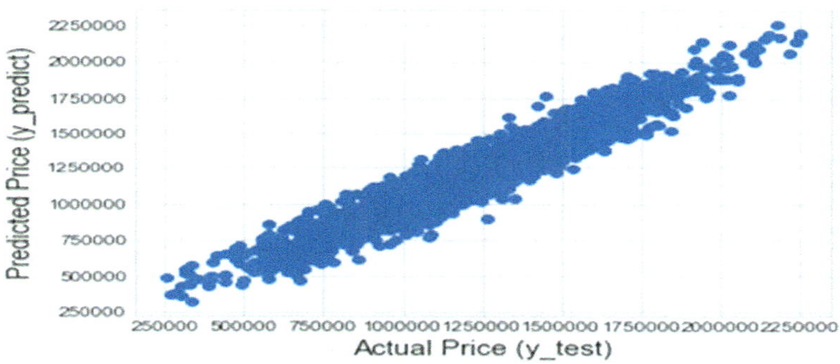

Fig. 15 ANN scatter plot

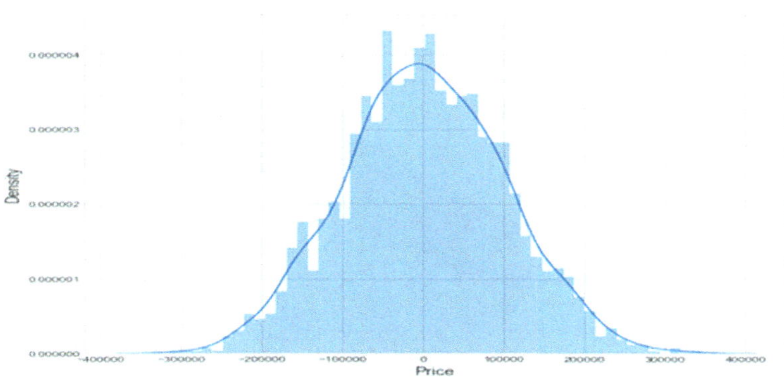

Fig. 16 Gradient boosting distplot

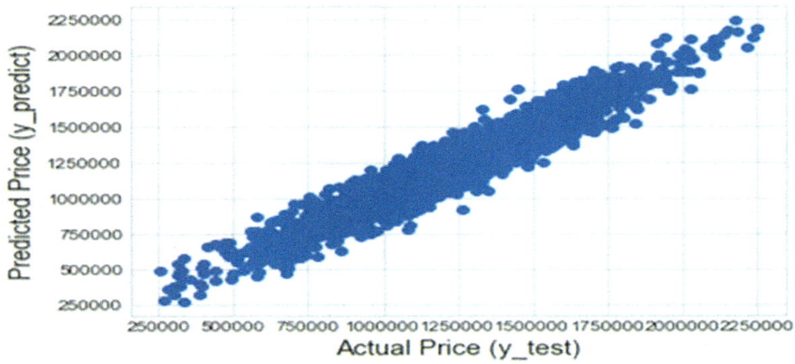

Fig. 17 Gradient boosting scatter plot

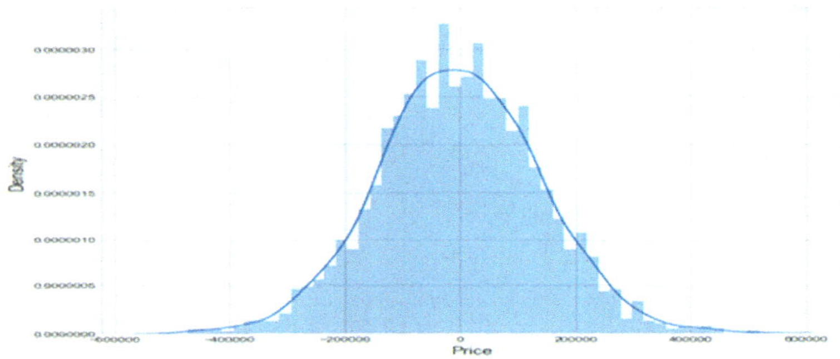

Fig. 18 Ada boosting distplot

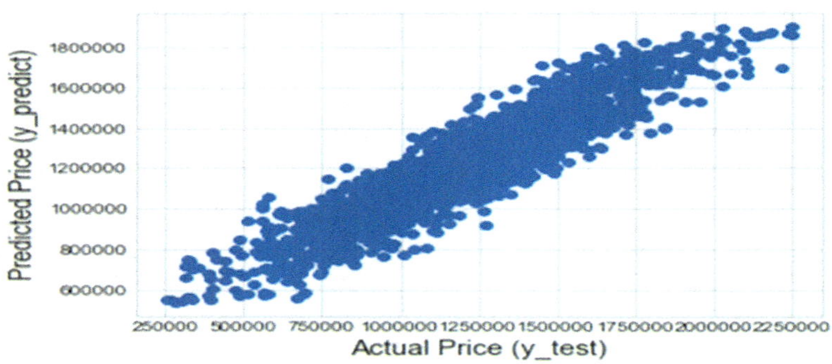

Fig. 19 Ada boosting scatter plot

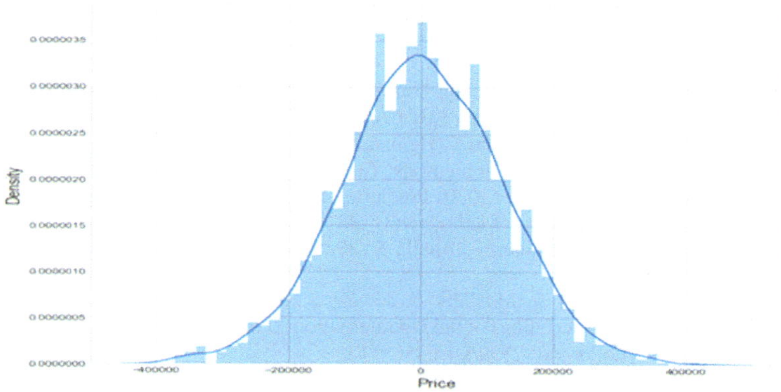

Fig. 20 Random forest regression distplot

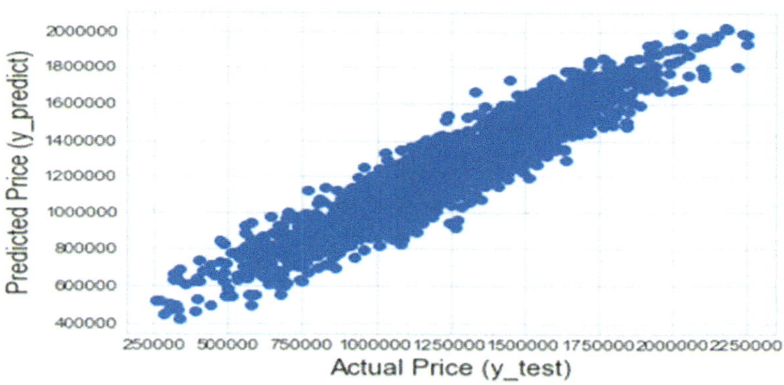

Fig. 21 Random forest regression scatter plot

8 Conclusion

For predicting the house cost price six models are used, namely Random Forest Regression, XGBoost Regression, Linear Regression, Artificial Neural Network, Gradient Boosting, and Ada Boosting. The results are compared, of which Linear Regression outperforms compare to the other five algorithms as per the investigation of the experimental outcomes. The results of the model help the customers to find the cost of the house which they are interested to purchase as per their budget.

References

1. Adyan Nur, Alfiyatin (2017) Modeling house price prediction using regression analysis and particle swarm optimization. Ind Int J Adv Comput Sci Appl 8:323–326
2. Monika R, Nithyasree J, Valarmathi V, Hemalakshmi MGR, Prakash NB (2021) House price forecasting using machine learning methods. Turk J Comput Math Educ 12:3624–3632
3. Erkek M, Cyril K, Hepşen ALİ (2020) Predicting house prices in Turkey by using machine learning algorithms. J Stat Econ Methods 9:31–38
4. Kamal N, Chaturvedi E, Gautam S, Bhalla S (2021) House price prediction using machine learning. Springer, LNNS
5. Kuvalekar A, Manchewar S, Mahadik S, Jawale S (2021) House price forecasting using machine learning. In: Proceedings of the 3rd international conference on advances in science & technology. https://doi.org/10.2139/ssrn.3565512
6. Varma A, Sarma A, Doshi S, Nair R (2018) House price prediction using machine learning and neural networks. https://doi.org/10.1109/ICICCT.2018.8473231
7. DanhPhan (2018) Housing price prediction using machine learning algorithms: the case of Melbourne City Australia. https://doi.org/10.1109/iCMLDE.2018.00017
8. Viktorovich PA, Aleksandrovich PV, Leopoldovich KI, Vasilevna PI (2018) Predicting sales prices of the houses using regression methods of machine learning. https://doi.org/10.1109/RPC.2018.8482191
9. Banerjee D, Dutta S (2017) Predicting the housing price direction using machine learning techniques. https://doi.org/10.1109/ICPCSI.2017.8392275
10. Ahtesham M, Bawany NZ, Fatima K (2020) House price prediction using machine learning algorithm—the case of Karachi City Pakistan. https://doi.org/10.1109/ACIT50332.2020.9300074
11. Ramya GC (2020) House price prediction. Dissertation, Kuvempu University
12. Truong Q, Nguyen M, Dang H, Mei B (2020) Housing price prediction via improved machine learning techniques. Sci Direct Procedia Comput Sci 174:433–442
13. Yan Z, Zang L (2020) Spatial prediction of housing prices in beijing using machine learning algorithms, pp 64–71. https://doi.org/10.1145/3409501.3409543
14. Ravi Kumar AS (2017) Real estate price prediction using machine learning. Dissertation, National College of Ireland
15. Ho WKO, Tang BS, Wang SW (2020) Predicting property prices with machine learning algorithms. J Property Res 38:1–23. https://doi.org/10.1080/09599916.2020.1832558
16. UdayDeo (2021) House price prediction using various regressions: a comparative study. Shri Mata Vaishnodevi University
17. Thamarai M, Malarvizhi SP (2020) House price prediction modeling using machine learning. Int J Inf Eng Electron Bus 12:15–20. https://doi.org/10.5815/ijieeb.2020.02.0
18. Park B, Bae JK (2015) Using machine learning algorithms for housing price prediction expert systems with applications. Int J 42:2928–2934.https://doi.org/10.1016/j.eswa.2014.11.040
19. Rossiter D (2015) Real estate prediction. Dissertation
20. Fan C, Cui Z, Zhong X (2018) House prices prediction with machine learning algorithms, pp 6–10. https://doi.org/10.1145/3195106.3195133
21. Vineeth N, Ayyappa M, Bharathi B (2018) House price prediction using machine learning algorithms. Soft computing systems. Springer, Singapore
22. Wang F, Zou Y, Zhang H, Shi H (2019) House price prediction approach based on deep learning and ARIMA model. https://doi.org/10.1109/ICCSNT47585.2019.8962443
23. Roman V (2019) Predicting boston house prices with regression. Project. www.towrdsdatascience.com
24. Guptha R, Narahari NS, Manasa J (2020) Machine learning based predicting house prices using regression techniques. In: Proceedings of the second international conference on innovative mechanisms for industry applications
25. Jain M, Rajput H, Garg N, Chawla P (2020) Prediction of house pricing using machine learning with Python. https://doi.org/10.1109/ICESC48915.2020.9155839

26. Mohd T, Jamil NS, Johari N, Abdullah L, Masrom S (2020) An overview of real estate modelling techniques for house price prediction. Charting a sustainable future of ASEAN in business and social sciences. Springer, Singapore
27. Begum S, Siddique FA, Tiwari R (2021) A study for predicting heart disease using machine learning. Turk J Comput Math Educ 12:4584–4592
28. Begum S, Siddique FA (2021) A study to predict home loan defaulter using machine learning. In: Samaroh Sampathy's memorial international conference technology innovation and quality management

Part II
Computational Intelligence for Marketing, Business Process and Human Resource Applications

Chapter 8
SDN-Based Network Resource Management

João Carlos Marques Silva, José André Moura, and Nuno Manuel Branco Souto

Abstract In recent years there has been a growing demand for network resources. However, fixed contracts between users and providers tend to result in network use inefficiencies and high costs. To promote the best accommodation for high network demand and usage, a setup where every user has the most amount of network resources at his disposal is paramount—this way users minimize the risk of not having sufficient resources to meet their service needs, and providers maximize the usage of their networks. In this chapter, we consider a setup based on Software Defined Networking (SDN), where connections between users' devices and providers' nodes are defined according to resource needs and pricing. The adoption of an SDN-based approach is detrimental of other more distributed control alternatives is since the scenario under investigation is very specific and dynamic, which is more efficiently managed in a logical centralized way than in a decentralized way. In this direction, an auction SDN-based broker is proposed, so that both users and providers get the best deal for every resource-allocation procedure, according to all players' needs and network restrictions. We present and discuss evaluation results taken from our auction business model. Our results suggest that the best bidding strategy depends on several aspects, namely: (i) the competitor's bidding strategy; (ii) the operating cost of each participant; or (iii) the available resources of all participants and the broker's requisites.

Keywords SDN · Broker · Distributed resources · Configurable computer networks · Auctions · 5G

J. C. M. Silva (✉) · J. A. Moura · N. M. B. Souto
Instituto Universitário de Lisboa (ISCTE-IUL), ISTAR, Lisboa, Portugal
e-mail: joao.silva@iscte-iul.pt

J. A. Moura
e-mail: jose.moura@iscte-iul.pt

N. M. B. Souto
e-mail: nuno.souto@iscte-iul.pt

S. Kautish et al. (eds.), *Computational Intelligence for Modern Business Systems*, Disruptive Technologies and Digital Transformations for Society 5.0,
https://doi.org/10.1007/978-981-99-5354-7_8

1 Introduction

The Internet has had massive growth in the XXI century, with people and businesses using online data and tools; in fact, the recent Covid-19 pandemic promoted these resources even more by imposing online working routines, which in some cases boosted network traffic by more than 50% [1].

To cope with this increase in demand, network and cloud service providers are installing more equipment to increase connection and routing capacity. However, this increase in capacity has not been able to keep up with the real-time stress that the increase in global demand has caused [2]. To solve this problem, edge computing comes into play, allowing network service providers to provide on-demand resources, hence avoiding over and under-provisioning (e.g., on-peak and off-peak times), and speeding-up up online services by allocating computational assets in proximity to the user [3–6]. This data resource proximity to the users decreases the data access latency and diminishes the amount of traffic on the backhaul links.

For this to work effectively, the user (person, organization, broker, middleware, etc.) will have to choose an edge provider to connect to. However, using a sole provider results in the previous shortcomings, and does not resolve anything. The solution is to use a broker to procure the resource needs (based on a variety of attributes such as price and quality) from any edge computing provider, on a case-by-case basis, so that the capacity from all providers is available to the user.

The advantages of having a broker are many since the user delegates the weighed choice between price and quality of different service providers that keep changing their offers continually. Furthermore, it is not practical to optimize for the most suitable vendor manually (or if a fixed provider is chosen then the user will not always get the best deal).

In the context of aforementioned challenges, this chapter aims at investigating scalable and automated mechanisms for users to acquire the resource services they need from edge computing providers. To achieve this, SDN (Software Defined Networking) comes into play, since it is the best solution to implement a broker that manages heterogeneous network resources from different providers, creating a federated market of network resources. This federated market can run efficient negotiations among all the players, enabling win–win situations between providers and users. In this way, the providers can increase their profit and the users can afford a high quality of experience when they run their applications.

In this chapter, we propose a multi-attribute bidding strategy for users to interconnect with an edge computing resource, via an SDN-based broker, for their application needs. Since this broker is in contact with a multitude of edge providers, several bids will be proposed to cover the network requests, allowing the broker to choose the best option on behalf of its users. This way, users can diminish their usage costs, receive quicker responses, and sustain the quality of their services above specific expected thresholds.

We have opted to use an auction business model because in our opinion the auctioneer role seems very suitable to be assumed by an SDN-based system. In addition, a system operating in a more centralized way seems better than others, which operate in a more distributed way, because the former normally converges faster than the latter towards the optimum solution. The deals can then be safely registered in a blockchain-based system.

This chapter aims to:

(a) Propose an auction model, intermediated by an SDN-based broker, for clients to use the edge computing resources made available by the network providers in the most cost-efficient way. This model permits users to procure resources and incorporate price and QoS attribute ratings to determine the best allocation of resources that address their on-demand networking traffic. The added value of this model lies in the way that the best offerings are assessed, ranking potential vendors to achieve maximum utility.

    a. In order to promote truthful bidding, two main strategies are suggested: a one-step Vickery auction [7] and, as an alternative, a two-step bidding model, providing the providers an incentive to bid using their own valuation of price and quality attribute.
    b. If the two-step bidding model is chosen, then the providers will have two stages to propose their offers. Initially, in the first stage, the automatism uses the bids to rank providers based on the broker's reservation quality and determines the cost value of the various attributes. Nest, in the second stage, the broker shares the maximum score obtained from the first stage and permits the providers an opportunity to increase their bids. The significance of this two-stage model is that it deters the competing service providers from deliberately submitting unfavourable/market-deficient bids (a.k.a. expensive bids), but instead encourages them to submit bids that meet the truthful market expectation.

(b) Describe the SDN-based solution for the broker, highlighting the main aspects and layers associated with such control, namely the data transfer, control, and management layers. Since all providers' resources will be interconnected via the SDN mechanism, the use of the management layer to act as a broker whose decisions will posteriorly control which connections to activate is simply the application of the normal operation of an SDN-based system with an auction process running on top of that system.
(c) Use blockchain technology to securely register and validate all deals
(d) We present and discuss some results extracted from our auction business model. Our results suggest that the best bidding strategy depends on several aspects, namely: (i) the competitor's bidding strategy; (ii) the operating cost of each participant; or (iii) the available resources of all participants and the broker's requirements.

2 Literature Review

The need for network resources has been growing exponentially, and the current networking infrastructure is lacking in resources. As stated before, part of the solution comes from making available edge computing resources to their closest users, alongside a flexible and resource-efficient pricing model. We will review studies related to edge computing resource allocation using bidding models and design mechanisms supported by SDN.

2.1 Edge Computing Resources Allocation via Auction Models

Today's edge computing resource providers define their costs and deal directly with the users, with fixed prices and service levels, although the promised QoS attributes are only informative and not contractually obligated—this motivated many studies to investigate the design of models that could induce providers to fulfil on their promises.

An initial auction model for users to bid on some quality features was proposed in [8], considering a combinatorial auction for limited resources, but had the shortcoming of being unable to rank such bids. Work done in [9–11], and [12] compares different types of pricing structures concerning fairness for edge computing resources, considering different settings such as dynamic environments, single or double auction models, credit borrowing, network economics, etc. The scale of the application was also considered in [12], where the authors proposed micro and macro auctions for different types of traffic. Most of the works do not consider the different quality attributes of each edge computing provider [13] but focus instead on pricing and resource allocation, to simplify the overall problem.

A reverse auction model was considered in [14, 15]. The work in [15] discussed a centralized strategy (similar to a broker) to allow third parties to lease spare/unused bandwidth and storage [16, 17], allowing users to decide which access point they would prefer to use based on a set of criteria. This work was then evolved in [18], considering an evolutionary mechanism to distribute resources based on market efficiency, user satisfaction, and QoS. A penalty scheme for bidders that cheated was also considered in [19]. The work in [20] considers a two-stage auction and takes into account the multi-attribute characteristics of the required parameters for a specific service, alongside its pricing functions, and thus is one of the main bases for the work developed in the current chapter.

2.2 Use of Software Defined Networking

We have recently assisted in an increase in the number of research contributions that aim for automated selection and management of edge computing resources [20, 21] to assist novel network demands. Within these recent works in the literature, some propose auction mechanisms [15, 20–22] to manage the network resources, mostly in mobile access cases [23, 24]. To deal with the high complexity of managing network resources in a more efficient way, machine learning techniques have also been used [25]. In this case, the authors discuss network network-slicing traffic framework for a Reinforcement Learning-based 5G Network Slice Broker (RL-NSB) building on: (i) traffic forecasting; (ii) slice admission control; and (iii) slice traffic scheduling.

Despite the existence of a few auction-based solutions in the literature, as far as we are aware, the usage of the SDN paradigm to control auction-based solutions, where the SDN controller acts as a broker that supervises the auction process, has not been comprehensively studied yet. The need for our current contribution came from previous work, such as [23], where the authors offer a comprehensive literature revision on recent solutions related to 5G network slicing using SDN and NFV. In their work, they acknowledge that further extensive investigation is required for new business models for network slices based on novel pricing and auction mechanisms that consider joint resource and revenue optimization in 5G networks. In addition, they claim for further research to investigate fairness problems during resource allocation in 5G network slices that are requested by distinct mobile virtual network operators (MVNOs) [14].

2.3 Blockchain Technology

Blockchain is a concept for database of digital contracts. The concept was introduced alongside the cryptocurrency Bitcoin [26], to promote secure transactions between (anonymous if desired) participants using a decentralized system, where all operations were validated by different entities (known as miners, which are participants of the blockchain system that earn certain types of resources as a reward; in the bitcoin system they naturally earn bitcoins). Blockchain got its name basically because it's a distributed chain of blocks, where each block contains information about a certain number of made deals. The miners compile blocks from the aggregation of several digital transactions, which are added sequentially to the whole ever-growing blockchain after validation from peers. Once a block is added to the chain, it is not possible to change its content or remove it (immutability), since it becomes part of the chain's global hash value.

Blockchain can operate in one of two different modes: permission-less or permissioned [27]. The permission-less mode, also known as a public Blockchain network, allows any user to create and add transactions to the chain; whereas the permissioned blockchain is a private centralized Blockchain network that requires authentication and authorization to use.

2.4 Basis for the SDN-Based Broker Implementation

As mentioned previously, our work builds on [20], where multi-attribute bidding was considered, intermediated by a broker that manages the payment allocation to the provider. To be truthful and to encourage providers to provide fair bidding, [20] suggested a 2-stage approach in which only the providers that met the broker's reserved quality would be considered for final bidding. The second stage was inspired by [8], using the bid density as a factor.

Our work considers both the 1- and 2-stage auction processes and assumes that user satisfaction will always be guaranteed if the broker provides a solution for every request. Table 1 summarizes the added contributions of our work, comparing it to other references.

3 Auction Model

In the auction model, a broker working on behalf of several clients/users wants to procure resources from edge computing service providers. We considered a broker in this model because it is not favourable for a user to negotiate directly with the service provider, since it lacks negotiation weight. Thus being, if several users contact the service provider with the intermediation of a broker, they should be able to negotiate for a global better offer that benefits everyone. This way, a common market negotiation approach was employed, which benefits all users and promotes, fair-market-value negotiations.

The model assumes that the broker is contacted by several clients/users (via applications), all with different requests. These requests are defined by the characteristics

Table 1 Differences between research chapter contributions

Research paper	Multi-attribute	Auction model	Utility function	User aggregation	Bidding strategies	SDN approach
Li et al. [10], Hossain and Muhammad [13]	No	Yes	Yes	No	No	No
Habiba and Hossain [15]	Yes	Yes	Yes	No	No	No
Our work	Yes	Yes	Yes	Yes	Yes	Yes

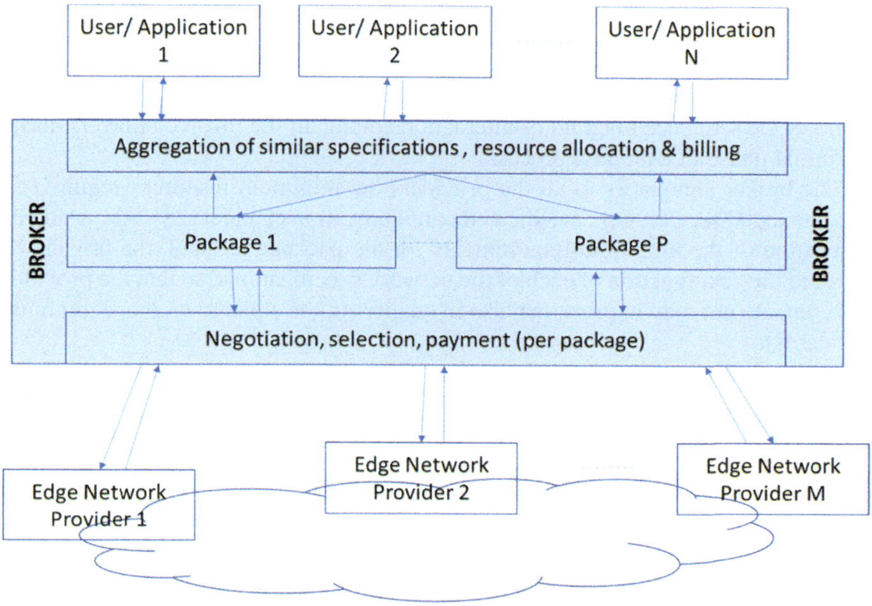

Fig. 1 Interaction between users, broker and providers

of the network traffic they desire, namely their minimum requirements. These characteristics are sent to the broker, who converts the client's minimum requirements into the appropriate network performance variables which are then used in the negotiation protocol between the broker and edge providers. Figure 1 illustrates the high-level process between *N* users, the broker, and the *M* edge network providers.

After a pre-determined window of time, all registered requests are gathered and aggregated into packages. These packages incorporate one or more requests from the users and are used to concentrate all similar requests so that they can be negotiated by the broker in bulk mode. The broker will then start negotiating with the edge network providers for the resources needed by each package. So being, *P* ($P =$ number of packages) requests are formulated by the broker and passed to all edge network providers, starting the negotiation process.

Once the winning bid is selected, the broker will bill its users/applications accordingly and pay directly to the edge network provider. The deal will then be registered and validated by all peers (edge network providers) using blockchain technology.

It will also be responsible for serving as an intermediate for the distribution of the negotiated resources, from the edge providers to the users (resource allocation).

3.1 Negotiation Process

The negotiation process between the broker and the providers is depicted in Fig. 2. It shows the sequence diagram of interaction among all the involved players during the initial phase of a bidding process.

The broker announces to all the providers the minimum resources required for the demanded service, e.g., a dedicated communication channel [28]. Alongside the translation of the minimum requirements for the package at hand, the broker will also add the cost function of each of the network specifications, so that the providers can compete not only in price, but also in quality (if it is equal to or above minimum standards).

The broker's cost function is basically made of two vectors, namely

$$v = [v_1 \dots v_m], \lambda = [\lambda_1 \dots \lambda_m] \tag{1}$$

where v_k is the broker's valuation (per unit) of the excess quality attribute k over the broker's posted minimum requirements and λ_k is the description of the quality attribute k (alongside the considered unit for valuation). The providers get these cost vectors and know exactly how the broker values each amount of excess quality attribute.

The providers then reply by offering their bids, including their price and (a pre-determined number of) their quality attributes related to the required service, to

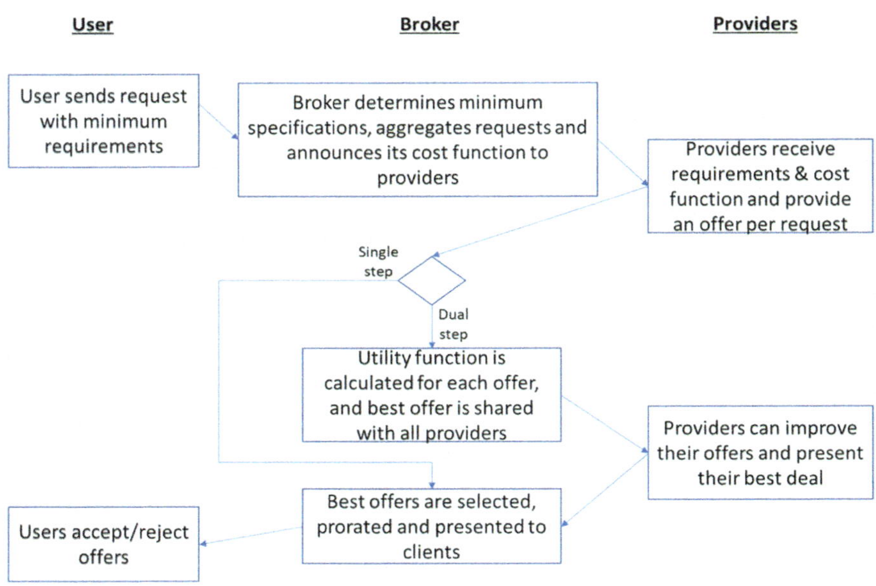

Fig. 2 Negotiation process between broker and providers

address all aggregate requests. It is expected that all bids address all minimum requirements.

The broker's contract for each package will be awarded to one winner only. In our model, the broker initiates the auctioning process and, after receiving the bids from the different providers, needs to decide on the best offer. The participating providers are determined to fulfil the request using their resources (or third-party resources that they have access to) and specify that in their bids. For the broker to be able to compare all offers, a normative will have to be followed—this normative will ensure that the quality attributes are enumerated with an associated dimension (kbit/s; MHz, ms, etc.); each provider will have to provide a fixed coefficient associated to the quality attributed of their offerings. Each quality attribute is independent, and the broker is free to favour certain attributed over other, depending on the user's requirements. The broker selects the provider that maximizes its utility function, based on the quality parameters and a reservation price. The utility function can be defined as

$$U_{broker}(p_i) = \left[\sum_{k=1}^{m}\left(v_k \cdot \lambda_{i,k}^{exc}\right) + \beta_{\mathbf{min}}\right] - \beta_{\mathbf{i}} \tag{2}$$

where v_k is the broker's valuation of the excess quality attribute k over the broker's posted minimum requirements, $\lambda_{i,k}^{exc}$ is the excess quality attribute k of provider i over the minimum requirements, β_{min} is the price value that the broker awards to the network performance variables that fulfil the client's minimum requirements and β_i is the price proposed by provider i with incremental network performance variables $\lambda_{i,k}$.

The provider i also has its own utility function, given

$$U_{p_i} = \beta_{\mathbf{i}} - \sum_{k=1}^{m}\left(\gamma_{i,k} \cdot \lambda_{i,k}^{total}\right) \tag{3}$$

where $\gamma_{i,k}$ is the provider's valuation of the quality attribute k and $\lambda_{i,k}^{total}$ is the proposed quality attribute k of provider i. The provider must make such a bid such that its utility function is ideally ≥ 0 (to avoid the winner's curse [29, 30], which in this case would be to undercharge for a service, yielding a loss). The bid will ideally make both the provider's and the broker's utility function positive.

The bidding can be done via a single or a dual step. The number of bidding steps is previously defined by the broker. In the case of a dual-step bidding mode, the broker announces its best offer after the first iteration, by presenting two values for each provider; the utility value of the best offer and the utility value of the provider's offer, which is different for each provider. Using all this received information from the broker, each provider can estimate what is the broker's valuation for $\beta_{\mathbf{min}}$, and the provider can decide to either maintain or change its previous offer. Naturally, the providers' strategy for the auction will differ depending on the type of auction

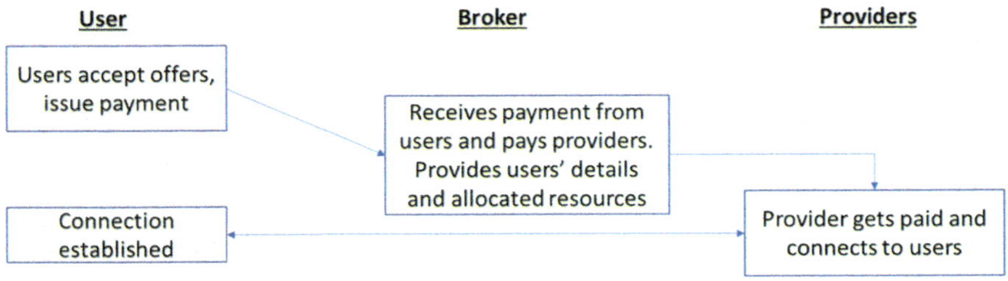

Fig. 3 Broker's final intermediation

that the broker is providing, and on the provider's best interests at hand; this will be discussed at a later section of this chapter.

As shown in Fig. 3, once the broker selects the best offer and transmits it (prorated) to the different users, the last functional part of our proposal is initiated, with the acceptance (or not) of the conditions by the user. Once the user accepts the conditions suggested by the broker, the user acknowledges the conditions (alongside payment[1]) to the broker. The broker will then contact the provider, issue payment, and deliver the user's details and its prorated quality attributes. Henceforth, the provider and user connect, and the requested network service is provided to the user through the network using the available connectivity resources from the diverse network providers.

3.2 SDN-Based Broker

This sub-section discusses how an SDN-based broker can manage the available resources of a heterogeneous access network. The deployment of this proposal is made using three layers: data transfer, control, and management, as shown in Fig. 4. The data layer is mainly responsible for data transfer among the network nodes. The control layer is mainly concerned with deciding about the used paths through the data layer to divert in an efficient way the diverse data flows among the selected network nodes. The management layer is responsible for initially accepting or not the network admission of a new client (see Fig. 4, steps 1 and 4) and use the available resources of the entire network infrastructure. Basically, it consists of a server (with secondary servers for backup) that is connected to the SDN controllers in the control pane. The main processing and management of the whole admission and network attribution procedure will be done by the server's broker application.

In the case the new client is accepted then, the algorithm running in the NFV of the management layer runs an auction with the diverse SDN controllers of the control layer (see Fig. 4, step 2). The expected outcome of the auction will be a set of selected switches from diverse network players that will be coordinated among them

---

[1] The payment will most probably be part of a monthly subscription, which will include both the broker's and provider's fees.

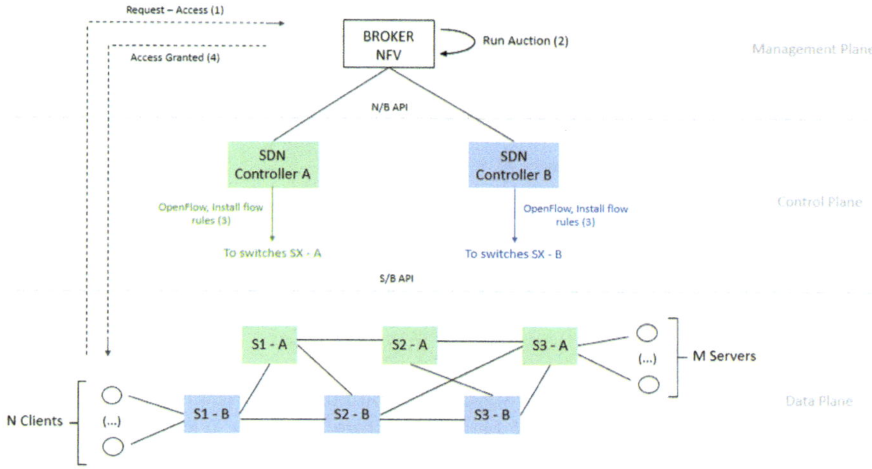

Fig. 4 Proposal of an SDN-based broker

to establish an end-to-end network path (see Fig. 4, step 3). Using this network path, the data flow of the new client can be transferred through the heterogenous network access.

Figure 5 is presented in a more detailed way the interaction among the diverse entities of our proposed system. To explain this interaction, we assume the scenario of a new client arriving at the system. This client initially sends the first message of a data flow. This message contains a list of network requisites the client data flow is requesting. The message is received by the switch of the network infrastructure which is the nearest one from that client. This switch has a default OpenFlow flow rule for any "unknown" data flow. The action of this default rule is to send a copy of the received message to the SDN controller responsible for controlling that switch. In this way, the received message of the new data flow is encapsulated by the last switch in a PacketIn message of the OpenFlow protocol and, the PacketIn message is sent to the controller. Then, the controller, in reaction to the received PacketIn, sends a notification message via the NorthBound API to the Broker (i.e. NFV). As a next step, the Broker initiates a round of negotiation with the group of SDN Controllers of the diverse network providers of the geographical location where the scenario is in operation. This negotiation considers the list of network requisites and an auction mechanism, as was previously explained in the auction model.

The outcome of the auction is a selected path through the heterogenous network for the data flow of the new client. The selected path is a set of nodes (switches) and edges (links) that should be used for implementing end-to-end communication between the new client device and the server offering the service, which the client aims to use.

The next operating step of our proposal involves each SDN controller sending in a proactive way FlowMod messages to the selected switches responsible to support

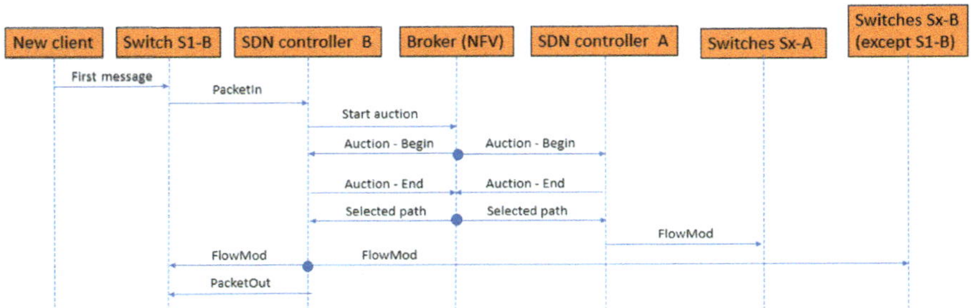

Fig. 5 Message exchange in the SDN-based broker system

the path previously negotiated during the auction phase. These FlowMod messages transport data flow rules that will be locally stored in each selected switch. In a more particular way, the switch that has initially received the first message of the new client receives not only FlowMod messages but also a PacketOut message. The second message type is necessary so that the switch sends out the first message correctly, which was initially buffered in that switch before the same switch sent the initial PacketIn to the SDN controller.

To assure that the broker is active most of the time, there should be one or more backup servers running the broker application. This means that the broker protocol should include mechanisms to transfer the data between the master and slave servers, much like the DNS mechanism of having secondary servers [31].

3.3 Blockchain for Registering Smart Contracts

Since in enterprise ecosystems such as telecoms, all parties involved are known to each other (participants and miners/managers are the sub-slice operators and verticals), there is no need for anonymity, and as such, permissioned blockchains are considered a better fit. This is because permissioned blockchains are designed for enterprise ecosystems and use simpler and less resource-consuming consensus protocols (such as Raft [32]) than public blockchains.

Each network resource deal attribution is a smart contract that should have a unique identifier and data fields that represent what resources were transitioned. The broker should issue a smart contract for block placement; the miners will then compose blocks out of contracts that are still "free" (not yet placed in a block) and try to issue a block (composed of a number of "free" contracts) to place it in the blockchain so that it is validated by all peers. This validation is necessary to guarantee transparency to all parties involved, especially since they are competing among each other.

3.4 Bidding Strategies

As discussed earlier, having a bidding strategy employed is the best way to guarantee the best results for both the clients and service providers. An overloaded service provider can raise its prices for its scarce resources, and a provider with many spare resources can lower the price to put its resources to use—the market thus becomes very dynamic and efficient in this way as such bidding is paramount. Of course, this does not mean that the providers can't reserve some of their resources for long-term, pre-defined contracts with some clients; best business practices state that the best way to manage the resources is to have some permanent deals and some dynamic deals, to maximize resource usage and revenues.

The bidding strategy of each provider varies, depending if the broker implements a single ou dual step voting mechanism. In the single step voting, the providers usually do not have feedback on their competitor's voting, and thus the information available for bidders participating in these auctions is relatively limited, rendering useless any analysis of bidding strategies [7]. Surely, providers can keep testing the broker and lower their price/increase their excess quality attributes in every auction until they start being awarded as the winners, to assess the current utility level being used. Nevertheless, the providers engaging in the previous strategy can reach a non-interesting trade-off price vs. offered quality. Alternatively, the providers could try to ask if the broker can make its records accessible, but in most cases that request probably will not be accepted by the broker.

Since every player expects to get the best deal, which might not be a fair deal for the client/ broker, the Vickery auction [33] is suggested for such cases, where the bidder that provides the best offer (yielding a higher value for the broker's utility function) will be making use of the utility value from the second-best bidder. This promotes truthful bidding [33], but some adjustments must be made—the broker's second-best utility function will be lower in value than its best value, and thus the simple suggestion is to increase the winner's price so that its utility function matches the second-best utility function—the winner will thus receive more than what he suggested (as with all Vickery auctions)! Note that there may be a case where the second-best proposal has a lower price than the winner, but in that case, it means that the second-best proposal also had less excess quality attributes than the winning bid. The winner could also match the second-best offer by reducing its offer of excess quality attributes while maintaining the price, or through a combination of increasing the price and reducing some excess quality attributes. Still, but to keep it manageable, we assume that all adjustments are done at the price level. In our opinion, this assumption still holds for real cases.

The dual step voting is different from the single step since it involves two bidding sessions. This allows for the providers to improve (or maintain) their offers, once the broker announces the best utility function after the first round, alongside the utility function of each provider individually (each provider will only get these two values so that they know exactly how much "value" they must increase their bid, if any). With this information, each provider can infer all variables, including the broker's

β_{min} (which can vary between auctions, according to the broker's algorithm for the determination of the minimum price), and increase their bid if they think it is worth it. This dual step voting allows for the bidders to bid less aggressively in the first round and adjust their strategy according to their goals.

Realistically, the providers' strategies may differ from each other, especially in the case of a provider that has many resources to offer and no clients, which may even consider bidding at a negative utility value, to gain market share or simply opt for the best choice between having no clients or having a few low-paying ones. However, these situations would not be an issue in a consolidated market, and as such that is what we will consider henceforth.

We will assume a scenario where bidders act independently, without any cooperation or sharing of information. So being, each provider prepares its bid according to its best interest and its perception of market values, expecting their competition to do the same, acting rationally. In this context, the question in each provider's mind is how it should behave (bid aggressively or not), if nothing is known from its competitor's behaviour?

When rational bidders (with incomplete information) are considered, we can analyse if a dominant bidding strategy is possible by using economic game theory. The optimum bid for a two-person auction game has been derived by means of restrictive probability distribution function forms for bids [34, 35]. Basically, the idea is to calculate the estimated payoff table for each bidder and evaluate where lies the equilibrium point between two bidders—this is known as the Nash equilibrium in game-theory jargon. At the point of (Nash) equilibrium, no player can unilaterally increase its gain by moving to a different position/strategy [36].

Referring to the case at hand, and assuming a consolidated market, we know that each provider wants to maximize their utility function. A zero value means that the provider is getting just enough to cover the costs of its resources, and a negative value means that the provider is charging less than its costs, losing money. The only way to make profit is the provider to have a positive value on the utility function, and the hard part is knowing what the highest profit margin one could have. If each provider has its own network (even if some providers may share different portions of the same networks) with different specifications (or quality attributes), we know that each provider's cost to meet the minimum quality attributes is different. So being, the providers with lowest network costs will be able to aspire for greater profit margins than those with higher network costs, and thus the optimum profit margin for each operator will be different. Yet they will all meet at a similar utility value for the broker.

Finding this similar (broker's) utility value is equivalent to letting the market work and see where it settles. Ideally, it will be in a Nash equilibrium, where the change of strategy from either provider will yield to lower profits. As an example, this equilibrium may state that provider 1 reaches its equilibrium state with a profit margin of 20% and provider 2 reaches the equilibrium with a profit margin of 35%, in the case that provider 2's cost for the minimum quality attributes is lower than provider 1's. The broker's utility function will yield the same numerical result for all at equilibrium.

4 Simulations

The benchmark model, as defined in [37] will be used. This model offers a generalization of auction formats, and is based on four assumptions:

1. All providers/ bidders are risk neutral. A provider being risk-neutral means that risk (e.g. running out of resources for a special future need) is not considered when bidding (realistic situations may take some risk into account).
2. Each bidder has a private valuation for the item independently drawn from some probability distribution. This is equivalent to saying that, since the cost structure is different among providers, the bids of a provider will vary from those from other providers, turning the situation equivalent to independent private valuations.
3. The bidders possess symmetric information, meaning that all have access to the same data made available by the broker.
4. The payment is represented as a function depending on only the bids. Even though this is not quite true, since the broker assumes a quantitative measure for both the minimum requirements and the excess quality attributes, if we replace all these with a specific value, then the previous condition holds.

Considering this as a mathematical game represented by a set of players, where each player has a set of actions (strategies) available to him, and a payoff vector corresponding to each combination of strategies, we can apply a theoretical game model. In this case, given the realistic scenarios and multitude of providers, it is quite difficult to effectively simulate a real-life situation. However, if we restrict our analysis to a couple of providers and variables, then some simple but interesting conclusions can be obtained.

We assumed a case of two providers, PA and PB, each with a limited number of resources, RA and RB, and a single round bidding session. We also assumed that the broker would run 15 bids for 15 resources, one at a time (in this basic example there is only one type of resource). We also considered that PA charges a slightly lower cost to cover the minimum quality requirements than what PB charges. Next, we will deem that PA and PB will suggest bids with a profit margin of 10% and 50% from their base value (assumed to be 100).

Finally, we will analyse three different strategies and will comment on the results of each one.

The analysed strategies were:

1. Static—each provider maintains the value of its bids, independently of winning or not.
2. Drop and stay (Dynamic)—each provider decreases its profit margin to 10% if it loses one of the rounds.
3. Drop and raise (Dynamic)—each provider decreases its profit margin to 10% when it loses a round and increases to 50% if it wins a round.

In the "static" strategy, portrayed in Fig. 6, we have 15 runs from the broker. Looking at the first case where RA = 10 and RB = 20, when PA & PB is 10%, PA

	15 Runs - static							
	PA - 10%	PB - 10%	PA - 50%	PB - 10%	PA - 10%	PB - 50%	PA - 50%	PB - 50%
RA=10, RB=20	100	50	0	150	100	250	500	250
RA=20, RB=10	150	0	250	100	150	0	750	0
RA=10, RB=10	100	50	50	100	100	250	500	250
RA=20, RB=20	150	0	0	150	150	0	750	0

Fig. 6 Financial results for different profit margins and resources—static strategy

will win every turn, since PA's final bid will always be lower than PB's (they are both competing with the same profit margin, and PA's base cost is smaller). However, since RA is 10, then PA will only compete for the first 10 rounds, after which it is left depleted of resources, allowing PB to win the remaining 5 rounds. A similar case holder for the case where PA = 10%, PB = 50%, with PB winning the remaining 5 rounds and making a larger profit than PA, since PB sold his 5 resources with a 50% profit margin each! It is interesting to assess that, as expected, PA will maximize its earnings if PB decides to have a high profit margin, although this strategy yields 0 for PB in the cases of RA = 20 (this happens for all cases, except PA = 50%, PB = 10%). It all comes down to PB knowing how many available resources PA has. If PA has limited resources, then PB should charge a 50% margin, otherwise PB should charge a 10% margin.

The "drop and stay" strategy is depicted in Fig. 7. In this setting, the provider lowers its profit margin once it loses in one of the rounds. Once the provider drops its margin, it never raises it. Let us look at the case RA = RB = 10 and PA = PB = 50%, for example. Here, PA will win the first round, and thus PB will drop its margin to 10%. Since PB dropped, PB will win round 2 and cause PA to drop its margin to 10%. The next 9 rounds are won by PA at 10% margin, until PA depletes all its available resources. Once PA is out of the bidding, PB wins the remaining 4 rounds also with a 10% margin. Looking at the overall results, we see that the best situation for both providers is when PA = PB = 50%. In this situation both providers cash in and it is clearly a Nash equilibrium.

	15 Runs - drop and stay when lose							
	PA - 10%	PB - 10%	PA - 50%	PB - 10%	PA - 10%	PB - 50%	PA - 50%	PB - 50%
RA=10, RB=20	100	50	100	50	100	50	140	50
RA=20, RB=10	150	0	140	10	150	0	180	10
RA=10, RB=10	100	50	100	50	100	50	140	50
RA=20, RB=20	150	0	140	10	150	0	180	10

Fig. 7 Financial results for different profit margins and resources—"drop and stay" strategy

	15 Runs - drop when lose, raise when win							
	PA - 10%	PB - 10%	PA - 50%	PB - 10%	PA - 10%	PB - 50%	PA - 50%	PB - 50%
RA=10, RB=20	80	70	70	80	80	70	120	70
RA=20, RB=10	80	70	70	80	80	70	120	70
RA=10, RB=10	80	70	70	80	80	70	120	70
RA=20, RB=20	80	70	70	80	80	70	120	70

Fig. 8 Financial results for different profit margins and resources—"drop and raise" strategy

Finally, the "drop and raise" strategy is depicted in Fig. 8. Here, each provider lowers its profit margin to 10% when it lose a round and raises it to 50% when they win a round. For the two-player case, this means that from round 2 onwards, both players will bid with different profit margins: one with 10% and the other with 50%. This also means that each provider will alternate winning the round with the other, always at 10% from round 2 onwards. Of course, if the competing provider ran out of resources prematurely, then the surviving provider would be able to charge a 50% profit margin for all the remaining rounds.

In Fig. 9 we have a summary of the profit margins yielding the highest returns for each strategy. We also have a column NE that states if the pair of the best profit margins is at Nash Equilibrium by itself or not. As expected, we see that the best profit margin for PA is always 50%, whereas it differs for PB, since PB cannot compete directly with PA using the same margins, due to its higher operating costs.

To summarize, we can infer from the given example that the best bidding strategy depends on many factors, namely:

(a) The competitor's bidding strategy.
(b) The operating cost of each participant.
(c) The available resources of all participants and the broker's needs.

Note that, for simplicity, we assumed only one type of resource, but in a more realistic scenario we will typically have many of them. The broker's valuation of the "excess" resources will also come into play, allowing for providers with "valuable" excess quality attributes to be able to bid a higher offer than its competitors, since their offered package is deemed more valuable. The two-step bidding process is also

	Static			Dynamic- drop and stay			Dynamic- drop and raise		
	PA	PB	NE?	PA	PB	NE?	PA	PB	NE?
RA=10, RB=20	50%	50%	yes	50%	50%	yes	50%	10%	no
RA=20, RB=10	50%	10%	yes	50%	50%	yes	50%	10%	no
RA=10, RB=10	50%	50%	yes	50%	50%	yes	50%	10%	no
RA=20, RB=20	50%	10%	no	50%	50%	yes	50%	10%	no

Fig. 9 Highest returns for the 3 strategies

another factor that will alter the strategy of all players: when the best current offer is made known, all competitors will have the chance to improve on their offer, and thus minimize differences between all bids.

5 Conclusions

This chapter addressed the use of an SDN-based broker to auction network resources to users, to allow both users and providers to make the most of the available network resources. Through an auction model, the network resources are attributed to competition between providers and users, allowing for the market of supply and demand to function accordingly. This is possible through a programmable SDN, in which all providers are interconnected, and that serves to procure the best deals for its users. This SDN will manage the negotiation process and handle all payments between users and providers.

6 Future Work

Future work will have us dig deeper into the different quality parameters of the network (most aren't linearly additive nor continuous) and for the possibility of collusion between providers. Elaborate broker strategies to promote fairness and truthfulness must also be addressed, alongside Artificial Intelligence mechanisms (such as neural networks) that will allow the broker and providers to learn from past experiences. To promote security and anonymity, blockchain technology [26] can also be used during the bidding procedure.

The interaction with 5G and post-5G networks will also be studied in the future, since 5G uses the concept of Network Slicing, which basically consists in using several network substrates, known as Virtual Networks, to share the same physical infrastructure. Using 5G nomenclature, the verticals/ slice owners (those who will pay for the services) will bid for slices from slice providers/ network brokers [38, 39], in a context where the broker will have to negotiate several sub-slices (either from the same provider or multiple providers) in order to compose (or "stitch") an end-to-end (full) slice carrying a 5G network service [40]. Hybrid strategies [41] can be also used for smoother deployment of new solutions, without disrupting the service supply and controlling the global cost of system changes.

Acknowledgements The authors acknowledge the support given by Instituto de Telecomunicações, Lisbon, Portugal.

Funding This publication/research was partially supported by Fundação para a Ciência e Tecnologia throught project grants FCT UIDB/04466/2020, UIDP/04466/2020 and UIDB/50008/2020.

References

1. Heaven WD (2020) Why the coronavirus lockdown is making the internet stronger than ever. MIT Review
2. Hossain MS, Muhammad G, Guizani N (2020) Explainable AI and mass surveillance system-based healthcare framework to combat COVID-I9 like pandemics. IEEE Network 34(4):126–132
3. Yassine A, Shirehjini AAN, Shirmohammadi S (2016) Bandwidth on-demand for multimedia big data transfer across geodistributed cloud data centers. IEEE Trans Cloud Comput. https://doi.org/10.1109/TCC.2016.2617369
4. Yassine A, Singh S, Hossain MS, Muhammad G (2019) IoT big data analytics for smart homes with fog and cloud computing. Future Gener Comput Syst 0167–739X, 91, pp 563–573
5. Chen M et al (2018) Edge-CoCaCo: toward joint optimization of computation, caching, and communication on edge cloud. IEEE Wirel Commun 25(3):21–27
6. Hao Y et al (2019) Smart-edge-CoCaCo: AI-enabled smart edge with joint computation, caching, and communication in heterogeneous IoT. IEEE Network 33(2):58–64
7. Vickery W (1961) Counterspeculation, auctions, and competitive sealed tenders. J Finance 16:8–37
8. Bahreini T, Badri H, Grosu D (2018) An envy-free auction mechanism for resource allocation in edge computing systems. In: 2018 IEEE/ACM symposium on edge computing (SEC). Seattle, WA, pp 313–322.https://doi.org/10.1109/SEC.2018.00030
9. Baek B, Lee J, Peng Y, Park S (2020) Three dynamic pricing schemes for resource allocation of edge computing for IoT environment. IEEE Internet Things J 7(5):4292–4303. https://doi.org/10.1109/JIOT.2020.2966627
10. Li Z, Yang Z, Xie S, Chen W, Liu K (2019) Credit-based payments for fast computing resource trading in edge-assisted Internet of Things. IEEE Internet Things J 6(4):6606–6617
11. Sun W, Liu J, Yue Y, Zhang H (2018) Double auction-based resource allocation for mobile edge computing in industrial Internet of Things. IEEE Trans Industr Inf 14(10):4692–4701
12. Tasiopoulos AG, Ascigil O, Psaras I, Pavlou G (2018) EdgeMAP: auction markets for edge resource provisioning. In: 2018 IEEE 19th international symposium on "A World of Wireless, Mobile and Multimedia Networks" (WoWMoM). Chania, pp 14–22
13. Hossain MS, Muhammad G (2019) Emotion recognition using secure edge and cloud computing. Inf Sci 504(2019):589–601
14. Tun YK, Tran NH, Ngo DT, Pandey SR, Han Z, Hong CS (2019) Wireless network slicing: generalized kelly mechanism-based resource allocation. IEEE J Sel Areas Commun 37(8):1794–1807. https://doi.org/10.1109/JSAC.2019.2927100
15. Habiba U, Hossain E (2018) Auction mechanisms for virtualization in 5g cellular networks: basics, trends, and open challenges. In: IEEE communications surveys & tutorials, vol 20, no 3, pp 2264–2293. https://doi.org/10.1109/COMST.2018.2811395
16. Zhang Y et al (2019) COCME: content-oriented caching on the mobile edge for wireless communications. IEEE Wirel Commun 26(3):26–31
17. Hossain MS, Muhammad G (2020) A deep-tree-model-based radio resource distribution for 5G networks. IEEE Wirel Commun 27(1):62–67
18. Sandholm T, Suri S, Gilpin A, Levine D (2005) CABOB: a fast optimal algorithm for winner determination in combinatorial auctions. Manage Sci 51(3):374–390
19. Wang XW, Sun JJ, Li HX, Wu C, Huang M (2013) A reverse auction based allocation mechanism in the cloud computing environment. Appl Math Inf Sci 7(1):75–84
20. Abdulsalam Y, Hossain MS, COVID-19 networking demand: an auction-based mechanism for automated selection of edge computing services. In: IEEE transactions on network science and engineering. https://doi.org/10.1109/TNSE.2020.3026637
21. Le THT et al (2020) Auction mechanism for dynamic bandwidth allocation in multi-tenant edge computing. IEEE Trans Veh Technol 69(12):15162–15176. https://doi.org/10.1109/TVT.2020.3036470

22. Kim DH et al (2020) Pricing mechanism for virtualized heterogeneous resources in wireless network virtualization. In: 2020 international conference on information networking (ICOIN). Barcelona, Spain, pp 366–371. https://doi.org/10.1109/ICOIN48656.2020.9016477
23. Barakabitze AA, Ahmad A, Mijumbi R, Hines A (2020) 5G network slicing using SDN and NFV: a survey of taxonomy, architectures and future challenges. In: Computer networks, vol 167, 106984, pp 1–40
24. Samdanis K, Costa-Perez X, Sciancalepore V (2016) From network sharing to multi-tenancy: the 5G network slice broker. IEEE Commun Mag 54(7):32–39. https://doi.org/10.1109/MCOM.2016.7514161
25. Sciancalepore V, Costa-Perez X, Banchs A (2019) RL-NSB: reinforcement learning-based 5G network slice broker. IEEE/ACM Trans Networking 27(4):1543–1557. https://doi.org/10.1109/TNET.2019.2924471
26. Feng Q, He D, Zeadally S, Khan MK, Kumar N (2019) A survey on privacy protection in blockchain system. J Netw Comput Appl 126:45–58
27. Salman T, Zolanvari M, Erbad A, Jain R, Samaka M (2019) Security services using blockchains: a state of the art survey. IEEE Commun Surveys Tuts 21(1):858–880
28. Wang XY, Ho P (2011) Gossip-enabled stochastic channel negotiation for cognitive radio ad hoc networks. IEEE Trans Mob Comput 10(11):1632–1645
29. Capen E, Clapp R, Campbell W (1971) Competitive bidding in high-risk situations. J Petroleum Technol 23:641–653
30. Thaler R (1988) Anomalies: the winner's curse. The J Econ Perspect 2:191–202
31. Elz R, Bush R, Bradner S, Patton M (1997) Selection and operation of secondary DNS servers. BCP 16, RFC 2182. https://doi.org/10.17487/RFC2182, https://www.rfc-editor.org/info/rfc2182
32. Ongaro D, Ousterhout J (2014) In search of an understandable consensus algorithm. In: USENIX conference
33. Rothkopf M, Harstad R (1994) Modeling competitive bidding: a critical essay. Manage Sci 40:364–384
34. Friedman L (1956) A competitive bidding strategy. Oper Res 4:104–112
35. Dougherty E, Nozaki M (1975) Determining optimum bid fraction. J Petrol Technol 27:349–356
36. Smith B, Chase J (1975) Nash equilibria in a sealed bid auction. Manage Sci 22:487–497
37. McAfee R, McMillan J (1987) Auctions and bidding. J Econ Literature 25(2):699–738
38. Afolabi I, Taleb T, Samdanis K, Ksentini A, Flinck H (2018) Network slicing and softwarization: a survey on principles, enabling technologies and solutions. IEEE Commun Surveys Tuts 20(3):2429–2453
39. Ksentini A, Frangoudis PA, Amogh PC, Nikaein N (2018) Providing low latency guarantees for slicing-ready 5G systems via two-level MAC scheduling. IEEE Netw 32(6):116–123
40. 3GPP "Study on management and orchestration of network slicing for next generation network, v15.0.0," 3GPP SA5, Sophia Antipolis, France, Rep. 28.801, 2017
41. Sandhya Y, Sinha K (2017) Haribabu, "A survey: Hybrid SDN." J Netw Comput Appl 100(2017):35–55

Chapter 9
The Future of Digital Marketing: How Would Artificial Intelligence Change the Directions?

Khan Md. Raziuddin Taufique and Md. Mahiuddin Sabbir

Abstract Technological advancements have made the most disruptive change in marketing and consumer behavior in the last few decades. The history of change suggests that technology has entirely transformed the media from cable TV to more personalized technologies. In particular, the Internet and other relevant information technologies and platforms such as social media, powerful search engine, big data, mobile apps, and augmented reality are redefining marketing theories and practices. These advancements, on the one hand, have enabled marketers to enhance customer relationship and engagement. On the other hand, customers are becoming more powerful than sellers in creating and controlling the information content. Artificial intelligence (AI), the use of computerized programs and machinery that exhibit human intelligence, is expected to have even much greater impact on marketing and customer behavior than social media and other recent advancements. Applying a desk research method, the primary purpose of this chapter is to highlight the present state of the application of AI in marketing with a focus on digital marketing. The chapter also aims to identify the future directions of digital marketing with AI as a potential major driver. The major hotspots identified for future research include future marketing jobs and relevant skills, change in consumer decision making, AI-driven social media marketing and new product development, enhanced recommender engine, and augmented reality marketing. Practically, the findings will help marketers better prepare for designing marketing strategies for ever-emerging and more empowered digital consumers.

Keywords Artificial intelligence · Consumer behavior · Digital marketing · Information technology

---

K. Md. R. Taufique (✉)
Oxford Brookes Business School, Oxford Brookes University, Headington Campus, Oxford OX3 0BP, UK
e-mail: ktaufique@brookes.ac.uk

Md. Mahiuddin Sabbir
Department of Marketing, Faculty of Business Studies, University of Barishal, Barishal 8254, Bangladesh
e-mail: mmsabbir@bu.ac.bd

S. Kautish et al. (eds.), *Computational Intelligence for Modern Business Systems*, Disruptive Technologies and Digital Transformations for Society 5.0,
https://doi.org/10.1007/978-981-99-5354-7_9

1 Introduction

The accelerated emergence and extensive adoption of information and communication technologies (ICTs) and digital media are rapidly changing the way consumers buy products as well as the way marketers communicate, reach, and engage with customers. The proliferation of ICT capabilities and tools such as big data, artificial intelligence (AI), powerful analytics, GPS, and other emerging mobile and internet tools are reshaping business models. Among the recent advancements in ICTs, AI is gaining special attention in the marketing domain due to enhanced computing power with relatively low cost, emergence of big data, and the advancement of machine learning algorithms and models [1]. As a disruptive technology, AI is continually changing the way business is running across a diverse array of industries from public healthcare to retail settings. The rapid advancement of AI-driven technology signals that future marketing, specifically digital marketing strategies, will require integrating new models for the sales process, online retailing, customer service and relationship management, tracking customer behavior, and so on.

Undoubtedly, marketing is a practical science that continuously integrates new approaches (e.g., digital marketing) to targeting, reaching, and analyzing customers. Digital marketing however could provide the best possible outcomes if it can systematically process the massive amount of data it is increasingly generating where AI has already proven its significance. AI has revolutionized the digital marketing spectrum by facilitating marketers in making data-based marketing decisions more quickly and precisely. For example, AI-enabled image-recognition technology enables firms to identify the most likely approach to developing and placing ads across multiple social media platforms (e.g., Facebook, Instragram), and thus, boost firms' digital marketing strategies. Therefore, many companies are already using different AI-based services such as Amazon Lex, Google Assistant, and Microsoft Cognitive Services as part of digital marketing strategies. From marketing perspective, the adoption of AI-based platforms helps firms better interact with customers and other channels as well as enhance market forecasting and overall automation. The importance and relevance of AI in marketing, especially in digital marketing is already well recognized. For example, a recent survey of business leaders reports that sales and marketing is a priority area of AI's application [2]. The report also suggests that 24% of US firms are already using AI for different business functions and 60% are expected to use by 2022. The effect of AI on marketing activities is on the rise [3], and by 2025, it is estimated to have a $40 billion impact on marketing [4]. However, it is argued that the vast potentials of AI capabilities, including personalizing campaigns and predictive models are minimally explored and implemented [5], while such capabilities are offering significant positive impact on firms' growth.

The importance, emergence, and future advancements of AI capabilities in marketing have already been well documented in recent marketing and ICT literature [6–8]. In order to advance the literature and relevant practices, the purpose of this chapter is to highlight the current state of digital marketing and its potential advancement with AI-driven capabilities. The chapter also aims to identify the hotspots for immediate future research in the areas of AI-driven digital capabilities and its potential impacts on advancing digital marketing strategies.

2 Digital Marketing Today

In its current and ever-advancing capabilities, digital marketing is used as an umbrella term that refers to the application of information and communication technologies to the entire marketing process for customer acquisition, retention and engagement, brand building, and customer relationship management [9, 10]. Precisely, digital marketing is "an adaptive, technology-enabled process by which firms collaborate with customers and partners to jointly create, communicate, deliver, and sustain value for all stakeholders" [9] (p. 23).

The history of digital marketing suggests it first emerged in the 1990s, primarily as a channel of advertising to customers that was extended significantly with the emergence of mobile technologies and social media technologies during the 2000s and around 2010, respectively [11]. Today's digital marketing offers various new ways to communicate, reach, sell, and engage customers. It does this very successfully with ever-changing new digital marketing tools. A list of currently used major digital marketing tools with a short description is presented in Table 1.

3 AI and Its Capability

Like any other discipline or concept, there are many different definitions of AI, and perhaps no single one is universally accepted by practitioners. In computer science, "artificial intelligence (AI) refers to any human-like intelligence exhibited by a computer, robot, or other machine" [17]. A more general definition refers AI to "a computerized system that exhibits behavior that is commonly thought of as requiring intelligence" [18] (p. 16).

One distinct feature of AI compared to previous generation information technology relates to AI's capabilities to learn and update using data (e.g., text, audio, and video) as input [19] where AI performance is the output [20]. AI performance can be qualitatively described at three different levels [19]: (1) *Mechanical intelligence* refers to the capability of performing standardized mechanical tasks that require limited extent of learning or adaptation (e.g., factory automation). In digital marketing, for example, machine learning (ML) can be used to micro-segment online retail customers based on their preferences for personalized recommendation [21]. (2) *Thinking intelligence* is the performance of making rational decision with systematic learning and adapting capabilities (e.g., autonomous cars). Using natural language processing chatbots, thinking AI can now handle very diversified customer information such as customer accents and other context-specific customer issues [1]. (3) *Feeling intelligence* refers to the capability of recognizing, mimicking,

Table 1 Major digital marketing tools

SN	Digital marketing tools	Description
1	Search Engine Optimization (SEO)	A process of optimizing the rank of digital platform (e.g. website) in search engine results pages (SERP), increasing the chance of organic (free) traffic to the website or other relevant platforms such as blogs, infographics. For example, by incorporating long-tail keyword (using air purifier instead of air filter), AIRESERV received a better spot in Google rankings. A similar result was also achieved by Sun Auto Service as they created a web page that is less cluttered with a more structured visual design [12]
2	Search engine marketing (SEM)	The main difference between SEM and SEO is that SEM is a *paid* strategy and SEO is an *organic* strategy. SEM uses paid tactics to gain visibility in SERPs. SEM is often referred to as paid search or pay-per-click (PPC) marketing. *Google Ads* (formerly Google Adwords) is the most commonly used tool for this strategy. SEM allows brands to conduct keyword research and create campaigns that target the best keywords for their industry, products, or services. When users search for those keywords, they see the custom ads at the top or bottom of SERPs. The brand is charged each time a user clicks on the ad. For example, using Google Ads service, Zomato (a renowned Indian delivering foodservice brand) has acquired greater customer awareness of their brand [13]
3	Social Media Marketing (SMM)	SMM involves creating and sharing tailored content (e.g. text, image, and video) on social media to drive audience engagement and promote business. *Paid social media* is another word for advertising, when brands pay social media to have their content shared with specific audiences through boosting their organic content or designing unique advertisements. A social media campaign "The Best Men Can Be" (depicting modern elucidation of manhood) by Gillette in 2019 has been viewed by more than 30 million viewers with over 150 million Netizen has shared #GilletteAd hashtag [14]. Such a successful campaign helped the brand to positively reshape its image globally

(continued)

Table 1 (continued)

SN	Digital marketing tools	Description
4	Email Marketing	Email is used to promote content, discounts, and events, follow up sales calls as well as to direct people toward the firm's website. For example, Uber usually sends very brief and simple but tasteful texts, allowing busy subscribers to have a quick look. Yet there is always a link to details in case anyone is interested to learn more [15]
5	Content Marketing	Content marketing involves creation and sharing of online content (e.g. blogs, video, social media posts) on any topic of general interest that does not directly promote the brand, but is intended to stimulate interest in the product. For example, REI (an outdoor retailer) publishes many videos according to its audience's demands, even sometimes answers common questions (e.g., the process of using a compass). This further helps REI in finding the most potential customers for their products
6	Sponsored Content	Sponsored content appears on a publisher's website, which is usually written by the publisher's staff to make the content consistent with other content on the site. For example, Taco Bell sponsored Snapchat filter has reached over 224 million audience and thus, generated augmented customer engagements [16]
7	Native Advertising	It contains a headline and description that induce users to click on the link, leading to an article on the brand's website or other sponsored content. It appears "native" to the publisher's website. For example, Allbirds (a shoe company) sponsored a link on the New York Times website. A click on the link would lead a reader to an article about the importance of birds in our environment. This content is "native" to Allbirds as it is committed to sustainability and has the word "bird" in their brand name Both native ads and sponsored content are paid media. The main difference is that native advertising is more like a traditional ad and sponsored content is more like a media placement

(continued)

Table 1 (continued)

SN	Digital marketing tools	Description
8	Affiliate Marketing	A process by which a third party (affiliate) earns commission by promoting another individual's or firm's products. For example, X is an affiliate who has an attractive website where Y's sales link is displayed. If a customer buys any product from Y by clicking on a link displayed on X's site, X will be paid commission for that sale. Every Day Carry offers analyses of notebooks, flashlights, backpacks, and related products used by adventurous people with links to each product which will lead a user to the Amazon website
9	Chatbot Marketing	A way to promote products or assist customers using a chatbot—a computer application enabling conversations with customers by programming or with the help of AI. Chatbots assist customers in solvingreal-time problems or queries by offering live chat while customers surf and/or shop online. Booking.com, Wingstop, Louis Vuitton are examples of some websites that have successfully implemented chatbot marketing

Source Created by the authors

and responding to human emotions. In marketing, feeling AI (e.g., feeling analytics) can be used to develop strategic brand positioning statement as well as brand slogans to enhance brand association with customers' feelings [1]. The last capability is still largely work in progress and AI has a long way to go with human emotions [22].

The significant amount of data produced today by machines and humans outstrips humans' competency to captivate, organize, analyze, and deduce to make inferences based on such data [23], where AI is offering the best possible solution with vast potentials yet to be unleashed [24]. AI outperforms the conventional statistical methods in the sense that AI can use both deductive and inductive reasoning while conventional statistical methods use deductive reasoning. Deductive reasoning is primarily based on the preceding knowledge about data and it makes rigid inferences relating to the problem [25]. In contrast, with inductive reasoning, AI methods do not make any firm assumptions rather enable learning based on the available data [24].

With its persistent enhanced capabilities and applications, AI has gone from a buzzword to an essential business competence over the last decade. AI's application is penetrating nearly every industry with various new opportunities for businesses operating in the digital marketing domain. For instance, virtual assistant, such as online chatbots is transforming customer service into self-service [26]. Similarly, AI applications are doing the job of a portfolio manager [27] as well and AI-enabled robots are substituting humans to greet customers once they enter any online or offline service platform [28]. Indeed, using both ML algorithms and natural language processing, online chatbots can better comprehend users' requirements and provide feedbacks accordingly. Moreover, these activities are done at a faster rate than humans with lower costs. AI-infused predictive analytics and big data enable recommender engines to help customers in selecting the best products or services that match their preferences and stimulate buying decisions. AI-powered technology has made changes in the field of offline targeted marketing as well. Several firms are merging technologies such as geospatial software and facial recognition, for better profiling their customers which is followed by personalized promotion of products and/or services [29]. Supply chain firms are also increasingly interested in AI-driven solutions as it is estimated that from 2020 to 2024 almost 50% of such firms will make investments in AI and associated technologies [30]. In fact, ML algorithms with predicting optimal supplies are facilitating supply chain firms to be more efficient and effective by lessening or to some extent, abolishing overstocking.

IT operations are also capitalizing on AI as some AI-associated anomaly detection applications can signal any ongoing hacking undertakings and malware attacks. Some AI technologies offer self-healing solutions to software and hardware failures. Another interesting application of AI is its use in maximizing safety protocols. For example, AI-driven applications allow managers to observe workers' behavior in ascertaining their compliance with required safety procedures for goal achievement [31]. Managers therefore are immediately warranted about any undesirable behaviors, supporting essential resolutions to be made appropriately. Though some AI applications have raised some ethical and emotional concerns, their use in different sectors including business and marketing cannot be overlooked. For example, a general ethical concern is that human capabilities are now compared with machines rather than animals, questioning human superiority [32]. There are also other more specific ethical concerns such as increasing surveillance of AI on individual's presence and activities in different digital platforms [33], raising the question of privacy and interruption.

4 Applications of AI in Digital Marketing

AI is rapidly changing today's digital marketing platform. With powerful analysis of customers' data as input, AI can now understand many customer needs or trends and make human-like decisions such as Amazon's personalized recommendations. Essentially, AI is enabling marketers to save time and resources through making many marketing tasks automated digitally. Some major current applications of AI in digital marketing are briefly outlined below.

4.1 Online Chatbots

While online retailers are experiencing an unprecedented rise in sales, the dearth of real-time customer service together with imprecise product-related data presented on seller's website is one of the major obstacles to online shopping [34]. The deficiency of social communication and pleasing buying experience are putting off many customers from adopting online shopping. In satisfying customers' need for hedonic and utilitarian facilities during online shopping, retailers can rely on online chatbots which in its augmented version now include anthropomorphic features.

Chatbots are "any software application that engages in a dialog with a human using natural language" [35] (p. 813), which are the most common example of the application of AI in marketing. In particular, with AI-assisted technology, chatbots can "understand natural language and respond in natural language to a user request" [36] (p. 220). Chatbots are considered to be an impeccable demonstration of the advancement and application of customer-centric AI [37]. One study reports that the users' acceptance rate of chatbots ranges from 41 to 57% across six different countries including the US, UK, Australia, France, Japan, and Germany, indicating a promising trend for marketers [38]. Chatbots can play an important social function as they are capable of communicating with customers by imitating human-to-human interaction. In addition, with AI-enabled technology, chatbots can scrutinize and influence customers' behavior by receiving questions and answering them like a real person [39].

Chatbots can assist customers along the complete buying process, particularly from their first appearance on the website to the checkout stage [40–43]. Specifically, AI chatbots can provide support to customers at three different stages including pre-purchase, purchase, and post-purchase. Chatbots act as real-time shopping assistants for customers in the virtual environment round the clock. Accordingly, AI chatbots generate augmented customer service by ensuring support beyond regular office hours. Amazon Alexa is an example. Unlike traditional one-to-one customer service, AI chatbots can deliver one-to-many customer service to multiple customers simultaneously in different time zones. AI chatbots also offer localized customer service by enabling customers to choose their preferred language. Currently used messenger applications (e.g., WhatsApp, Facebook Messenger) by different companies can produce better outputs in communicating with their customers if such applications are combined with AI chatbots. One such example could be the Starbucks Barista bot with Facebook Messenger. The bot allows customers to order coffee using either voice instruction or a messaging gateway.

A recent study reports that the use of AI chatbots by online retailers in assisting customers with product recommendations and answering relevant queries generates higher level of positive customer responses [37]. In particular, interactions with virtual assistants resulted in greater level of customer satisfaction with incremental purchase and patronage intentions. Another encouraging finding in this regard is that AI-enabled anthropomorphized female chatbots demonstrated more significant results in stimulating positive customer responses compared to those of male chatbots. Interestingly, this finding remains consistent despite some errors occurring in the process. Indeed, female chatbots are more frequently excused in the event of any errors as opposed to male chatbots.

4.2 AI-Assisted Customer Insights

Understanding customer is the key to successful marketing. This requires real-time accurate insights from customers that help marketers customize the content for the targets. AI can track customer behavior on different digital marketing platforms and provide marketers with quality customer data for designing and customizing marketing campaigns and other relevant marketing strategies precisely for specific target markets. For example, AI-powered programs are providing marketers with data that exceeds conventional survey- or interview-based data to include vast amounts of unstructured customer data generated by mining customer preferences and associated data from social media, web, and mobile activity. Medallia is one such provider that develops customer experience software [44], allowing firms reshaping their decisions instantaneously and providing feedbacks to customers in real time. Crimson Hexagon's AI-enabled user-insights platform is engaged by Samsung to study user-generated discussions and related images in different digital marketing platforms (e.g., social media). This facilitates the company in gathering knowledge about how customers are interacting about their products which allows the company to modify their campaigns with the contents to which customers can best relate to [45].

Using image recognition algorithms, Pinterest provides its users with more tailored image content [46]. Moreover, with AI-optimized image mining technology, analysts are now able to generate insights that are supposed to be valuable in developing ads. For example, [47] applied a deep learning mechanism to explore how the different car images impact on their popularity. The authors gathered about 800,000 car images of 17 brands, which were posted on different social media platforms. Drawing on relevant theories, they further determined three specific image design characteristics that affect popularity. This insight can guide digital marketers in incorporating the most appropriate image to digital banner ads. Besides, American scented candle manufacturer and retailer Yankee Candle effectively developed and launched new variety of scented candles using AI-driven text mining techniques. In so doing, they leveraged the information available online, specifically text data on diverse social media platforms, and then investigated the data to determine what specific scents customers link with particular seasons [46].

4.3 Product Recommenders

Product recommenders are applications allowing firms to provide users (existing or new) with the best relevant product or service recommendations online in the process of leading them to act that generates value to both ends (firm and user). Nowadays, *personalized product recommendations* are done by AI through collecting and analyzing customer data of their past engagements with the brand. Amazon uses AI-powered recommendation engines to effectively provide personalized suggestions. AI-enabled product recommendation is typically designed to facilitate customers' buying decision by easily identifying products that match their tastes and preferences. AI-driven technologies also help customers through the checkout stage with recommendations that better reflect their personality and habits, making customers' shopping journey more relaxed.

Moreover, in many industries, customer acquisition and account management are still done by salespeople using telephone, email or other modes of communication. AI is predicted to assist salespeople via AI-enabled agents through monitoring teleconversations in real time [6]. The authors also indicate that AI agent might be able to provide real-time recommendations to the salesperson for approaching the customer by detecting customer's tone. While such capabilities will supplement salesperson's job, possible negative consequences might involve customers being uncomfortable with AI's monitoring of conversation [6].

4.4 Augmented Retailing

Augmented reality (AR) can be defined as an interactive experience of a natural environment in which real-world objects are augmented by computer-produced information, sometimes across different sensory properties such as haptic, visual, somatosensory, olfactory, and auditory. AR is often termed as mixed reality (MR) as in both cases the real-world environment is augmented through technologies [48]. However, AR and virtual reality (VR) are different in a sense that AR changes a person's continuing perception of a natural environment, while VR entirely substitutes its user's real-world environment with an artificial one. Furniture retailer IKEA's "Place" app (helps customers in choosing best fitting furniture), paint specialist Dulux's "Visualizer" (helps customers in choosing suitable wall colors), Home Depot's "Project Color" app (helps customers in choosing suitable paint colors in home), and Timberland's virtual fitting room (helps customers in selecting best fitting garments) are some examples of AR application in the current business world.

Application of augmented reality in marketing is termed as *Augmented Reality Marketing* (ARM) [49] that stimulates customers' *situated cognition* where customers enthusiastically interact with a digitally-enhanced environment to regulate their decision making [50]. It is argued that ARM has the ability to allow customers "adjust" their thinking (e.g., by picturing a furniture into their drawing room), offering firms a certain level of influential advantage over customers' decision making process [50, 51]. In this context, customers might be willing to purchase more and pay premium prices for products chosen through ARM [52]. The proliferation of augmented reality is so vigorous that the tech giants, such as Microsoft, Apple, Google, and Facebook are factually in a race in order to make AR omnipresent across different digital marketing platforms [49]. This is better reflected in their investments in AR which are predicted to touch $100 billion mark by 2024 [53]. Renowned academician Michael Porter thus asserted that "every company needs an AR strategy" [54] (p. 6).

In fact, many online retailers including 1–800-Flowers.com and Amazon.com are already utilizing different advanced AI-driven tools to better understand customers' behavior and provide more pleasant and customized customer experience. *Augmented retail shopping experience* can now be implemented by AI through enhancing the capability of augmented reality (AR). With the help of AI-enabled technology, shoppers can now "try" products (e.g., clothing, furniture in the room) before buying them, instead of just browsing product images with AR. This is already implemented by the UK-based TopShop for clothing and Swedish IKEA for furniture. These facilities strengthen consumers' consumption experience and form strong brand identifications which further create a loyal customer base.

As an early adapter of AR technology, IKEA launched IKEA Place App in September 2017 and has been enjoying a dominant first-mover advantage in terms of AR technology's implementation in service provision since then [55]. With as much as 98% accuracy rate of scaling the selected products to dimension according to buyers' kitchen, guest, or dining room's sizes, this AI-driven technology offers some additional benefits in relation to firm's existing marketing and customer service policy. For example, the app increasingly minimizes the existing multi-channel gap by coupling the virtual and real world at a smartphone's touch. Second, it can reduce stress relating to in-store furniture shopping experience by allowing its users to pick out preferable items using AI-enabled fit experimentation with the digital duplications of furniture. Lastly, it can lessen product returns and the associated troubles.

More importantly, the app attempts to translate the practice of selecting a furniture item digitally into a fairly enjoyable online engagement by customers, which is particularly crucial for formulating digital marketing strategies [56].

4.5 Market Segmentation

One of the key strategic decisions in marketing involves segmentation, targeting, and positioning which is commonly known as STP strategies. Segmentation is to divide the customers into different segments where each segment is supposedly to have different needs and wants corresponding to their profiles. The very traditional bases of segmentation include geographic, demographic, psychographic, and behavioral segmentation. Recent studies [57–60] on these traditional segmentation variables have raised some challenges that AI can potentially solve [61]. Specifically, AI has been proven to uncover the patterns of consumer behavior through powerful data mining that are difficult for human analysis. For example, data mining can segment the tourist markets based on the different meaning of destinations to different customers, which is proven to be better than the classic cluster analysis [62]. Similarly, AI-enabled personalized recommendations can be used to micro-segment retail customers based on their preferences for recommendations [21].

4.6 Market Targeting

Market targeting is to select specific segment(s) on which the firm focuses its marketing actions and offers its products. In digital marketing domain, various technological tools and analytics are already in use for targeting customers. Some commonly used tools include search engines for targeting customers based on users' keyword search and browsing history as well as social media platforms for targeting social media consumers based on users' interests, content, and connections [63]. AI-enabled recommendation engine is another powerful tool that can recommend target markets as well as predictive modeling that can be used to determine specific segment(s) to target [1].

4.7 Marketing Mix Decisions

Marketing mix decisions involve strategies for fundamental elements of marketing programs including product, price, place, and promotion. Three alternative forms (i.e., mechanical, thinking, and feeling) of AI are already used in different capacities with more future potentials for some aspects of marketing mix elements [1]. For example, online chatbots as a *Mechanical AI* is successfully implemented for different routine service designs. Mechanical AI is also widely used as payment gateway for digital marketing such as Apple Pay, Amazon Payment, Google Pay, PayPal, etc. UPS and Amazon Prime Air's drone delivery systems are other examples of *Mechanical AI*'s application in distribution/forward logistics of marketing mix decisions.

Thinking AI can be used for personalized product and branding decision through big data analytics on consumer trends and preferences [21]. Nestle used Nespresso machines and coffee pods in order to maintain direct brand relationship with their customers and observe their preferences. This therefore led Nestle to curate the idea of developing one of their bestselling coffee flavors as well as providing a variety bundle offers customized to match specific customer preferences [64]. Another example is Coca-Cola's freestyle vending machines which in due course led the company to launch its Cherry Sprite drink that was produced based on the data on customers mixing flavors of their choice on the freestyle vending machines [65]. Some aspects of retailing as part of place decision can be implemented using *Thinking AI*. Amazon Go is an example of retailing where AI is used as a facial recognition technology that identifies and remembers each shopper. Interactive nature of digital marketing facilitates *Feeling AI* to track real-time customer sentiment responses (e.g., like, dislike) to promotional messages. Sky suggests programs to viewers based on their mood and Kia identified social media influencers for one of its Super Bowl campaign in 2016 using machine learning technology. More applications of *Feeling AI* include the use of "Affectiva" that senses customer's feeling and customizes advertisement messages accordingly [1].

4.8 Enhanced e-Mail Marketing

Email marketing refers to sending persuasive messages to customers in large quantities that contain information about any specific product. If executed properly, such messages create a decent connection with customers' feeling leaving them with trust and positive attitudes toward the advertised brand. Now, AI is capable of more personalizing email marketing campaigns based on user preferences and behaviors. The power of machine learning can more precisely determine the most appropriate time and frequency to send emails to the targets. AI is also capable of suggesting the most suitable content and subjects of email that are likely to generate more clicks. Some AI-enabled e-mail marketing tools such as Boomtrain, Persado, and Phrasee have already been proven to be more powerful than human in generating clicks [66].

4.9 Digital Advertising

Digital advertising refers to "a message of persuasion (regarding products, services, and ideas) that interacts with consumers through digital media" [67] (p. 4). Such digital media includes internet, smartphones, in-game advertising, AI speakers, smart TV, digital signage, VR/AR platforms, and over-the-top services. The often-used digital advertising platforms include Google, Instagram, and Facebook which examine users' demographic information (e.g., age gender, location) to generate the best possible results by delivering consumers with the most relevant advertising content. Adding to that, many tasks of digital advertising are now performed by AI. One of the most powerful and widely used digital advertising platforms is Google AdWords which already implements an AI-enabled auction system enabling advertisers to pay based on cost per conversion. AI-enabled technology helps marketers track current market trends and predict future trends. Marketers can then better allocate their funds to target the right customers with right platforms and contents, allowing them to minimize advertising waste. With the help of AI-enabled image-recognition technology, for example, Coca-Cola tracks when pictures of its products, or those of its competing brands, are uploaded online, and then uses algorithms to find out the best alternative approach to develop and position their advertisements. One study suggests that this type of advertising is likely to have a four times greater probability of being clicked on compared to other means of targeted advertising used by the company [68].

Current AI applications also include AI-enabled interactive advertisement placements [69] and displaying video ads in relation to the content just viewed by the users [70]. For example, on Black Friday 2018, LEGO used AI-enabled interactive ads where the company empowered the AI system with the information of a vast array of its products [69]. This enabled the system with ads customized specific to customers' interests and needs, allowing the brand to ensure persuasive, one-on-one communications with customers along their entire journey to purchase. Back in 2016, 20th Century Fox teamed up with IBM Watson to develop the first-ever AI-powered promo for its movie Morgan [24]. Studying hundreds of past thriller and horror promos, IBM Watson suggested what specific thrilling moments to be included in the promo for the movie Morgan to make the promo more suspenseful. In 2015, the NBA collaborated with WSC Sports to deliver its audiences with region-specific highlight clips through NBA websites. Specifically, with AI-powered technology, they created highlight clips for each player in a particular game and delivered custom-made clips to its global viewers. For example, an Australian viewer would be exposed to highlight clips of Australian-born NBA stars [71]. These examples clearly signal the promising dominance of AI-enabled technologies in digital advertising.

4.10 AI-Enabled Website Builders

AI-enabled website builders can be used to design website based on customers' data on how they interact with the shopping site. The AI algorithm makes the decisions entirely based on the user data where the algorithm has access to number of website content, designs, layouts, and navigation options. The AI then selects the most suitable combinations to create a unique website for the user. Examples of providing such services include Grid, Squarespace, Weebly, and Wix.

4.11 New Online Retailing Model

Online retailing is a process enabling customers to explore, compare, and select, negotiate prices, and eventually buy products from any location using internet-based applications. The current progression of online retailing is greater than ever, allowing marketers to devise the best competitive retailing models to reach and retain the targets. Now more firms are entirely shifting to or complementing their traditional retailing with online retailing thanks to incremental internet access and rise of online shopping from the customer's perspective. Amazon, eBay, Rakuten, and Walmart are among the most recognized current online retailers.

The core tenet of marketing is to shape customer needs and wants. That is, the job of a marketer is not just following the information a customer is providing rather analyzing customers' mind to discover needs that have not been surfaced yet. This subsequently helps marketers match their products or services to customer preferences. Aligning with this concept, AI may enable retailers to transform the current online retailing of "shopping-then-shipping" model into "shipping-then-shopping" model [72, 73]. In shopping-then-shipping model, customers first order the product and then retailers make the shipment, whereas shipping-then-shopping model works in other way around. In the second retailing model, AI might enable retailers to predict customers' preferences instead of waiting for customers to order the product. More specifically, AI will help retailers to identify customers' preferences (with certain level of accuracy) and accordingly, retailers will ship items to customers without a formal order from customers, providing customers the option to return the items that they do not want [72, 73]. Some businesses have already initiated such retailing models in the USA (e.g. Birchbox, Stitch Fix).

4.12 Analyzing Online Customer Engagement Behavior Data

Customer engagement is a psychological event involving customer's cognitive, emotional, and behavioral properties that is induced by shared, co-creative customer experiences with a focal agent and/or object (e.g., brand) in a specific service relationship [74]. Simply put, customer engagement is the continual communication between the firm and its customer, mostly initiated by the firm, and chosen by the customer. Observing customer engagement behavior signifies remarkable significance for the company as it can exert favorable and unfavorable impacts on the company [75]. More precisely, customer engagement is supposed to produce preferred outcomes for the firm [76]. With technological advancement, online customer engagement behavior has been accelerated significantly. For example, customers can now easily engage with the brand and the firm through its presence on Facebook and other social media platforms, leaving firms with ample opportunity to gather much more data on what customers think, share, and talk about their products and services.

Firms can now depend on machine learning and AI to investigate and generate insights from big data relating to customers' online engagement behavior [77], where Machine learning is a subfield of AI that enables machines to learn from historical data to make predictions and some decisions. AI-enabled information processing systems [94] can use three methods to process customer data [78]. First, *content-based* data processing system uses customers' metadata (e.g., likes, dislikes) and real-time data (e.g., keywords used while searching on Google). Second, *collaborative recommendations* compare different customers by analyzing customers' ratings or their past purchase records. Lastly, *hybrid recommendations* combine the essentials of the previous two systems. Moreover, it is now possible through AI to classify customers' messaging to determine which employees in the organization should respond to which messages [79].

In terms of firms' response to customer feedback, AI can determine whether it should be manual or automatic [79]. In e-commerce, favorable product feedback usually does not need immediate reaction to the customer, while negative feedback requires prompt response by firm with tailored answer and further probing. As such, the company's response can be made absolutely automated when the customer feedback is positive. In contrast, if AI senses any sort of unusual behavior or unfavorable feedback in the case of online customer engagement behavior, it can instantaneously alert the company to make necessary reconciliations and act accordingly. Prediction and sensing of this kind is considerably impossible for humans.

4.13 Enhancing Online Pricing Strategies

Pricing has always been an important marketing tool for the firms. With different pricing approaches including fixed pricing, dynamic pricing, seasonal pricing, and geographical pricing, firms persistently strive to gain competitive advantage. The rise of digital marketing and the optimization of AI-powered technology have accelerated this endeavor further than before. AI-driven technologies allow firms to determine consumer price sensitivity, estimate pricing errors, and track related purchasing trends. This enables firms to formulate the best competitive pricing strategy which they can apply to nudge customers at the stage of final purchase decision [80].

Amazon, for example, collects customer data at various customer touchpoints, including pre-purchase (situations when customers view or search for products, read product reviews, or navigate retailers' page), purchase (includes tracking past purchase records and shopping cart lists), and post-purchase (includes analyzing product returns and post-purchase service requirements). AI-enabled technology then lets Amazon analyze those customer data to comprehend what these particular customer segments are searching for and the prices they are more likely to pay

[81]. AI and big data are also crucial in the hotel context as AI enables hotels to offer dynamic pricing as well as geographical pricing for customers around the globe. This can more precisely address the under-occupancy concerns of the hotels by altering pricing to stabilize supply and demand [82].

4.14 Emotional Support to Customers

Emotional support to customers may involve helping customers in picking the right products for themselves, carefully handling their pre- and post-purchase complaints about products or services, which are crucial for building long-term and trustworthy relationships with customers. Though AI's capabilities in recognizing and responding to human emotions are still being challenged, it aims to provide some emotional support to customers in different ways in various digital marketing platforms. For example, AI-enabled technology can detect customer sentiment and accordingly suggest alternatives. AI can also provide emotional support to digital customers by asking relevant questions as well as adjusting customers' linguistic syntax [6]. Marketing is now more analytics- and data-driven than ever before where AI plays the driving role. More and more firms are realizing this reality, integrating AI in different marketing strategies, and planning to enhance AI-enabled marketing models.

5 Research Opportunities in AI-Driven Digital Marketing

5.1 Future Marketing Jobs and Skills

It is predicted by scholars and practitioners that the advancement of AI will create many job displacements [22]. Some marketing jobs have already been identified to be replaced by AI such as telemarketing, market research analysts, and retail salespeople. However, much research attention is needed to pinpoint the types of marketing jobs to be vulnerable due to the advancement of AI and types of marketing jobs that are safe or relatively safe. Advancement of AI will also create opportunities for jobs that will require new skills. As such, future research should look into the skills that marketing jobs will demand in AI-driven marketplace.

5.2 Change in Consumer Decision Making

As AI is performing part of human job, it would make decisions including consumer decision [22], which is already partly in existence [83]. This is very critical for marketers, because this may call for significant redesigning of marketing strategies.

Hence, future research should investigate how AI is changing consumer behavior and what decisions AI might take on behalf of consumers. We also need to know how much consumers are likely to be comfortable to allow AI to make consumer decision.

5.3 AI-Driven Social Media Marketing

While social media is increasingly making its strong hold in digital marketing, researchers predict that AI will liberate the power of social media data and enhance the customer digital journey experience [7]. In fact, social media marketing can be enhanced by AI-driven strategies by using widespread user-generated data available on social media. However, the use of AI in social media platforms has raised ethical concerns, causing consumer distrust [7]. This requires future research to focus on better understanding consumer perception of AI intervention in social media environment, especially in accessing user data. From marketing decision perspective, future research should investigate more on the quality and reliability of widespread and vast amount of social media data as inputs for marketing strategies [7].

5.4 Privacy and Data Security

Users are becoming more and more concerned about their privacy and data security on social media platform. This is more so after public outcry due to Facebook's unauthorized licensing of millions of user accounts to Cambridge Analytica which was a British political consulting firm. Similar concern has also been documented in a recent empirical study [84] suggesting that the acceptance of chatbots, specifically the usage frequency is negatively influenced by the users' privacy concerns. Thus, this opens up ample opportunities to do further research on how AI can be used to access users' data for marketing purposes while maintaining users' privacy and security. Some scholars specifically highlighted the need for research on how privacy and data security might affect marketing's use of AI-driven data collection [1] and how AI governance may facilitate companies to retrieve privacy in social media [7]. This is even more important in the context of AI-driven marketing strategies that require lot of customer data for personalizing different market offerings. Future research should investigate on how AI-driven customer personal data can be used for personalizing marketing programs by limiting the loss of privacy.

5.5 AI-Driven New Product Development

Considering its rapid advancement in capabilities, AI is to some extent capable of recommending new product development. However, the question is whether such products might be able to meet customers' needs and wants. This is another potential avenue for further research on AI's capabilities on new product suggestions and matching such product ideas to customer needs and wants. [1] suggest that AI's capabilities to detect the patterns and regularities in data can be used to discover consumers' implicit needs and wants and to match products at different life cycle stages, which necessitates further research.

5.6 Enhanced Recommender Engine (RE)

RE is an AI-enabled system that is used in different virtual platforms such as email, social media, and online shopping sites to recommend relevant content/items to users. While RE is already used by firms (e.g. Amazon Recommender Engine) to recommend products to customers based on their preferences, future research should focus on AI-powered data analytics tools that can be leveraged for more powerful predictive analysis of customer future needs and wants or preferences.

5.7 AI-Driven Brand Positioning

As a vital element of STP strategies, positioning is to place the brand in the mind of the customers distinctively from the competitors. In other words, positioning is to feel the customer and speak to the customers in their language. While *mechanical AI* can be used for segmentation and targeting, positioning requires the use of *feeling AI* (e.g. feeling analytics) that can sense the behavior and potentially the emotion of the customers. Such feeling AI will then be able to recommend some compelling positioning themes or slogans by understanding what echoes with target market's feelings [1]. This is an interesting future research area in AI-driven digital marketing domain which is still sparse [1].

5.8 Better Comprehension of Augmented Reality Marketing (ARM)

It is argued that though ARM is engaged across various stages of customers' buying process [85, 86] or exploring its impact on brand attitudes [87], extensive implementation of ARM remains ambiguous [88, 89]. Such ambiguity is mostly hinged on scant conceptualization of ARM and lack of comprehensive discussion on ARM's distinctiveness in relation to traditional marketing tactics. Indeed, many companies are gradually downsizing their investment in AR, considering they lack adequate knowledge and competencies to target and involve their customers seamlessly using AR [90]. Specifically, some marketers find it challenging to make AR as a commonplace technology with augmented user experience through most relevant content offerings [49]. Within the broad scope of future digital marketing opportunities, this therefore necessitates further study to understand how to create and deliver superior user experience through ARM that will be valued by customers in a manner that is unique to conventional marketing tactics.

5.9 Tracking Consumer Five Senses

The capabilities of AI in guiding online businesses are already well documented [91]. Digital consumers spend significant amount of time online for various purposes including shopping, sharing, browsing, entertainment, and other activities. In response, AI capabilities are already used by marketers to better track and understand consumer insights. To advance the capabilities and applications, future research should specifically focus on how consumers' five senses (sight, hearing, taste, smell, and touch) can be more accurately tracked in order for marketers to customize market offerings and enhance customer-brand association based on more precise customer insights.

5.10 Integrating Psychological Theories to Sentiment Mining

Various psychological theories applied in marketing and consumer behavior, especially relevant to cognitive and affective components need to be more integrated with AI capabilities in understanding and segmenting customers. One promising area where psychological theories require more integration is sentiment mining which refers to "the use of natural language processing and computational linguistics to find and extract subjective information from text data" [92] (p. 31). Essentially, sentiment mining identifies the emotional tone by analyzing the texts that involve the use of data mining, machine learning, and AI. Future research should especially focus on how sentiment mining can be used to precisely classify sentiments by integrating relevant psychological theories in marketing.

Figure 1 presents a snapshot of different AI branches, their relevant applications in digital marketing, and future research opportunities to enhance AI-enabled digital marketing strategies.

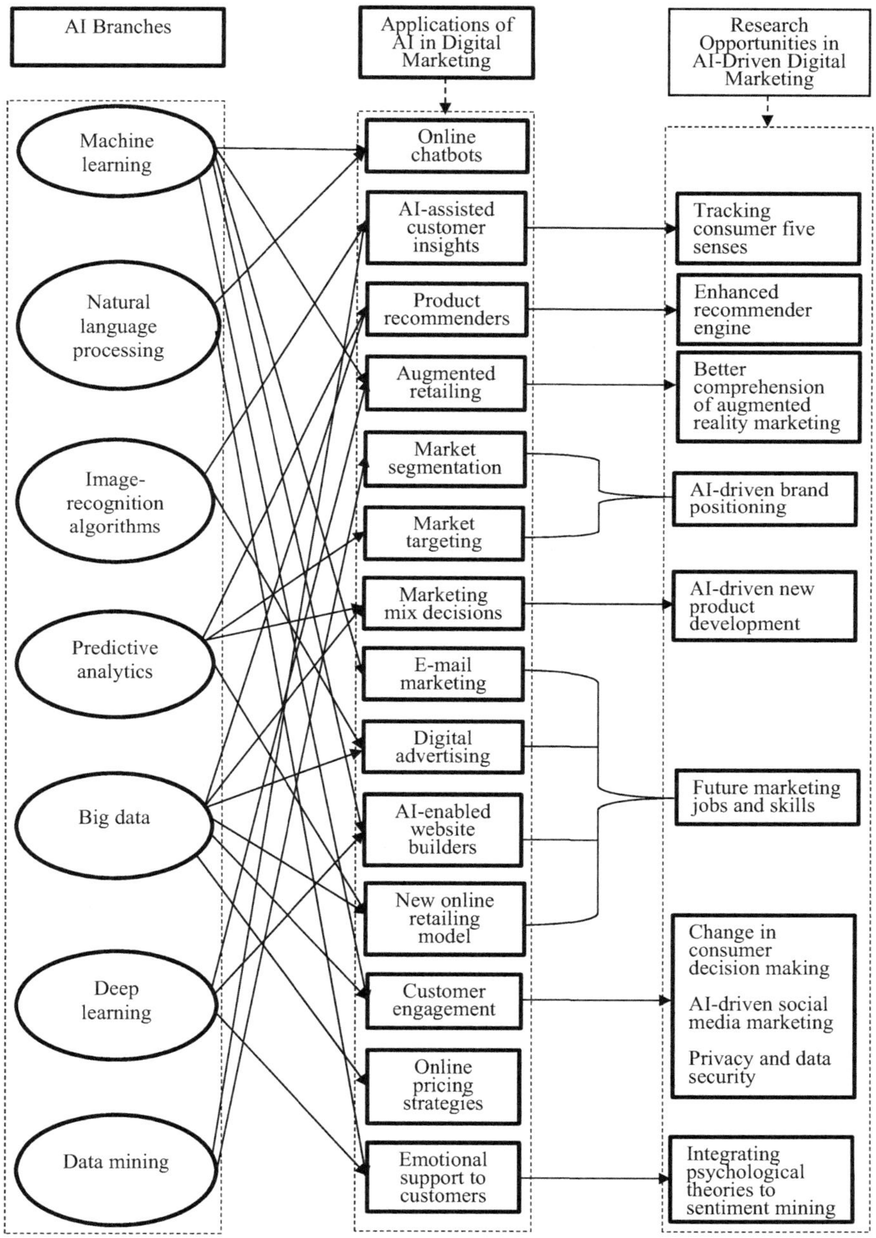

Fig. 1 AI applications and research opportunities in digital marketing

6 Conclusion

The rapidity of the advancement of digital marketing platforms does not need to be overstated. Some advancement occurs too fast to be predicted. Yet, in order for businesses to leverage on the best of digital advancement, it is important to know the current state of advancement relevant to business functions. It is also critical to predict future advancement in order for businesses to be prepared to adopt and capitalize on the advancement. In line with the aims of this chapter, it outlines the present state of digital marketing with the major ICT tools currently being used including search engine marketing, social media marketing, content marketing, affiliate marketing, chatbot marketing, and other tools. The chapter also highlights the AI and its capabilities in enhancing digital marketing performance in different marketing functions such as product recommender, augmented retailing, enhanced e-mail marketing, AI-enabled website builders, and emotional support to customers.

With its rapidly changing and improved capabilities, AI is predicted to offer a lot more to advance future digital marketing efforts. This requires marketers to be prepared for such changes and advancement which calls for extensive research. Accordingly, the chapter identifies ten different critical areas where research attention is called for leveraging AI capabilities in advancing future digital marketing strategies. While these ten areas are not inclusive, in line with many other researchers cited in this paper, the authors believe that these research areas should be given priority at this point in time to take the best that AI has to offer for future of digital marketing.

Glossary

Machine learning involves letting computer programs or applications to think or act automatically without much human interferences. Machine learning is a subset of artificial intelligence.

Deep learning emulates the human brain in generating specific types of information by analyzing data and producing patterns to make decisions. Deep learning is a subset of machine learning.

Natural language processing divides human languages in different parts and analyze sentence structure and words' meaning to better comprehend the written or spoken languages.

Big data is a gathering of data that is large in volume, persistently fast in growing, and widely diverse in types, generating unprecedented insights about customers.

Predictive analytics or **predictive modelling** includes a wide range of statistical techniques such as machine learning and data mining that examine available facts to make forecasts about the future.

Data mining is a practice of extracting and analyzing the large datasets to produce new insights.

References

1. Huang MH, Rust RT (2021) A strategic framework for artificial intelligence in marketing. J Acad Mark Sci 61(4):43–65. https://doi.org/10.1007/s11747-020-00749-9
2. MIT Technology Review Insights (2020) North America—The global AI agenda. https://mittrinsights.s3.amazonaws.com/AIagenda2020/NAAIagenda.pdf. Accessed 2 July 2021
3. Columbus (2018) 77% Of marketing execs see AI adoption growing this year. https://www.forbes.com/sites/louiscolumbus/2018/05/14/77-of-marketing-execs-see-ai-adoption-growing-this-year/?sh=2137b3b37ef8. Accessed 2 July 2021
4. Reavie V (2018) Do you know the difference between data analytics and AI machine learning? https://www.forbes.com/sites/forbesagencycouncil/2018/08/01/do-you-know-the-difference-between-data-analytics-and-ai-machine-learning/?sh=20e0bdd5878d. Accessed 4 July 2021
5. Blueshift (2018) Activating customer data for AI-powered marketing. https://blueshift.com/library/activating-customer-data-for-ai-powered-marketing-report/. Accessed 2 July 2021
6. Davenport T, Guha A, Grewal D, Bressgott T (2020) How artificial intelligence will change the future of marketing. J Acad Mark Sci 48(1):24–42.https://doi.org/10.1007/s11747-019-00696-0
7. Dwivedi YK, Ismagilova E, Hughes DL et al (2021) Setting the future of digital and social media marketing research: perspectives and research propositions. Int J Inf Manage 59.https://doi.org/10.1016/J.IJINFOMGT.2020.102168
8. Mustak M, Salminen J, Plé L, Wirtz J (2021) Artificial intelligence in marketing: topic modeling, scientometric analysis, and research agenda. J Bus Res 124:389–404. https://doi.org/10.1016/j.jbusres.2020.10.044
9. Kannan PK, Li H (2017) "Alice" digital marketing: a framework, review and research agenda. Int J Res Mark 34(1):22–45.https://doi.org/10.1016/J.IJRESMAR.2016.11.006
10. Mogaji E, Soetan TO, Kieu TA (2021) The implications of artificial intelligence on the digital marketing of financial services to vulnerable customers. Australas Mark J 29(3):235–242. https://doi.org/10.1016/j.ausmj.2020.05.003
11. Fierro I, Cardona Arbelaez DA, Gavilanez J (2017) Digital marketing: a new tool for international education. Pensam Gest 43:241–260. https://doi.org/10.14482/PEGE.43.10594
12. Gibbons M (2021) 6 SEO examples to imitate on your business website. https://www.webfx.com/blog/web-design/seo-examples/. Accessed 24 August 2021
13. Juneja D (2021) Top 11 benefits of search engine marketing in 2021. https://iide.co/blog/benefits-of-search-engine-marketing/. Accessed 24 August 2021
14. Bredava A (2020) 15 awesome examples of social media marketing. https://www.searchenginejournal.com/social-media-marketing-examples/380202/#close. Accessed 24 August 2021
15. Watts E (2018) 10 of the best email marketing campaign examples You've ever seen. https://mailbakery.com/blog/best-email-marketing-campaign-examples/. Accessed 24 August 2021
16. Curle M (2020) 10 Examples of sponsored content to inspire your marketing. https://www.fool.com/the-blueprint/sponsored-content/. Accessed 24 August 2021
17. IBM Cloud Learn Hub (2020) What is artificial intelligence (AI)? https://www.ibm.com/cloud/learn/what-is-artificial-intelligence. Accessed 2 July 2021
18. NSTC (2016) Preparing for the future of artificial intelligence. https://obamawhitehouse.archives.gov/sites/default/files/whitehouse_files/microsites/ostp/NSTC/preparing_for_the_future_of_ai.pdf. Accessed 2 July 2021
19. Huang MH, Rust RT, Maksimovic V (2019) The feeling economy: managing in the next generation of artificial intelligence (AI). Calif Manage Rev 61(4):43–65. https://doi.org/10.1177/0008125619863436
20. Darwiche A (2018) Human-level intelligence or animal-like abilities? Commun ACM 61(10):56–67. https://doi.org/10.1145/3271625
21. Dekimpe MG (2020) Retailing and retailing research in the age of big data analytics. Int J Res Mark 37(1):3–14. https://doi.org/10.1016/J.IJRESMAR.2019.09.001
22. Rust RT (2020) The future of marketing. Int J Res Mark 37(1):15–26. https://doi.org/10.1016/J.IJRESMAR.2019.08.002

23. Hurwitz J, Kaufman M, Bowles A (2015) Cognitive computing and big data analytics. Wiley, New York
24. Campbell C, Sands S, Ferraro C, Tsao HY (Jody), Mavrommatis A (2020) From data to action: how marketers can leverage AI. Bus Horiz 63(2):227–243. https://doi.org/10.1016/j.bushor.2019.12.002
25. Teboul W (2018) Why use machine learning instead of traditional statistics? https://towardsdatascience.com/why-use-machine-learning-instead-of-traditional-statistics-334c2213700a. Accessed 1 July 2021
26. Fluss D (2017) The AI revolution in customer service. https://www.destinationcrm.com/Articles/Columns-Departments/Scouting-Report/The-AI-Revolution-in-Customer-Service-115528.aspx. Accessed 3 July 2021
27. Javelosa J (2017) Major firm announces it's replacing its employees with A.I.. https://futurism.com/major-firm-announces-its-replacing-its-employees-with-a-i. Accessed 2 July 2021
28. Choudhury SR (2016) Softbank's Pepper robot gets a job waiting tables at Pizza Hut. https://www.cnbc.com/2016/05/24/mastercard-teamed-up-with-pizza-hut-restaurants-asia-to-bring-robots-into-the-pizza-industry.html. Accessed 2 July 2021
29. Pratt MK (2021) 9 Top applications of artificial intelligence in business. https://searchenterpriseai.techtarget.com/tip/9-top-applications-of-artificial-intelligence-in-business. Accessed 25 August 2021
30. Hippold S (2021) How supply chain technology will evolve in the future. https://www.gartner.com/smarterwithgartner/gartner-predicts-the-future-of-supply-chain-technology/. Accessed 25 August 2021
31. Moore L (2020) 10 AI use cases in manufacturing. https://searcherp.techtarget.com/feature/10-AI-use-cases-in-manufacturing?_gl=1*lrnn00*_ga*MTg4NDA2Nzk4Mi4xNjI5ODgyNTk4*_ga_TQKE4GS5P9*MTYyOTg4OTE5NS4yLjAuMTYyOTg4OTE5NS4w&_ga=2.80216146.498479587.1629882598-1884067982.1629882598. Accessed 25 August 2021
32. Belk R (2019) Machines and artificial intelligence. J Mark Behav 4(1):11–30. https://doi.org/10.1561/107.00000058
33. Belk R (2021) Ethical issues in service robotics and artificial intelligence. Serv Ind J 41(13–14):860–876. https://doi.org/10.1080/02642069.2020.1727892
34. Sivaramakrishnan S, Fang W, Zaiyong T (2007) Giving an 'e-human touch' to e-tailing: the moderating roles of static information quantity and consumption motive in the effectiveness of an anthropomorphic information agent. J Interact Mark 21(1):60–75. https://doi.org/10.1002/dir.20075
35. Dale R (2016) The return of the chatbots. Nat Lang Eng 22(5):811–817. https://doi.org/10.1017/S1351324916000243
36. Lester J, Branting K, Mott B (2004) Conversational agents. In Singh MP (ed) The practical handbook of internet computing. Chapman & Hall/CRC, New York
37. Toader DC, Boca G, Toader R, Măcelaru M, Toader C, Ighian D, Rădulescu AT (2020) The effect of social presence and chatbot errors on trust. Sustain 12(1):1–24. https://doi.org/10.3390/SU12010256
38. LivePerson (2017) How consumers view bots in customer care. https://docsend.com/view/826nkc4. Accessed 2 July 2021
39. Abdul-Kader SA, Woods DJ (2015) Survey on chatbot design techniques in speech conversation systems. Int J Adv Comput Sci Appl 6(7):72–80
40. Copulsky J (2019) Do conversational platforms represent the next big digital marketing opportunity? Appl Mark Anal 4(4):311–316
41. Sotolongo N, Copulsky J (2018) Conversational marketing: creating compelling customer connections. Appl Mark Anal 4(1):6–21
42. Forrest E, Hoanca B (2015) Artificial intelligence: marketing's game changer. In Tsiakis T (ed) trends and innovations in marketing information systems. IGI Global
43. Marinchak CMD, Forrest E, Hoanca B (2018) Artificial intelligence: Redefining marketing management and the customer experience. Int J E-Entrepreneursh Innov 8(2):14–24. https://doi.org/10.4018/IJEEI.2018070102

44. Dunwoodie B (2018) How AI Is impacting the voice of the customer landscape. https://www.cmswire.com/customer-experience/how-ai-is-impacting-the-voice-of-the-customer-landscape/. Accessed 3 July 2021
45. Sentance R (2018) How Samsung uses social listening for product marketing & sentiment analysis. https://econsultancy.com/how-samsung-uses-social-listening-for-product-marketing-sentiment-analysis/. Accessed 4 July 2021
46. Ordenes FV, Zhang S (2019) From words to pixels: text and image mining methods for service research. J Serv Manag 30(5):593–620. https://doi.org/10.1108/JOSM-08-2019-0254
47. Wulf J, Mettler T, Ludwig S, Herhausen D (2019) A computational visual analysis of image design in social media car model communities. The 27th European conference on information systems (ECIS). Stockholm & Uppsala, Sweden
48. Siriwardhana Y, Porambage P, Liyanage M, Ylianttila M (2021) A survey on mobile augmented reality with 5G mobile edge computing: architectures, applications, and technical aspects. IEEE Commun Surv Tutorials 23(2):1160–1192. https://doi.org/10.1109/COMST.2021.3061981
49. Chylinski M, Heller J, Hilken T, Keeling DI, Mahr D, de Ruyter K (2020) Augmented reality marketing: a technology-enabled approach to situated customer experience. Australas Mark J 28(4):374–384. https://doi.org/10.1016/j.ausmj.2020.04.004
50. Hilken T, de Ruyter K, Chylinski M, Mahr D, Keeling DI (2017) Augmenting the eye of the beholder: exploring the strategic potential of augmented reality to enhance online service experiences. J Acad Mark Sci 45(6):884–905. https://doi.org/10.1007/s11747-017-0541-x
51. Dror IE, Harnad S (2008) Offloading cognition onto cognitive technology, 1–23. https://doi.org/10.1075/BCT.16.02DRO
52. Heller J, Chylinski M, de Ruyter K, Mahr D, Keeling DI (2019) Let me imagine that for you: transforming the retail frontline through augmenting customer mental imagery ability. J Retail 95(2):94–114. https://doi.org/10.1016/J.JRETAI.2019.03.005
53. Grand View Research (2016) Augmented reality (AR) market size, industry report 2024. https://www.grandviewresearch.com/press-release/global-augmented-reality-market. Accessed 2 July 2021
54. Porter M, Heppelmann J, Morse G (2017) A Manager's guide to augmented reality. Harv Bus Rev 95(6):46–57
55. Williams R (2018) Ikea Place ranks as No. 2 free ARKit app. https://www.marketingdive.com/news/ikea-place-ranks-as-no-2-free-arkit-app/520761/. Accessed 1 July 2021
56. Ozturkcan S (2020) Service innovation: using augmented reality in the IKEA Place app. J Inf Technol Teach Cases 11(1):8–13. https://doi.org/10.1177/2043886920947110
57. Belanche D, Casaló LV, Flavián C (2019) Artificial Intelligence in FinTech: understanding robo-advisors adoption among customers. Ind Manag Data Syst 119(7):1411–1430. https://doi.org/10.1108/IMDS-08-2018-0368
58. Belk R (2016) Understanding the robot: comments on Goudey and Bonnin. Rech Appl en Mark 31(4):83–90. https://doi.org/10.1177/2051570716658467
59. Poria S, Cambria E, Winterstein G, Bin Huang G (2014) Sentic patterns: Dependency-based rules for concept-level sentiment analysis. Knowl Based Syst 69(1):45–63. https://doi.org/10.1016/J.KNOSYS.2014.05.005
60. Wu CH, Ho GTS, Lam CHY, Ip WH (2015) Franchising decision support system for formulating a center positioning strategy. Ind Manag Data Syst 115(5):853–882. https://doi.org/10.1108/IMDS-10-2014-0291
61. Vlačić B, Corbo L, Costa e Silva S, Dabić M (2021) The evolving role of artificial intelligence in marketing: a review and research agenda. J Bus Res 128:187–203.https://doi.org/10.1016/J.JBUSRES.2021.01.055
62. Valls A, Gibert K, Orellana A, Antón-Clavé S (2018) Using ontology-based clustering to understand the push and pull factors for British tourists visiting a Mediterranean coastal destination. Inf Manag 55(2):145–159. https://doi.org/10.1016/J.IM.2017.05.002
63. Liu X (2020) De-targeting to signal quality. Int J Res Mark 37(2):386–404. https://doi.org/10.1016/J.IJRESMAR.2019.10.003

64. Kumar V, Rajan B, Venkatesan R, Lecinski J (2019) Understanding the role of artificial intelligence in personalized engagement marketing. Calif Manage Rev 61(4):135–155. https://doi.org/10.1177/0008125619859317
65. Miller B (2016) Sprite cherry and sprite cherry zero launching in early 2017. https://www.chewboom.com/2016/10/28/sprite-cherry-and-sprite-cherry-zero-launching-in-early-2017/. Accessed 2 July 2021
66. Adext AI (2019) 9 Applications of artificial intelligence in digital marketing. https://blog.adext.com/applications-artificial-intelligence-ai-digital-marketing/. Accessed 22 June 2021
67. Lee H, Cho CH (2020) Digital advertising: present and future prospects. Int J Advert 39(3):332–341. https://doi.org/10.1080/02650487.2019.1642015
68. Marr B (2017) The amazing ways coca cola uses artificial intelligence and big data to drive success. https://www.forbes.com/sites/bernardmarr/2017/09/18/the-amazing-ways-coca-cola-uses-artificial-intelligence-ai-and-big-data-to-drive-success/?sh=5051244f78d2. Accessed 2 July 2021
69. Sweeney E (2018) IBM's interactive AI ads reach more sites, brands. https://www.marketingdive.com/news/ibms-interactive-ai-ads-reach-more-sites-brands/538558/. Accessed 1 July 2021
70. Caygill D (2017) Six trends brands should know about for 2018 and the tech they need to craft responses. https://www.campaignlive.co.uk/article/six-trends-brands-know-2018-tech-need-craft-responses/1450488. Accessed 2 July 2021
71. NBA (2015) NBA teams up with WSC Sports Technologies to provide next-gen video highlights to fans. https://pr.nba.com/nba-wsc-sports-technologies-partnership/. Accessed 2 July 2021
72. Agrawal A, Gans J, Goldfarb A (2018) Prediction machines: the simple economics of artificial. Harvard Business School Press
73. Agrawal A, Gans J, Goldfarb A (2017) How AI could change amazon: a thought experiment. https://hbr.org/2017/10/how-ai-will-change-strategy-a-thought-experiment. Accessed 2 July 2021
74. Brodie RJ, Hollebeek LD, Jurić B, Ilić A (2011) Customer engagement: Conceptual domain, fundamental propositions, and implications for research. J Serv Res 14(3):252–271. https://doi.org/10.1177/1094670511411703
75. van Doorn J, Lemon KN, Mittal V, Nass S, Pick D, Pirner P, Verhoef PC (2010) Customer engagement behavior: theoretical foundations and research directions. J Serv Res 13(3):253–266. https://doi.org/10.1177/1094670510375599
76. Pansari A, Kumar V (2017) Customer engagement: the construct, antecedents, and consequences. J Acad Mark Sci 45(3):294–311. https://doi.org/10.1007/s11747-016-0485-6
77. Akter S, Wamba SF (2016) Big data analytics in E-commerce: a systematic review and agenda for future research. Electron Mark 26(2):173–194. https://doi.org/10.1007/s12525-016-0219-0
78. Isinkaye FO, Folajimi YO, Ojokoh BA (2015) Recommendation systems: principles, methods and evaluation. Egypt Informatics J 16(3):261–273. https://doi.org/10.1016/J.EIJ.2015.06.005
79. Perez-Vega R, Kaartemo V, Lages CR, Borghei Razavi N, Männistö J (2021) Reshaping the contexts of online customer engagement behavior via artificial intelligence: a conceptual framework. J Bus Res 129:902–910. https://doi.org/10.1016/j.jbusres.2020.11.002
80. Arevalillo JM (2019) A machine learning approach to assess price sensitivity with application to automobile loan segmentation. Appl Soft Comput J 76:390–399. https://doi.org/10.1016/j.asoc.2018.12.012
81. Ke W (2018) Power pricing in the age of AI and analytics. https://www.forbes.com/sites/forbesfinancecouncil/2018/11/02/power-pricing-in-the-age-of-ai-and-analytics/?sh=1752b1c2784a. Accessed 2 July 2021
82. O'Hear S (2017) The price is right! Pace raises £2.5M to automate hotel room pricing based on demand. https://techcrunch.com/2017/12/13/pace-pricing/. Accessed 4 July 2021
83. Dawar N, Bendle N (2019) Marketing in the age of alexa. Harv Bus Rev 96(3):80–86
84. Rese A, Ganster L, Baier D (2020) Chatbots in retailers' customer communication: How to measure their acceptance? J Retail Consum Serv 56.https://doi.org/10.1016/j.jretconser.2020.102176

85. Hilken T, Heller J, Chylinski M, Keeling DI, Mahr D, de Ruyter K (2018) Making omnichannel an augmented reality: the current and future state of the art. J Res Interact Mark 12(4):509–523. https://doi.org/10.1108/JRIM-01-2018-0023
86. Javornik A (2016) Augmented reality: research agenda for studying the impact of its media characteristics on consumer behaviour. J Retail Consum Serv 30:252–261. https://doi.org/10.1016/j.jretconser.2016.02.004
87. Rauschnabel PA, Felix R, Hinsch C (2019) Augmented reality marketing: how mobile AR-apps can improve brands through inspiration. J Retail Consum Serv 49:43–53. https://doi.org/10.1016/j.jretconser.2019.03.004
88. Davis B (2019) AR fails on its only selling point—escaping reality. https://www.marketingweek.com/ar-fails-on-its-only-selling-point-escaping-reality/. Accessed 2 July 2021
89. Haque U (2015) Google glass failed because it just Wasn't cool. https://hbr.org/2015/01/google-glass-failed-because-it-just-wasnt-cool. Accessed 2 July 2021
90. Willersdorf S, Barton C, Dupreelle P (2019) A new (Augmented) reality for brands and retailers. https://www.bcg.com/publications/2019/new-augmented-reality-ar-brands-retailers. Accessed 1 July 2021
91. Nazim SS, Rajeswari M (2019) Creating a brand value and consumer satisfaction in E commerce business using artificial intelligence with the help of Vosag technology. Int J Innov Technol Explor Eng 8(8):1510–1515
92. Salehan M, Kim DJ (2016) Predicting the performance of online consumer reviews: a sentiment mining approach to big data analytics. Decis Support Syst 81:30–40. https://doi.org/10.1016/J.DSS.2015.10.006

Chapter 10
Business Process Reengineering in Public Sector: A Case Study of World Book Fair

M. A. Sikandar, M. Razaulla Khan, and Anita Sikandar

Abstract This chapter examines the World Book Fair (WBF) organised in New Delhi by the National Book Trust (NBT), India, a public sector entity. The World Book Fair was initially set up as a biennial event but was changed into an annual event from 2012 onwards in an announcement by the Cabinet Minister concerned. The case takes through the challenges that the then NBT Director and his team faced in making this change and outlines how they tackled these challenges. The case study aims to take through an example of how to bring about changes in the functioning of the public sector, with its inherent constraints in terms of resources, bureaucratic inertia, lack of coordination and competing interests. This case study broadly sets to examine the significant presence of the public sector across the world and its dominance in some parts of the world.

Keywords Business process reengineering · Business strategy · Change management · Organisational change · International book fair · Public sector

1 Background

The top management of the National Book Trust (NBT), India, an autonomous organisation, under the then Ministry of Human Resource Development (MHRD), Govt. of India was tense after attending the inaugural function of the World Book Fair (WBF)

M. A. Sikandar (✉) · M. R. Khan
School of Commerce & Business Management, Maulana Azad National Urdu University, Hyderabad, India
e-mail: masikandar@manuu.edu.in

M. R. Khan
e-mail: razakhan@manuu.edu.in

A. Sikandar
Shyamlal College, University of Delhi, Delhi, India
e-mail: anitasikandar@shyamlal.du.ac.in

S. Kautish et al. (eds.), *Computational Intelligence for Modern Business Systems*, Disruptive Technologies and Digital Transformations for Society 5.0,
https://doi.org/10.1007/978-981-99-5354-7_10

2012, a biannual edition held on February 25, 2012, at India Trade Promotion Organization (ITPO), New Delhi (India). The reason behind such tension was that the Cabinet Minister, HRD, announced that the WBF would be an annual event from the following year in his inaugural address. There was a mixed reaction among the audience present in the inaugural session of the event. The audience included diplomats, ambassadors, publishers, booksellers, eminent writers, academics, school children, senior government officials, members of literary societies, and officials of ITPO.

1.1 National Book Trust, India

The National Book Trust, India is an apex body established by the Government of India in 1957 to promote books and develop a reading culture in the country. The NBT books are subsidised and priced at affordable rates to reach the masses. It organises Book Fairs/ Exhibitions throughout the country, including the prestigious World Book Fair at New Delhi in India. It receives over US$ 1 million as a grant-in-aid per annum from the government of India for its book promotional activities. NBT is a leading multifaceted institution engaged in promoting, publishing, production, marketing books, book exhibition, providing grant-in-aid, etc. NBT also represents India in several International Book Fairs and regularly participates in prominent international book fairs worldwide. NBT being a government agency, must work as per the Government regulations and control. The government funds 60% of expenses to meet staff salaries, and the remaining 40% is meted out from revenues accrued from the sale of books. The uniqueness of the NBT is that it functions as a government department. At the same time, it is expected to operate as a corporate entity because of its publishing business. At times NBT had to compete with many private English and vernacular publishers in the market. The Government of India provides special funds to NBT for organising the WBF each financial year.

The Director is the executive head of the NBT. He is responsible for the entire functioning of the organisation, and he was to function under the general guidance of an Honorary Chairperson appointed by the government. Both the Chairman and Director were responsible to the Board of Directors, nominated by the government. The director was assisted by Divisional heads of Administration, Finance, Editorial, Production, Exhibition, Art, and Sales & Marketing (refer to the Organization chart of NBT at Exhibit-1 (Fig. 1)). The government appointed the current Director in 2011, overlooking candidates from the bureaucracy. He had spent a decade managing a public university and had firsthand experience of organising the WBF in Feb 2012. He was perplexed by the sudden announcement by the Minister, which means he has to undertake the tedious exercise of organising the WBF every year in addition to the core functioning of publishing and marketing of NBT publications with limited resources.

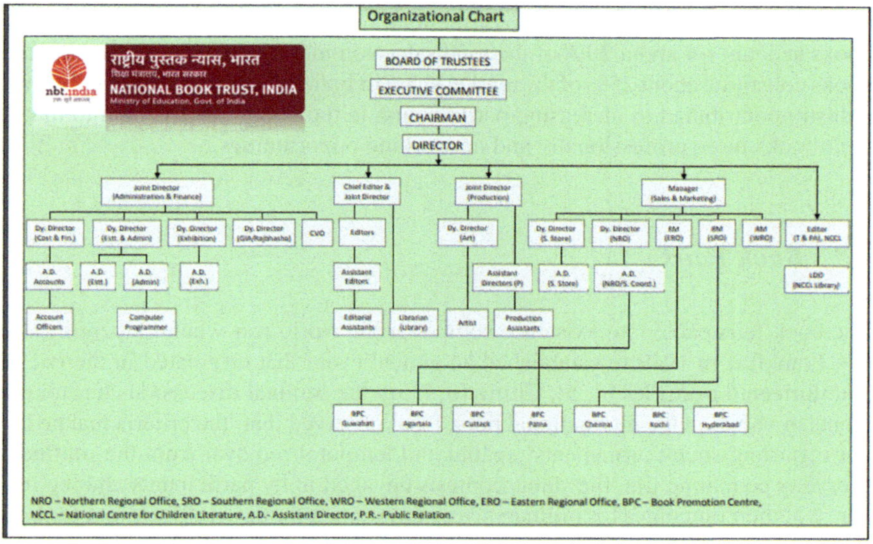

Fig. 1 Organization chart of NBT at Exhibit. *Source* [17] Annual Reports of National Book Trust, India. https://www.nbtindia.gov.in/ (Exhibit-1)

1.2 Publishing Industry

Publishing is one of the creative industries well-acknowledged globally. It is known for creativity and brings together people to share their intellectual capabilities. Publishers have been playing a pivotal role in disseminating the ideas of writers to reach the readers. 'Book publishing is the commercial activity of putting books in the public domain' (Feather 2006). It is also an economic activity for creative people and provides for expanding their horizons. 'Publishers are more than just intermediaries between authors and readers. They commission manuscripts and finance their production, marketing, promotion, and sales. In the process, they confer authority and add value to authors' works' [1]. Publishing companies are 'content-acquiring and risk-taking organisations oriented towards the production of a particular kind of cultural commodity' [2]. Books publishing attracts authors who want to communicate and seek recognition of their ideas among the readers.

The publishing industry in India is one of the fastest-growing industries. Neilson BookScan statistics, India is the third-largest in English language publication ranking, next to the U.S. and U.K. India is also the 6th largest publishing industry in the world. India produces over 21 million books annually [3]. The past decade has witnessed a booming expansion of the book industry in India, facilitated by the growth of publishing houses, booksellers, and distributors. The development of higher education in India was one of the reasons for the growth of the publishing industry in India. According to an estimate, the demand for books in India (book retail) grew 15% per annum. The Indian publishing industry is dominated by academic and children's books, accounting for 40 and 30% of the total market share. The English language books account for about 20% of the total sales volume, whereas the Hindi language books constitute about 25% of the market share in India. The growth in the publishing industry contributed to increasing participation in the book fairs by them to market their books more professionally and networking opportunities.

1.3 Book Fairs

The book fairs came into existence soon after the world war when Leipzig in 1946 and Frankfurt in 1949 re-established an annual event that originated in the twelfth and thirteenth centuries [4, 5], Citing [6, 7] on the seminal discussions on 'tournaments of value' in the international book fairs, observed that 'the criteria that he lays out regarding such tournaments' spatial and temporal removal from the routine of everyday economic life, the status contests engaged in by participants and the relevance of their outcomes for ordinary every day among them, are all relevant for book fairs'. International Book fairs 'bring together all members of the supply and value chains in the publishing industry, and provide a unique occasion for them to interact face-to-face'. Further, such book fairs have provided visible events and structures to publishing, putting together the economic, human, symbolic and intellectual capacity of writers, etc. [7].

The international publishers' Association (IPA) carried out a customer satisfaction survey to assess the facilities available in the venues of the 55 different international book fairs across the world in 2014. Responses were taken from the trade and general visitors to these book fairs. Several factors have been included in the survey form on the facilities, location, layout plan, digital or I.T. services, availability of hotel or leisure services, media promotion, and overall attendance and perception on a five-point Likert scale. The survey indicated that 69% of the participants rated the book fairs either very good or good. The survey also suggested that the publishing business was slowly shifting to online, but attending book fairs was highly valued by publishers worldwide to understand the new trends and network and purchase copyrights [8]. Exhibit 4 (Table 3) shows the total number of copies of books sold and sales revenue, 2018 of the top 10 countries globally. India is one of these ten countries where the demand for books is high [3].

The growing urban middle class and their hunger for books are reasons for the ever-increasing demand for book events like WBF in India. There are three prominent book events in India viz. NBT organises the World Book Fair in New Delhi with the generous fund provided by the government. There are other regional level book Fairs—Kolkata Book Fair, Chennai Book Fair, and many small book fairs organised by either publisher's associations or through private initiatives across India. Some of them receive funding from the NBT as per the government scheme.

1.4 The World Book Fair (WBF)

The World Book Fair (WBF), which was started in 1972, is considered one of the largest International Book Fairs, and the venue for the WBF has been provided by the ITPO, a Government agency since 1974 on a rental (license fee) basis. ITPO did not give any preferential treatment to NBT till 2012 for booking halls, and it was required to deposit around 25% of the license fee in advance for every booking towards the rental of Exhibition Halls. The inaugural event is usually a spectacular one, sometimes presided over by the Head of the State or the Union Education Minister. The WBF spread about 40 to 45 thousand Sqr. Meters over 2000 stalls with nearly 1300 exhibitors, including foreign exhibitors. Exhibit-2 (Table 1) shows the number of participants, stalls/stands, cost of the stalls, area of the WBF from 2010 to 2017. Exhibit-4 (Table 3) shows the steady increase in footfalls to the Fair each year since 2012. Each edition of WBF is usually a nine-day affair, including two Saturdays & two Sundays and organised in February or March.

The WBF has been managed by an in-house team consisting of editorial, publishing, design, marketing, and administrative sections of the NBT. The team was involved in all stages of the event, right from choosing the dates in consultation with publishers' associations, planning the event, preparing the drawings for the bookstalls, booking and organising draw of lots of bookstalls/stands, promoting the event in different parts of the country, inviting tenders for setting up of pavilions, finalising the themes, designing the pavilions, preparation of advertisement materials, organising literary events, bringing out souvenirs, mobilising visitors, and attending Very Important Persons (VIPs). ITPO charged a heavy amount as hall charges from NBT for the event.

Table 1 Total number of copies of books sold and sales revenue, 2018 (Exhibit-2)

Country	Total number of books sold (million)	Distribution	Total sales revenue (in USD million)
Australia	61.2	22.3	44.3
Brazil	44.4	26.6	23.3
India	21.0	19.7	20.1
Ireland	11.8	26.0	37.9
Italy	85.6	34.3	29.4
Mexico*	8.4	16.5	17.1
New Zealand	6.2	21.0	43.6
South Africa	9.2	19.8	36.7
Spain	64	26.3	42.4
United Kingdom	190.9	26.8	33.2

Source [3] Neilson Book Scan, September 2019

1.5 India Trade Promotion Organization (ITPO)

ITPO is a public sector company under the administrative control of the Ministry of Commerce, Govt. of India. It provides services to promote Indian exports by organising trade fairs, buyer–seller meets and giving information on produce and market. As a part of its mandate, it has been maintaining substantial exhibition grounds—a landmark in Delhi, for more than five decades. As per the booking policy of ITPO, the venue is open for booking well in advance, sometimes even before two years and was hosting some famous events like Auto Expo, Defense Expo, Trade Fair, and other Business-to-Business (B2B) events every year. ITPO is also the organiser of the 'Delhi Book Fair (DBF)' on the annual mode in association with the Federation of Indian Publishers' Association (FIP) in the peak summer month (Sept/October) every year. FIP used to receive funding from NBT through a scheme introduced by the government, and the funds were used to subsidise the bookstalls for their members.

2 The Dilemma of Converting the WBF from Biennial to Annual Mode

The immediate challenge before the NBT administration was the lack of financial resources. Due to prevailing sluggish economic conditions, the Ministry of Education officials expressed their inability to provide additional funds for the Fair. The dilemma of the NBT administration was:

(a) whether to continue with the biennial Fair with status quo citing financial constraints to the Minister concerned or
(b) switching over to annual mode by taking a risk with several cost-cutting measures and optimum utilisation of resources.

However, to honor the public announcement made by the Minister, the NBT Board asked its Director to explore various options to organise the Fair within the financial outlay allocated for the activity. The Director was in a fix to take multiple cost-cutting measures to organise the event with half of the financial resources compared to previous editions of the biennial Fairs. The NBT Board also disagreed with the proposal of the Director to engage a few event management specialists and professionals to manage the world book fair event citing a shortage of funds from the government.

2.1 The Predicament of the Director

The Director had left with no option but to look for a managerial team to deliver their best to meet the new challenge of organising the event professionally. The predicament of the Director was:

(a) whether to continue the Fair with the present venue at ITPO by paying high organising costs or
(b) moving the Fair to the Expo grounds at the sub-city considering the budgetary constraints enforced by the government.

2.2 Consultative Process

As a first step, the Director took an open brainstorming session for idea generation. He shared the board's views to support the Minister's decision and sought their valuable input to chalk out a plan to manage the event effectively. The kneejerk reaction of the staff was not very optimistic about the decision. Some felt that organising the World Book Fair was one of the many activities undertaken by NBT, and making the event annual would derail its core publishing activities. However, the Director started convincing them that it was an opportunity for the NBT officials to demonstrate the capability and expertise of NBT in delivering this challenging task. After that, suggestions started pouring into the session.

The Finance team initially suggested shifting the ITPO venue to other suitable locations with low-cost options. But this was immediately opposed by the Marking and Exhibition teams because the present venue was an iconic one and was retained by NBT for a long time since its inception. The ITPO venue also provides easy access to visitors and publishers.

The third suggestion was to discontinue offering the stalls/stands at a subsidised rate or free of cost to a section of foreign participants. (NBT was offering stalls free of charge to some participants from Asian countries under regional cooperation) Such an arrangement could earn higher revenue to offset the reduction in grants from the government. Consequently, the NBT had to increase the stall rates substantially up to 100%, making them very costly for the publishers. There was apprehension in the FIP may thwart such moves by NBT as they were not very comfortable about the announcement by the Minister of conversion of the biennial Fair into annual mode. After discussions, there was a consensus to increase the stall/stand prices proportionate to the increase by ITPO, i.e. 10% every year.

Director held informal consultations with the President and Secretary of FIP to create consensus on the conversion of WBF as an annual event from 2013. But there was an indication from the FIP side that some multinational publishing houses may not be ready to participate because of budgeting issues at their parent company.

2.3 *Identification of Factors for Evaluation of Venue Options*

Based on the first round of discussions the NBT Exhibition and Administration teams had come up with some inputs regarding (i) the factors to be considered for selecting the exhibition venue and (ii) the Identification of stakeholders' needs and expectations.

Similarly, some of the significant factors identified for evaluating the two available venues were (a) rental cost; (b) quality of basic exhibition infrastructure, (c) distance from the venue, (e) conferencing or meeting facility; (f) I.T. infrastructure; (g) Food courts; (h) Hotel/stay arrangement inside or outside the venue; (i) entry fee; (j) business hours etc. The two of the available exhibition venues were evaluated based on factors (a) to (j) by the NBT Exhibition team (refer to Exhibit-6 (Table 5)).

The NBT team identified six stakeholders viz. (a) NBT/Government (b) Indian publishers; (c) Foreign publishers; (d) Authors/writers; (e) Domestic visitors; and (f) International business visitors. Stakeholder's needs and expectations were also identified carefully after intensive discussions with the stakeholders (refer to Exhibit-7 (Table 6)).

2.4 *Idea Generation Stage*

The Director wanted to have focused discussions with the above data to arrive at a decision. He again called a second brainstorming meeting with his officers, and the details were shared with the other members for their reaction. He explained to his officers the venue constraint and his predicament about inadequate resources to organise the annual Fair at the ITPO venue. There was a pressing need to cut down the cost to the tune of 40–50%, which seems impossible for him unless NBT takes some drastic steps to control the cost factors. Discussions indicated that a cost reduction of 10 to 15% was possible but not 40–50% as about 70% of the total budget for WBF was going to ITPO as rent. The Director told his team that he was considering himself a newcomer to NBT, and he hardly spent seven months at NBT. Therefore, he requested his team members to apply their wisdom and develop solutions to overcome the financial constraints of organising two book fairs at the cost of one. But nothing concrete was forthcoming from them.

Then, the Director casually put forth a question 'how will it be if NBT joins hands with ITPO to organise the Fair jointly through an MoU?' The question was sudden, and none of them was able to digest it immediately. As the discussions advanced by the Director, a good number of them opined that it might not be feasible as the ITPO itself was the organiser of another book fair, 'Delhi Book Fair', in collaboration with the Federation of Indian Publishers (FIP) from the last two decades. Some of them recalled the past unpleasant relationships between the ITPO and successive Directors of NBT during the Fair. There was a consensus that NBT may not gain from the proposed tie-up with ITPO, and NBT may lose ground gradually.

However, the proposal was somewhat supported by a few from the Administration and Exhibition division. According to them, a proper negotiation with the ITPO and an MoU may help the NBT save considerable money long-term. However, they were sceptical about the proposal because of the fact ITPO being a Public Sector Enterprise. However, the Director firmly believed that collaboration with the ITPO was the key to coming out of this predicament. If he succeeded in his efforts, it would be a game-changer for NBT and may create a 'win–win' situation for both the organisation. However, some of the venue issues shall remain the same.

The Director has given serious thought to the idea and started meeting some government nominees on the NBT board and seeking their endorsement and active support. The Director also called the Chairman (Honorary), a former banker, and informed him about the outcome of the lengthy deliberations and his subsequent meetings with some board members. The Chairman expressed his complete support of the idea and readily agreed to accompany the Director to hold a meeting with the CMD of ITPO at the earliest opportunity.

3 Collaborative Exercise with ITPO by NBT Administration

NBT had to adapt to the new situation compulsorily after the change announcement made by the Minister concerned in 2012. The initial dilemma of the Director was to find a new low-cost exhibition venue or to continue with the existing one at high cost. After two brainstorming meetings with his NBT team, he decided to retain the current ITPO venue for the next annual book fair. However, he had to find a way to cut down the cost drastically to the extent of 45–50%. One of the ways to reduce the price further was to negotiate a deal with the ITPO as co-organiser of the event. The Director and Chairman, NBT, decided to meet the new CMD to explore the possibilities of having the ITPO on Board as co-partner. At that time, ITPO was slowly losing prominent exhibition events to the new expo ground at the sub-city because of high rental rates and inadequate services at the venue.

The CMD also saw a long-term business opportunity of retaining NBT India, an old client. She also noted the remarkable improvements in the just concluded 2012 edition of the WBF on the theme of '100 years of Indian Cinema'. The Fair also saw an increased footfall and brought some popular cine-artists. The CMD has shown interest in the proposal of NBT offering the ITPO as co-organiser of the event. After initial consultations with her team, she decided to take up the matter before the ITPO Board for a final decision.

According to the proposal, the NBT shall be the event's main organiser with total funding. NBT shall provide hall charges as per actual utilisation. The ITPO, being the co-organiser, has to offer specific facilities other than Exhibition space for the Book Fair free of charge. In addition, the ITPO has to deploy some of its professionals and managerial staff to oversee the peripheral activities of the event. Both NBT and

ITPO sides agreed to scale up the Business-to-Business (B2B) events to promote the book trade and make joint efforts to bring more international participants. The ITPO has been organising an International Trade Fair at the same venue for more than five decades, attracting sizeable foreign participants.

The CMD, ITPO took full two months to get a green signal for the tie-up from the board. But some of the senior executives from ITPO were sceptical about the deal as they thought it was a revenue loss for the ITPO. NBT and ITPO entered into an MoU for five years for organising the world book fair jointly. NBT saved about 30–35% of expenses on hiring charges of conference, meetings, auditoriums for the programmes and some fees towards availing professional services to set up the stalls, pavilions, and security in the first annual edition of the WBF in 2013. The same ITPO executives who used to create hindrances to NBT officials during the World Book Fair have started acting as facilitators.

4 Business Process Re-engineering Interventions in NBT

After entering into a partnership with ITPO, the Director felt the need for gearing up the organisation and decided to implement the Business Process Re-engineering (BPR) exercise to scale up the business activities to meet the emerging challenges of the organisation. NBT had to take several measures—like cutting costs, optimum utilisation of human resources, sharpening their skills, improved services, image makeover, active engagement with the stakeholders and scaling up other literary activities to attract more visitors, particularly young ones, to the Fair.

The Director during his tenure observed there were several gaps- (a) Highly bureaucratic work culture and rule-bound; (b) resistance to change; (c) centralisation and tight control even for routine decisions; (d) people work in silos; (e) limited communication restricted to downwards; (f) no role clarity as the work was not reviewed for several years; (g) less customer focus; (h) no digital intervention; and (i) Not using the social media and digital platforms to promote the WBF.

4.1 Lack of Coordination Between Government Agencies

The Board of the NBT had two eminent authors, publishers, academicians, President of the National Academy of Letters, Senior Bureaucrats from three Ministries of the Government—Ministry of Education, Information & Broadcasting and Finance. However, somewhat, the successive administrators missed the opportunity of having a continuous dialogue with them to achieve the aims and objectives of NBT India. Two government stakeholders, namely the National Academy of Letters and Publication Division of the Information and Broadcasting Ministry, were also publishing books parallel to NBT India since the 1950s. These stakeholders' role was critical in mobilising resources viz. authors and literary events for the successful organisation of the world book fair. Nevertheless, the new team of NBT India started engaging with them continuously and was able to get sponsorship of authors and writers' programmes which added value to the event. It was a win–win situation for both NBT and these Government agencies.

4.2 Nature of Interventions

The NBT gradually introduced the following intervention measures related to process improvement in the organisation:

4.2.1 Developing a Vision

The Director has aligned the mission and objectives of the organisation with the actual activities and created an eco-system where the top management and the employees share common goals and objectives. Employees felt self-motivated and proud to be a part of the organisation. These were achieved through continuous communication at all levels and articulation in the meetings by the Administration. Employees started realising their natural strengths and potentials.

4.2.2 Understanding the Existing Process of WBF

The Director evaluated each division/function of the organisation by involving internal and external stakeholders.

4.2.3 Identification of Process for the Redesign of WBF

Through continuous engagement, the Director opened up feedback channels with internal and external stakeholders, Government officials, authors, publishers, intellectuals, media, civil society, and foreign participants. These steps provided the much-needed hindsight and understood the expectations from the stakeholders. In the process, many of the employees opened up with the management team.

4.2.4 Identification of Change Levers

The exhibition division of NBT engaged itself in organising book fairs at the national and international levels. They have to be innovative as well as showcase the Indian ethos outside the world. The process helped the organisation identify the competencies needed in the services and strategies to synergise within the divisions, external players, and government departments.

4.2.5 Implement the New Processes

Cost-effective measures adopted and implemented through Alliances with stakeholders and Government Agencies. Joining hands with ITPO resulted in about 30% cost reduction through various free services at the venue—implementing new activities in the Fair through sponsorships. To give a truly international flavour, France was the first one to accept the request by NBT. Guest of honour presentations made the event genuinely global and encouraged trade visitors from outside India. Authors' corners, Meet-the-Author programmes, literary seminars, cultural and musical shows by the publishers/participants. Mobilisation of School children and College/University students to the Fair. Student volunteers in place of the Fair hostess to guide the visitors and escorting the prominent persons. Collaborative arrangements with the Federation of Indian Chambers of Commerce & Industry (FICCI) to create a forum of Indian and International publishers to exchange ideas to promote business. Tie up with National Broadcasting agencies (T.V. and Radio) to run free publicity programmes to promote the Fair. Use of social media platforms and creating a separate website for the WBF to disseminate the information of the Fair. Celebrity endorsement of the Fair to attract young visitors. In-house Fair News Letter to cover the events and receive feedback from the visitors.

4.2.6 Make New Process Operational

Main stakeholders of the event were brought on board to get the desired results—continuous engagement of a small team of officers to ensure implementation of new processes and practices/policies. The Director continuously monitored and ensured the operationalisation of new processes to yield better results. Because of the active involvement of officials of NBT, the latest systems and processes were put in place and made operational. The joint team of NBT and ITPO in the new experiment was a big success. Both organisations' strengths and expertise were handy in putting a good show of force in each successive edition of WBF. Saving precious government resources had sent a good message among book lovers, readers, intellectuals, government officials, and envoys. The networking with the various stakeholders and the government agencies yielded good results and helped to reduce the advertisement budget of subsequent editions of WBF.

4.2.7 Evaluating the New Processes

NBT received good feedback from the stakeholders at the end of each successive book fairs.

4.3 Leveraging the BPR Interventions

Sustained efforts by the NBT team ensured the new initiatives and changes brought in the annual mode of the Fair were successful. After the initiation of BPR interventions, several changes were visible in the Organization—(a) Employees have more goal-oriented; (b) Board Meetings were used for brainstorming and exchange of ideas by the Members and active contribution by the members; (c) working in teams, exchange of ideas both internal and external, sharing of opinions and data across the divisions; (d) Employees become receptive to new idea generation, innovation and creativeness at work; (e) clarity in roles and responsibilities at each level through written policies; (f) delegation of work among managers and supervisors; (g) Customer and stakeholder orientation; (h) Digital intervention through online services to the publishers and visitors; (i) Dynamism and professional outlook with objectivity and; (j) extensive social media coverage for Fairs and other book promotional activities.

Over a period, notable good organisational practices introduced by NBT were—Multi-dimensional communication—Free flow of Peer to peer and inter-departmental communications at the workplace—Timely appraisal and upward career mobility within the norms; Participative management practices brought more involvement of employees in the decision making; Open door policy by the Administration. National Institute of Design (NID) Ahmedabad (India) was engaged in redesigning a new logo project for the WBF to give a contemporary look to NBT. A tie-up was made with the 'Delhi Metro Rail' to sell the entry passes through select Metro stations to avoid crowds at ticket counters managed by ITPO. The Hostess arrangement with ITPO was replaced with Student Volunteers drawn from prominent city colleges. They were controlling the information desk, escorting the VIPs and authors. 'Show-Daily' was introduced during the book fair, which covered the events happening across various halls and pavilions.

Exhibit-3 (Table 2) shows the Stakeholder participation in the World Book Fair from 2010 to 2017. The numbers were stabilised after 2013. Similarly, Exhibit-4 (Table 3) shows multiple events organised in the World Book Fair from 2010 to 2017. There was a steep increase in the number of activities at the Fair from 2013 onwards.

The NBT Director and the Head of the Exhibition team had an opportunity to visit some prominent international book fairs to understand the business models and processes. The learnings from such trade visits helped him and his team bring vast improvements in the WBF quickly.

The revenue generated, financial support received from the government, cost incurred for the organisation of the successive World Book Fairs from 2010 to 2017 have been tabulated at Exhibit-5 (Table 4) and Graph 1. The NBT, over a period, was able to maintain its viability through various cost-cutting measures and process improvement. From 2013 onwards, the revenue generation started improving, and at the same time, the dependability on government assistance for WBF is reduced. NBT started showing judiciousness in space utilisation in the Fair. The government

Table 2 Stakeholder participation in the World Book Fair from 2010 to 2017 (Exhibit-3)

Year of event	No. of publishers participated	No. of foreign publishers participated	Area of the venue in SQM	No. of stalls and stands (approx.)	Charges per stall (36 SQM) for english language publishers	Charges per stall (36 SQM) for indian language publishers (50% less)	Costs for booth (9 SQM)
2010	1234	10	41,069	2450	INR 34,000/-	INR 17,000/-	INR 8500/-
2012	1297	17	45,913	2600	INR 34,000/-	INR 17,000/-	INR 8,500/-
2013	1092	25	46,258	2250	INR 45,000/-	INR 22,500/-	INR 11,250/-
2014	1023	29	37,724	1950	INR 49,500/-	INR 24,750/-	INR 12,500/-
2015	1047	34	36,549	1875	INR 54,500/-	INR 27,250/-	INR 13,750/-
2016	886#	32	36,548	1850	INR 60,000/-	INR 30,000/-	INR 15,000/-
2017	950#	25	36,548	1850	INR 66,000/-	INR 33,000/-	INR 16,500/-

Note
1. Bare space: 10% reduction in the charges)
2. Stand to Booth ratio: 8: 2
3. Rental costs for nine days of the Book Fair

increasing trends were seen in booking more bare space to construct a pavilion for branding by English and Indian language publishers

Source [17] Annual Reports of National Book Trust, India. https://www.nbtindia.gov.in/

Table 3 Details of various events organised in the World Book Fair from 2010 to 2017 (Exhibit-4)

Year of the event	No. of B2B events organised	Number of literary/ book-related events organised	No. of popular music and cultural shows sponsored by other Govt. agencies and local government	Number of trade visitors (approx.)	Number of general visitors (approx.)
2010	–	15–20	–	200	200,000
2012	5	25–30	15–20	500–600	350,000
2013	12	150–175	20–25	1050	450,000
2014	19	240–260	25–30	1080	650,000
2015	21	250–275	30–35	1150	700,000
2016	20	240–250	20–25	1100	750,000
2017	18	350–360	15–17	850	800,000

Source [17] Annual Reports of National Book Trust, India. https://www.nbtindia.gov.in/

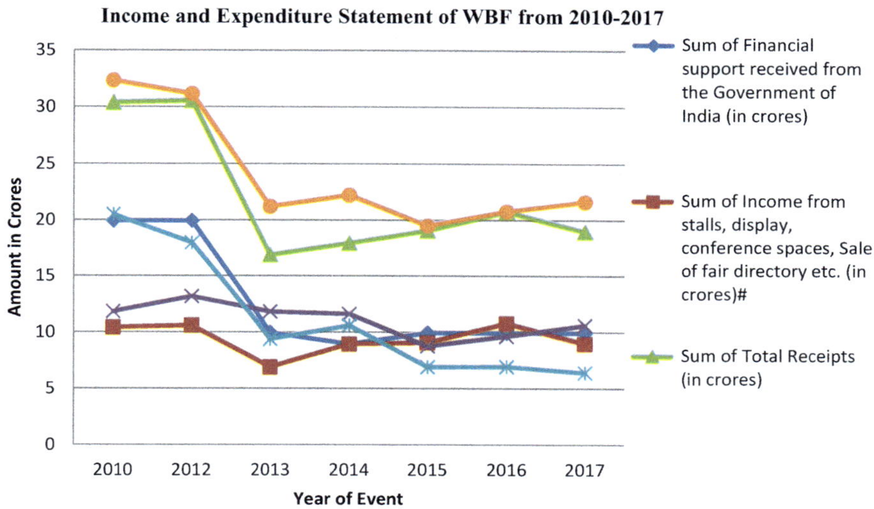

Graph 1 Income and expenditure statement of WBF from 2010–2017

of India, since 2013, has reduced the financial support from INR 20 crore to INR 10 crore.

The management of NBT did not face many challenges to bring changes in specific areas–Employee development through training; Employee Engagement and Empowerment in the decision making process; Clarity in roles and responsibilities prescribing standard operating procedure (SoPs) and through delegation; Digital intervention and adapt to the new work environment; Improving customer relationship; Synergy between various divisions and breaking the silos and; Free flow of Vertical and horizontal communication in the Organization. The Director also introduced a matrix-type reporting structure among the WBF activity teams to take out the best in both efficiency and responsiveness.

However, there were specific issues that took a lot of effort to implement in NBT. Some of them related to bringing cultural changes, viz. convincing employees for their new roles; promoting collaborative work culture; negating with the ITPO team to clinch the deal; convincing FIP to get on board; bringing agility in the decision making process, etc., were found more challenging for the NBT management to adopt and implement successfully.

The initiative helped NBT undertake an aggressive advertisement campaign through print and digital media to promote the World Book Fair. As a part of good practice and to engage foreign visitors/participants in the WBF, new features such as 'CEO-Speak' in collaboration with trade bodies like FICCI and Rights Table to sell/acquire copyrights of books between the Indian and foreign publishers were introduced.

Table 4 Income and expenditure statement of world book fair editions from 2010 to 2017 (Exhibit-5)

Year of event	Financial support received from the Government of India	Income from stalls, display, conference spaces, sale of fair directory etc (in crores)#	Total receipts	Payment to ITPO towards rent in the form of the license fee (after adjusting the advances)	Additional expenditure incurred for the construction of stalls and promotional activities of the event	Total expenditure incurred for the event	Shortfall or surplus met out from the other income of NBT India	Area hired for the event (including walking space inside the pavilions)	Number of stalls and stands
	INR in crore (1 crore = 10 million)							In SQM	
2010	20.00	10.45	**30.45**	11.85	20.55	32.40	−1.95	41,069	2450
2012	20.00	10.63	**30.63**	13.20	18.02	31.22	+0.59	45,913	2600
2013	10.00^	06.94	**16.94**	11.84	9.43	21.27*	−4.33	46,258	2250
2014	09.00	09.00	**18.00**	11.65	10.64	22.29*	−4.29	37,724	1950
2015	10.00	09.13	**19.12**	8.82	6.97	19.57	+0.45	36,549	1875
2016	10.00	10.83	**20.83**	9.70	7.00	20.86	−0.03	36,548	1850
2017	10.00	09.05#	**19.05**	10.67	6.47	21.72	−2.67	36,548	1850

Source [17] Annual Reports of National Book Trust, India. https://www.nbtindia.gov.in/

Note

^From 2013 onwards, the event became annual, and the Financial support from the Government of India was reduced to 50% after that

*Advertisement cost was high in the years 2013 & 2014 to marketing the annual event

#Participation of Indian language publishers was very high in the year 2017, which reduced the income

Due to redevelopment work started at the ITPO Exhibition venue, the book fair was scaled down and restricted to only a few venues. Hence the data from 2018 onwards have not been taken for comparison purposes

5 Change in the Perception

The BPR intervention introduced by the NBT management helped reduce the costs of organising a mega event like WBF. NBT has become increasingly competitive by embracing newer technologies at the workplace. NBT reinvented itself, which helped it to be more relevant to the business environment. It emerged as a winner and perhaps the only Government Agency that directly engages itself in event management at this scale.

The public image of the organisation improved vastly. Because of continuous book promotional activities in the WBF, its popularity among the masses also increased. A substantial increase in the sale of NBT publications due to the annual mode of the Fair. NBT was not over-dependent on Government funds which became a performing organisation. Publishing houses benefited out of B2B events, increased business visitors, general visitors. Similarly, authors were also helped through improvements. In the same way, visitors had an opportunity to interact with eminent authors and writers and more book-buying prospects.

The overall reputation of the WBF has slowly improved among the foreign participants as the brochure was promoted in different international book fairs participated by NBT right from Frankfurt Book Fair to Beijing Book Fair. Several friendly countries, one by one, were starting to show their keenness to become Guest of Honour country in WBF. The liaison between foreign embassies in India also improved, and the external affairs ministry has recognised NBT as an institution that promotes soft diplomacy through book fairs. In 2015, to the surprise of the Ministry of External Affairs, Govt. of India, a proposal was received from the Chinese foreign affairs ministry just before the state visit of the Chinese President.

Despite all improvements and initiatives made by NBT and ITPO in organising the WBF, there were a few venue facilities-related issues such as no parking space for visitors and traffic jams around the ITPO bookfair venue. All the limited Car parking space was either used by NBT officials or publishers. The visitors had to park their cars at a one-mile distance and avail of shuttle buses to reach the venue. Visitors also complained about the high prices of food items made available through the Food Courts inside the venue. The vendors were justifying the high prices due to high rental charges by the ITPO. Now the ITPO is coming up with a state-of-the-art exhibition-cum-convention Centre.

Yet another problem faced by the Business visitors was some of the prestigious B2B events like CEO Speak etc., were organised by NBT outside the venue because of the non-availability of quality infrastructure at the venue.

In April 2017, the ITPO floated a redevelopment plan with a world-class iconic state-of-the-art integrated Exhibition and Convention Centre (IECC) with modern amenities to address these issues. The new venue which has been made operatioal now by the Prime Minister during the recent G20 event and the IECC will have a five-star hotel in the fure. It has a state of art conference and exhibition facilties now with underground tunnels to decongestion traffic with a basement parking facility for 4800 vehicles [9].

6 Teaching Note

1. Case Summary

This case describes various Business Process reengineering (BPR) and change management interventions undertaken by National Book Trust, India. The case study presented the effective mobilisation of resources to bring synergy in organising the World Book Fair (WBF), one of Asia's biggest International Book fairs in the Indian context. Thousands of people visit the Fair every year.

WBF is the largest International Book Fair that started in 1972. The Fair is being organised by the National Book Trust (NBT), India, an apex body established by the Government of India, Ministry of Education, Department of Higher Education. The government substantially funds it through grant-in-aid. The peculiarity of the NBT was it functions as a Government department. At the same time, NBT engaged itself in the publishing business and has to compete with many private publishers in the market. The NBT staff headed by Director managed the WBF. The tasks involved event planning, drawings for the bookstalls, booking and organising drawing of lots of Book Stalls/Stands, promoting the event in different parts of the country, inviting tenders for setting up of pavilions, finalising the themes, designing the pavilions, preparation of advertisement materials, organising literary events, bringing out souvenirs, mobilising visitors and attending VIPs/VVIPs.

The Government of India provided special funds to NBT for organising the WBF every alternative year to the tune of INR 20 crore. A chunk of the funds was going to ITPO as Hall charges for the event. NBT used to charge staff at lower rates than English language publishers and offered a 50% further subsidy to Indian language publishers, including Hindi. There was a shortfall of funds every year from the charged amount vis-à-vis the publishers' actual amount charged by NBT. On the one hand, there was a steady increase of 10%, or so in the hall charges and on the other hand, NBT was unable to increase the hall charges due to pressure and affordability issues from the publishers' and booksellers' lobby. One of the other challenges faced by the NBT was that the same ITPO was the organiser of another event, 'Delhi Book Fair' charged less for the stalls and stands than the WBF.

The WBF became an annual event in 2013 with a 50% cut in the budget. In one budget, two yearly fairs had to be managed by the NBT, which was another challenge before the Director and his team.

In this case, the Director introduced several BPR and change management processes to energise the organisation and sustain the changes. It also narrates how the organisation used all the resources of the organisations to bring improvement and changes in the delivery of services through WBF.

2. Learning Objectives

The case is intended to help the participants, students of Business Administration, recognise the challenges of managing a public sector organisation and various change management processes adopted and implemented successfully to re-energise it as one of the spectacular events amidst all adversities by government agencies. The case study can develop a deeper understanding of the dilemma and challenges experienced by a Government organisation and help the participants/students understand the situational issues and decision-making process in uncertainties and adversities. The case can help the participants understand the nuances of organisational velocity to improve its speed, efficiency, and effectiveness to organise a prestigious international event.

Course(s) where this case can be taken for discussions include:

(a) Business Reengineering Process
(b) Change Management
(c) Organizational Change & Development
(d) Business Strategy
(e) Business Environment

3. Suggested Assignment Questions

Q1. What were the initial challenges encountered by the top management of NBT in organising the World Book Fair?

Q2. What were the risk factors involved in the WBF event of NBT continuing with its WBF as an annual event without partnering with ITPO?

Q3. What was the predicament of the Director of NBT in the selection of venue?

Q4. What could have been the advantages in choosing the new Expo venue at Noida sub-city to organise the event?

Q5. How the NBT benefited from the partnership with ITPO?

Q6. What were the process improvement intervention measures introduced by the management of NBT?

4. Analysis of the Six Case Assignment Questions

Q1. What were the initial challenges encountered by the top management of NBT in organising the World Book Fair?

The Director was new, and he, with great difficulties, was able to put things together and organised the WBF 2012 edition. When the Minister, in his inaugural address, justified holding the WBF on an annual basis, it added a burden on his shoulders. The Director knew his staff would not be willing to do additional work for holding such an event. He knew that most of the human resources available at his disposal in NBT were pressed into WBF work for the last few months, and the core publishing and other books promotional work, including organising book fairs in other parts of the country, may suffer. He also knew that the government would provide no additional posts. The immediate cause of worry was additional funding from the government to organise the event. The other challenges faced by NBT were the NBT being an old institution established in the 1950s, was working in the old ways in silos and organising a public event at this scale requires multitasking and teamwork, which were lacking at that point.

Further, there was no motivation for shouldering the additional burden by the staff. Further, there were not many synergies between different stakeholders, especially with the publishing industry, the exhibitors for WBF. In a nutshell, there was a lack of financial resources and bureaucratic inertia to go ahead with organising the World Book Fair in annual mode from 2013. A section of the employees genuinely believed that their core function—publication work in the organisation (NBT) would suffer if a decision is taken by NBT management to organise WBF on an annual mode.

Q2. What were the risk factors involved in the WBF event of NBT continuing with its WBF as an annual event without partnering with ITPO?

The NBT may have lost face before the Cabinet minister announced that the WBF should be organised on annual mode from 2013 through a public announcement during the inauguration of the last biennial WBF in 2012. The reputation of the top management may have taken a beating for non-performance or cooperation with the government. The popularity of the NBT and its WBF may have gone down because of similar book fair activities undertaken by the ITPO itself on the annual mode in the same venue in collaboration with the Federation of Indian Publishers (FIP). Finally, the NBT may not have gone through a needed Business Process Reengineering process and may not have seen substantial improvement in its systems and processes. The general public and authors/writers may have been deprived of an annual event at this scale.

Q3. What was the predicament of the Director of NBT in the selection of venue?

The Director had left with no option but to look for a managerial team to deliver their best to meet the new challenge of organising the event professionally. The predicament of the Director was:

(a) Whether to continue the Fair with the present venue at ITPO by paying high organising costs or
(b) Moving the Fair to the Expo grounds at the sub-city considering the budgetary constraints enforced by the government.

He engaged himself in a series of internal and external meetings to arrive at the most suitable decision on the venue. His team also analysed various factors connected with the two exhibition venues (advantages and disadvantages) and stakeholders' needs and expectations (refer to exhibit 6 & 7 (Tables 5 and 6)). These inputs helped him arrive at a sound decision about the venue and lead to a collaborative arrangement with ITPO and other Government Agencies to reduce the cost of the WBF.

Q4. What could have been the advantages in choosing the new Expo venue at Noida sub-city to organise the event?

One of the prime advantages of choosing the new Expo venue at Noida sub-city, as perceived by the Finance team of NBT, was low cost, and the NBT could have saved a substantial amount in the payment of hall charges to ITPO. The quality of infrastructure was far more superior to the one at ITPO. B2B events could have been arranged within the Expo grounds rather than going outside. Another advantage was

Table 5 Major factors used for the evaluation of the Exhibition venue for WBF in 2013 by NBT (Exhibit-6)

S no	Factors/ parameters of the venues	Venue 1	Venue 2
1.	Rental cost for the venue	High with 10% incremental every year	About 50% less than the venue-1
2.	Quality of basic infrastructure	Basic standard	Excellent with international expo standards
3.	Conferencing and meeting halls	Available at a moderate level with additional costs	Excellent. Cost included on the rental
4.	I.T. infrastructure	To be hired by the organiser at own cost	Plug-and-play facility with international standards
5.	Food courts	Excellent	Kiosks are available
6.	Hotel/stay arrangement in the Expo venue	Not available	Available in a moderate scale
7.	Entry fee	Negligible	High
8.	Business hours	Till 8 pm	Till 7 pm
9.	Distance from the main city	0 km (centrally located)	Located at Noida Sub-city (30 km away from the Venue-1)
10.	Connectivity (by metro train, bus services)	• Metro station directly opens at the venue • Bus connectivity is excellent • 15 km from the International Airport • Connectivity available through metro train	• No metro connectivity • Need to depend on the costly transportation means and occasional inter-city bus services • The organiser has to arrange special bus services to ferry the visitors from the main city to the expo venue • 50 km from the Airport and need to hire a taxi to reach the venue
11.	Visitor's car parking facility	Limited car parking facility	Ample visitors' parking facility

ample parking space for the visitors and not having to face the traffic jams as in the ITPO venue. There was always a possibility of improvement in the connectivity between the main city hub with such sub-cities.

Q5. How the NBT benefited from the partnership with ITPO?

The partnership with ITPO brought NBT several advantages and benefits, both monetary and non-monetary. NBT had to hire several services and non-exhibition spaces at the venue to organise several activities in the background of the WBF. Under the MoU, most of such benefits were made available by ITPO free of cost as co-organiser.

Table 6 Stakeholder Interests involved in the world book fair (Exhibit-7)

S. no	Stakeholder	Needs	Expectations
01	NBT and government	• To promote a reading culture among the citizen of the country • To make available quality books in all Indian languages at an affordable price • To promote the sale of NBT publications through its exclusive pavilions	• To popularise books through participation by the publishers and booksellers in large numbers in the world book fair • To organise book promotional events- literary to promote the books • To organise and support B2B meets for promotion of the sale of publishing trade • Facilitate networking opportunities for Indian and Foreign publishers • To support budding authors • To provide an opportunity to the visitors towards foreign authors and publishers • To attract young readers, students, school children and visitors to the Fair to achieve the objectives of the NBT India • To improve the public image of the organisation • To provide an opportunity for young authors to interact with the readers
02	Publishers (Indian)	• Quality of exhibition space with a good layout design for the display of publications • To get an opportunity to meet the business visitors, librarians, prospective authors • to achieve a decent volume of retail sales and more profit margins	• Hassle-free bookings of exhibition space online to avoid the last-minute draw of the lot as was done earlier by NBT • Subsidy in the stall rates • Author Corner space for holding book release functions, Meet-the-Author programme as promotional events for their publications free of cost • publicity through print and Digital media • Facilitating business visitors to the venue • Lounge facility for meetings • Ample parking space • Extended fair timings • Good wifi connectivity, ATMs • Getting a facilitation role by the organisers in B2B meetings • An increased footfall of visitors

(continued)

Table 6 (continued)

S. no	Stakeholder	Needs	Expectations
03	Publishers (foreign)	• Quality of exhibition space with a good layout design to put up their stands at international standards • Arranging meetings with the Indian publishing trade for copyright exchanges of publishing contents • Facilitation of visitors to their stands • Inviting Diplomats from their home country to the inaugural events	• Date of the next edition of the Fair to be announced in advance. The same would help the publishers in planning and making their travel arrangements • Visa assistance services by the organisers • Facilitation of Diplomats and Embassy staff to the venue • Encouragement of participation of the Indian publishers in international fairs of their home country (e.g. Frankfurt, Germany, Seoul, South Korea, London, U. K, etc.) • Providing more opportunities for participation and activities in B2B events • Networking dinners to meet the Indian publishing trade • Media interaction to promote their events in the Fair • Social media promotion • Decent amenities at the Fair site (Hotel, Bar, Restaurant, good connectivity to Airport, travel desk to visit some places of interest)
04	Authors	• Get a fair opportunity to meet the readers to understand what they want to read • To interact with the publishers for their future publications	• A good number of diverse audience at their programmes organised at the venue during the book fair (Authors'/ Reading corners, literary activities) • Easy access to the venue • Free passes
05	Visitors (domestic)	• To buy text and general books at an affordable price • A large number of Indian and foreign books • To meet the famous authors at the authors' corners • To take part	• More discounts on books • E-contents • To get a feel of books • To watch the guest of Honour presentations by foreign countries in the fair • Hang out with other friends • Cultural shows by prominent performers
06	Visitors (international)	• Better display of their stands/stalls • Conferencing and meeting facility at the venue • Easy access from the Airport, Hotels and Diplomatic Mission	• More B2B visitors for exchanging buying copyrights of books • Showcasing the rich contents of books from their home country and introduction to the Indian visitors and readers

Some of the professional services from ITPO were also made available to NBT. It brought a much-needed unity of efforts to make the event a success. ITPO has also taken care of protocol and security management services at the venue through its experienced pool of officers. Such help by ITPO allowed the NBT official to concentrate on the professional work of holding the book fair. There was teamwork between NBT and ITPO in making the event more vibrant and thriving. NBT need not divert all its staffing for organising the WBF.

Q6. What were the process improvement intervention measures introduced by the management of NBT?

The NBT management, mainly the Director, carefully studied the situation and evaluated various options and action plans. He had foreseen the possible hindrance within and outside the organisation. He used Business Process Reengineering (BPR) and change management principles to overcome the challenges to realise the vision of the Minister and the NBT in this case. Firstly, he developed the business vision and evaluated existing processes in the organisation. He identified the weak areas where he had to intervene for redesigning the processes. The Director also clearly identified the change levers, such as repositioning the WBF with new features and good practices followed in other International Book Fairs. His visits to some international book fairs to understand the business models and processes helped him bring vast improvements quickly. He involved all the stakeholders at the decision-making level and got regular feedback from them.

The Director also used a set of change management principles, including forming an informal group coalition, recognising the resistance to change by his own staff, and convincing his team of officers about the need for change by articulating in different meetings/interactions. He slowly brought various H.R. interventions in the BPR interventions in employee training and development programmes, holding stakeholders' meetings before and after the WBF. Such feedback sessions helped the employees to understand the issues correctly and reduced the gaps in the decision making. It brought openness and transparency in the day-to-day business operations and shed away the shyness or reservations being a government official among the employees of NBT.

Continuous engagement with the staff and other stakeholders brought visible improvement and helped the organisation improve its public image. Employees became motivated and committed to their work. The Director himself involved in the change management process helped the organisation to transform and perform. The Director introduced a matrix-type reporting structure among the WBF activity teams to take out the best in both efficiency and responsiveness. The leadership styles used by the Director were quite different from the bureaucratic way of getting things done in the Government sector.

The NBT director being the protagonist in this case study, was able to identify the gaps in the processes and take adequate management intervention through BPR systematically. After the BPR intervention, the results were terrific. The success of the event can be attributed to NBT and ITPO team members and the stakeholders of WBF.

5. Teaching/Whiteboard Plan

Discussion points/questions would include the following:

5.1. Write down some critical background information on the whiteboard, including important dates and events and names of key people or agencies involved in organising the WBF. Such steps will save time and enable the students to grasp and refresh their memory on the case-study facts.

5.2. The instructor may open the case with a whiteboard and then start a discussion with questions such as—What were the initial challenges encountered by the NBT management? How the NBT benefited from the partnership with ITPO? etc. Students may identify the sequence of important events. The discussion may be on the issues about financing the event.

5.3. Some of the other points the instructor can emphasise are: (a) Over the period improvements that were brought in the conduct of WBF. (b) The Business environment is changing very fast, and the importance of agility in the decision-making process in public sector bodies. (c) Advertisement campaigns became high-tech using digital and social media platforms. (d) Perhaps the NBT officials ignored such external changes and relied upon the old pattern of holding the event and (e) it seems; they were not taking a risk in the decision-making process. They may be comfortable working in a bureaucratic working style and not bothered with the results/situations for holding such a public event of national interest.

6. Learnings

Business process reengineering (BPR) has been extensively used by several organisations and falls in management and information systems. The NBT Director, in this case, used a mixture of Business Process Re-engineering (BPR) and principles of change management process suggested by Vakola and Rezgui [10]. In his case study, Covert [11] argued that BPR means not only change, but was intended to bring dramatic change. The change in such a situation constitutes a complete overhauling of organisational structure, reporting system, employee responsibilities and performance measures, skill development, introducing technology, etc.

The literature indicates that the BPR is one of the successful reengineering processes adopted by several companies across the world. The model of BPR by Vakola and Rezgui [10] has been discussed in detail for the benefit of the teachers and readers. Simsion [12] argued that 'the widely accepted methodologies were based on how the business processes should change and how the organisation should adapt itself in this change, rather than on evaluating current practices and on the codification of successful practical experiences'. Companies operate according to unarticulated rules. Every company operates according to a great many unarticulated rules. Hammer argued that BPR could not be planned in a precious manner thread cautiously and 'it's an all-or-nothing proposition with an uncertain result'. However, companies have been left with little or no choice in experimenting with it [13].

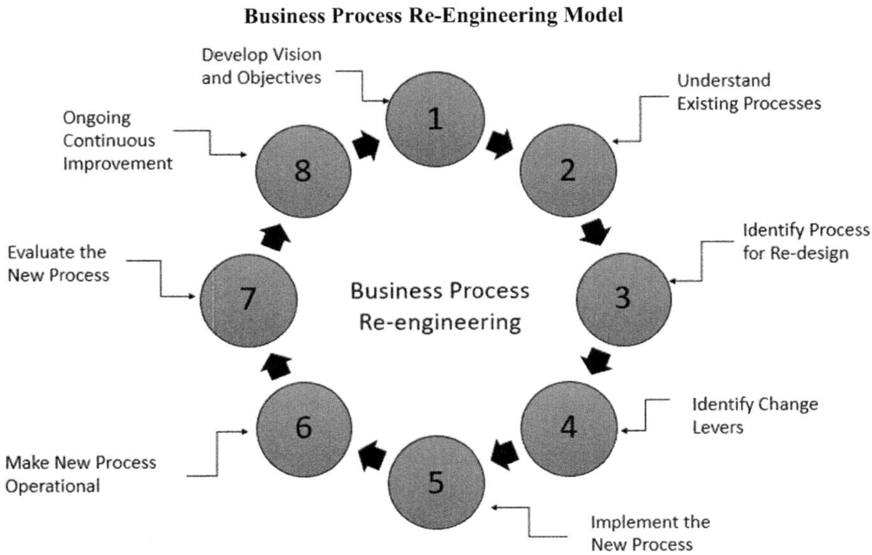

Fig. 2 A Generic model for business process Re-engineering adapted from Vakola et al. (1998)

Business Process Re-engineering Model

See (Fig. 2).

The 8 Stage Business Process Reengineering by Vakola

1. **Developing business vision and process objectives**: Development of vision and objectives of the organisation is the first and foremost stage of BPR intervention wherein the organisation should have a clear understanding of its strength, weaknesses, area of service or operation and any innovative practices adopted in its business operations. After evaluating the same, the new Director has aligned the activities of WBF with the core objectives of NBT India. For example, to promote books and reading, the WBF events were modified to bring the readers and authors closer. Authors' corners added. At the same time, CEO Speak and Rights Table, etc., were introduced in the WBF.
2. **Understanding existing processes**: In the second stage of the model, the focus has to be made to understand the existing process in an organisation before attempting to redesign. This process helps the organisation facilitate information sharing and improved communication among the divisions or participants in the BPR engagement. The new Director evaluated the functioning of each division/ function of the organisation by involving internal and external stakeholders.
3. **Identifying processes for the redesign**: The third stage in the model was the identification of bottlenecks in the existing process has been considered an essential process for the redesign of workflow. Through continuous engagement, the Director opened up feedback channels with internal and external stakeholders,

Government officials, authors, publishers, intellectuals, media, civil society and foreign participants. These steps provided the much-needed hindsight and understand the expectations of the stakeholders. In the process, many of the employees opened up with the management team.

4. **Identifying change levers**: The fourth stage in the model involves identifying the change lever process consisting of relying on the organisation's intellectual capital and creative engagement of the human capital. NBT is essentially a knowledge-based organisation where editors and production officers engaged in the production of books. Similarly, the Exhibition division has been involved in organising book fairs at the national and international levels. They have to be innovative as well as showcase the Indian ethos outside the world. The process helped the organisation identify the competencies needed in the services and processes to bring synergies within the divisions, external players, and government departments. Such steps provided ways to modernise the organisation through digital intervention and helped bring a transparent and friendly atmosphere against the bureaucratic attitude. The continuous feedback and informal meetings with stakeholders also helped the Director identify the gaps in the systems and change levers. The organisation sought the occasional help of specialists and consultants' services to determine the change levers and best practices.
5. **Implementing the new processes**: In the third stage, the approach was to identify the process for redesign, whereas the focus in the fifth stage has to be shifted to implementing those identified processes. Implementing the new processes and methods is considered a critical phase for any organisation, and due to a lack of will or support from the top management, the strategy fails. In some cases, the implementation was partial, which fails to yield the desired results.
6. **Making the new processes operational**: In the present case, the Director took a keen interest and brought the main stakeholders on board to get the desired results. A small team of officers has been continuously engaged in implementing new processes and practices/policies. Innovation and transparency were the keys to these processes. The Director constantly monitored and ensured these new processes were appropriately implemented to get better results. Because of the active involvement of officials of NBT, the latest systems and processes were put in place and made operational. The joint team of NBT and ITPO in the new experiment was a big success. The expertise available with both organisations was handy in putting a good show of strength in each successive edition of WBF. The government's precious resources were also saved, which sent a good message among the book lovers, readers, intellectuals, government officials, and envoys.
7. **Evaluating the new processes**: The seventh stage of evaluation of new methods is equally crucial in the model. The assessment involves the potential achievements and advantages of the new approaches and systems devised by the organisation through the BPR model.
8. **Ongoing continuous improvement**: The BPR model is essentially a successive as well as an ongoing process. Hence it has to be regarded as a new business model

innovation and intervention by the organisation. Such interventions help organisations to make the right moves by deviating from the traditional functional-based models. Once the organisation overcomes initial challenges and hiccups and commits to its new business process change, it can grow with its innovative ways with the help of digital technologies. I.T. intervention and human resources of the organisation may prove its principal enabler for change management. The BPR process is considered an ongoing process and may reach some saturation level, but it would continuously engage itself by reinventing itself in the ever-changing business environment. Such practices would lead to the transformation of business strategy to achieve the organisations' vision, mission, and objectives.

The authors in this case study were also guided by Cunningham and Kempling [14] who argued that 'Public Sector changes may not be more difficult than those in the private sector, but they are different. Change is not simply an exercise in convincing the various stakeholders to get on the side; it is an exercise in negotiation and compromise'. 'Doing what is right in government is a matter of responding to conflicts and negotiating with various interests much more than it is for a corporate executive trying to implement a strategy'. Cunningham and Kempling [14] argued in favour of nine principles for changing public sector organisations viz. '(i) Forming a guiding coalition; (ii) Recognising and responding to resistance; (iii) Establishing a need for change; (iv) Articulating envisioned outcomes; (v) Establishing a process to implement planning; (vi) Focusing on continuous improvement; (vii) Developing a commitment plan and; (viii) Managing by walking around ad (ix) Changing structures and H.R. Systems'.

According to [11] 'Business Process Reengineering (BPR) can potentially impact every aspect of how we conduct business today. Change on this scale can cause results ranging from enviable success to complete failure'. But at the same time, successful BPR can bring enormous cost reduction and substantial improvements in quality, customer service and business objectives. Successful BPR can result in considerable reductions in cost or cycle time [13]. BPR strives for a dramatic process improvement in an organisation, and it should aim to think differently from the conventional wisdom and beyond the boundaries. It should use technologies to bring innovations.

The Director in this case also recognised the arguments put forth by Waterman and Peters [15], 'management by walking around' as a process where leaders and managers were seen and involved in face-to-face interactions with front-line staff. Lewin [16] identified resistance to change as a force, like inertia that is preventing the disruption of an old equilibrium.

References

1. Clark G (2001) Inside book publishing, 3rd edn. Routledge, London
2. Thompson J (2005) Books in the digital age. Polity, Cambridge
3. Nielson Book Research: Data and Insight for the Book Industry (2020) https://nielsenbook.co.uk/wp-content/uploads/sites/4/2020/09/Research_Brochure.pdf
4. Flood J (2007) 'Omnium totiusorbisemporiorum compendium': the Frankfurt fair in the early modern period. In: Myers R, Harris M, Mandelbrote G (eds) Fairs, markets and the itinerant book trade. Oak Knoll Press, London; The British Library, Delaware, pp 1–42
5. Weidhaas P (2007) A history of the Frankfurt book fair. Dundurn, Toronto
6. Appadurai A (1986) Introduction. In: Appadurai A (ed) The social life of things. University Press, Cambridge, pp 1–72
7. Moeran B (2010) The book fair as a tournament of values. J Royal Anthropol Inst 16(1):138–154. http://www.jstor.org/stable/40541809. Accessed 15 June 2021
8. The Future of Book Fairs: An IPA Special Report (2014). Retrieved 15 June 2021. http://www.international-publishers-association.org/images/news/2015/BookFairs.pdf
9. Press Note, ITPO (2017) https://architexturez.net/pst/az-cf-182242-1493000817
10. Vakola M, Rezgui Y (2000) Critique of existing business process reengineering methodologies. Bus Process Manag J
11. Covert M (1997) Successfully performing business process reengineering. Vis Syst Corp 201
12. Simsion G (1994) A methodology for business process reengineering. In: Paper presented at the IFIP information systems international working conference, business process reengineering-information systems opportunities and challenges. Queensland Gold Coast, Australia
13. Hammer M (1990). Reengineering work: don't automate, obliterate, harvard business review (July–August)
14. Cunningham JB, Kempling JS (2009) Implementing change in public sector organisations. Manag Decis
15. Waterman RH, Peters TJ (1982) In search of excellence: lessons from America's best-run companies. Harper & Row, New York
16. Lewin K (1947) Quasi-stationary social equilibria and the problem of permanent change. Organ Change Compr Read 73–77
17. Annual Reports of National Book Trust, India. https://www.nbtindia.gov.in/

Chapter 11
Improved Machine Learning Prediction Framework for Employees with a Focus on Function Selection

Kamal Gulati, T. S. Ragesh, K. Bhavana Raj, Bhimraj Basumatary, Ashutosh Gaur, Gaurav Dhiman, and Uma S. Singh

Abstract Companies are constantly looking for methods to keep their workers with them to minimize further recruitment and training expenses. Predicting whether a specific staff member can depart helps the business to take preventative measures. Unlike physical systems, a scientific and analytical formula cannot explain human resource issues. Machine learning methods are thus the ideal instruments for this purpose. This chapter offers a three-stage paradigm for the prevention of attrition (up to processing, processing, post-processing). IBM HR dataset for the case study

K. Gulati (✉)
Associate Professor, Amity University, Noida, Uttar Pradesh, India
e-mail: drkamalgulati@gmail.com

T. S. Ragesh
Assistant Professor, Business Analytics, Prin. L. N. Welingkar Institute of Management Development and Research, Mumbai, India
e-mail: ragesh.ts@welingkar.org

K. B. Raj
Assistant Professor, Department of Management Studies, Institute of Public Enterprise, Hyderabad, India
e-mail: bhavana_raj_83@yahoo.com

B. Basumatary
Assistant Professor, Mathematical Sciences, Bodoland University, Kokrajhar, Assam, India
e-mail: brbasumatary14@gmail.com

A. Gaur
Assistant Professor, Mangalamay Insitute of Management and Technology, Greater Noida, India
e-mail: me@ashutoshgaur.co.in

G. Dhiman
Department of Computer Science, Government Bikram College of Commerce, Patiala, India
e-mail: gdhiman0001@gmail.com

U. S. Singh
Associate Professor, Department of Commerce, ARSD College, University of Delhi, New Delhi, India
e-mail: umasinghdu@gmail.com

S. Kautish et al. (eds.), *Computational Intelligence for Modern Business Systems*, Disruptive Technologies and Digital Transformations for Society 5.0,
https://doi.org/10.1007/978-981-99-5354-7_11

is selected. As there are many functions in the dataset, the selection technique for the "maximum-out" feature is suggested for the extent of decreasing the to-processing phase. In the planning retrogression model, the coefficient of each variable indicates the significance of the feature in attrition predictions. The findings indicate an improvement in the F1 score using the "maximum-out" feature selection technique. Finally, through learning the model for many bootstrap datasets, the validity of data is verified. The average deviation of the parameters is then evaluated to verify the confidence and stability of the model parameters. The modest average deviation of the data shows the model is stable and generalized.

Keywords Machine learning · Management of human resources · Characteristic selection · Logistic regression · Prediction of attrition · Bootstrap

1 Introduction

Human resources are any company's source and the most important element. Managers spend significant effort hiring skilled staff. In addition, extra resources are routinely spent on training personnel. Each employee attrition, leaving the job without being replaced, costs the business a new employee to recruit and train. Consider two examples (a) and demonstrate the concept of attrition. (b). It is not attrition; the employer chooses to substitute an employee with another qualified individual. If (b) is attrition, the worker quits the enterprise [1]. The second scenario involves the business being faced with delays in its project timeline because the substitute employee is recruited and trained. It is simpler for decision-makers to take appropriate preventative measures to predict attrition. Some variables, like age, income, distance, level of education, etc. help to determine whether or not an employee leaves a business. Because the attrition of employees and all these variables will have no deterministic analytical relationship, machine learning techniques that rely on experience to improve performance or accurately forecast can be used [2]. Several literature works seek to develop a classification to forecast whether a certain employee quits. Mohebi and Kumar have trained for this job in the irregular woodland. Accuracy, reminder, F1 values indicate that the logistic regression has been done in the forecast task and that certain indicators in this model are greater than in other classifiers. A logistic regression classification is used for the attrition prediction of IBM HR database employees. But this study did not choose influential characteristics. There are many features in the database. Only eight workers out of 70 attritions were expected to leave their employment. The remaining samples were incorrectly projected like depreciation. On how these factors might be optimally blended to benefit both customers and enterprises, taking into consideration criteria including the degree to which technologies have been included and its client satisfaction rate via relationship building [3]. The author reviewed and analysed the literature, paying special attention to the characteristics of wireless connections for energy conservation and data aggregation [4]. However, many characteristics are binary variables in attrition databases that don't function well for the PCA method [5]. PCA also lowers the size based on the linear connection between characteristics. The possible relationship between candidate characteristics and production is not examined. The authors removed those factors that link poorly to extra variables. A feature selection was then conducted based on the random forest [6]. The logistic Regression Model was finally trained for attrition prediction. Recall measures, essential for performance analysis in the event of severe imbalances of the dataset, were not evaluated in papers [7]. On the other side, numerous studies have calculated and attempted to increase these metrics. Before separating the training and test sets, they resampled the data set. Some samples may be repeated in the check dataset before the training and test data set are separated. For these specific test samples, the model would be successful [8]. The ability of the model to generalize may therefore be overstated. A few linear models are employed to predict whether an employee wants to quit, taking linear regression and linear discriminant analysis. The article calculated recall measures and precise measurements depending on the workers that were with the business. This class is more crowded and the classifier's performance is overestimated. Usually, classifiers are more accurate in the prediction of the crowded class. The major difficulty is to accurately forecast the minority class. An attrition predictor for the (ANN) was reported. Despite the significant accomplishments of this research, the model's performance was investigated with average mistakes and confirmation [9]. Mean square error is not a suitable classification task performance measure, and accuracy is not enough to conduct an unbalanced dataset analysis. The authors tested various gradients including logistical regression, Ada Boost, irregular woodland and gradient increase for prediction of attrition. Though this study is exhaustive, the recollections and accuracy of the copies are not compared. It also utilized the correlation-based feature selection, which did not provide a satisfactory selection of the most important characteristics [10]. The connection between the characteristics may be seen in. This accommodation gives a good insight into the data set.

Each article in the current literature lacks a few of the succeeding aspects:

- Proper technique of feature selection.
- Informative classifier performance assessment.
- Conviction extent with the value of the coefficient in the logistic regression copy for each feature. An accurate examination of the forecast step should comprise up to-processing, processing and after-processing to provide a realistic, trustworthy predictor that human resources organizers may trust. In the pre-processing phase, characteristics are chosen that are most important for this job. Redundant functions are also removed. A correct selection of features increases the performance of the models learned during processing [11]. The forecast copy is ready, checked and compared with other models at the performing step. Lastly, a great conviction in the model must be guaranteed at the post-processing step.

For the first time, this article offers an attrition prediction job to our knowledge which deals with all three phases of up to-processing, processing's and post-processing. The author claims that a sequence of CNN architectures having different depths produces diverse semantic features of the picture. The max-out algorithm is first and foremost a naval feature selection technique for increasing the performance of the attrition prediction classifier. Then for the new set of characteristics for the processing step, a logistic regression model is trained. Next, confidence analysis is added at the post-processing step, to measure how certain we are about the parameters of our models. The approach is finally checked using attrition statistics from IBM. The figure shows the overall structure of the proposed framework. This graphic represents the pre-processing, processing's and post-processing phases of yellow, green and red blocks. The primary purpose of these procedures is to ensure.

The overall structure of the suggested framework for attrition prediction. The remainder of this paper is structured accordingly. It presents the "max-out" approach and analyses the algorithm complexity. Next, the confidence analysis of the parameters is presented in Sect. 3. Logistic regression, which is the attrition predictor, is then examined briefly in Sect. 4. Section 5 then examines this algorithm (Fig. 1).

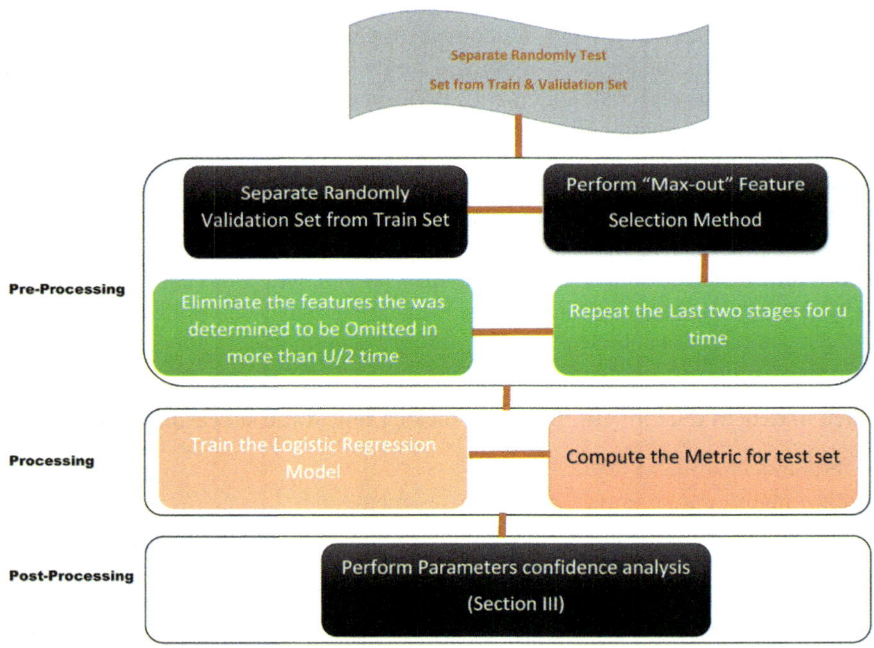

Fig. 1 The basic structure of the suggested framework for attrition predictions

2 Selection of Feature During Pre-processing

Many unwanted characteristics may be removed. One approach is to train the model for all feature sub-sets, compare the validation metrics. This method needs 2n times a model training to a dataset of n functionality. It is thus extremely time intensive. Several approaches for selecting features for this specific process have been previously described. Filter techniques are based on the correlation of feature values.

On the other hand, wrapper approaches focus on the choice depending on the performance of the classifier with the characteristics chosen. Embedded techniques include the selection of functions that help in the algorithm. Hybrid techniques are the ways that combine them.

Table 1 provides a comparison between these groups. Note that the technique for selecting the feature must agree to the characteristics.

2.1 Maximum is Known Selection Characteristic Data

Depending on the nature of the feature set of this issue, which comprises double and continuous features, a selection of "max out" features, it is from the covering category, is created. The method is shown in Algorithm 1. First, every below set to n-m features is trained according to this method. The subset with the most important measure is selected as the function. For the next set of features, the procedure is repeated. If the metric is less than the preceding task, the characteristic class is not updated. As the copy is only trained for a fraction of all available features, the method is considerably quicker than testing all potential feature combinations. When m is equal to 1, it is termed 1-max-known, and when m is 2, it is named 2-max-out. The backward option is the 1-max-out algorithm. However, each one of them may not have a major impact on selection performance. Thus, 1-max-out may remove these characteristics one by one. In certain situations, maximum-known ($m > 1$) works better than 1-max-out. The choice of an appropriate m is thus equally reliant on the available computing resources. Shows an example in which the optimum option chosen by the "1-max-out" approach is to remove (X2, X3, X9 and X8) characteristics of the total 15. The first feature to be removed requires 14 rounds for this procedure,

Table 1 Characteristic selecting methods

Category	Merits	Demerits
Seep	High Classifier	Few accurate
Cover	Very valid	Prone to overfitting
Inserted	More accurate	Selective specific
Crossbred	1. More validity than seep 2. Less computational Complexity than covers	Selective Specific

thirteen iterations, 12 for the third one, 11 for the fourth and 10 for the realization that further elimination is not helpful. If the brute-force search is 215, equivalent to 32,768 times the model is trained.

Data 1: maximum-known.

1. Create Set F = Set of n characteristics.
2. Calculate variable M as a classifier metric using Set F3 characteristics. Take n of the size S 1. Create Set F = Set of n characteristics.
3. Calculate variable M as the classifier metric using Set F 3 characteristics. Take n equal to Set F.
4. Sizes. Set Sub1, Sub2, ..., Subk all sub-sets of Set F size n-m.
5. Calculate Sub1 metric, ..., Sub.
6. Take M' equal to the largest collection of line 4 metrics (for the corresponding subset j).
7. If M' Alto M: M = M, and set F = Subj, proceed to 3.

2.2 Example of Illustrative 1-Maximum-Out

A model to the fish market issue is presented in this part to demonstrate further the Max-Out technique. Note that this is not the paper's primary case study. This benchmarking issue aims to forecast fish weight. The seven variables are binary, "Bream", "Pike", "Roach", "Smelt", "Whitefish" and five variables are constant, "Longitude1", "Longitude2", "Length", and "Height" and "Width". After they are indexed from 0 to 11, the 1-max-out is executed. The R2-score measurement for the validation set improves from 0.9384 to 0.9396 as a consequence of new functionality and from 0.9145 for the Test.

3 Confidence Analysis Parameter

Figure 2 depicts the procedure for verifying our model parameters' values. Our model's parameters changed every time we bootstrapped the original training dataset. A large parameter fluctuation occurs if the model's training set exceeds [12]. If not, parameters differ somewhat, indicating that the parameters assess the true trend and not simply save the training set. On each parameter, several statistical analyses may be done.

A toy example is to demonstrate this. Consider that we have a dataset including six samples, three in the star class and others in the circles. A classifier of logistic regression is trained. If the bootstrap model has been trained in which are and green star circles are removed, the parameters will vary significantly. It may thus be inferred that we are unsure about the parameters of the model which are significantly dependent on the training data [13]. The parameters may be regularised to make the parameters more stable (Fig. 3).

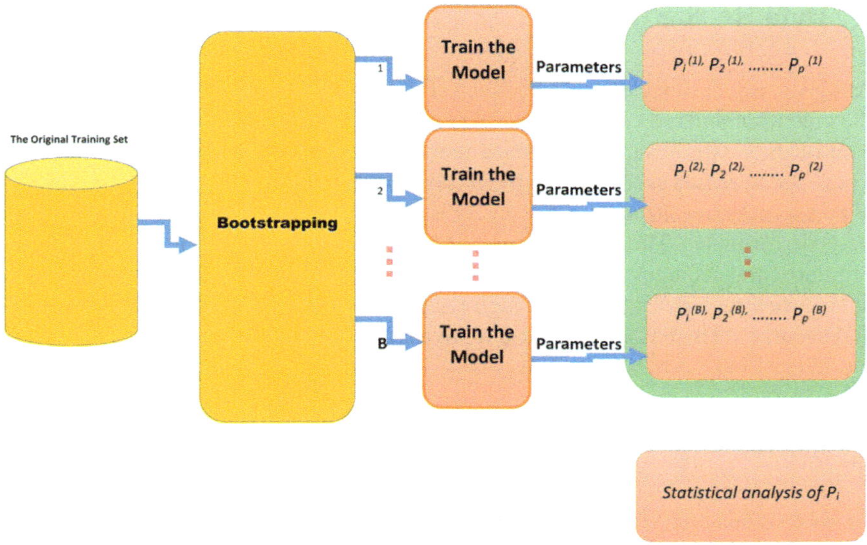

Fig. 2 The total confidence analysis parameter structure

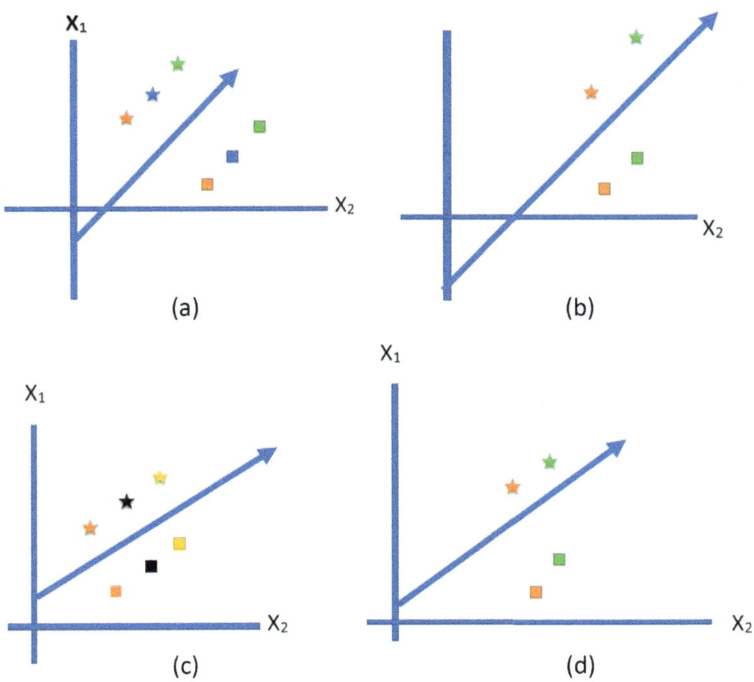

Fig. 3 The total confidence analysis parameter structure

4 Data Retrogression

A logistic regression classifier is trained (Fig. 2). A bootstrap model trained without the green star and red circle has radically different parameters. So we can't be confident of the model's parameters since they are strongly reliant on the training data. They can be regularised to make them more stable. If a weak classifier classifies all copies as negative, its accuracy is 99.99%. Thus, accuracy may overstate the classifier's performance. The reminder measures estimate the correct samples tagged [14]. Similarly, the accuracy measure measures the proportion of positive samples in the data. A few partials can result in high recall. So the F1-score represents both recall and accuracy. The F1-Score is little if any recall or accuracy is low.

5 Chronology

As case studies, the IBM attrition dataset is used. This dataset contains 35 columns per worker. "Abrasion", is the classifier's work output. The remaining 34 columns are functional. The following features are listed in other categories like: "period", "business travel", "daily rate", "department of employment", "home distance", "academic area", "Environmental satisfaction", "sex" "hourly rate" "jobs extent", "employment satisfaction" "jobs role" "matrimonial income". "Monthly rate", "number of businesses working" these characteristics are either numerical or categorical. From machine mode of learning.

Either category of numerical is these characteristics. Since copy cannot deal directly with categorical features in machine learning, categorical data will be transformed into binary functions using dummy coding. For instance, as shown in Table 2 the category "education field" characteristic may be transformed into five binary variables.

Table 2 Conversion to binary features of categorical feature education

Academic sphere	EF-HR	EF-LS	EF-Ma	EF-Me	EF-Oath
Living being resources	1.4	0.2	0.1	0.2	0.1
Biology	0.42	1.5	0.2	0	0
Merchandise	0.5	0.2	1.3	0.3	0.1
Medication	0.33	0.1	0.2	1.5	0.2
Additional	0.2	0.1	0.1	0.2	1.4

5.1 Selection of Feature

To determine the most significant characteristics the data set was first split into the learning and confirmation group and the check sec. The validation set was then removed from the training set. To decide what characteristics should be missed, the 1-max-out method was used. After that, the random separation process for the permanence test and the 1-max-out method is repeated. After seven cycles, characteristics that were more than four times excluded were deemed deleted.

"Hourly price", "Academic Area-Human Resource", "Month wise income", "Gender women", "Service-Research and Development", "Over18-Yes", "Education", "Educational level", "Department-Human Resources", "Business Travel-Rarely", "Performance", "Month wise price", "Academicsection-more", "Business Travel-Non-Travel", "Education Field-Maritime" "Education Field-Marine Education". These are not always the least essential characteristics for predicting attrition. Some are selected to be removed since they are fully linked with other data set characteristics. For example, "sex female" is one characteristic minus "sex male". For categorical characteristics transformed into binary features, being removed implies that the likelihood of attrition in this category is neither increased nor decreased.

5.2 Model Final

According to the coefficients, "periods from the previous development", "extra time", performing like sales representative and "the number of businesses employed" are the most significant reasons that leave an employee. As one of these numbers rises, the likelihood of the employee leaving work increases. The most important criteria in which an employee is allowed to remain at the business are his performance as research director, "whole work years" and "years with present managers" and "job involvement". The model exhibits 81% accuracy of the check database Precision, recall, and F1 are all 0.43. This would be %, 0.39, 0.82 and 0.53 without the 1-max-out option.

According to the coefficients, the main reasons for an employee's abandonment are "period as final promotion", "overtime", work like a sales representative and "number of businesses operating". The chance that the employee quits the job rises with the increase in any of these variables. The criteria for an employee to remain with the business as a director of research, whole functioning years, years with the present management and "job involvement" is the most important. The model exhibits 81% accuracy in the test database. Accuracy, reminder and F1 score are respectively 0.43, 0.82 and 0.56. If the selection of the 1-maximum-out function were done, the validity, precision, reminder and F1- score would be 78%, correspondingly 0,39, 0,82, 0,53. Figure 4, shows a comparison of the suggested technique performance and the classification utilized in prior studies. The findings indicate a significant improvement in the F1 score of the suggested approach.

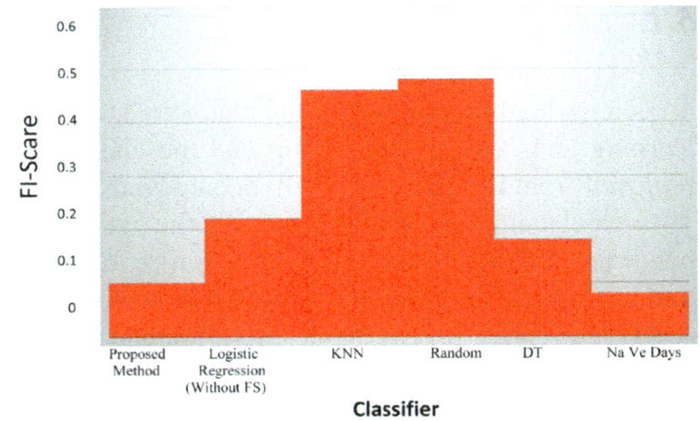

Fig. 4 Comparison of the performance measures of the various selectors for the HR prediction task

It should be noted that these findings are only valid for this data set. These factors may change for other businesses with distinct cultural and economic condition in another nation.

5.3 Analysis of Confidence Parameters

The model is then trained on each dataset. It shows the average, standard deviation and correlation (average deviation to the absolute value of the average ratio). Standard deviations show average trust. We can be measured in domains where the coefficient of determination is low. This shows the fluctuation of coefficients related to the most important characteristics mentioned in the preceding paragraph. This shows that in the years after the previous promotion coefficient all bootstrap training datasets had a value between 2 and 4. We may thus trust its significant impact on attrition. The "Over Time Yes" coefficient changes little. We can thus be certain of the value of this coefficient. By comparison, the coefficient "Years with Current Manager" fluctuates across a broad range. Therefore, we can't be sure about this parameter. However, this parameter is negative in all bootstrap datasets. It may thus be concluded that this feature has a positive effect on keeping the employers with the business.

6 Findings

This article aims to provide a machine learning approach to forecast the attrition of employees. A feature selection technique was originally proposed to reduce the feature space dimension. Then reasonable model for prediction was trained. A comparison of the findings with current techniques shows that the selection of features suggested improves the prediction performance. The model showed that the reasons for quitting the work are "years after the previous promotion", "Overtime—yes", "Job Role Representative" and "Number Companies Working". Higher values for these characteristics lead to a greater likelihood of attrition. On the other hand, "total years of employment", "years with present managers" and "job participation" are the main consideration for the business. 300 hundred bootstrap datasets have been created to see if the parameters are valid. A model was fitted for each of them. Statistical analysis was then conducted of the coefficient of each characteristic. The fluctuation of coefficients was generally acceptable. Variations in parameters related to the most important characteristics were negligible in. We are thus confident that the above characteristics are important to forecast attrition.

Compared with prior work, this article provides up to processing, processing and post-processing methodology for creating an accurate forecast model for attrition and verifying the validity of the model parameters. The m-max-out method is presented for the pre-processing feature selection. Because of the limitations of the computer equipment presently faced by the writers, the 1 max-out (a particular situation in which m is equal to one) is utilized. Bigger m may also be utilized for greater computational resources available. By evaluating the parameters' changes when trained across numerous bootstrap data sets, the validity of logistic regression model parameters is verified. These pre-processing and post-processing steps may be used to create precise and reliable models for any broad issue. The max-out function selection process may be applied for any feature set, including binary and continuous functions. The statistical examination of the models' parameters on several bootstraps may infer if we are confident of the model for any type of parametric machine learning model. Psychological variables relating to employee turnover are recommended for study for future research on attrition forecasting. The impact of the number of openings available for each company also takes into account its specification. In future studies, situational variables regarding his/her likelihood of attrition may also be studied.

References

1. Mohri M, Rostamizadeh A, Talwalkar A (2012) Foundations of machine learning. The MIT Press, Cambridge, USA
2. Mohbey KK (2020) Employee's attrition prediction using machine learning approaches. In: Machine learning and deep learning in real-time applications. IGI Global, pp 121–128
3. Ponnuru S, Merugumala G, Padigala S, Vanga R, Kantapalli B (2020) Employee attrition prediction using logistic regression. Int J Res Appl Sci Eng Technol 8(5):2871–2875
4. Frye A, Boomhower C, Smith M, Vitovsky L, Fabricant S (2018) Employee attrition: What makes an employee quit? SMU Data Sci Rev 1(1):9
5. Alduayj SS, Rajpoot K (2018) Predicting employee attrition using machine learning. In: 2018 International conference on innovations in information technology (IIT). IEEE, pp 93–98
6. Bhuva K, Srivastava K (2018) Comparative study of the machine learning techniques for predicting the employee attrition. Int J Res Anal Rev (IJRAR) 5(3):568–577
7. Qutub A, Al-Mehmadi A, Al-Hssan M, Aljohani R, Alghamdi HS (2021) Prediction of employee attrition using machine learning and ensemble methods. Int J Mach Learn Comput 11(2):110–114

8. Fallucchi F, Coladangelo M, Giuliano R, William De Luca E (2020) Predicting employee attrition using machine learning techniques. Computers 9(4):86
9. Zebari R, Abdulazeez A, Zeebaree D, Zebari D, Saeed J (2020) A comprehensive review of dimensionality reduction techniques for feature selection and feature extraction. J Appl Sci Technol Trends 1(2):56–70
10. Venkatesh B, Anuradha J (2019) A review of feature selection and its methods. Cybern Inf Technol 19(1):3–26
11. Dhiman G, Kumar V (2018) Emperor penguin optimizer: a bio-inspired algorithm for engineering problems. Knowl-Based Syst 159:20–50
12. Kaur S, Awasthi LK, Sangal AL, Dhiman G (2020) Tunicate swarm algorithm: a new bio-inspired based metaheuristic paradigm for global optimization. Eng Appl Artif Intell 90:103541
13. Dhiman G, Kumar V (2019) Seagull optimization algorithm: theory and its applications for large-scale industrial engineering problems. Knowl-Based Syst 165:169–196
14. Gulati K, Kumar Boddu RS, Kapila D, Bangalore SL, Chandnani N, Saravanan G (2021) A review paper on wireless sensor network techniques in Internet of Things (IoT). Mater Today Proc. https://doi.org/10.1016/j.matpr.2021.05.067

Chapter 12
Applications of Data Science and Artificial Intelligence Methodologies in Customer Relationship Management

E. Fantin Irudaya Raj

Abstract Customer relationship management (CRM) is a set of technologies, methods, and practices that companies adopt to analyze and manage customer interactions and data throughout their engagement with them. The primary purpose behind using CRM is to enhance the business and improve interaction with the customer at the service point. It also helps in increasing the revenue of suppliers or manufacturers and is more useful in customer retention. The efficient operation of the CRM system needs to collect information from several contact points with the consumer, such as direct phone calls, live chat, marketing materials, the company's website, chatbot, social media, and mail communication. In addition, the CRM system also collects and provides detailed information about the customers' personal information, preferences, history of purchase, and various concerns with the employees from the marketing side. Thus, the CRM system rapidly processes a huge volume of data from the consumer and uploads numerous data from the supplier or manufacturer. In such cases, an efficient way of handling this voluminous data is highly needed. The advent of Data Mining techniques and their rapid growth makes it easier to handle this large volume of CRM data more effectively. The present work discusses various Data Mining techniques and their application to CRM. It also outlines how effective it is to make a contemporary CRM from a conventional one.

Keywords Customer relationship management · Data mining · Modern business system · Data science · Artificial intelligence

E. F. I. Raj (✉)
Department of Electrical and Electronics Engineering, Dr. Sivanthi Aditanar College of Engineering, Tiruchendur, Tamilnadu, India
e-mail: fantinraj@gmail.com; fantinirudhayaraj@drsacoe.com

S. Kautish et al. (eds.), *Computational Intelligence for Modern Business Systems*, Disruptive Technologies and Digital Transformations for Society 5.0,
https://doi.org/10.1007/978-981-99-5354-7_12

1 Introduction

Customer relationship management (CRM) is a process through which a company or other organization manages its contacts with customers, usually analyzing vast volumes of data via data analysis. Normally, the CRM systems collect information from various sources, including marketing materials, live chat, email, phone, the official website, and social media. They enable organizations to understand more about their target consumers and how to best respond to their demands, resulting in increased sales and customer retention. Customer satisfaction was measured using annual surveys or front-line questions in the early 1970s when the notion of customer relationship management was born. Businesses had to rely on separate mainframe systems to automate sales at that time, but they could now categorize clients in spreadsheets and lists thanks to advancements in technology.

Data Mining (DM) is the subfield of Data science that evolved in the early of this millennium. Adopting the Data Mining techniques in CRM makes it more renovative and makes the business more profitable and customer friendly. DM is the process of predicting outcomes by looking for anomalies, patterns, and correlations in massive data sets. Organizations can foresee future trends in markets and customer behavior by employing DM approaches. DM enables enterprises to better satisfy the needs of their customers by providing extensive knowledge of current market trends and consumers. The following sections explain CRM and DM and their types and approaches and the adoption of DM in CRM in detail.

2 Literature Review

Buttle [1] explained Customer Relationship Management (CRM) and its various underlying principles in his work. He also explained the contexts in which it can be used, how it can be implemented, and its important strategies and objectives. Kumar [2] provides various methodologies to implement CRM and describes the data collecting points, which are the key section in CRM implementation. Payne and Frow [3] discussed different strategic frameworks for implementing CRM in the marketing environment. He also insisted that customer satisfaction is the key element to be considered and the basis for successful CRM implementation. The authors [4–7] discussed various CRM frameworks in different domains like banking, shopping mall, E-Business, Stock Market, etc. The authors [8, 9] explained how CRM can be effectively implemented and creates a good impact in the healthcare environment as well as how it improves social customer relationship management from the customer's point of view. Gil-Gomez et al. [10] pointed out the importance of digital transformation and sustainable business model innovation in CRM. He also mentioned that any organization must come out of the traditional system and adapt to this new environment.

Whenever we talk about digitalization, we come across the term electronics as well as Artificial Intelligence (AI). Mitchell et al. [11] mentioned that AI is a branch of computer science, but it can influence every domain in the present world. By employing AI in the business environment, we can take our business to new heights [12–15]. AI gets introduced and implemented around the year the 1960s. Then after the development of processors and fast computing, Machine Learning (ML) was introduced around the 1990s [16]. ML is mainly focused on making computers think independently of a particular task and take or predict the decisions. It consists of various algorithms [17–20]. Deep Learning is the subfield of Machine Learning, and it is introduced and implemented around 2000 [21]. DL mainly has Artificial Neural Networks (ANN) for its effective operation and highly accurate prediction [22]. It can also be implemented in a business environment used for classification and pattern recognition. It also provides great yields to the modern business world [23–26]. There are various deep ANN configurations used in the deep learning process. The efficiency of the deep ANN can be measured using multiple parameters [27–30]. Data science is one of the evolving fields in managing numerous data. Data Mining is the subfield of Data science. DM fields get the inferences of AI, ML, and DL [31, 32].

Data Mining is the data management tool for effective business management. It can be useful for extracting useful data from the raw database [33]. It can be used to improve the business environment and improve the profit of the manufacturer. Application of Data Mining (DM) in CRM makes it more profitable [34]. There are various DM techniques used in CRM for effective system management. It can be used for marketing, sales, and manufacturing [35]. The authors [36–40] explained various Data Mining frameworks for CRM in the business management system. There are numerous works listed in the literature. The present work discusses more in detail Data Mining and its applications in Customer Relationship Management. It also discusses various DM strategies and their various advantages.

3 Customer Relationship Management (CRM)

CRM refers to a company's systems and processes for measuring and improving how it communicates with others and manages consumers. CRM's major goal is to gather sufficient information about the customer and then use it effectively enough to boost its positive experiences with the company, leading to increased sales [41]. The CRM system includes the following process. They are 1. Lead Generation and Conversion, 2. Relationship Building, and 3. Customer Service. Figure 1 shows the overview of the CRM system.

Locating and contacting a company's ideal customers, producing high-quality sales prospects, developing and defined objectives and goals in marketing initiatives are all examples of Lead Generation and Conversion. Creating frequent communication channels, building and improving customer connections, and delivering specialized attention to the most profitable consumers are examples of relationship building.

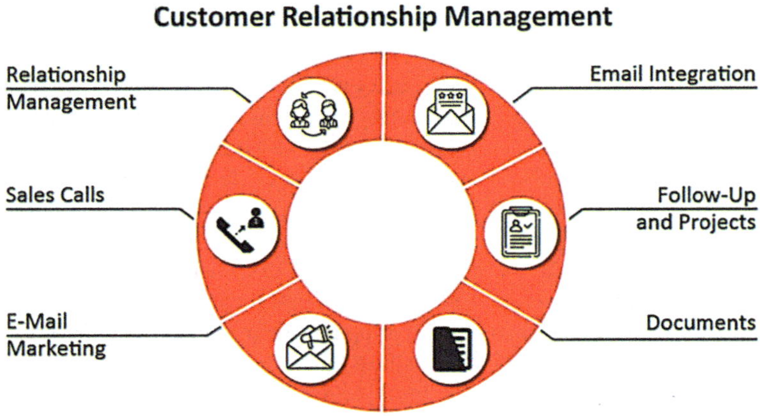

Fig. 1 The overview of customer relationship management (CRM) system

Customer service gives employees the knowledge they need to understand their customers' delight, handle issues, organization goals, and objectives, and respond to complaints. CRM systems provide a clear view of customer engagement, enabling business owners and managers to make sensible decisions to improve it and boost revenue.

CRM systems use desktop computers, mobile apps, cloud apps, and browser-based software to collect and organize customer data. This CRM software collects information on customers and how they engage with your company, which can be used to 1. Drive Product Development, 2. Increase in the sales, 3. Finding new customers, 4. Personalize advertising and Marketing, 5. Guide the consumer's buying journey and their experience, and 6. Improve customer service. Collecting specific data about business clients is the key to a successful CRM system. There should be data capturing points at each point in which the organization interacts with its consumers.

Depending on when and how our staff communicates with clients, there are various opportunities to collect CRM data. Depending upon which knowledge regarding our clients is valuable to our business, the information we require will differ. Figure 2 shows the different types of data normally collected and processed in the CRM system.

Name of the customer, preferred method of communication, and client contact information (social media accounts, website address, phone, physical address, email, etc.), and how the customer discovered your organization is all included in the Contact Details section. The personal profile includes group associations or memberships (which might be leveraged to generate new business), hobbies (which can be used to give high-level clients bonuses or prizes), and family information (which includes birthdays, anniversaries, and other milestones). When you create personal ties with particular customers, this type of information is usually obtained over time. Response to advertising campaigns, promotions, and other marketing efforts are coming under

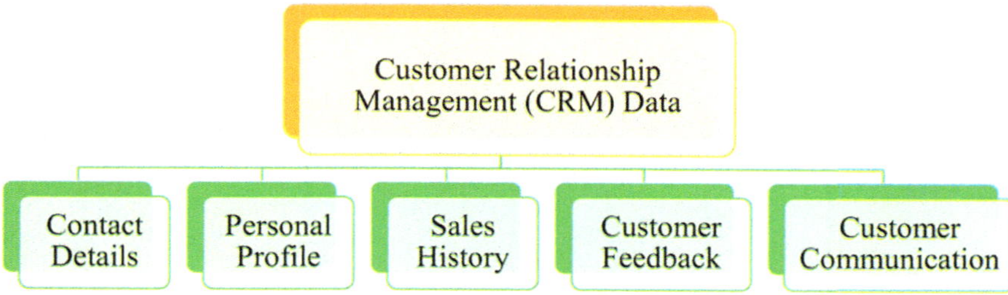

Fig. 2 Types of CRM data

Sales history data. It also includes details about credit terms and payment history, method of payment (Credit card, Debit card, cheque, net banking, or cash), services obtained and products purchased, and transaction amounts and time or date, and if purchases are made on credit. This CRM data is quite important for data analysis. Purchasing habits can be utilized to customize product portfolios to the interest of customers. Customer feedback on advertisements and promotions might help you fine-tune your marketing plan. When it comes to late payment concerns, a credit payment history can be helpful.

Collecting information on client communication allows you to contact customers via their chosen communication method, sending out notifications to encourage more sales in the future and more. You may also want to acquire the following information: 1. Sales calls, emails, or promotions; 2. Likely responses to various contact methods (phone calls, text messaging, email, etc.), and 3. Preferred method of communication. Responses to customer surveys and how customer service concerns are resolved are examples of consumer feedback (Follow-up correspondence, such as whether the consumer was satisfied, whether a refund was issued, and so on.), Product returns, Customer complaints, and information about support calls, social media reviews or Online ratings, as well as clients that have departed (who leaves, why, and which competitor they chose).

All these processes in the CRM system collect an enormous amount of data continuously. Dealing with this large volume of data is quite complicated by using a traditional system. The automation process may also get stagnate due to the improper handling of data. Artificial Intelligence (AI) techniques like Data Mining (DM) can be adopted to overcome these problems. DM approaches work on this huge data and process accordingly and make it suitable for further process.

4 Data Mining (DM)

Data mining (DM) is one of the most helpful approaches for extracting valuable information from large data sets by individuals, researchers, and enterprises. In layman's terms, mining is the process of extracting precious minerals. Data is the

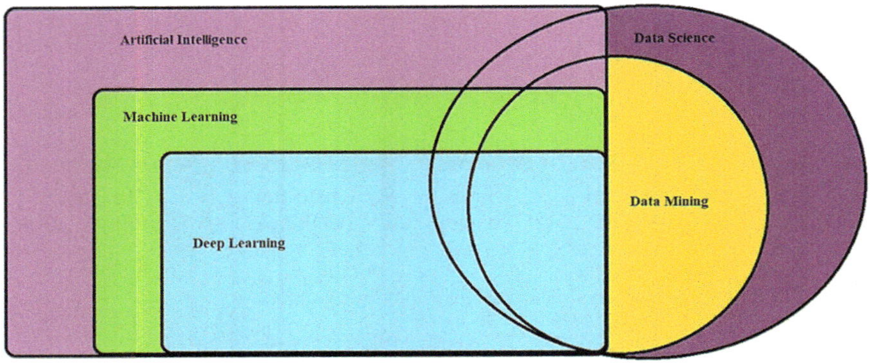

Fig. 3 Relation between DM with AI, ML, DL, and data science

most valuable mineral in the twenty-first century. We utilize DM approaches to extract usable data from a set of raw data. It is also known as Database Knowledge Discovery. Data cleansing, Integration of data, selection, and transformation of data, data mining, pattern evaluation, and knowledge presentation are all part of the knowledge discovery process from the database [42]. Organizations use it to extract particular data from large datasets to solve business problems. DM's main function is to convert raw data into valuable information. DM is the subfield of Data Science. It is also part of or a mixture of Artificial Intelligence (AI), Machine Learning (ML), Deep Learning (DL). Figure 3 shows the connection of DM with Data Science, AI, ML, and DL. The main roles of DM are 1. Developing and predicting foreseen models, 2. Discovering new and hidden patterns, and 3. Extracting the useful information from the raw data from the database.

4.1 Types of Data Mining

The following categories of data can be used for data mining: 1. Transactional Database, 2. Object-Relational Database, 3. Data Repositories, 4. Data Warehouses, and 5. Relational Database.

4.1.1 Transactional Database

This type of database refers to a database management system (DBMS) that has the ability to undo a database transaction if it isn't completed correctly. Despite the fact that this was once a unique function, most relational database systems now offer transactional database activities.

4.1.2 Object-Relational Database

An object-relational model is a hybrid of an object-oriented database model with a relational database model. It supports Classes, Objects, and Inheritance, among other things. One of the main goals of the object-relational data model is to bridge the gap between relational databases and the object-oriented model techniques common in many programming languages, such as C++ , Java, and C#.

4.1.3 Data Repositories

The Data Repository is a broad term for a data storage location. On the other hand, many IT professionals use the phrase to refer to a certain type of arrangement within an IT organization. For example, a collection of databases in which a company has stored numerous types of data.

4.1.4 Data Warehouses

It is the technology that gathers data from various sources within a company to deliver useful business insights. The massive volume of data comes from a variety of sources, including Marketing and Finance. The retrieved data is used for analytical purposes and assists a corporate organization in making decisions. Therefore, rather than transaction processing, the data warehouse is intended for data analysis.

4.1.5 Relational Database

It is a collection of many data sets that are officially organized by tables, records, and columns that can be accessed in various ways without needing to know the database tables. Tables let people find and share information, making data search, reporting, and organization easier.

4.2 Steps in Data Mining

Extracting useful information and knowledge discovery are the most important parts of the Data Mining process. The important steps which are involved in Data Mining are shown using the flowchart in Fig. 4.

The data is cleaned in the data cleansing process to make no noise or irregularity in the data. We merge various data sources into one during the Data Integration process. The next step is known as Data Selection. In this, the required data is extracted from the database. In the next step of Data Transformation, the data is transformed to execute the summary analysis and inclusion actions. Data Mining is the next step in

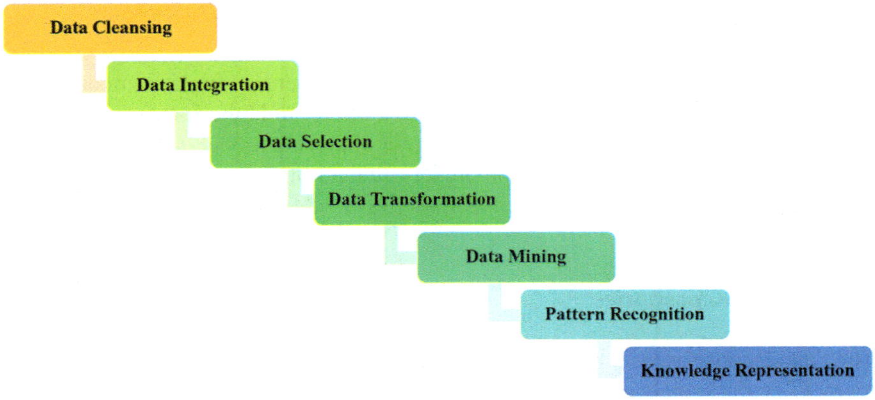

Fig. 4 Steps involved in data mining process

this useful data extracted from the pool of existing data in the database. Several hidden and meaningful patterns are recognized in the next step of Pattern recognition [43]. In the last and final step in which as per the pattern recognition and useful information extracted from the database, the knowledge information is represented to the user in the form of matrices, tables, trees, and graphs. The entire process of knowledge extraction from the raw and unstructured database using Data Mining is shown in Fig. 5.

4.3 Important Data Mining Techniques

The important techniques of Data Mining are as follows, 1. Clustering Analysis, 2. Classification Analysis, 3. Regression Analysis, 4. Choice Modelling, 5. Rule Induction, 6. Neural Network, and 7. Association Rule Learning.

4.3.1 Clustering Analysis

Clustering analysis helps marketers discover both differences and similarities in data. It is a form of DM method that is characterized as identifying DM instruments that are comparable to each other. Clusters can be utilized to improve targeting algorithms because they have common properties. For instance, if a specific group of customers is buying a specific brand of goods, a campaign can be developed to promote that product [44]. Understanding this can assist brands in raising their sales conversion rates, resulting in increased brand power and engagement. Clustering analysis also leads to the formation of personas. Personas, which are a crucial component of

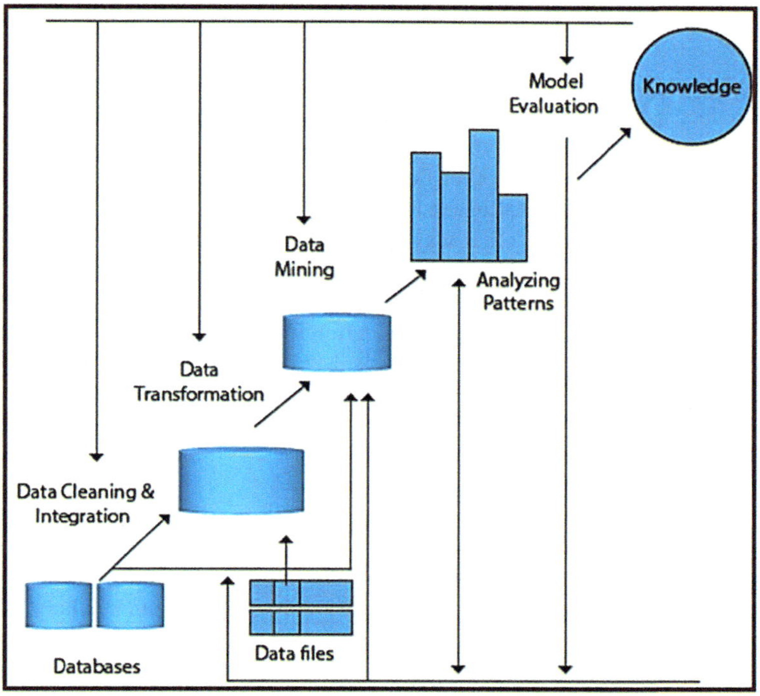

Fig. 5 Knowledge extraction from the unstructured raw database using data mining

clustering analysis, assist firms in making wise marketing decisions and creating effective campaigns.

4.3.2 Classification Analysis

This data mining strategy uses a systematic approach to gather vital and relevant information about metadata (information about data) and data; classification analysis aids brands in identifying distinct types of data mining approaches [45, 46]. This analysis is closely related to cluster analysis because they both help you make better data mining tool choices. Email is a well-known example of classification analysis since it employs algorithms to sort emails into legitimate and spam categories. It's done via data mining software on the message, which looks for terms and attachments that signal whether an email is spam or not.

4.3.3 Regression Analysis

Regression analysis is another method that helps brands determine the relationship between factors. This is based on the assumption that there is a one-way causal influence from one variable to the response of another variable. While independent variables might influence each other, dependency is rarely affected in both directions, as it is in correlation analysis. Regression analysis can reveal if one variable is dependent on another, but not the other way around. Because regression analysis is good at determining customer satisfaction, it may help brands learn new and interesting things about customer loyalty and how external elements like weather disturb service stages. This data mining technique's application in matrimonial website matching is an excellent example of regression analysis. Many websites use variables to match users based on their interests, hobbies, and preferences.

4.3.4 Choice Modelling

Choice modeling is a data mining skill that allows brands to make probabilistic predictions about their customers' judgment behavior. Choice modeling enables firms to focus their data mining strategies on customers who are most likely to make a genuine purchase. It is also utilized to figure out the most important factors that assist a customer in making a decision. It aids brands in determining the possibility of customers attaining a marketing decision depending upon variables such as attitudes, preceding purchases, and locations. Investing in choice modeling can help brands boost their revenues quickly and comprehensively.

4.3.5 Rule Induction

Rule induction is another data mining technique that facilitates the generation of formal rules consisting of observations. The rules of this DM technique can be used to generate local trends in data or a scientific model for a data-mining program. The association rule is also part of the induction paradigm. The technique of determining convincing links between variables, especially in huge databases, is known as the association rule. Data mining software technology aids brands in identifying patterns between particular products. When a customer purchases butter, for example, there's a significant chance they'll also purchase bread. The association rule's main purpose is to detect that if a client performs a specific function, such as X, the likelihood of achieving function Y is high. This knowledge can assist firms in forecasting sales and developing smart marketing strategies such as promotional pricing and improved product positioning in stores and malls.

4.3.6 Neural Network

Neural networks, a formative stage in data mining technology, have their own set of benefits and advantages. The most important benefit of a neural network is that it produces highly accurate predictive models that may solve a variety of issues. Artificial and neural networks are the two types of networks. Real neural networks, or human brains that can create patterns and predictions, are biological. It makes decisions about the circumstances during this process. Artificial program is those that are installed on computer systems. The term "artificial neural networks" comes from a historical trend in which scientists attempted to enable computer software to think like the human brain. Even though the brain is a far more complex organism, neural networks can accomplish many of the same activities like the human brain. Although it is impossible to pinpoint when neural networks were first used for data mining, a study of this data mining technique was identified during WWII. Since then, a neural network has come a long way, and many data analysts have used it to address real-world prediction problems and improve algorithm outcomes.

Furthermore, many of the most notable neural network innovations have been in applying challenges like enhancing consumer prediction or fraud detection. They can assist organizations in discovering new and improved ways to engage with customers. Neural networks have effectively assisted brands and organizations with a variety of issues, such as detecting credit card theft. They've also been used in the military to automate uncrewed vehicles' operations and fix the English language in written documents. One of the most difficult tasks for a company is determining which option is the best. The ideal technique to adopt is determined by the sort of problems that the brand is trying to tackle through data mining. Occasionally, trial and error will assist a brand in better resolving this issue. However, it is also a reality that markets and clients are always changing and completely dynamic. Because it is practically difficult to anticipate the future accurately, these dynamics have ensured that there will never be a flawless data mining technique.

Nevertheless, they are essential because they can assist scientists and organizations in using appropriate data mining software and better adapting to the changing environment and economy. Because the more models there are for data mining approaches, the more business value can be created for the brand, which can help construct models that will help anticipate a shift in a much more concentrated and enhanced manner. Overall, it assists brands in a far more scientific and systematic understanding of data mining tools, enabling and ensuring greater brand connect on the one hand and a better growth story on the other.

4.3.7 Association Rule Learning

The goal of this sort of data mining is to find intriguing relationships between variables in massive databases. The goal of this data mining technique is to find hidden patterns in the data. They can be used to find variables in the data as well as co-occurrences of distinct variables with the highest frequencies. The association rule data mining technique is widely employed in retail businesses to detect trends in point-of-sale data. This data mining software can suggest new items, particularly for determining which products individuals refer to others or identifying new products to propose to customers. Association rule learning, a useful data mining technique, can help boost a brand's conversion rate. In 2004, Walmart presented a good example of association learning's effectiveness. Strawberry pop-starts sales jumped by seven times before a hurricane, according to these data mining algorithms. Walmart has been placing this product near checkouts before hurricanes since this discovery, resulting in higher sale conversions.

4.4 Advantages of Data Mining

- Data mining aids an organization's decision-making process.
- It allows for the automatic finding of hidden patterns as well as trend and behavior prediction.
- It's a rapid technique that allows new users to examine large amounts of data in a short amount of time.
- Data mining is a cost-effective alternative to conventional statistical data applications.
- It is possible to introduce it into both new and existing systems.
- Data mining enables businesses to make profitable changes to their operations and production.
- Organizations can collect knowledge-based data using the Data Mining technique.

5 Application of Datamining in Customer Relationship Management

Customer relationship management (CRM) is becoming a more significant aspect of E-Business success. It is the process of managing consumer relations with the company. It entails segmenting customers according to their purchasing power to uncover high-profit opportunities from which marketing strategies are developed.

5.1 Customer Segmentation

It's the process of categorizing clients into homogeneous groups based on common characteristics. It is the technique of segmenting a client base into groups of similar marketing-related factors such as spending habits, interests, gender, and age. In Data Mining, the clustering technique is used to find a group of clients based upon these attributes [47]. Customers with comparable attribute values are grouped together in each cluster. One of the most significant advantages of clustering in this fashion is that the firm or product developer may better understand the needs and behaviors of each customer group.

5.2 Click Stream Analysis

In addition to classifying individuals based on transaction or personal data, the paths visitors, or Click streams, take through a website provide valuable knowledge. Click-stream information is used to identify how customers navigate an online store by the retailers. It's useful for determining the efficiency of marketing campaigns, such as how customers locate the online store, what things they look at, and what they buy. The data reveal that the time and urgency of the transaction are essential aspects to consider for individuals who rarely utilize online shopping facilities. On the other hand, price and human resources were emphasized as the most important aspects for individuals who purchase online regularly. Findings like these help businesses better understand their clients and meet their requirements and preferences.

5.3 Customer Profitability

Although segmentation and clickstream analysis provide detailed information about the consumers, Data Mining can also predict customer profitability. In other words, a company can determine its most potential consumers and focus its marketing efforts around them. Neural Network, Choice Modelling, Association Rule Learning are all very important Data Mining tools to predict the future market and are useful in creating precautionary steps to improve the profitability of the companies. In addition, it allows the product developers to develop a suitable product for their valuable customers based on their future needs.

5.4 Marketing Recommendations

Single target groups are identified using cluster analysis. To make marketing forecasts, regression analysis is used. Analysis of classification to determine spam and other things. To recognize any abnormalities, anomaly detection is used. Detection of intrusion for better system security to discover links between data, learn the association rule. To optimize project risk management, use decision trees. Neural networks are utilized for learning automation predictive data-based analysis and induction rule data warehousing to process large amounts of information.

5.5 *Targeted Marketing*

Marketing management revolves around targeting. It's all about giving the right product to the right customer at the right time and through the appropriate channel. Indeed, marketing had progressed from the days of mass marketing, when everyone was exposed to the same product, to today's fragmented and diverse marketplaces. The emphasis has shifted away from the product and onto the customer. The goal has moved from gaining market share to increasing customer share and improving customer loyalty and happiness. Recent advances in computer and database technology are assisting these goals by utilizing database marketing, data mining, and, more recently, CRM technologies to understand the client better and only approach her with items and services that she is interested in.

6 Summary and Conclusion

The present work discussed the importance of Customer Relationship Management (CRM) in the present world. It also explains various ways to collect data at the customer meeting points and store it in the database. Normally, these data are unlabeled, uncategorized raw data. Processing this much data and extracting valuable information from it is a very complicated and time-consuming process. The conventional methods to extract useful information from the actual data are not precise and accurate. Artificial Intelligence-based Data Mining tools are employed to overcome this. The current work discussed various Data Mining approaches and the processing steps involved in the Data Mining Process. The various advantages of adopting Data Mining techniques in a modern business environment are also listed. In addition to that, the Application of Data Mining in Customer Relationship Management and its significant impact on E-Business is also being discussed. With this, we can conclude that adopting Data Mining Tools in the modern business environment makes it more profitable and successful.

References

1. Buttle F (2008) Customer relationship management. Routledge
2. Kumar V (2010) Customer relationship management. Wiley international encyclopedia of marketing
3. Payne A, Frow P (2005) A strategic framework for customer relationship management. J Mark 69(4):167–176
4. Winer RS (2001) A framework for customer relationship management. Calif Manage Rev 43(4):89–105
5. Knox S, Payne A, Ryals L, Maklan S, Peppard J (2007) Customer relationship management. Routledge

6. Chen IJ, Popovich K (2003) Understanding customer relationship management (CRM). Bus Process Manag J. https://doi.org/10.1108/14637150310496758
7. Bhat SA, Darzi MA (2016) Customer relationship management. Int J Bank Mark. https://doi.org/10.1108/IJBM-11-2014-0160
8. Baashar Y, Alhussian H, Patel A, Alkawsi G, Alzahrani AI, Alfarraj O, Hayder G (2020) Customer relationship management systems (CRMS) in the healthcare environment: a systematic literature review. Comput Stand Interfaces 71:103442
9. Dewnarain S, Ramkissoon H, Mavondo F (2021) Social customer relationship management: a customer perspective. J Hosp Mark Manag. https://doi.org/10.1080/19368623.2021.1884162
10. Gil-Gomez H, Guerola-Navarro V, Oltra-Badenes R, Lozano-Quilis JA (2020) Customer relationship management: digital transformation and sustainable business model innovation. Eco Res-Ekonomska Istraživanja 33(1):2733–2750
11. Mitchell R, Michalski J, Carbonell T (2013) An artificial intelligence approach. Springer, Berlin
12. Schoen S, Sykes W, Little AD (1987) Putting artificial intelligence to work: evaluating and implementing business applications
13. Reitman WR (Ed) (1984) Artificial intelligence applications for business: proceedings of the NYU symposium, May, 1983. Intellect Books
14. Min H (2010) Artificial intelligence in supply chain management: theory and applications. Int J Log Res Appl 13(1):13–39
15. Muthusamy V, Slominski A, Ishakian V (2018) Towards enterprise-ready AI deployments minimizing the risk of consuming AI models in business applications. In: 2018 first international conference on artificial intelligence for industries (AI4I), IEEE, pp 108–109
16. Jordan MI, Mitchell TM (2015) Machine learning: trends, perspectives, and prospects. Science 349(6245):255–260
17. Bose I, Mahapatra RK (2001) Business data mining—a machine learning perspective. Inf Manag 39(3):211–225
18. Finlay S (2021) Artificial intelligence and machine learning for business: a no-nonsense guide to data driven technologies, 4th ed. Relativistic
19. Canhoto AI, Clear F (2020) Artificial intelligence and machine learning as business tools: a framework for diagnosing value destruction potential. Bus Horiz 63(2):183–193
20. Khan WA, Chung SH, Awan MU, Wen X (2019) Machine learning facilitated business intelligence (Part I): neural networks learning algorithms and applications. Ind Manag Data Syst 120(1):164–195
21. Goodfellow I, Bengio Y, Courville A, Bengio Y (2016) Deep learning, vol 1. MIT press, Cambridge
22. LeCun Y, Bengio Y, Hinton G (2015) Deep learning. Nature 521(7553):436–444
23. Kraus M, Feuerriegel S, Oztekin A (2020) Deep learning in business analytics and operations research: models, applications and managerial implications. Eur J Oper Res 281(3):628–641
24. DeLotell PJ, Millam LA, Reinhardt MM (2010) The use of deep learning strategies in online business courses to impact student retention. Am J Bus Educ 3(12):49–56
25. Howard J (2013) The business impact of deep learning. In: Proceedings of the 19th ACM SIGKDD international conference on Knowledge discovery and data mining, pp 1135–1135
26. Mehdiyev N, Evermann J, Fettke P (2017) A multi-stage deep learning approach for business process event prediction. In: 2017 IEEE 19th conference on business informatics (CBI), vol 1, IEEE, pp 119–128
27. Raj EFI, Balaji M (2021) Analysis and classification of faults in switched reluctance motors using deep learning neural networks. Arab J Sci Eng 46(2):1313–1332
28. Raj EFI, Kamaraj V (2013) Neural network based control for switched reluctance motor drive. In: 2013 IEEE international conference on emerging trends in computing, communication and nanotechnology (ICECCN), IEEE, pp 678–682
29. Gampala V, Kumar MS, Sushama C, Raj EFI (2020) Deep learning based image processing approaches for image deblurring. Materials Today: Proceedings
30. Agarwal P, Ch MA, Kharate DS, Raj EFI, Balamuralitharan S (2021) Parameter estimation of COVID-19 second wave BHRP transmission model by using principle component analysis. Annals of the Romanian Society for Cell Biology, pp 446–457

31. Hand DJ, Adams NM (2014) Data mining. Wiley StatsRef: statistics reference online, pp 1–7
32. Chung HM, Gray P (1999) Data mining. J Manag Inf Syst 16(1):11–16
33. Larose DT, Larose CD (2014) Discovering knowledge in data: an introduction to data mining, vol 4. John Wiley & Sons
34. Chen MS, Han J, Yu PS (1996) Data mining: an overview from a database perspective. IEEE Trans Knowl Data Eng 8(6):866–883
35. Rygielski C, Wang JC, Yen DC (2002) Data mining techniques for customer relationship management. Technol Soc 24(4):483–502
36. Berry MJ, Linoff GS (2004) Data mining techniques: for marketing, sales, and customer relationship management. John Wiley & Sons, Hoboken
37. Sharma S, Goyal DP, Mittal RK (2008) Data mining research for customer relationship management systems: a framework and analysis. Int J Bus Inf Syst 3(5):549–565
38. Chen Y, Zhang G, Hu D, Wang S (2006) Customer segmentation in customer relationship management based on data mining. In: international conference on programming languages for manufacturing. Springer, Boston, MA, pp 288–293
39. Shokouhyar S, Shokoohyar S, Raja N, Gupta V (2021) Promoting fashion customer relationship management dimensions based on customer tendency to outfit matching: mining customer orientation and buying behaviour. Int J Appl Decis Sci 14(1):1–23
40. González-Serrano L, Talón-Ballestero P, Muñoz-Romero S, Soguero-Ruiz C, Rojo-Álvarez JL (2021) A big data approach to customer relationship management strategy in hospitality using multiple correspondence domain description. Appl Sci 11(1):256
41. Pacha NH, Khebazi FZ, Mazouz N (2021) Data mining and its contribution to decision-making in business organizations. In: Sedkaoui S, Khelfaoui M, Kadi N (eds) Big data analytics. Apple Academic Press, NE Palm Bay, pp 67–80
42. Hernández-Nieves E, Parra-Domínguez J, Chamoso P, Rodríguez-González S, Corchado JM (2021) A data mining and analysis platform for investment recommendations. Electronics 10(7):859
43. Baloch S, Muhammad MS (2021) An intelligent data mining-based fault detection and classification strategy for microgrid. IEEE Access 9:22470–22479
44. Zhao Y, Chang C, Hannum M, Lee J, Shen R (2021) Bayesian network-driven clustering analysis with feature selection for high-dimensional multi-modal molecular data. Sci Rep 11(1):1–11
45. Omuya EO, Okeyo GO, Kimwele MW (2021) Feature selection for classification using principal component analysis and information gain. Expert Syst Appl 174:114765
46. Neelakandan S, Rene Beulah J, Prathiba L, Murthy GLN, Irudaya Raj EF, Arulkumar N (2022) Blockchain with deep learning-enabled secure healthcare data transmission and diagnostic model. Int J Model Simul Sci Comput. https://doi.org/10.1142/S1793962322410069
47. Pradana MG, Ha HT (2021) Maximizing strategy improvement in mall customer segmentation using K-means clustering. J Appl Data Sci 2(1):19–25

Chapter 13
AI Integrated Human Resource Management for Smart Decision in an Organization

S. B. Goyal, Pradeep Bedi, Anand Singh Rajawat, Deepmala Singh, and Prasenjit Chatterjee

Abstract Employees are viewed as an asset that leads to an organization's growth. As competition grows in the market, the attrition of employees increases. Consequently, employee attrition can be a major problem for businesses, particularly when there is mobility or movement of technically skilled employees, in search of better opportunity. This results in financial loss for a company or organization. Therefore, in this chapter, advantage of machine learning algorithm is presented to predict the behavior of current employees' employees. In this research, Long Short-Term Memory (LSTM) algorithm and fuzzy rules are used to analyzes attrition actions of workers by forecasting a shortage of qualified employees by department and sending out a warning message about recruiting new employees or distributing the workload among existing employees. This template also ensures that incentive and promotion decisions are made without bias, assisting the organization's growth indirectly.

Keywords Supply chain management · Industry 4.0 · IoT · Machine learning · Big data · Blockchain · Production planning

---

S. B. Goyal (✉)
City University, Petaling Jaya, Selangor 46100, Malaysia
e-mail: drsbgoyal@gmail.com

P. Bedi
Galgotias University, Greater Noida, Uttar Pradesh, India

A. S. Rajawat
School of Computer Sciences and Engineering, Sandip University, Nashik 422213, India

D. Singh
Symbiosis Centre for Management Studies (SCMS), Symbiosis International (Deemed University) (SIU), Mouza-Wathoda, Nagpur, Maharashtra, India

P. Chatterjee
Department of Mechanical Engineering, MCKV Institute of Engineering, Howrah 711204, West Bengal, India

S. Kautish et al. (eds.), *Computational Intelligence for Modern Business Systems*, Disruptive Technologies and Digital Transformations for Society 5.0,
https://doi.org/10.1007/978-981-99-5354-7_13

1 Introduction

Nowadays, one of the most significant or important problems that an organization faces is employee attrition/churn. This condition is quite unpredictable and causes an unnoticeable void in an organization [1]. The unexpected attrition of employees causes a delay in services or the productivity of an organization gets compromised [2].

Though a corporation's workers are the most important assets, managers need all the resources they can get to better execute their employees' skills and abilities [3]. An inventory of skills is a tool that managers can use. The use of a skill inventory can become very helpful for achieving company objectives and making smarter choices on personnel management. A staff, also known as Skills, is a list of existing employees' credentials, training, and experience. The method of building a skills inventory varies depending on the size and complexity of an organization. Commercial software is used by several businesses. The database programs are used by other companies. Some small firms are creating the skills stock through the use of simple sheets or a paper-based system [4].

Regardless of the terminology used, a collective stock of skills, knowledge, and professional experience of an employee of the enterprise will be made available, once completed correctly [5]. This data on the inventory of abilities can then be used by the organization to enhance decision making in several areas, including:

- Employing personnel to better meet the needs of the various business units both today and future [15].
- Assign the right staff to the right tasks.
- Staffing the best talented internal project teams to ensure organizational success.
- To close existing ability gaps, focus training and growth efforts.
- Identify key staff for future business requirements.
- Creating an internal talent pipeline to replace key workers and managers who leave the business.
- Create a workforce plan for the company's future strategic needs.

Human resources can benefit from an improved expertise database by reviewing and accurately evaluating the information on different workers within the company. This information can then be used to make decisions about things like promotions, training, changes, and the organization's potential recruiting needs [6].

However, the people management stock must be maintained on a regular interval to maintain that it remains relevant and useful. Precise information collected from every company employee is often a tedious and time-consuming activity on paper. The static human resources inventory, also known as a "snapshot" of an organization, has no value [7].

The following are a few of the most commonly used methods for predicting employee attrition:

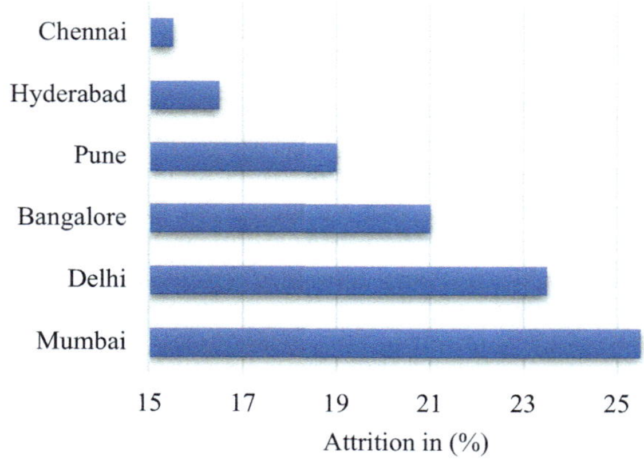

Fig. 1 Employee attrition trends by cities

IBM HR Analytics: IBM Kenexa HR Analytics, provided by IBM Watson Analytics, is one of IBM Software's products in the Talent Management Software (TMS) category. Employee turnover prediction is one of the IBM Workforce Analytics solutions [8]. It analyzes the key wear variables and predicts employee wear.

SAP Workforce Analytics: Predictive analytics are used by SAP Success Factors to address questions about sales and to detect and assess the risks of fraud. IBM and SAP, for example, have created a solution and their technology to anticipate workplace friction so that the organization can find a substitute for an employee ahead of time.

A. Attrition Trends in India

At 25.7%, Mumbai has the highest annual attrition rate in analytics. The city has a thriving data science ecosystem, which is dominated by the banking and finance industries. Mumbai has moved to the top of the list in terms of employee turnover due to high wages and demand for data scientists [9] (Fig. 1).

On the other hand, Chennai has the lowest attrition rate, which is only 15.5%. Large cities with larger analytical employees (Bangalore, Mumbai & Delhi/NCR) have higher rates of attrition than 20%. Meanwhile, towns with a lower staff base have lower attrition rates, below 20%. Besides, current analytics personnel in Mumbai remain for an average of 4.2 years in their existing organizations. We found no correlation between attrition and tenure, as mentioned above. Chennai has the highest tenure of analytical professionals, at 4.5 years.

The paper is dedicated to discussing about application of machine learning-based human resource management for employee attrition. It was observed from a brief literature review that machine learning techniques show their efficiency to benefit the HRM of an organization [10, 13]. Finally, an AI-integrated HRM framework is proposed for the development of organization which is focused toward its adaptability with machine learning algorithms.

2 Related Work

Alduayj and Rajpoot [1] studied employee attrition using machine learning model. Soni et al. [2] examined the worker features and different organizational variables that could result in the turnover of employees. Product inventions and associated product parameters can be duplicated, but the unity of an organization's workers can never be duplicated, so it is critical. The Artificial Neural Network (ANN) and Adaptive Neuro-Fuzzy Inference System are the two classification methods used to compare the prediction accuracy and generalization capabilities (ANFIS).

Sisodia et al. [3] proposed a model that will predict employee churn rate. The correlation matrix and heatmap are created to demonstrate the relationship between attributes. The histogram is created in the experimental section, which demonstrates the comparison between left employees vs. compensation, department, satisfaction level, and so on. We use five different machine learning algorithms for prediction. This article suggests the factors that help an organization's employee turnover rate to be as low as possible.

The Economic Order Quantity Model (EOQ), introduced in 1913 by Ford W. Harris, was the first statistical model for inventory management [4]. It was created for the manufacturing process. EOQ is a deterministic and dynamic mono-product model that is very basic. The model identifies the best method for determining the inventory system's actions. It's also easy to measure the closed solution.

The inventory management and stock forecasting method were the subjects of [5]. A new approach has been established for a stock management system to manage and locate food additives in the Codex Alimentarius Commission's global food additive database. To forecast the stock of food additives, four machine learning models were used: Naïve Bayes, Decision Tree, Linear Regression, and Support Vector Regression. Data mining techniques such as clustering, grouping, and association rule were used by [6, 14]. For the success of any retention attempt, precision and accuracy of the technique used are essential. After all, if a customer who will leave the business is unaware of the company, that customer cannot be adequately dealt with.

Dolatabadi and Keynia [7] addressed how necessary it is to predict the propensity of customers to retain customers in the face of increasing rivalry among service companies. For businesses, the effects of brand loyalty and the loss of a customer base, as well as the difficulties of attracting a new customer to replace each lost customer, are extremely difficult. To prepare and process this sort of scenario, having a forecast theory of consumer behavior can be extremely useful.

Franciska and Swaminathan [8] mentioned the aggregation of particles based on the similarity of their properties to provide user information patterns and knowledge. Partitioning, hierarchical, and grid-based clustering approaches are examples of different types of clustering algorithms. K-means clustering, k-medoids clustering, Hierarchical clustering, DBSCAN, and Fuzzy c-means clustering are all used in this study. Customer churn analysis employs clustering algorithms for a variety of reasons, one of which is that the cost of acquiring a new customer is much greater than the cost of maintaining a current customer.

Zhu et al. [9] intends to boost employee turnover forecasting with the impact of or without the economic indicators being taken into account. Several time series modeling techniques have been used to identify optimal models for effective predictions of staff turnover. The proposed models were constructed and validated over 11 years of monthly turnover data. The R2 training turnover = 0.77 and holdout R2 = 0.59 have been predicted in contrast to the other models by a dynamic regression model with additive tends, seasonality, interventions, and very important economic indicators. This study is aimed at discovering well-trained and skilled people for bonuses or replacements in any organization. The hiring of new staff always takes some huge costs for the organization. The Departments of Human Resources provide a huge number of daily data: leaves, societal disputes, yearly assessments, payments, and benefits, etc. Nevertheless, the main issue that arises here is to whom this appreciation can be provided. Further, apply artificial intelligence techniques to detect autism spectrum disorder [11] also.

Some challenges that are generally faced in an organization are [12]:

1. The attrition of employees is a trivial problem for losses of companies like financial loss, replacement and hiring costs and time, retrain new employees as well as customer dissatisfaction.
2. Somehow, a company may bear the turnover of workers that are not as seasoned as others who have worked for a long time and whose attrition often results in substantial losses.
3. Attrition of employees can give existing employees a negative impression.
4. In the manual assessment of employees' performance, the HR department can be partial.

3 Methodology

The proposed methodology works in different levels termed multi-agent architecture which performed their function as discussed below:

1. In the very first step, input dataset is taken in which records of employees are mentioned.
2. The raw data is passed to Information Selection Agent (ISA).
3. Attrition Detection Agent (ADA) decides the behavior of employees using a machine learning tool i.e., LSTM.
4. Further, such employees that don't show attrition behavior are processed to Appreciation Providing Agent (APA) to provide appreciation/bonus in the future.

A. Information Selection Agent (ISA)

Co-relation is used to determine whether there is an association between features or details about employees. It's a feature selection technique of some kind. Feature selection is a technique of optimization to reduce the data dimensionality in various areas such as machine learning, pattern recognition, and data extraction. Its primary goal is to eliminate features that are obsolete or irrelevant. These aspects surely improve the classification algorithms' performance in terms of time and cost and give the data mining systems little or no predictive information. It is not quite easy to remove unnecessary properties because the relevance or irrelevance of a specific feature in the target class is difficult to determine. An evaluation criterion measures the relevance of a particular feature between two features or among many features.

Once this algorithm is used, Information Selection Agent (ISA) is retrieved to determine the attrition between employees that helps the Attrition Detection Agent (Fig. 2).

B. Attrition Detection Agent (ADA)

The Attrition Detection Agent (ADA) uses an LSTM Classifier to predict employee attrition. Long short-term memory is one kind of recurring neural network (RNN) (LSTM). It's a convolutional neural network technique made up of several neural network modules. There are four units in the LSTM network: memory cell, input unit, output unit, and forget unit. The storage device is the device that stores the data values over time and controls the flow of data values for the output value evaluation by the three remaining units. Both classification and regression are performed using the LSTM deep network and train the features concerning time instance t (Fig. 3).

The output of hidden layer h_t is evaluated as following equations:

$$\text{fun}_t = \sigma(Wt_{\text{forget}} * x_t + Ut_{\text{forget}} * h_{t-1} + \text{bais}_{\text{forget}}) \tag{1}$$

$$i_t = \sigma(Wt_{\text{input}} * x_t + Ut_{\text{input}} * h_{t-1} + \text{bais}_{\text{input}}) \tag{2}$$

$$\widehat{\text{Cell}}_t = \tanh(Wt_{cell} * x_t + Ut_{\text{cell}} * h_{t-1} + \text{bais}_{\text{cell}}) \tag{3}$$

$$\text{Cell}_t = \text{input}_t * \widehat{\text{Cell}}_t + \text{forget}_t * \text{Cell}_{t-1}) \tag{4}$$

$$\text{cell\_output}_t = \sigma(Wt_{\text{output}} * x_t + Ut_{\text{output}} * h_{t-1} + \text{bais}_{\text{output}}) \tag{5}$$

$$\text{Finaloutput}_t = \text{cell\_output}_t * \tanh(\text{Cell}_t)) \tag{6}$$

where, σ = activation (sigmoid) function.

tanh = activation (tangent).

x_t = input variable at time instance t.

$Wt_{\text{input}}, Wt_{\text{Cell}}, Wt_{\text{forget}}, Wt_{\text{output}}, Ut_{\text{input}}, Ut_{\text{Cell}}, Ut_{\text{forget}}, Ut_{\text{output}}$ = weight vectors.

$\text{bais}_{\text{input}}, \text{bais}_{\text{Cell}}, \text{bais}_{\text{forget}}, \text{bais}_{\text{output}}$ = bias vectors.

C. Appreciation Providing Agent (APA)

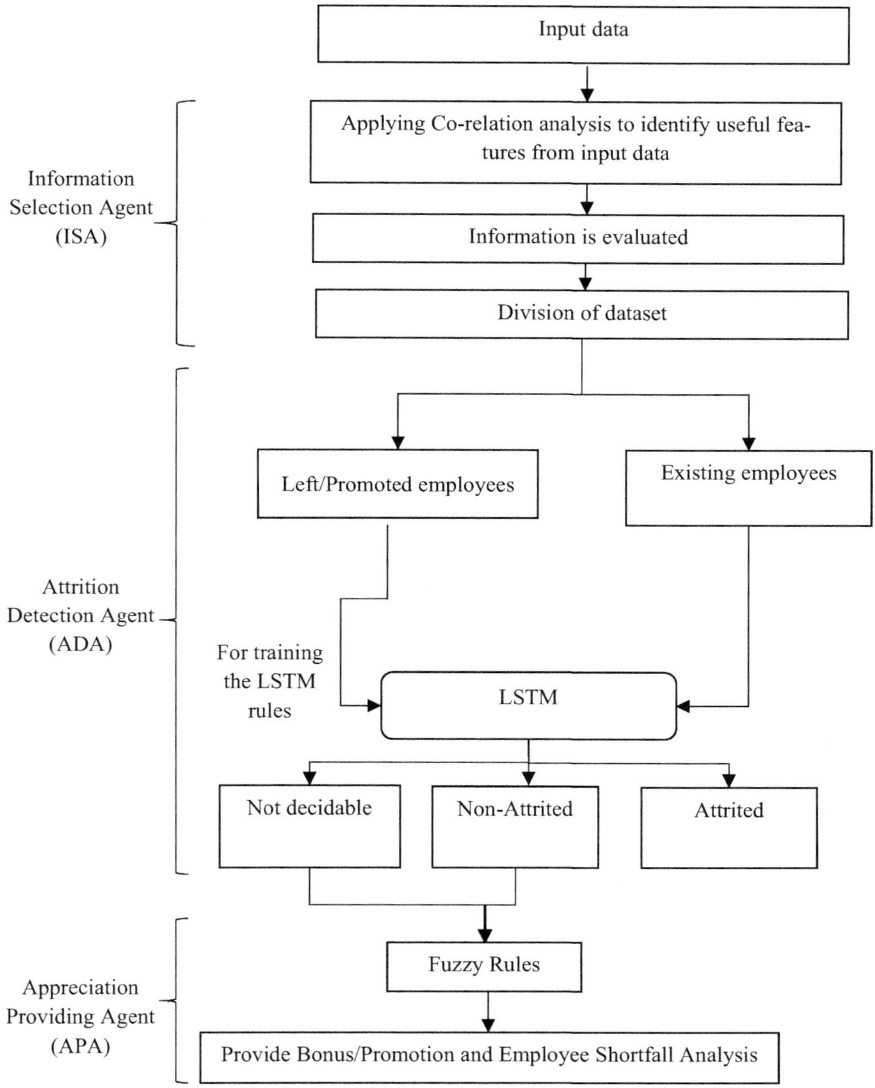

Fig. 2 Methodology for employees attrition detection

With designing of fuzzy model, bonus/promotion is provided to such employees that doesn't show attrition behaviour. This model is also compatible to predict the staff shortfall and their arrangement.

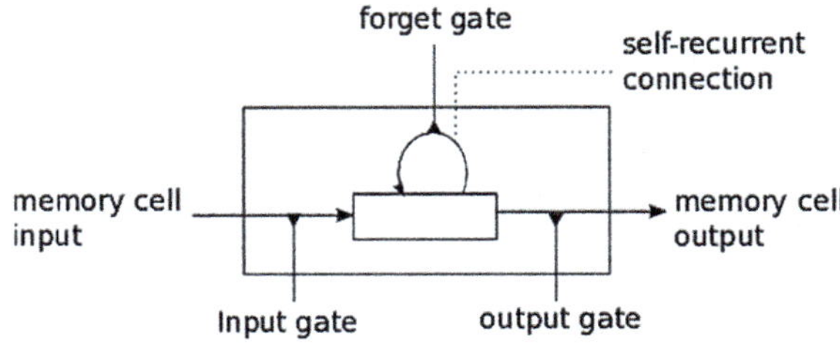

Fig. 3 Long short-term memory (LSTM) units

4 Results and Discussions

In this section, result analysis is presented on the co-relation feature extraction method that is used further to predict the attrition behavior of employees. Simulation of this work is presented over the IBM employee dataset that contains details of 1470 employees. Some of the features that contribute to the visualization step of the feature selection process are illustrated in Fig. 4.

In this section, a correlation analysis is performed to visualize how features determine employee attrition. Table 1 illustrates the correlation analysis and it has been concluded that demographic features such as educational background or gender don't

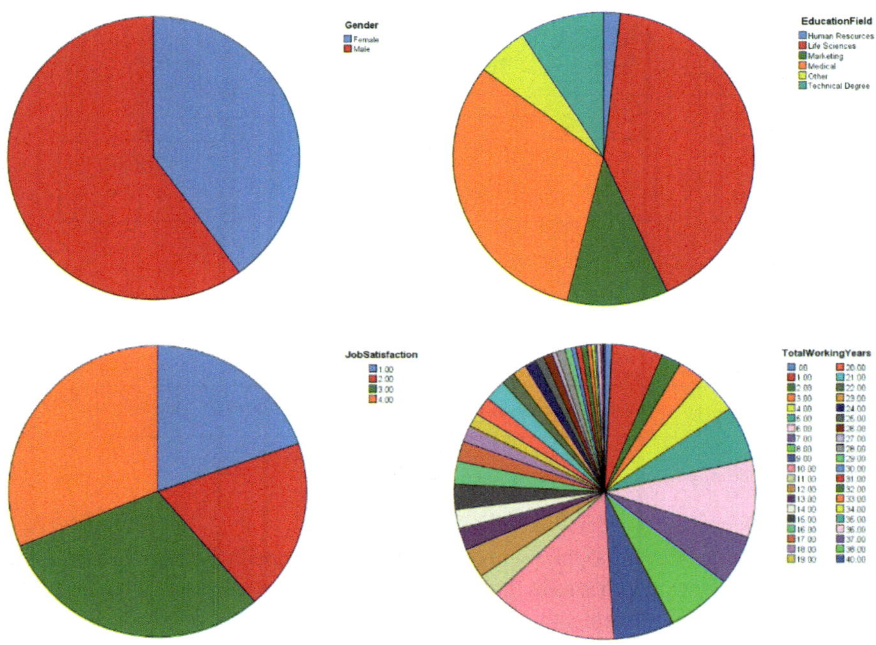

Fig. 4 Pie-Chart visualization of features for selection

Table 1 Employee attrition correlation

Features	Employee attrition correlation
Age	0.1592
Education	−0.0093
Job satisfaction	0.1035
Gender	−0.0326
Working years	0.1711
Total years of working in current position	0.1605

have an impact on employee attrition whereas other features are correlated to determine the employee's attrition behavior. So, while feature selection non-correlated features are removed to reduce further classification complexity. These unnecessary features also decrease the classification performance.

5 Conclusion

Staff turnover can have several consequences for a business, including a loss of reputation, sales, and expense concerning time and money. To solve the existing manual HR system's problem multi-agent artificial architecture is proposed that contributes to the development of smart HRM with a predictive approach that can make better decisions for the complete growth of an organization. The efficiency is assessed using the IBM HR analytical dataset. The approach employs AI techniques such as LSTM and Fuzzy logic. The system architecture is referred to as multi-agent architecture since it operates on several levels. The very first agent is called an ISA, which hands data to the second agent, called an attrition detection Agency (ADA). Co-relation is evaluated for feature selection by the Attrition Detection Agent (ADA), which finds relationships between features or employees' knowledge. LSTM is used by the Attrition Detection Agent (ADA) to classify employees for their attrition actions. Last but not least, Appreciation Providing Agent (APA) is activated for deciding the list of employees for bonus/promotion distribution. The following conclusions are derived from this chapter:

1. This framework is very effective at identifying and filtering the most qualified and worthy applicants.
2. This framework also provides the existing employees a motivating message to engage themselves and deserving applicants get a promotion/bonus without bias.
3. This model can also forecast a shortage of qualified employees by department and send out a warning message about recruiting new employees or redistributing workload between current workers.
4. This decision-making method is also effective in terms of time.

5. This model ensures that incentive and promotion decisions are made without bias and, as a result, the organization's growth is aided indirectly.

Future studies should be conducted to add to the current study's findings, as variables such as employee knowledge and information about the strategies could change over time. Researchers in the IT Service Industry will be encouraged to build and implement new constructs relevant to employee retention.

References

1. Alduayj SS, Rajpoot K (2019) Predicting employee attrition using machine learning. In: Proceedings of the 2018 13th international conference on innovations in information technology, pp 93–98. https://doi.org/10.1109/INNOVATIONS.2018.8605976
2. Soni U, Singh N, Swami Y, Deshwal P (2019) A comparison study between ANN and ANFIS for the prediction of employee turnover in an organization. In: 2018 international conference on computing, power and communication technologies, pp 203–206. https://doi.org/10.1109/GUCON.2018.8674886
3. Sisodia DS, Vishwakarma S, Pujahari A (2018) Evaluation of machine learning models for employee churn prediction. In: Proceedings of the international conference on inventive computing and informatics, pp. 1016–1020. https://doi.org/10.1109/ICICI.2017.8365293
4. Lippman SA (1971) Economic order quantities and multiple set-up costs. Mgmt Sci 18(1):39–47. https://doi.org/10.1287/MNSC.18.1.39
5. Tangtisanon P (2018) Web service based food additive inventory management with forecasting system. In: 2018 3rd International conference on computer and communication systems, pp 72–79. https://doi.org/10.1109/CCOMS.2018.8463339
6. Mitkees IMM, Badr SM, Elseddawy AIB (2018) Customer churn prediction model using data mining techniques. In: ICENCO 2017-13th international computer engineering conference: boundless smart societies, pp 262–268. https://doi.org/10.1109/ICENCO.2017.8289798
7. Dolatabadi SH, Keynia F (2017) Designing of customer and employee churn prediction model based on data mining method and neural predictor. In: 2nd International conference on computer and communication systems, ICCCS, pp 74–77. https://doi.org/10.1109/CCOMS.2017.8075270
8. Franciska I, Swaminathan B (2017) Churn prediction analysis using various clustering algorithms in KNIME analytics platform. In: Proceedings of 2017 3rd IEEE international conference on sensing, signal processing and security, ICSSS, pp 166–170. https://doi.org/10.1109/SSPS.2017.8071585
9. Zhu X, Seaver W, Sawhney R, Ji S, Holt B, Sanil GB, Upreti G (2016) Employee turnover forecasting for human resource management based on time series analysis. J Appl Stats 44(8):1421–1440. https://doi.org/10.1080/02664763.2016.1214242
10. Bedi P, Goyal S, Kumar J, Ritika (2021) Integration of knowledge management and web 3.0 for human resource management. In: Kautish S, Singh D, Polkowski Z, Mayura A, Jeyanthi M (eds) Knowledge management and web 3.0: next generation business models. De Gruyter, Berlin, Boston, 65–84. https://doi.org/10.1515/9783110722789-005
11. Bedi P et al (2021) Application of image processing for autism spectrum disorder. In: Kautish S, Dhiman G (eds) Artificial intelligence for accurate analysis and detection of autism spectrum disorder. IGI Global, pp 1–24. https://doi.org/10.4018/978-1-7998-7460-7.ch001
12. Sustrova T (2016) A suitable artificial intelligence model for inventory level optimization. Trends Econ Manag 10(25):48–55. https://doi.org/10.13164/TRENDS.2016.25.48
13. Reyana A, Kautish S, Vibith AS, Goyal SB (2021) EGMM video surveillance for monitoring urban traffic scenario. Int J Intell Unmanned Syst

14. Upadhyay H, Juneja S, Juneja A, Dhiman G, Kautish S (2021) Evaluation of ergonomics-related disorders in online education using fuzzy AHP. Comput Intell Neurosci
15. Uppal M, Gupta D, Juneja S, Dhiman G, Kautish S (2021) Cloud-based fault prediction using IoT in office automation for improvisation of health of employees. J Healthc Eng

Chapter 14
A q-ROF Based Intelligent Framework for Exploring the Interface Among the Variables of Culture Shock and Adoption Toward Organizational Effectiveness

Sanjib Biswas, Dragan Pamucar, Poushali Dey, Shreya Chatterjee, and Shuvendu Majumder

Abstract Nowadays, Culture shock is very much common to work life and this is an important issue in identification of the variables to develop morale of the employees towards an aim to increase productivity. The term 'cultural shock' refers the emotional state of uncertainty, confusion, anxiety that people may experience when transforming to a new state of affairs or feeling a new work culture or social status. So, it happens when an individual is censored from acquainted surroundings and culture after transferring to a new environment. It brings the logic that Culture shock tends to mean a bustle of emotions, including excitement, emotional labor, job stress, and job satisfaction and helplessness. Social scientists are of the opinion that culture shock is treated as 'mental illness' as common men are suffering as they are distracted from the cultural environment. In this context, it is pertinent to identify the impact of cultural shock during the pandemic time especially in the service sector management as people are constrained to work from home and that is also to some extent detrimental to their mental health. In this present study, the researchers are trying to explore the possible attributes that are responsible for cultural shock

S. Biswas (✉)
Decision Sciences and Operations Management Area, Calcutta Business School, South 24 Parganas, Kolkata 743503, West Bengal, India
e-mail: sanjibb@acm.org

D. Pamucar
Faculty of Organizational Sciences, Department of Operations Research and Statistics, University of Belgrade, 11000 Belgrade, Serbia

P. Dey
Indian Institute of Foreign Trade, Kolkata Campus, Kolkata 700107, West Bengal, India

S. Chatterjee · S. Majumder
HR and OB Area, Calcutta Business School, South 24 Parganas, Kolkata 743503, West Bengal, India

S. Kautish et al. (eds.), *Computational Intelligence for Modern Business Systems*, Disruptive Technologies and Digital Transformations for Society 5.0,
https://doi.org/10.1007/978-981-99-5354-7_14

and also try to measure the possible impact of the same. This study will provide a working model for the stakeholders to frame a strategy to get rid of the crisis and develop an employee retention policy with respect to stress coping behavior as well, in the days of New Normal. The study is essentially focused on tech-based gaming industry. The present chapter proposes a new q-Rung Orthopair Fuzzy (qROF) based computational intelligence framework of psychological assessment. The procedural steps of forced field analysis (FFA) is followed in this chapter wherein Level Based Weight Assessment (LBWA) is applied for calculating the weights of the attributes. Further, it carries out stability analysis of the results obtained. Finally, it puts forth policy recommendations for formulating effective employee motivation and retention strategy.

Keywords Covid-19 · Culture shock · Emotional labor · Job stress · Helplessness · q-rung orthopair fuzzy · Forced field analysis (FFA) · Level based weight assessment (LBWA) · Gaming industry

1 Introduction

Individuals, in general, face a series of unknown situations, and they suffer from a crisis like emotional distress that may result in a sudden shift into a culture or environment that is different from their own. In this context, people face difficulty coping with the situation, which can be considered culture shock. Oberg [1] first used the term culture shock to describe an individual who becomes unable to adjust to the society or culture as he might experience anxiety, discomfort, and other forms of psychological distress, which often becomes chronic and enfeebling. The recent pandemic has significantly changed the working conditions. Working in a different country, not being able to return home, working from home are some of the reasons that have made people experience some culture shock (CS). In simpler terms, CS is defined as the trauma due to an unprecedented or unfamiliar cultural transition.

Taft [2] opined that there are six aspects of CS such as psychological strain, sense of loss and deprivation, feelings of being isolated, role ambiguity, agony or anxiety due to cultural differences, and feeling if impotence. According to the model developed by Oberg [1], there are four stages or periods of CS. It begins with the honeymoon period where a person encounters a new culture or environment, and he/she feels excited, euphoric, and pleased by the new things he/she encounters [3]. After this stage, the individual enters a period of CS where he experiences negative emotions or psychological disturbances like stress, anxiety, and depression. A sense of uncertainty regarding the self, environment, and future may also arise. After a point of time, the individual slowly shifts into the stage of recovery where he becomes aware of the problems he/she is facing and starts working on them. The feeling of frustration slowly subsides as the person moves into the adjustment period. Here the individual has understood the foreign environment in which he/she has been placed. They can deal with the problems successfully and accept the new culture positively. As a result, the individual becomes flexible and works effectively in new ways with the required skills. The duration of each of the stages may vary from individual to individual, but it is expected that the person goes through the overall process of adjustment within a year [4–6]. Kathirvel and Febiula [7] mentioned some of the impacts of CS such as mental health of the employees, interpersonal communication, productivity of the individuals, lifestyle, and morale among others. As a result, organizational performance is largely affected by CS.

The recent outbreak of Covid-19 has affected the socio-economic state of the countries across the globe significantly and has brought about a radical change in the organizational culture. The sudden change in communication and working style, social distancing, increasing level of anxiety and fear psychosis due to high infection and death rates, job losses and salary reduction, and pressure to adopt new technologies have posited a transformational shift in the organizational culture [8]. In other words, the challenge of absorbing the CS and building a stress-free enduring culture in the post-Covid phase has become a critical success factor for the organizations [9, 10].

It is, therefore, a matter of paramount importance for organizations to show resilience on cultural dimensions for keeping the sound mental health of the employees for combating the effect of environmental disruptions and ensure performance. Both physical and mental well-being of the employees and the health of the organizational culture have become essential parts of the organizational strategy now. Amidst the highly turbulent, uncertain, and gloomy environment, shock leadership has emerged as one of the essential qualities of corporate governance [11]. In this context, the present study aims to answer the following questions: What are the critical manifestations of the CS vis-à-vis Covid-19? Does coping behavior (CB) withstand the impact of CS?

We attempt to understand the interplay of various manifested attributes of CS and behavioral resilience in this chapter. We utilize the dimensions and psychological assessment scale for measuring CS and CB as obtained from the extant literature and apply a Force Field Analysis (FFA). The manifested attributes of CS and CB are treated as restraining forces (RF) and driving forces (DF) respectively. Our initial assumption is that CS (resulted due to Covid-19) outperforms the natural CB of the individuals. To test this null hypothesis, we have conducted a study on the employees of gaming industry. The gaming industry is a representative of technology-based innovative firms that require creative employees. We are inquisitive about the impact of CS on the mental health of the employee that affects the performance and how CB counters the CS to show resilience. In this sense, CS is associated with mind and creativity is dependent on mental conditions. Therefore, it is justified to select the gaming industry. Now, responses to psychological manifests suffer from a considerable amount of subjective bias. Further, we use a group decision-making framework with small sample study. Hence, to combat the subjective bias and work with imprecise information, we use a fuzzy framework here. We utilize qROFS-based FFA (qROF-FFA) in our study.

The concept of qROFS was proposed in the work of Yager [12] as a generalization of Intuitionistic Fuzzy Sets (IFS) [13] and Pythagorean Fuzzy Sets (PyFS) [14].

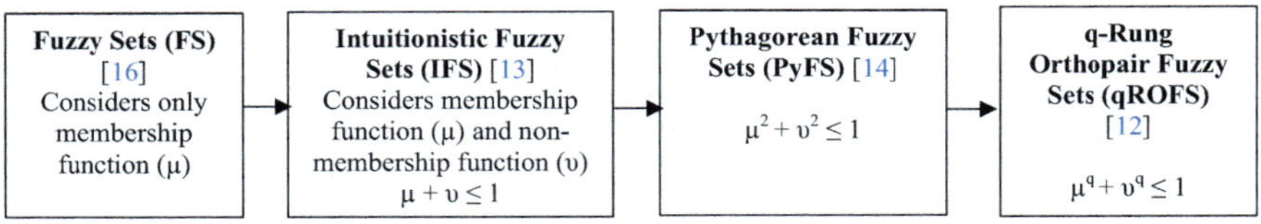

Fig. 1 Evolution of qROFS

Figure 1 shows the chronological evolution of qROFS. The need for developing qROFS stems from a common psychological behavior of human beings known as "bounded rationality" which talks about the preference of human beings for the most obvious selection. To illustrate further, there are incidents where people may opine for favorable choices whereas in other cases, opinions are made related to non-favorable perspectives. As a result, the development of IFS was necessitated which takes into account both membership (the favorable element the decision maker wants to be included in the selection set) and non-membership (the element that the decision maker does not want to have) functions. The researchers felt the need to have a generalization of IFS in terms of qROFS due to the fact in some occasions where the sum of favorable and non-favorable memberships is not bounded to unit interval as explained with an investment decision making scenario in [15]. In a typical investment decision making scenario, an investor may give opinion on incremental (membership) and detrimental (non-membership) outcomes as per his/her choice. In effect, the sum of both membership and non-membership may exceed 1. The generalized version of IFS, aka qROFS enables the analyst in similar kind of real-life situations. The advantages of using qROFS in the multi-criteria-based group decision making (MAGDM) scenarios (as compared to IFS and PyFS) include greater precision and flexibility (of having a wide range of values for μ and υ) in capturing impreciseness under uncertainty and carrying out granular analysis with accurate and stable solutions.

In the proposed framework of qROF-FFA, we calculate the aggregated score of RF- and DF-based on the opinions of the respondents (from gaming industry) on the various attributes of CS and CB respectively. To find out the aggregate scores, we use LBWA algorithm for the calculation of the weights of the individual attributes. LBWA algorithm is used for deriving the criteria weights by level (according to the relative priorities of the criteria)-based partitioning [16]. LBWA provides the following advantages:

- Ability to work with a large number of criteria. In our chapter, we deal with a large number of attributes pertaining to CS and CB. Therefore, LBWA is assumed to be apt for data analysis.
- LBWA requires only $(n - 1)$ number of pairwise comparisons for n criteria under consideration. Unlike the other popular algorithms like AHP, PIPRECIA among others, LBWA therefore is less susceptible to subjective bias and computational complexity and provides a reasonably accurate, reliable, and stable solution.

However, if the degree of comparative priority of one criterion with respect to others, LBWA suffers from a difficulty in level-based partitioning on some occasions. To sum up, our proposed qROF-FFA framework provides the decision-makers (DM) a number of benefits such as

- A better representation and capture of vagueness associated with subjective opinions.
- More granular analysis with reasonable accuracy and flexibility to the DMs.
- Ability to consider a large number of criteria and variations in the values.
- A precise comparison of DF and RF in the process of carrying out FFA.

The present chapter, to our limited knowledge, is the first of its kind that extends the growing strand of literature in the following ways

- In this chapter, a new framework of FFA using qROFS and LBWA is presented.
- No prior study has considered qROFS-based assessment of CS and CB.
- In the context of Covid-19, measuring of CS and subsequently CB to counter the impact of the shock is seen as less evidenced in the literature.
- The extant literature is found to be inclined towards measuring CS for students. We have not found plentiful applications of CS measurement in the industry, especially in the gaming industry.

The remaining part of the chapter is presented as follows. In Sect. 2, some of the related work are described. Some preliminary concepts and definitions of qROFS are discussed in Sect. 3. Section 4 sheds light on the research methodology. In Sect. 5 the results are presented and discussions are summarized. Section 6 provides some of the research implications. The concluding remarks and some of the future scope of work are given in Sect. 7.

2 Literature Review

In this section, we present a summary of some of the related past work. Evans [17] highlighted in his research work on escapism, which is a "dysfunctional avoidance coping response" to unfavorable situations that occur. Escaping from such a negative environment helps a person feel 'happy, relaxed and pleased by engaging or without engaging in any sort of activities. In this context, Warmelink et al. [18] stated escapism also often leads to feelings of unhappiness, dissatisfaction with life, high levels of anxiety, isolation, and addiction. Escapism is considered as an important factor in culture shock as many people found it difficult to adjust to the sudden and unprecedented shift to remote work and hybrid work environment. This unexpected shift and work from home culture has caused mental distress among many employees due to workload, long hours of working, and reduced rest periods. This can make people find ways to escape to get and maintain their mental and emotional stability. In the case of role shock, researchers found that many individuals undergo a stress reaction when there is a disparity between the expected role and the encountered roles one has to play in a new environment [19–22]. The society has experienced that employees need to change their work pattern from physical to online, which is directed towards ambiguity and is certainly related to culture shock. A number of factors were identified that are related to culture shock like role novelty, role ambiguity, role conflict, and role overload [23]. These also important elements are associated with culture shock in the days of new normal. According to [24], stress is a particular relationship between the person and the environment. Stress depends on how threatening the situation is perceived by the individual and what resources he has to overcome the stressful situation and protect his wellbeing. Anxiety is also related to stress characterized by feelings of tension, worrying, restlessness, and physical changes, which in turn affects blood pressure, hyperventilation, and trembling. Stress and anxiety often lead to depression, which is a serious mental illness that negatively affects how we feel, think, and act. The common symptoms of depression include sadness, learned helplessness, hopelessness, feeling of worthlessness, difficulty in concentrating, and others. This is evident that researchers have found significant results where people experienced culture shock that gave a sense of anxiousness, confusion, and apathetic which affects their normal behavior; until they learn a new set of cognitive constructs to adjust to the environment [25].

Mio [26] experimented that psychological disorientation increases with the increase in accumulation of stress, anxiety, and depression, which may impair the individual's problem solving skills and decision-making process. This situation might decrease the individual's motivation to adapt to a new environment. According to Searle and Ward [27] adjustment is the behavioral process by which human beings maintain a balance among their various needs and the obstacles in their environment. Adjustment or adaptation during culture shock is into two categories—psychological and socio-cultural. Psychological adjustment is the feeling of wellbeing and satisfaction whereas socio-cultural adjustment is ability to 'fit in' in the host culture. By getting familiar with the new culture, people can encounter obstacles. This helps them to accept and adjust with their surroundings [28–30]. However, social support was found to be a key element in reducing stress that helps the person in adjusting to the new environment [31]. Hamid and Durmaz [32] in their study noted that support from co-workers is an important determinant of employee's performance rather than the training he is provided with. In this context, culture fatigue is behavioral response of culture shock. It refers to the weariness and overwhelming experience that result from the big or small adjustments one has to learn to function in a new environment.

It is amazing to note that technostress was first defined by Brod [33] as "a modern disease" which is caused due to inability to cope with computer technology. There are other potential causes of stress and anxiety such as "application multitasking", "constant connectivity", "information overload", "frequent system upgrades" and "consequent uncertainty", "continual relearning" and "consequent job related insecurities", and "technical problems" [34]. Due to work from home employees have work faster; stay connected, available and operative all the time. According to World Health Organization, increased use of technology has fostered expectations about individuals being constantly available and working faster and better. In one of the studies, it was found that during the Covid-19 crisis, social media communication in workplace has increased which led to higher amount of technostress [35].

Adaptation to a new culture is the key to success of an individual, as people need to be constantly motivated towards performance. This is referred to as cultural intelligence—to adapt behaviors where one can work effectively [36]. There are two types of cultural quotient (CQ)—the motivational CQ and behavioral CQ. Individuals with higher motivational CQ are eager to accept challenges in a new environment and have a strong will to tolerate frustration which helps the person to adapt to the new situation. On the other hand, individuals with high behavioral CQ have the capability to perform in a different culture, which makes them feel accepted by the host group. Therefore, higher motivation CQ are more likely to accept challenges but on the other hand high behavioral CQ are more prone to perform better. The researcher [37] in a recent study noted that there exists a positive relationship between the culture of the organization and the employees' performance, engagement, teamwork, and productivity. In this framework, emotional labor is another component, related to an employee's inclination towards desired behavior related to interpersonal relationships in a given situation. Performance at work can be judged based on an employee's emotional dissonance, which is an inconsistency between felt and enacted emotions.

Sengupta and Manjumdar [38] in their research work on emotional labor identified a few factors like Machiavellianism, self-esteem, self-monitoring, risk taking behavior, etc., which are essential components to emotional labor and emotional dissonance. In a study, it was found that culture shock positively affects emotional labor but it has a negative impact on job satisfaction and this leads to employee turnover. However, the perceived managerial role plays a crucial role in lessening culture shock and increasing job satisfaction [39]. Xia [40] opined that to overcome culture shock "self-confidence and optimism", "accepting new culture", "seeking social support" are three effective ways to get rid of psychological discomfort. He also added that a certain amount of stress could also help an individual to learn and know about the culture, which may speed up the adjustment and adaptation process and help the person work effectively and efficiently.

Pantelidou and Craig [41] examined the psychological impact of the culture shock and social support structure (in terms of size, diversity of the social network, and quality of support received), helping the migrant Greek students in the UK overcome the shock. The authors observed that social support is an essential factor associated with culture shock. Quality of support as an aspect of social support is the most closely related to the psychological distress of migrant students. The research of Cullen et al. [42] offered insights regarding how organizations may help improve perceptions of organizational support by reducing perceived uncertainty as well as identifying employees who may need assistance to adapt to workplace changes. The study reported the role of perceived organizational support as a mediator of the relationship between employees' adaptability and perceptions of change-related uncertainty and employees' satisfaction and performance.

The recent pandemic has brought about an unprecedented cultural shock as compared with the other major catastrophizes in the last century. The work of Lai et al. [39] studied the effect of cultural shock on emotional labor, job satisfaction, and turnover intentions of service industry employees. The empirical study revealed that higher level of cultural shock the foreign employees perceived, higher is the emotional labor, lower is the job satisfaction, and hence have a higher intention of employee turnover. However, if perceived managerial support is higher than cultural shock may be reduced having moderating effect on job satisfaction, emotional labor and hence lower would be the intention of employee turnover.

Creative industry has been affected severely by the pandemic, with large cities often covering the greatest share of jobs at risk. The most severe damage from the pandemic as a whole was suffered by micro-enterprise and the self-employed population. The latter forms the backbone of the labor force in the most affected industries, culture, leisure, tourism, and retail [43]. Covid 19 and the large scale social and economic shock have profoundly transformed organization culture. The underlying values and assumptions have shifted from exploration creativity towards safety and resilience. It is a major challenge for managers as to how to build a culture where everyone is working from home and this practice has evolved as a new norm. It further studies how large scale transitions in society unsettle organizational culture and how those cultures might adapt [8]. The work [44] found substantial evidence for an association between cultural consumption from pre-pandemic periods and individual levels of happiness during the pandemic. The chapter addresses two aspects of the cultural impact on mental health resilience: the cultural effect on mental health prevention (i.e., the effects of past consumption of culture on the levels of happiness during the COVID-19 pandemic) and the cultural impact on the resilience of the community spirit approximated by the change in the social capital propensity in human behavior due to cultural consumption during the lockdown period.

3 Preliminaries

Since its proposal qROFS has garnered attention from many researchers and practitioners resulting in a plethora of applications in complex real-life problems. For instance, selection of collaborating firms for setting up food processing plants in India using interval-valued qROFS [45]; selection of laptop using a novel knowledge measures for qROFS [46]; company selection for investment using complex interval-valued qROFS based AHP-TOPSIS [47]; selection of all-rounder players for cricket team using axiomatically supported divergence measures of qROFS [48]; comparison of agri-farming choices using m-polar qROFS based MCDM [49]; enterprise risk planning and management [50]; investment decision making and software selection problem using power neutral aggregation operators of qROFS [51] among others. In this section, we present some of the primary definitions related to concepts and operators of qROFS.

Definition 1 The Pythagorean Fuzzy Sets (PyFS) is defined as [14]

$$\tilde{A}^p = \left\{\langle x, \mu_{\tilde{A}^p}(x), \vartheta_{\tilde{A}^p}(x)\rangle : x \in U\right\}; U \text{ is the universe of discourse}$$

where,

$\mu_{\tilde{A}^P}(x) : U \to [0, 1]$ and $\vartheta_{\tilde{A}^P}(x) : U \to [0, 1]$ are the degree of membership and degree of non-membership respectively, where $0 \leq \left(\mu_{\tilde{A}^P}(x)\right)^2 + \left(\vartheta_{\tilde{A}^P}(x)\right)^2 \leq 1; \forall x \in U$; the degree of indeterminacy is being given by

$$\pi_{\tilde{A}^P}(x) = \sqrt{1 - \left(\mu_{\tilde{A}^P}(x)\right)^2 - \left(\vartheta_{\tilde{A}^P}(x)\right)^2}; \forall x \in U \tag{1}$$

Definition 2 A q-ROFS is defined as [12, 52]

$$\tilde{A}^Q = \left\{\langle x, \mu_{\tilde{A}^Q}(x), \vartheta_{\tilde{A}^Q}(x)\rangle : x \in U\right\};\ U \text{ is the universe of discourse}$$

where,

$\mu_{\tilde{A}^Q}(x) : U \to [0, 1]$ and $\vartheta_{\tilde{A}^Q}(x) : U \to [0, 1]$ are the degree of membership and degree of non-membership respectively, where $0 \leq \left(\mu_{\tilde{A}^Q}(x)\right)^q + \left(\vartheta_{\tilde{A}^Q}(x)\right)^q \leq 1; \forall x \in U$; the degree of indeterminacy is being given by

$$\pi_{\tilde{A}^Q}(x) = \sqrt[q]{1 - \left(\mu_{\tilde{A}^Q}(x)\right)^q - \left(\vartheta_{\tilde{A}^Q}(x)\right)^q} \forall x \in U \tag{2}$$

When $q = 1$, it $\tilde{A}^Q$ becomes an Atanassov's Intuitionistic Fuzzy Sets (IFS) and for $q = 2$ it represents PyFS.

In this chapter, in the same way as defined by expression (2), we shall use the notation $\mathbb{Q} = (\mu, \vartheta)$ for representing a q-Rung Orthopair Fuzzy Number (q-ROFN).

Definition 3 Basic operations on q-ROFN

Let, $\mathbb{Q}_1 = (\mu_1, \vartheta_1)$, $\mathbb{Q}_2 = (\mu_2, \vartheta_2)$ and $\mathbb{Q} = (\mu, \vartheta)$ are the three q-ROFNs. Some of the basic operators are defined as follows [12, 52]

$$\mathbb{Q}^c = (\vartheta, \mu) \tag{3}$$

$$\mathbb{Q}_1 \boxplus \mathbb{Q}_2 = (\sqrt[q]{\mu_1{}^q + \mu_2{}^q - \mu_1{}^q \mu_2{}^q}, \vartheta_1 \vartheta_2) \tag{4}$$

$$\mathbb{Q}_1 \boxtimes \mathbb{Q}_2 = (\mu_1 \mu_2, \sqrt[q]{\vartheta_1{}^q + \vartheta_2{}^q - \vartheta_1{}^q \vartheta_2{}^q}) \tag{5}$$

$$\alpha\mathbb{Q} = \left(\sqrt[q]{1 - (1 - \mu^q)^\alpha}, \vartheta^\alpha\right); \alpha \text{ is a constant} \tag{6}$$

$$\mathbb{Q}^\alpha = \left(\mu^q, \sqrt[q]{1 - (1 - \vartheta^q)^\alpha}\right) \tag{7}$$

Definition 4 Score and Accuracy Function

There are a number of definitions available in the extant literature for score and accuracy functions. For example, in [51] the Score Function (SF) is defined as

$$\mathfrak{H} = \mu^q - \vartheta^q; \mathfrak{H} \in [-1, 1] \tag{8}$$

Proceeding further, in [53–56] the alternative version of SF is given as

$$\mathfrak{H}\prime = (1 + \mu^q - \vartheta^q)/2 \tag{9}$$

Peng and Dai [57] defined SF as

$$\mathfrak{H}'' = \mu^q - \vartheta^q + \left(\frac{e^{\mu^q - \vartheta^q}}{e^{\mu^q - \vartheta^q} + 1} - \frac{1}{2}\right)\pi^q \tag{10}$$

Assuming that if $\mu = \vartheta$, for precise ranking of qROFNs an extended definition of SF is given in [57, 58] as given below. In this chapter, we use the definition given in [57, 58] for calculating score values.

$$\mathfrak{H}^* = \frac{\mu^q - 2\vartheta^q - 1}{3} + \frac{\lambda}{3}(\mu^q + \vartheta^q + 2), \lambda \in [0, 1] \tag{11}$$

The Accuracy Function (AF) is defined as [51]

$$H = \mu^q + \vartheta^q, H \in [0, 1] \tag{12}$$

If $\mathfrak{H}_1 > \mathfrak{H}_2$ then $\mathbb{Q}_1 > \mathbb{Q}_2$
If $\mathfrak{H}_1 < \mathfrak{H}_2$ then $\mathbb{Q}_1 < \mathbb{Q}_2$
If $\mathfrak{H}_1 = \mathfrak{H}_2$, then if $H_1 < H_2, \mathbb{Q}_1 < \mathbb{Q}_2; H_1 > H_2, \mathbb{Q}_1 > \mathbb{Q}_2$.

Definition 5 q-Rung Orthopair Fuzzy Weighted Averaging Operator (q-ROFWA) [51]

$$\text{q} - \text{ROFWA}(\mathbb{Q}_1, \mathbb{Q}_2, \ldots, \mathbb{Q}_r) = \langle\left(1 - \prod_{k=1}^{r}\left(1 - \mu_k^q\right)^{\alpha_k}\right)^{1/q}, \prod_{k=1}^{r}\vartheta_k^{\alpha_k}\rangle \tag{13}$$

Here, α_k is the corresponding weight.

4 Research Methodology

In this section, we briefly showcase the research methodology. In our study, a group of six professionals from the gaming industry participated. The profile of the respondents is given in Table 1.

We derive the broad dimensions and subsequent attributes of the CS (i.e., RF) from [1, 59] and the extant literature as mentioned in Sect. 2. The broad dimensions of CS are escapism, role shock/identity crisis, feeling of helplessness/acceptance, anxiety/stress, adjustment, cultural fatigue, technostress, performance and motivation, and emotional labor. Corresponding to different dimensions, there are a total 30 statements wherein each statement is a representative of an attribute of CS. It may be noted

Table 1 Respondent profile

Respondent	Gender	Qualification	Past experience	Present designation	Total experience (years)
R1	Female	Post graduate	Gaming, IT	AVP HR	18
R2	Male	Post graduate	Product Design, Innovation Lab, IT	General Manager	25
R3	Male	Graduate	Gamming	Businesses Head India	21
R4	Male	Post graduate	IT Industry	Manager	16
R5	Male	Post graduate	Information Technology	Service Delivery Head	19+
R6	Male	Graduate	IT	Senior Manager	21

that some attributes of CS are carrying reverse scaling (i.e., disagreement is the better representation of shock) which are marked as ©. A brief description of the dimensions of CS is given in the Table 2. The attributes of CB are constructed following the "Brief-Cope" scale [60, 61] as used in the experimentation [62]. There are 14 broad dimensions of CB such as active coping, planning, positive reframing, acceptance, humor, religion, emotional support, instrumental support, self-distraction, denial, venting, substance, behavioral disengagement, and self-blame. The attributes are mentioned in the questionnaire given in Appendix 1. We calculate the final weighted aggregated score for both CS and CB using qROFN-based operations and LBWA. We then compare the scores of CS and CB to see whether stress due to shock prevails over resilience or not. The steps of the research methodology (proposed qROF-FFA framework) are given in the Fig. 2 and described thereafter.

4.1 *Force Field Analysis (FFA)*

FFA is one of the fundamental and widely used frameworks for assessing the planned change management process that explains the DFs and RFs and their influences vis-à-vis the transition from the current state to a changed future state [63, 64]. FFA is one of the widely used frameworks enabling organizations to bring about strategic changes in the organizations to withstand market competition [65]. The extant literature shows various applications of FFA in social science, business management, and engineering, for instance, vulnerability analysis of drought [66], adoption of cloud computing [67], and adoption of environmental strategy [68] to name a few.

Table 2 Operational definitions of dimensions of CS

Dimension of CS	Description
Escapism	It is the individual's attempt to avoid the unpleasant aspects of life through engaging oneself into different set of activities, positive or negative, to gain mental and emotional stability
Role shock	It is a feeling of disorientation, uncertainty, and confusion when there is a sudden shift in role, which is different from the familiar role the employee has been playing for a long time
Feeling helplessness/ acceptance	The state of mind in utter disappointment, finding no solutions
Anxiety, depression and stress	Anxiety is the extreme and uncontrollable worry about everyday life events that leads the individual to experience different physiological (e.g., trembling and sweating), behavioral (e.g., restlessness, irritability, and cognitive symptoms (e.g., unwanted intrusive thoughts, lack of concentration) Depression is a continuous feeling of sadness, loss of interest in activities hopelessness, and worthlessness for a longer period Stress is a body's reaction to the strain experienced by a certain situation and the inability to cope with it, which in return affects the individual's wellbeing
Adjustment	It is the ability to adapt to a particular situation or environment by bringing a change in attitude and behavior to match a standard. Individuals who can adjust well have a high quality of life and on the other hand, individuals who are unable to adapt experience psychological disorders like clinical anxiety or depression
Culture fatigue	This happens when an individual is stressed by trying to deal and adjust to the new environment and information at once. It is the physical and mental exhaustion, which occurs due to the adjustments, which are required to survive in a different culture
Techno stress	It occurs due to constant learning and use of technology and the inability to cope with the demands of the technology
Performance and motivation	Performance is how an employee plays or fulfills the job role and achieves his goals. Motivation is the desire to achieve the goal. Motivation plays a key role in the effort that an employee will put in to improve his performance
Emotional labor	It is a way of satisfying or keeping others happy while interacting, by regulating and managing one's emotions to achieve the required task and professional goals

4.2 LBWA Method

LBWA finds applications in several complex problems pertaining to engineering, management, basic sciences, and social sciences. Some of the recent applications include defense equipment selection [69, 70], healthcare management [71], food supplier selection on sustainability dimensions [72], strategic feasibility analysis for

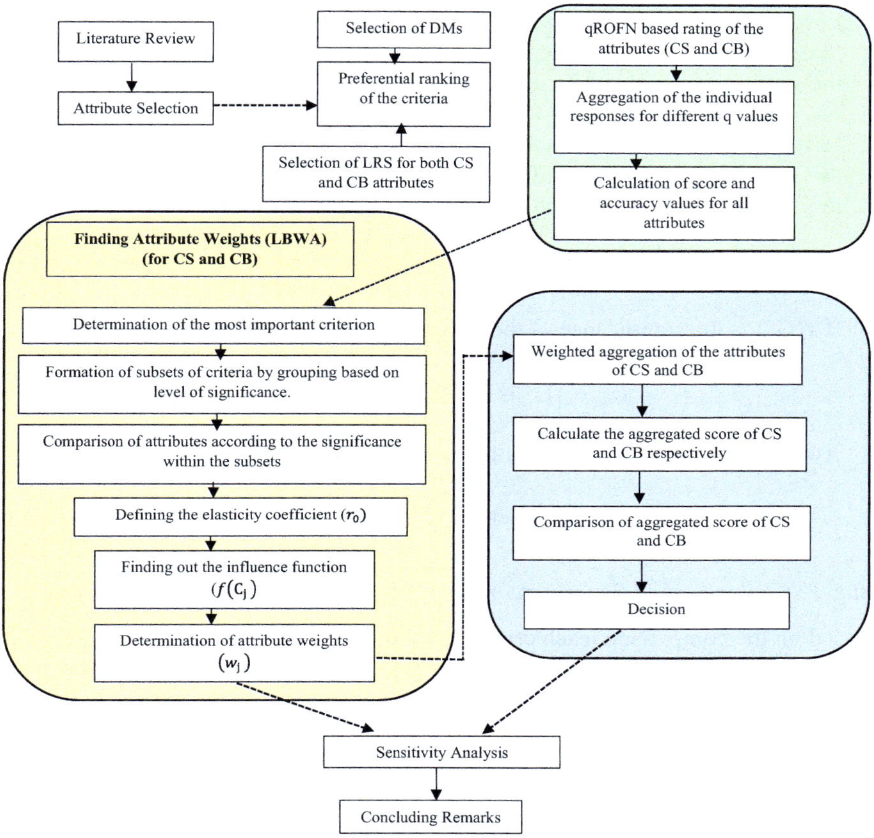

Fig. 2 Proposed framework of qROF-FFA

merger decision [73], deciding marketing mix using 7P [74], renewable energy selection [75] among others. Let, the criteria set is given by $\mathbb{C} = \{C_1, C_2, C_3 \ldots\ldots C_n\}$. Let, the ith criterion ($C_i \in \mathbb{C}$) is the most important one as opined by the respondents.

The computational steps of LBWA are as given below as mentioned in [16].

Step 1: Formation of subsets of criteria by grouping based on level of significance.

The grouping process is described below.

Level L_1: Group the criteria and form the subset with the criteria that are having equal to or up to twice as less as the significance of the criterion C_i.

Level L_2: Group the criteria and form the subset with the criteria having exactly twice as less as the significance of the criterion C_i or up to three times as less as the significance of the criterion C_i.

Level L_3: Group the criteria and form the subset with the criteria having exactly three times as less as significance of the criterion C_i or up to four times as less as the significance of the criterion C_i.

---------------- ----------------- ----------------- ----------------- ---------------------- ---

Level L_k: Group the criteria and form the subset with the criteria having exactly "k" times as less as the significance of the criterion C_i or up to "$k + 1$" times as less as the significance of the criterion C_i.

$$\text{Hence, } L = L_1 \cup L_2 \cup L_3 \cdots \cup L_k \tag{14}$$

If $s(C_j)$ is the significance of the jth criterion, we note that

$$L_k = \left\{C_j \in L : k \leq s(C_j) \leq k + 1\right\} \tag{15}$$

Also, the following condition holds good to appropriately define the grouping

$$L_p \cap L_q = \varnothing; \text{ where } p, q \in \{1, 2, \ldots k\} \text{ and } p \neq q \tag{16}$$

Step 2: Comparison of factors according to the significance within the subsets

Based on the comparison, each criterion $C_j \in L_k$ is assigned with an integer value $I_{C_j} \in \{0, 1, 2 \ldots r\}$; where, r is the maximum value on the scale for comparison and is given by:

$$r = \max\{|L_1|, |L_2|, |L_3| \ldots |L_k|\} \tag{17}$$

Conditions followed in this context are

(i) The most important criterion is assigned with an integer value of zero.
In other words, $I_{C_i} = 0$ (18)

(ii) If C_p is more significant than C_q then $I_{C_p} < I_{C_q}$ (19)

(iii) If C_p is equally significant with C_q then $I_{C_p} = I_{C_q}$ (20)

Step 3: Find out the elasticity coefficient

The elasticity coefficient r_0 is defined as any real number with the condition $r_0 > r$ and $\tau \in \mathbb{R}$; Where $\mathbb{R}$ represents a set of real numbers.

Step 4: Calculate the influence function of the criteria

For a particular criterion, $C_j \in L_k$; the influence function can be defined as $f : L \rightarrow R$

It is calculated as

$$f\left(C_j\right) = \frac{r_0}{\delta r_0 + I_{C_j}} \tag{21}$$

Here, ∂ is the number of levels or subsets to which C_j belongs and $I_{C_j} \in \{0, 1, 2 \ldots r\}$ is the value assigned to the criterion C_j within that level.

Step 5: Calculation of the optimum values of the priority weights of the criteria

$$\text{For the most significant criterion}: w_i = \frac{1}{1 + f(C_1) + f(C_2) + \cdots + f(C_n)} \tag{22}$$

where, $i \in j$; $j = 1, 2, \ldots, n$, the number of criteria

$$\text{For other factors}: w_{j \neq i} = f\left(C_j\right) w_i \tag{23}$$

Decision rule: rank the criteria in descending order of criticality based on the weight values.

4.3 The Proposed QROF-FFA Method

Suppose,

C_j, where $j = 1, 2, \ldots n$(n is finite and ≥ 2): The number of criteria. In our chapter, the criteria represent the list of attributes pertaining to CS and CB.

D_t, where $t = 1, 2, \ldots m$(m is finite and ≥ 2): The number of respondents. In our case $m = 6$.

The steps of the proposed qROF-FFA framework are described below.

Step 1. Selection of the attributes of CS (i.e., RF) and CB (i.e., DF)

In this chapter, the attributes (i.e., criteria) are given as $(C_1)_{\text{CS}}, (C_2)_{\text{CS}}, (C_3)_{\text{CS}}, \ldots, (C_{30})_{\text{CS}}$ and $(C_1)_{\text{CB}}, (C_2)_{\text{CB}}, (C_3)_{\text{CB}}, \ldots, (C_{28})_{\text{CB}}$.

Step 2. Selection of the linguistic rating scale

In our chapter, we modify the linguistic rating scale (LRS) in line with the work of [76] for CS and the qROFNs are defined for the scales used in [62] for rating CB. The LRS and corresponding qROFNs are given in Tables 3 and 4, respectively.

Step 3. Formation of qROFN-based rating matrix for CS and CB attributes

The responses of the respondents have been collected through telephonic conversation and then an online google sheet-based survey. The individual responses have been captured and collated and subsequently, as per Tables 3 and 4, corresponding qROFNs have been assigned. Appendix 2 provides the qROFN-based rating of the attributes of CS and CB as given by the respondents.

Table 3 Q-ROF linguistic scale for rating CS attributes

Linguistic scale	qROFN			
	Forward scaling		Reverse scaling	
	μ	ϑ	μ	ϑ
Strongly disagree (SD)	0.25	0.85	0.85	0.25
Disagree (D)	0.40	0.70	0.70	0.40
Moderately agree (MA)	0.55	0.55	0.55	0.55
Agree (A)	0.70	0.40	0.40	0.70
Strongly agree (SA)	0.85	0.25	0.25	0.85

Table 4 Q-ROF linguistic scale for rating CB attributes

Linguistic scale	qROFN	
	μ	ϑ
I have not been doing this at all	0.25	0.85
I have been doing this a little bit	0.45	0.65
I have been doing this a medium amount	0.65	0.45
I have been doing this a lot	0.85	0.25

Step 4. Aggregation of the individual responses for different q values

Here, we use the definition of q-ROFWA (see expression (13)) to aggregate individual responses while assigning equal priority (i.e., 1/6) to all respondents. This exercise we have done for both CS and CB separately.

Step 5. Calculation of score and accuracy values for all attributes

We use the definition to calculate the score values as given in Sect. 3 (see expression (11)) for attributes of CS and CB respectively.

Step 6. Obtain the weights of the attributes

We utilize the score values of the attributes (of CS and CB respectively) as obtained in step 5 and apply the steps of LBWA-based criteria weight calculation as given in Sect. 4.2 (see expressions (14)–(23)).

Step 7. Weighted aggregation of the attributes

We apply q-ROFWA for weighted aggregation of the qROFNs (representing various attributes of CS and CB respectively). The weights of the attributes as calculated in step 6 are utilized here.

Step 8. Calculate the aggregated score of CS and CB respectively

By applying the expression (11), the aggregated scores of CS and CB are calculated.

Step 9. Comparison of aggregated score of CS and CB

The decision rule is as follows

(i) If $\text{Score}_{\text{CB}} > \text{Score}_{\text{CS}}$; we conclude that the shock due to change is negotiated by coping behavior.
(ii) If $\text{Score}_{\text{CB}} < \text{Score}_{\text{CS}}$; we conclude that the shock has dominated.
(iii) If $\text{Score}_{\text{CB}} = \text{Score}_{\text{CS}}$; No adequate evidence to support the movement.

Step 10. Sensitivity analysis

We carry out the sensitivity analysis in two stages

(i) Varying the elasticity coefficient to check the stability of weight calculation by LBWA method.
(ii) Vary the values of q and λ to examine the stability of our proposed framework through sensitivity analysis.

5 Results and Discussions

In this section, we present the summary of the data analysis and findings in a step-by-step manner.

As mentioned in the previous section, the responses of the DMs are recorded and given in Appendix 2. For our primary analysis, we consider $q = 1$; $\lambda = 0.8$. Table 5 represents the aggregated qROF-ratings for the attributes of CS using expression (13) with equal weights for all DMs. We then move to calculate the score and accuracy values of the qROFNs given in Table 5 using expression (11) and find out the weights of the attributes of CS using the procedural steps of LBWA (see expressions (14)–(23)). Table 6 provides the score and accuracy values along with weights of the attributes of CS while Table 7 gives the supporting calculations for the determination of the weights of the attributes.

It may be noted here that a large number of attributes are involved. As a result, we notice that as many as 20 attributes are appearing in level 1 which results in the value of the elasticity coefficient, $r_0 = 21$. To check the stability in the calculated weights by using the LBWA method, we vary the values of r_0 to perform the sensitivity analysis (see Table 8) and plot the results graphically (see Fig. 3). The results obtained by using MCDM algorithms are vulnerable to changes in the given conditions. In our research, the weights of the attributes play a decisive role and therefore, we perform the sensitivity analysis. Figure 3 suggests that the result obtained by using LBWA is stable.

We then move to calculate the aggregated weighted score (considering all attributes together) of CS using expression (13). Table 9 provides the final weighted aggregated score of CS.

We then move to calculate the aggregated score value of CB following the same way as we have done for CS. Table 10 shows the qROFNs related to rating of

Table 5 Aggregated response in terms of qROFNs for the attributes of CS ($q = 1$)

N1		N2		N3		N4		N5	
μ	υ	μ	υ	μ	υ	μ	υ	μ	υ
0.3364	0.7653	0.7274	0.3803	0.5644	0.5423	0.7572	0.3516	0.5920	0.5186
N6		*N7*		*N8*		*N9*		*N10*	
μ	υ	μ	υ	μ	υ	μ	υ	μ	υ
0.3839	0.7174	0.4768	0.6278	0.3606	0.7410	0.5339	0.5718	0.3364	0.7653
N11		*N12*		*N13*		*N14*		*N15*	
μ	υ	μ	υ	μ	υ	μ	υ	μ	υ
0.7879	0.3162	0.3905	0.7118	0.6727	0.4337	0.5110	0.5953	0.5972	0.5257
N16		*N17*		*N18*		*N19*		*N20*	
μ	υ	μ	υ	μ	υ	μ	υ	μ	υ
0.5110	0.5953	0.6505	0.4535	0.4712	0.6327	0.5803	0.5329	0.4064	0.6945
N21		*N22*		*N23*		*N24*		*N25*	
μ	υ	μ	υ	μ	υ	μ	υ	μ	υ
0.5339	0.5805	0.6505	0.4673	0.6941	0.4175	0.7274	0.3803	0.7619	0.3420
N26		*N27*		*N28*		*N29*		*N30*	
μ	υ	μ	υ	μ	υ	μ	υ	μ	υ
0.7327	0.3754	0.7452	0.3606	0.6669	0.4422	0.5956	0.5083	0.6077	0.5054

the attributes of the CB which we get after aggregating the individual responses. Table 11 provides the calculation of the score and accuracy values of the attributes while Table 12 exhibits the derivation of the weights using LBWA method. It is seen that in the case of CB, the value of r_0 is 18. Likewise, the previous case, here also we consider $q = 1$; $\lambda = 0.8$.

Table 13 shows the sensitivity analysis of the weights calculated by using LBWA for the attributes of CB and Fig. 4 reflects the stability in the result. Finally, we obtain the aggregated weighted score considering all attributes of CB together (see Table 14).

We observe that Score_{CB} (= 0.396) is marginally greater than Score_{CS} (= 0.390) at $q = 1$; $\lambda = 0.8$ that suggests that the coping nature of human behavior has helped the professionals in the gaming industry just to absorb the stress imposed by CS. However, we do feel that there is a marginal difference. Therefore, it is imperative to examine whether, with the changing values of $q$andλ, our qROF-FFA indicates any change in the relationship between Score_{CB} and Score_{CS}. In other words, we perform a sensitivity analysis to examine whether our model provides a stable result given the changes in the underlying conditions (i.e., changes in the values of $q$andλ). Table 15 reflects that Score_{CB} and Score_{CS} are not substantially sensitive to the changes in the values of $q$andλ to contradict the relationship $\text{Score}_{\text{CB}} > \text{Score}_{\text{CS}}$. Further, the result of the Friedman test (see Table 16) indicates that there are not any significant changes in the score values of CS and CB given the changes in the values of q. Therefore,

Table 6 Score, accuracy, and weight values of the attributes of CS ($q = 1; \lambda = 0.8$)

Attribute	μ	ϑ	Score	Accuracy	Weight	Attribute	μ	ϑ	Score	Accuracy	Weight
N1	0.3364	0.7653	0.0957	1.1017	0.0111	N16	0.5110	0.5953	0.2685	1.1063	0.0268
N2	0.7274	0.3803	0.4843	1.1077	0.0454	N17	0.6505	0.4535	0.4089	1.1040	0.0381
N3	0.5644	0.5423	0.3217	1.1067	0.0311	N18	0.4712	0.6327	0.2296	1.1039	0.0257
N4	0.7572	0.3516	0.5136	1.1088	0.0513	N19	0.5803	0.5329	0.3350	1.1132	0.0319
N5	0.5920	0.5186	0.3478	1.1106	0.0328	N20	0.4064	0.6945	0.1660	1.1010	0.0184
N6	0.3839	0.7174	0.1434	1.1013	0.0179	N21	0.5339	0.5805	0.2882	1.1144	0.0295
N7	0.4768	0.6278	0.2350	1.1046	0.0262	N22	0.6505	0.4673	0.4034	1.1178	0.0369
N8	0.3606	0.7410	0.1200	1.1016	0.0139	N23	0.6941	0.4175	0.4494	1.1115	0.0421
N9	0.5339	0.5718	0.2916	1.1058	0.0303	N24	0.7274	0.3803	0.4843	1.1077	0.0437
N10	0.3364	0.7653	0.0957	1.1017	0.0110	N25	0.7619	0.3420	0.5203	1.1039	0.0536
N11	0.7879	0.3162	0.5462	1.1041	0.0562	N26	0.7327	0.3754	0.4895	1.1082	0.0472
N12	0.3905	0.7118	0.1496	1.1023	0.0182	N27	0.7452	0.3606	0.5029	1.1059	0.0492
N13	0.6727	0.4337	0.4301	1.1064	0.0407	N28	0.6669	0.4422	0.4232	1.1091	0.0393
N14	0.5110	0.5953	0.2685	1.1063	0.0274	N29	0.5956	0.5083	0.3540	1.1039	0.0347
N15	0.5972	0.5257	0.3480	1.1229	0.0337	N30	0.6077	0.5054	0.3625	1.1131	0.0358

Table 7 Calculation of the weights of the attributes of CS using LBWA

Level	Attribute	Score	CS	I	f	W
1	N11	0.5462		0	1.0000	0.0562
	N25	0.5203	1.0498	1	0.9545	0.0536
	N4	0.5136	1.0634	2	0.9130	0.0513
	N27	0.5029	1.0862	3	0.8750	0.0492
	N26	0.4895	1.1160	4	0.8400	0.0472
	N2	0.4843	1.1278	5	0.8077	0.0454
	N24	0.4843	1.1278	6	0.7778	0.0437
	N23	0.4494	1.2153	7	0.7500	0.0421
	N13	0.4301	1.2700	8	0.7241	0.0407
	N28	0.4232	1.2906	9	0.7000	0.0393
	N17	0.4089	1.3359	10	0.6774	0.0381
	N22	0.4034	1.3541	11	0.6563	0.0369
	N30	0.3625	1.5070	12	0.6364	0.0358
	N29	0.3540	1.5429	13	0.6176	0.0347
	N15	0.3480	1.5695	14	0.6000	0.0337
	N5	0.3478	1.5707	15	0.5833	0.0328
	N19	0.3350	1.6306	16	0.5676	0.0319
	N3	0.3217	1.6979	17	0.5526	0.0311
	N9	0.2916	1.8732	18	0.5385	0.0303
	N21	0.2882	1.8955	19	0.5250	0.0295
2	N14	0.2685	2.0344	1	0.4884	0.0274
	N16	0.2685	2.0344	2	0.4773	0.0268
	N7	0.2350	2.3244	3	0.4667	0.0262
	N18	0.2296	2.3787	4	0.4565	0.0257
3	N20	0.1660	3.2898	1	0.3281	0.0184
	N12	0.1496	3.6513	2	0.3231	0.0182
	N6	0.1434	3.8090	3	0.3182	0.0179
4	N8	0.1200	4.5534	1	0.2471	0.0139
5	N1	0.0957	5.7091	1	0.1981	0.0111
	N10	0.0957	5.7091	2	0.1963	0.0110
					Σ	1.000

the findings contradict our null hypothesis and we conclude that CB defeats CS for professionals belonging to high-tech industry like gaming.

Table 8 Sensitivity analysis for LBWA results (CS)

Attribute	W				
	$r_0 = 21$	$r_0 = 22$	$r_0 = 23$	$r_0 = 24$	$r_0 = 25$
N11	0.0562	0.0556	0.0551	0.0547	0.0542
N25	0.0536	0.0532	0.0528	0.0525	0.0521
N4	0.0513	0.0510	0.0507	0.0505	0.0502
N27	0.0492	0.0490	0.0488	0.0486	0.0484
N26	0.0472	0.0471	0.0470	0.0469	0.0467
N2	0.0454	0.0453	0.0453	0.0452	0.0452
N24	0.0437	0.0437	0.0437	0.0437	0.0437
N23	0.0421	0.0422	0.0423	0.0423	0.0424
N13	0.0407	0.0408	0.0409	0.0410	0.0411
N28	0.0393	0.0395	0.0396	0.0398	0.0399
N17	0.0381	0.0383	0.0384	0.0386	0.0387
N22	0.0369	0.0371	0.0373	0.0375	0.0377
N30	0.0358	0.0360	0.0362	0.0364	0.0366
N29	0.0347	0.0350	0.0352	0.0355	0.0357
N15	0.0337	0.0340	0.0343	0.0345	0.0348
N5	0.0328	0.0331	0.0334	0.0336	0.0339
N19	0.0319	0.0322	0.0325	0.0328	0.0331
N3	0.0311	0.0314	0.0317	0.0320	0.0323
N9	0.0303	0.0306	0.0309	0.0312	0.0315
N21	0.0295	0.0299	0.0302	0.0305	0.0308
N14	0.0274	0.0272	0.0270	0.0268	0.0266
N16	0.0268	0.0266	0.0264	0.0262	0.0261
N7	0.0262	0.0260	0.0259	0.0257	0.0256
N18	0.0257	0.0255	0.0254	0.0252	0.0251
N20	0.0184	0.0183	0.0181	0.0180	0.0178
N12	0.0182	0.0180	0.0179	0.0177	0.0176
N6	0.0179	0.0177	0.0176	0.0175	0.0174
N8	0.0139	0.0138	0.0136	0.0135	0.0134
N1	0.0111	0.0110	0.0109	0.0108	0.0108
N10	0.0110	0.0109	0.0108	0.0108	0.0107

6 Research Implications

In general, culture is an embodiment of shared norms, beliefs, and behaviors that provide long-term stability to organizations [77]. Organizational culture is one of the cornerstone of successful performance and building dynamic capabilities as it sets

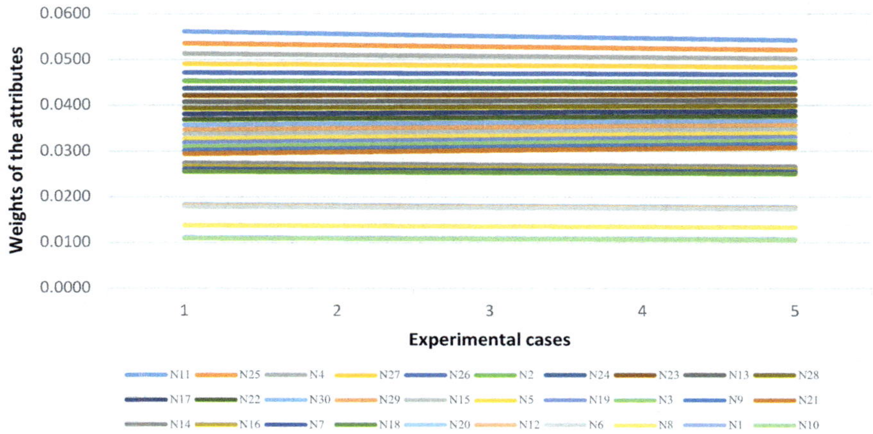

Fig. 3 Changes in the weights of attributes of CS with respect to changes in r_0

Table 9 Weighted aggregated score of CS ($q = 1; \lambda = 0.8$)

μ	υ	Score
0.6278	0.4660	0.390

Table 10 Aggregated response in terms of qROFNs for the attributes of CB ($q = 1$)

Attribute	μ	ϑ	Attribute	μ	ϑ
P1	0.7155	0.3933	P15	0.5613	0.5408
P2	0.7529	0.3566	P16	0.4949	0.6235
P3	0.3901	0.7108	P17	0.7361	0.3699
P4	0.3727	0.7311	P18	0.4043	0.6991
P5	0.4344	0.6685	P19	0.6517	0.4649
P6	0.4476	0.6576	P20	0.8010	0.3041
P7	0.7155	0.3933	P21	0.3577	0.7433
P8	0.3577	0.7433	P22	0.3727	0.7311
P9	0.4043	0.6991	P23	0.6467	0.4612
P10	0.6517	0.4649	P24	0.6082	0.5044
P11	0.3727	0.7311	P25	0.7361	0.3699
P12	0.8010	0.3041	P26	0.4043	0.6991
P13	0.5551	0.5608	P27	0.6932	0.4182
P14	0.7398	0.3729	P28	0.5203	0.5962

Table 11 Score, accuracy and weight values of the attributes of CB ($q = 1$; $\lambda = 0.8$)

Attribute	μ	ϑ	Score	Accuracy	Weight	Attribute	μ	ϑ	Score	Accuracy	Weight
P1	0.7155	0.3933	0.4720	1.1088	0.0486	P15	0.5613	0.5408	0.3204	1.1021	0.0354
P2	0.7529	0.3566	0.5091	1.1096	0.0584	P16	0.4949	0.6235	0.2475	1.1183	0.0307
P3	0.3901	0.7108	0.1497	1.1009	0.0201	P17	0.7361	0.3699	0.4937	1.1061	0.0531
P4	0.3727	0.7311	0.1312	1.1038	0.0160	P18	0.4043	0.6991	0.1630	1.1035	0.0208
P5	0.4344	0.6685	0.1932	1.1029	0.0292	P19	0.6517	0.4649	0.4050	1.1166	0.0417
P6	0.4476	0.6576	0.2055	1.1051	0.0299	P20	0.8010	0.3041	0.5590	1.1052	0.0614
P7	0.7155	0.3933	0.4720	1.1088	0.0467	P21	0.3577	0.7433	0.1173	1.1010	0.0152
P8	0.3577	0.7433	0.1173	1.1010	0.0154	P22	0.3727	0.7311	0.1312	1.1038	0.0156
P9	0.4043	0.6991	0.1630	1.1035	0.0212	P23	0.6467	0.4612	0.4035	1.1079	0.0403
P10	0.6517	0.4649	0.4050	1.1166	0.0432	P24	0.6082	0.5044	0.3632	1.1125	0.0389
P11	0.3727	0.7311	0.1312	1.1038	0.0158	P25	0.7361	0.3699	0.4937	1.1061	0.0508
P12	0.8010	0.3041	0.5590	1.1052	0.0649	P26	0.4043	0.6991	0.1630	1.1035	0.0205
P13	0.5551	0.5608	0.3088	1.1159	0.0343	P27	0.6932	0.4182	0.4487	1.1114	0.0449
P14	0.7398	0.3729	0.4947	1.1128	0.0556	P28	0.5203	0.5962	0.2737	1.1165	0.0315

Table 12 Calculation of the weights of the attributes of CB using LBWA

Level	Attribute	Score	CP	*I*	*f*	*W*
1	P12	0.5590		0	1.0000	0.0649
	P20	0.5590	1.0000	1	0.9474	0.0614
	P2	0.5091	1.0979	2	0.9000	0.0584
	P14	0.4947	1.1299	3	0.8571	0.0556
	P17	0.4937	1.1322	4	0.8182	0.0531
	P25	0.4937	1.1322	5	0.7826	0.0508
	P1	0.4720	1.1844	6	0.7500	0.0486
	P7	0.4720	1.1844	7	0.7200	0.0467
	P27	0.4487	1.2459	8	0.6923	0.0449
	P10	0.4050	1.3801	9	0.6667	0.0432
	P19	0.4050	1.3801	10	0.6429	0.0417
	P23	0.4035	1.3853	11	0.6207	0.0403
	P24	0.3632	1.5392	12	0.6000	0.0389
	P15	0.3204	1.7445	15	0.5455	0.0354
	P13	0.3088	1.8103	16	0.5294	0.0343
2	P28	0.2737	2.0422	1	0.4865	0.0315
	P16	0.2475	2.2581	2	0.4737	0.0307
	P6	0.2055	2.7199	3	0.4615	0.0299
	P5	0.1932	2.8934	4	0.4500	0.0292
3	P9	0.1630	3.4302	1	0.3273	0.0212
	P18	0.1630	3.4302	2	0.3214	0.0208
	P26	0.1630	3.4302	3	0.3158	0.0205
	P3	0.1497	3.7331	4	0.3103	0.0201
4	P4	0.1312	4.2602	1	0.2466	0.0160
	P11	0.1312	4.2602	2	0.2432	0.0158
	P22	0.1312	4.2602	3	0.2400	0.0156
	P8	0.1173	4.7645	4	0.2368	0.0154
	P21	0.1173	4.7645	5	0.2338	0.0152
					Σ	1.000

a conducive environment to facilitate innovation, ensure uninterrupted knowledge sharing through open communication and coordination, and fosters effective adoption of changes [78–80]. The recent pandemic has posited a deep impact on mental health and has been instrumental in causing burnout in several cases both at individual and organizational level [81]. The now-normal remote work culture and mental agony coupled with technological progress and disruptions have brought about a substantial impact on employee performance [82]. The employees are digitally connected but socially distant. Of course, technology enables them to have virtual experience and

Table 13 Sensitivity analysis for LBWA results (CB)

Attribute	W					Attribute					
	$r_0 = 18$	$r_0 = 20$	$r_0 = 22$	$r_0 = 24$	$r_0 = 25$		$r_0 = 18$	$r_0 = 20$	$r_0 = 22$	$r_0 = 24$	$r_0 = 25$
P12	0.0649	0.0636	0.0626	0.0618	0.0614	P13	0.0343	0.0354	0.0363	0.0371	0.0374
P20	0.0614	0.0606	0.0599	0.0593	0.0590	P28	0.0315	0.0310	0.0306	0.0303	0.0301
P2	0.0584	0.0579	0.0574	0.0570	0.0569	P16	0.0307	0.0303	0.0300	0.0297	0.0295
P14	0.0556	0.0553	0.0551	0.0549	0.0548	P6	0.0299	0.0296	0.0293	0.0291	0.0290
P17	0.0531	0.0530	0.0530	0.0530	0.0529	P5	0.0292	0.0289	0.0287	0.0285	0.0284
P25	0.0508	0.0509	0.0510	0.0511	0.0512	P9	0.0212	0.0209	0.0206	0.0203	0.0202
P1	0.0486	0.0490	0.0492	0.0494	0.0495	P18	0.0208	0.0205	0.0203	0.0200	0.0199
P7	0.0467	0.0471	0.0475	0.0478	0.0480	P26	0.0205	0.0202	0.0200	0.0198	0.0197
P27	0.0449	0.0455	0.0459	0.0463	0.0465	P3	0.0201	0.0199	0.0197	0.0195	0.0194
P10	0.0432	0.0439	0.0445	0.0449	0.0452	P4	0.0160	0.0157	0.0155	0.0153	0.0152
P19	0.0417	0.0424	0.0431	0.0436	0.0439	P11	0.0158	0.0155	0.0153	0.0151	0.0151
P23	0.0403	0.0411	0.0418	0.0424	0.0426	P22	0.0156	0.0153	0.0151	0.0150	0.0149
P24	0.0389	0.0398	0.0405	0.0412	0.0415	P8	0.0154	0.0152	0.0150	0.0148	0.0148
P15	0.0354	0.0364	0.0372	0.0380	0.0384	P21	0.0152	0.0150	0.0148	0.0147	0.0146

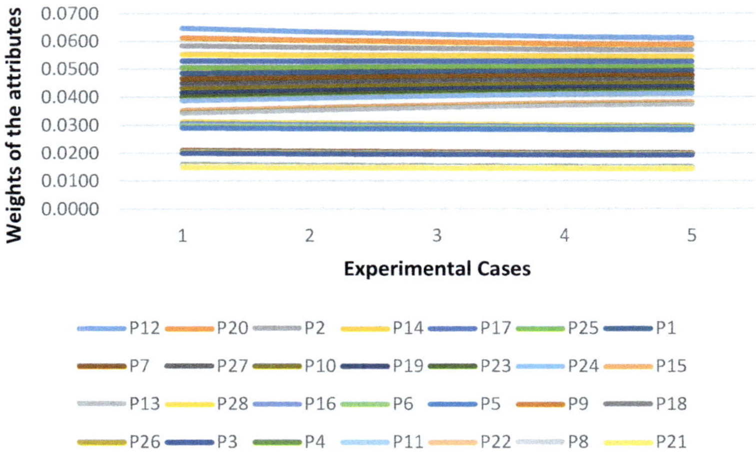

Fig. 4 Changes in the weights of attributes of CB with respect to changes in r_0

Table 14 Weighted aggregated score of CB ($q = 1; \lambda = 0.8$)

μ	ϑ	Score
0.6362	0.4633	0.396

Table 15 Sensitivity analysis of qROF-FFA results with respect to changes in $q$andλ values

Aggregate score	$\lambda = 0.8$				$q = 1$
	$q = 1$	$q = 2$	$q = 3$	$q = 5$	$\lambda = 0.9$
Culture shock (CS)	0.390	0.147	0.162	0.190	0.480
Coping behavior (CB)	0.396	0.392	0.362	0.296	0.482
Inference	CB > CS	CB >> CS	CB >> CS	CB >> CS	CB > CS

Table 16 Friedman test (H_0: there is no significant difference among the values of CB and CS with the changes in q values)

Test Statistic	Value
Chi-square	3.6
df	3
Asymp. sig.	0.308

social distancing allows them to stay close to family members. But, getting confined in home with mental stress due to Covid-19 and situations in the external environment may lead to an adverse impact on innovative work and performance. The researchers [32] argued that for efficient and innovative performance, training and development and support from co-workers are of critical importance which has got disrupted in Covid-19 phase. In this context, the findings of the present study reveal an interesting observation that CB marginally defeats CS. The present chapter is one of its kind that conducts a contrast analysis of RF (imposed due to stress related to CS) and DF (provided by the CB). To add further, on a technical note, the present research is a novel attempt in carrying out psychological analysis using imprecise information and subjective bias under q-ROF environment.

7 Conclusion and Future Direction

Aftermath of the recent pandemic the world has witnessed an unprecedented cultural shift. The transition from "no-mask" social connectedness to "with mask" social distancing phase, prolong lockdown, anxiety related to lives and livelihood and distress have taken normal life to a shocked state. The cultural shift has resulted in a strenuous condition known as CS. In this chapter, we have selected 30 manifested attributes of various dimensions of the CS. We have conducted a study on the impact of CS on the professionals belonging to IT-based gaming industry that provides innovative products. We aim to figure out the interplay among the RF (in terms of CS) and DF (natural coping behaviour) of the professionals using an extended FFA. There are 28 attributes that manifest the CB. To offset the subjectivity associated with the psychological attributes and responses of the six professionals (who have taken part in the study) we use qROFN-based analysis to propose a novel qROF-FFA framework. For the calculation of weighted aggregate score for both CB and CS, we use LBWA method to derive the weights of the attributes. Through our analysis, we observe that despite the impact of never seen before type of CS, the professionals show resilience as the CB defeats CS marginally. Our model shows stability in the result as the sensitivity analysis reveals that with changes in the values of q, CB remains as superior to CS.

However, our study has some scope for future extensions. For instance, in this chapter, only a handful number of professionals have taken part in the survey who represent a large section. One future study may think of a large scale empirical survey with professionals from different level. In that case, a distinctive analysis of impact of CS and counter effect of CB at different levels may be introspected. Further, in this study, we deal with the aggregate score of CS and CB. Our chapter remains silent about development of a new scale for psychological measurement of CS and CB. In addition, we do not focus on delving into the interrelationship of the attributes of the CS and CB. Next, we only consider CB as the DF that essentially is based on stress coping (Brief-Cope). There are other scales to cope up the shock. It may be an interesting study to utilize various coping scales and carry out a comparative analysis. As a technical extension, other fuzzy variants may be used for proposing new FFA frameworks.

Nevertheless, the present study is one of its kind that attempts to utilize uncertain analysis for assessing psychological impact of Covid-19 influenced CS and CB to counter the same. The proposed qROF-FFA may be useful for various other complex change management-related issues, for instance, adoption of new technology, new behavior, and social issues. We do believe that our model shall help policymakers and organizational leaders to formulate appropriate engagement and employee welfare strategies.

Appendix 1: Questionnaire

This questionnaire is part of a study on Workplace Behavior to confront with changing business environment in service sector management. Please fill-in the questionnaire and kindly respond to all the items mentioned in the questionnaire. Your cooperation is earnestly solicited in carrying out an effective research on the area. We solemnly affirm that the information collected from the questionnaire will be used for academic purpose only. The identity of the individual responding to the questionnaire and the name of the organization will not be disclosed in any form or manner.

Name: ....
Gender: Male/Female: ....
Qualification: ....
Past Experience: ....
Present Designation: ....
Total Experience (in years): ....

Section A. Assessment of Cultural Shock

Respondents are requested to select a specific option for each statements

Escapism

(1) I tend to adjourn tasks to avoid facing new challenges

(a) Strongly Disagree, (b) Disagree, (c) Moderately Agree, (d) Agree, (e) Strongly Agree

(2) I always feel like continue with my present assignment ©

(a) Strongly Disagree, (b) Disagree, (c) Moderately Agree, (d) Agree, (e) Strongly Agree

(3) I feel happy to be relaxed instead of taking tension in every moment

(a) Strongly Disagree, (b) Disagree, (c) Moderately Agree, (d) Agree, (e) Strongly Agree

Role Shock/Identity Crisis

(4) I am feeling depressed if no task is assigned to me by immediate superior

(a) Strongly Disagree, (b) Disagree, (c) Moderately Agree, (d) Agree, (e) Strongly Agree

(5) It is a state of great opportunity to work if I am performing individually ©

(a) Strongly Disagree, (b) Disagree, (c) Moderately Agree, (d) Agree, (e) Strongly Agree

(6) I am confused in wishing on the birthday of colleague as I am not invited

(a) Strongly Disagree, (b) Disagree, (c) Moderately Agree, (d) Agree, (e) Strongly Agree

Feeling of Helplessness/Acceptance, Anxiety/Stress

(7) I feel uneasy when I worry about unfavourable outcomes

(a) Strongly Disagree, (b) Disagree, (c) Moderately Agree, (d) Agree, (e) Strongly Agree

(8) I get utmost satisfaction when accepted by my critics ©

(a) Strongly Disagree, (b) Disagree, (c) Moderately Agree, (d) Agree, (e) Strongly Agree

(9) During unforeseen situation, I get upset

(a) Strongly Disagree, (b) Disagree, (c) Moderately Agree, (d) Agree, (e) Strongly Agree

(10) I feel confident in handling my personal problems ©

(a) Strongly Disagree, (b) Disagree, (c) Moderately Agree, (d) Agree, (e) Strongly Agree

Adjustment

(11) I am excited to receive support from my friends

(a) Strongly Disagree, (b) Disagree, (c) Moderately Agree, (d) Agree, (e) Strongly Agree

(12) It is not wise to trust people without judging in every steps ©

(a) Strongly Disagree, (b) Disagree, (c) Moderately Agree, (d) Agree, (e) Strongly Agree

(13) If I fall in trouble, seek help from my community

(a) Strongly Disagree, (b) Disagree, (c) Moderately Agree, (d) Agree, (e) Strongly Agree

(14) To me WFH has hindered performance at the work place ©

(a) Strongly Disagree, (b) Disagree, (c) Moderately Agree, (d) Agree, (e) Strongly Agree

Culture Fatigue

(15) I often feel stressed out due to lot of pressure

(a) Strongly Disagree, (b) Disagree, (c) Moderately Agree, (d) Agree, (e) Strongly Agree

(16) I do not feel exhausted in traveling across cities even during pandemic ©

(a) Strongly Disagree, (b) Disagree, (c) Moderately Agree, (d) Agree, (e) Strongly Agree

(17) I feel difficulty to start new initiatives without proper planning

(a) Strongly Disagree, (b) Disagree, (c) Moderately Agree, (d) Agree, (e) Strongly Agree

(18) I prefer not to keep distance with others during group work ©

(a) Strongly Disagree, (b) Disagree, (c) Moderately Agree, (d) Agree, (e) Strongly Agree

Technostress

(19) My workload has increased due to intricacy of technology

(a) Strongly Disagree, (b) Disagree, (c) Moderately Agree, (d) Agree, (e) Strongly Agree

(20) It is easier to adapt new technology related to our lifestyle ©

(a) Strongly Disagree, (b) Disagree, (c) Moderately Agree, (d) Agree, (e) Strongly Agree

(21) I am worried as technology is changing life every day

(a) Strongly Disagree, (b) Disagree, (c) Moderately Agree, (d) Agree, (e) Strongly Agree

(22) I have no inferiority when my peers can use technology better ©

(a) Strongly Disagree, (b) Disagree, (c) Moderately Agree, (d) Agree, (e) Strongly Agree

Performance and Motivation

(23) It is tough to complete daily task accurately on time ©

(a) Strongly Disagree, (b) Disagree, (c) Moderately Agree, (d) Agree, (e) Strongly Agree

(24) I am unable to solve problems in a creative way ©

(a) Strongly Disagree, (b) Disagree, (c) Moderately Agree, (d) Agree, (e) Strongly Agree

(25) I am able to submit assignments within deadline

(a) Strongly Disagree, (b) Disagree, (c) Moderately Agree, (d) Agree, (e) Strongly Agree

(26) It is mandatory to contribute always in joint project

(a) Strongly Disagree, (b) Disagree, (c) Moderately Agree, (d) Agree, (e) Strongly Agree

(27) I need not to express views in meetings as others are involved ©

(a) Strongly Disagree, (b) Disagree, (c) Moderately Agree, (d) Agree, (e) Strongly Agree

(28) Continuous improvement is a lifetime challenge to perform

(a) Strongly Disagree, (b) Disagree, (c) Moderately Agree, (d) Agree, (e) Strongly Agree

Emotional Labour

(29) In case of refusal of promotion, behavioral pattern should not be modest ©

(a) Strongly Disagree, (b) Disagree, (c) Moderately Agree, (d) Agree, (e) Strongly Agree

(30) In case of misbehavior shown by juniors, your reaction would be very submissive

(a) Strongly Disagree, (b) Disagree, (c) Moderately Agree, (d) Agree, (e) Strongly Agree

Section B. Assessment of Coping Behavior

The following questions ask how you have sought to cope with a hardship in your life. Read the statements and indicate how much you have been using each coping style. Respondents are requested to select a specific option for each statement. The options are:

1. I have not been doing this at all
2. A little bit
3. A medium amount
4. I have been doing this a lot.

Statements:

1. I've been turning to work or other activities to take my mind off things.
2. I've been concentrating my efforts on doing something about the situation I am in.
3. I've been saying to myself "this isn't real".
4. I've been using alcohol or other drugs to make myself feel better
5. I've been getting emotional support from others.
6. I've been giving up trying to deal with it.
7. I've been taking action to try to make the situation better.
8. I've been refusing to believe that it has happened.
9. I've been saying things to let my unpleasant feelings escape.
10. I've been getting help and advice from other people.
11. I've been using alcohol or other drugs to help me get through it.
12. I've been trying to see it in a different light, to make it seem more positive.
13. I've been criticizing myself.
14. I've been trying to come up with strategy about what to do.
15. I've been getting comfort and understanding from someone.

16. I've been giving up the attempt to cope.
17. I've been looking for something good in what is happening.
18. I've been making jokes about it.
19. I've been doing something to think about it less, such as going to movies, watching TV, reading, daydreaming, sleeping, or shopping.
20. I've been accepting the reality of the fact that it has happened.
21. I've been expressing my negative feelings.
22. I've been trying to find comfort in my religion or spiritual beliefs.
23. I've been trying to get advice or help from other people about what.
24. I've been learning to live with it.
25. I've been thinking hard about what steps to take.
26. I've been blaming myself for things that happened
27. I've been praying or meditating
28. I've been making fun of the situation.

Appendix 2: Response Sheet

See (Tables 17 and 18).

Table 17 Response for CS attributes

Respondent	N1	N2	N3	N4	N5
R1	Strongly disagree	Moderately agree	Agree	Strongly agree	Moderately agree
R2	Strongly disagree	Strongly disagree	Strongly disagree	Strongly agree	Strongly agree
R3	Strongly disagree	Strongly disagree	Strongly disagree	Strongly agree	Strongly disagree
R4	Disagree	Disagree	Agree	Moderately agree	Moderately agree
R5	Strongly disagree	Moderately agree	Moderately agree	Agree	Moderately agree
R6	Moderately agree	Disagree	Agree	Moderately agree	Moderately agree
Respondent	*N6*	*N7*	*N8*	*N9*	*N10*
R1	Disagree	Agree	Agree	Agree	Agree
R2	Strongly disagree	Strongly disagree	Strongly agree	Strongly disagree	Strongly agree
R3	Strongly disagree	Strongly disagree	Moderately agree	Strongly disagree	Strongly agree

(continued)

Table 17 (continued)

Respondent	N1	N2	N3	N4	N5
R4	Disagree	Moderately agree	Agree	Moderately agree	Moderately agree
R5	Moderately agree	Disagree	Strongly disagree	Moderately agree	Strongly disagree
R6	Disagree	Moderately agree	Strongly agree	Agree	Strongly agree
Respondent	*N11*	*N12*	*N13*	*N14*	*N15*
R1	Agree	Moderately agree	Moderately agree	Agree	Strongly agree
R2	Strongly agree	Strongly agree	Strongly agree	Strongly agree	Strongly disagree
R3	Strongly agree	Strongly agree	Moderately agree	Strongly agree	Strongly disagree
R4	Agree	Moderately agree	Moderately agree	Disagree	Moderately agree
R5	Agree	Strongly disagree	Agree	Disagree	Strongly disagree
R6	Strongly agree	Agree	Agree	Moderately agree	Strongly agree
Respondent	*N16*	*N17*	*N18*	*N19*	*N20*
R1	Disagree	Agree	Disagree	Strongly agree	Moderately agree
R2	Strongly agree	Agree	Agree	Strongly disagree	Agree
R3	Strongly agree	Strongly disagree	Strongly agree	Moderately agree	Strongly agree
R4	Moderately agree	Agree	Moderately agree	Disagree	Agree

(continued)

Table 17 (continued)

Respondent	N1	N2	N3	N4	N5
R5	Agree	Agree	Agree	Disagree	Agree
R6	Disagree	Agree	Agree	Agree	Agree
Respondent	*N21*	*N22*	*N23*	*N24*	*N25*
R1	Strongly agree	Agree	Moderately agree	Disagree	Agree
R2	Strongly disagree	Strongly disagree	Strongly disagree	Strongly disagree	Strongly agree
R3	Strongly disagree	Strongly disagree	Strongly disagree	Strongly disagree	Strongly agree
R4	Disagree	Strongly agree	Moderately agree	Disagree	Agree
R5	Moderately agree	Disagree	Disagree	Moderately agree	Agree
R6	Moderately agree	Agree	Agree	Moderately agree	Agree
Respondent	*N26*	*N27*	*N28*	*N29*	*N30*
R1	Agree	Moderately agree	Agree	Disagree	Strongly disagree
R2	Strongly agree	Strongly disagree	Strongly agree	Agree	Agree
R3	Strongly agree	Strongly disagree	Strongly disagree	Moderately agree	Strongly agree
R4	Disagree	Disagree	Moderately agree	Disagree	Disagree
R5	Agree	Disagree	Disagree	Agree	Agree
R6	Agree	Disagree	Disagree	Disagree	Disagree

Table 18 Response for CB attributes

Respondent	P1	P2	P3	P4	P5	P6	P7
R1	4	4	2	1	2	3	3
R2	3	4	2	3	2	3	4
R3	4	4	1	1	1	1	4
R4	3	3	1	1	2	1	3
R5	3	3	2	1	3	1	2
R6	2	2	2	2	1	2	3
Respondent	*P8*	*P9*	*P10*	*P11*	*P12*	*P13*	*P14*
R1	2	2	2	1	3	4	3
R2	1	1	4	3	4	3	4
R3	1	1	4	1	4	1	4
R4	1	3	2	1	3	1	3
R5	2	1	3	1	4	3	4
R6	2	2	1	2	4	1	1
Respondent	*P15*	*P16*	*P17*	*P18*	*P19*	*P20*	*P21*
R1	2	3	3	2	2	3	2
R2	2	4	4	1	2	4	1
R3	3	1	4	2	1	4	1
R4	2	1	3	1	4	4	2
R5	3	1	3	3	3	3	2
R6	3	1	3	1	4	4	1
Respondent	*P22*	*P23*	*P24*	*P25*	*P26*	*P27*	*P28*
R1	1	2	1	3	2	2	1
R2	2	4	4	4	3	4	1
R3	1	3	1	4	1	4	4
R4	1	2	3	3	1	3	1
R5	3	3	3	3	2	3	2
R6	1	3	3	3	1	2	3

References

1. Oberg K (1960) Cultural shock: adjustment to new cultural environments. Pract Anthropol 4:177–182
2. Taft R (1977) Coping with unfamiliar cultures. In: Warren N (ed) Studies in cross-cultural psychology, vol 1. Academic Press, London, pp 121–153
3. Bochner S (2003) Culture shock due to contact with unfamiliar cultures. Online Read Psychol Cult 8(1):1–12
4. Poster GM (1973) Traditional societies and technological change, 2nd edn. Harper and Row, New York
5. Ruben BD, Kealey DJ (1979) Behavioural assessment of communication competency the prediction of cross-cultural adaptation. Int J Intercult Relat 4:15–47
6. Zhou Y, Jindal-Snape D, Topping K, Todman J (2008) Theoretical models of culture shock and adaptation in international students in higher education. Stud High Educ 33(1):63–75
7. Kathirvel N, Febiula IC (2016) Understanding the aspects of cultural shock in international business arena. Int J Inf, Bus Manag 8(2):105
8. Spicer A (2020) Organizational culture and COVID-19. J Manage Stud 57(8):1737–1740
9. Wijaya OYA (2021) Risk management mitigation in the new normal era. Bp Int Res Crit Inst-J (BIRCI-J) 4(1):1088–1097
10. Amankwah-Amoah J, Khan Z, Wood G (2021) COVID-19 and business failures: the paradoxes of experience, scale, and scope for theory and practice. Eur Manag J 39(2):179–184
11. Shufutinsky A, Long B, Sibel JR, Burrell DN (2021) Shock leadership: leading amidst pandemics and other chaotic change. In: Global perspectives on change management and leadership in the post-COVID-19 era. IGI Global, pp 136–159
12. Yager RR (2016) Generalized orthopair fuzzy sets. IEEE Trans Fuzzy Syst 25(5):1222–1230. https://doi.org/10.1109/Tfuzz.2016.2604005
13. Atanassov KT (1986) Intuitionistic fuzzy sets. Fuzzy Sets Syst 20:87–96
14. Yager RR (2013) Pythagorean fuzzy subsets. In: 2013 joint IFSA world congress and NAFIPS annual meeting (IFSA/NAFIPS), June. IEEE, pp 57–61
15. Shaheen T, Ali MI, Toor H (2021) Why do we need q-rung orthopair fuzzy sets? Some evidence established via mass assignment. Int J Intell Syst 36(10):5493–5505
16. Žižović M, Pamucar D (2019) New model for determining criteria weights: level based weight assessment (LBWA) model. Decis Mak: Appl Manag Eng 2(2):126–137
17. Evans A (2001) This virtual life: escapism and simulation in our media world. Fusion Press
18. Warmelink H, Harteveld C, Mayer I (2009) Press enter or escape to play deconstructing escapism in multiplayer gaming. In: Proceedings of the Digital Games Research Association international conference, Dundee, Scotland, UK
19. Brislin RW, Pedersen P (1976) Cross cultural orientation programs. Gardner Press Inc., New York
20. Byrnes FC (1966) Role shock: an occupational hazard of American technical assistants abroad. Ann Am Acad Pol Soc Sci 368(1):95–108
21. Harris PR, Moran RT (1979) Managing cultural differences. Gulf Publishing Co, Houston
22. Higbee H (1969) Role shock—a new concept. Int Educ Cult Exch 4(4):71–81
23. Kocak MAREK (2014) Management of culture shock. CRIS-Bull Cent Res Interdiscip Study 2014(2):63–82
24. Lazarus RS (1999) Stress and emotion: a new synthesis. New York, NY
25. Eschbach DM, Parker GE, Stoeberl PA (2001) American repatriate employees' retrospective assessments of the effects of cross-cultural training on their adaptation to international assignments. Int J Hum Resour Manag 12(2):270–287
26. Mio JS (1999) Key words in multicultural interventions: a dictionary. Greenwood Publishing Group, Westport
27. Searle W, Ward C (1990) The prediction of psychological and sociocultural adjustment during cross-cultural transitions. Int J Intercult Relat 14(4):449–464

28. Harrison JK, Brower HH (2011) The impact of cultural intelligence and psychological hardiness on homesickness among study abroad students. Front: Interdiscip J Study Abroad 21(1):41–62
29. Shi L, Wang L (2014) The culture shock and cross-cultural adaptation of Chinese expatriates in international business contexts. Int Bus Res 7(1):23–33
30. Mundeza RS (2021) Process of student adaptation of culture shock. J Soc 2(2):26–31
31. Lafreniere KD, Cramer KM (2005) Applying social psychology to health. In: Schneider FW, Gruman JA, Coutts LM (eds) Applying social psychology: understanding and addressing social and practical problems. Sage Publications, London
32. Hamid D, Durmaz O (2021) Organizational culture impact on employee innovative behaviours in Kurdistan. Black Sea J Manag Mark 2(1):63–72
33. Brod C (1984) Technostress: the human cost of the computer revolution. Addison-Wesley Publishing Company, Reading, USA
34. Tarafdar M, Tu Q, Ragu-Nathan TS (2010) Impact of technostress on end-user satisfaction and performance. J Manag Inf Syst 27(3):303–334
35. Oksanen A, Oksa R, Savela N, Mantere E, Savolainen I, Kaakinen M (2021) COVID-19 crisis and digital stressors at work: a longitudinal study on the Finnish working population. Comput Hum Behav 122:106853
36. Chen ASY, Lin YC, Sawangpattanakul A (2011) The relationship between cultural intelligence and performance with the mediating effect of culture shock: a case from Philippine labourers in Taiwan. Int J Intercult Relat 35(2):246–258
37. Kamalakannan R (2021) Organization culture and its impact on work performance. Turk J Comput Math Educ (TURCOMAT) 12(7):61–67
38. Sengupta SK, Majumder S (2012) Challenge of emotional labor in present day work scenario. Parikalpana: KIIT J Manag 8(1&2):100–107
39. Lai HS, Hu HH, Chen ZYJ (2020) The effects of culture shock on foreign employees in the service industry. Serv Bus 14(3):361–385
40. Xia J (2009) Analysis of impact of culture shock on individual psychology. Int J Psychol Stud 1(2):97–101
41. Pantelidou S, Craig TK (2006) Culture shock and social support. Soc Psychiatry Psychiatr Epidemiol 41(10):777–781. https://doi.org/10.1007/s00127-006-0096-5
42. Cullen KL, Edwards BD, Casper W, Gue KR (2014) Employees' adaptability and perceptions of change-related uncertainty: implications for perceived organizational support, job satisfaction, and performance. J Bus Psychol 29(2):269–280
43. Bhowmik R, Debnath GC, Zafar RF, Lormon BL (2021) Creative industry in terms of covid-2019 pandemic: European countries responsive measures. Press Econ Rev 1(1):9–18
44. Tubadji A (2021) Culture and mental health resilience in times of COVID-19. J Popul Econ 34(4):1219–1259
45. Garg H (2021) A new possibility degree measure for interval-valued q-rung orthopair fuzzy sets in decision-making. Int J Intell Syst 36(1):526–557
46. Khan MJ, Kumam P, Shutaywi M (2021) Knowledge measure for the q-rung orthopair fuzzy sets. Int J Intell Syst 36(2):628–655
47. Garg H, Ali Z, Mahmood T (2021) Algorithms for complex interval-valued q-rung orthopair fuzzy sets in decision making based on aggregation operators, AHP, and TOPSIS. Expert Syst 38(1):e12609
48. Khan MJ, Alcantud JCR, Kumam P, Kumam W, Al-Kenani AN (2021) An axiomatically supported divergence measures for q-rung orthopair fuzzy sets. Int J Intell Syst 36(10):6133–6155
49. Riaz M, Hamid MT, Afzal D, Pamucar D, Chu YM (2021) Multi-criteria decision making in robotic agri-farming with q-rung orthopair m-polar fuzzy sets. PLoS ONE 16(2):e0246485
50. Cheng S, Jianfu S, Alrasheedi M, Saeidi P, Mishra AR, Rani P (2021) A new extended VIKOR approach using q-rung orthopair fuzzy sets for sustainable enterprise risk management assessment in manufacturing small and medium-sized enterprises. Int J Fuzzy Syst 23(5):1347–1369

51. Liu P, Wang P (2018) Some q-rung orthopair fuzzy aggregation operators and their applications to multiple-attribute decision making. Int J Intell Syst 33(2):259–280. https://doi.org/10.1002/int.21927
52. Zadeh LA (1965) Fuzzy sets. Inf Control 8(3):338–356
53. Wang H, Ju Y, Liu P (2019) Multi-attribute group decision-making methods based on q-rung orthopair fuzzy linguistic sets. Int J Intell Syst 34(6):1129–1157. https://doi.org/10.1002/int.22089
54. Wang R, Li Y (2018) A novel approach for green supplier selection under a q-rung orthopair fuzzy environment. Symmetry 10(12):687. https://doi.org/10.3390/sym10120687
55. Wei G, Gao H, Wei Y (2018) Some q-rung orthopair fuzzy Heronian mean operators in multiple attribute decision making. Int J Intell Syst 33(7):1426–1458. https://doi.org/10.1002/int.21985
56. Peng X, Dai J, Garg H (2018) Exponential operation and aggregation operator for q-rung orthopair fuzzy set and their decision-making method with a new score function. Int J Intell Syst 33(11):2255–2282
57. Peng X, Dai J (2019) Research on the assessment of classroom teaching quality with q-rung orthopair fuzzy information based on multiparametric similarity measure and combinative distance-based assessment. Int J Intell Syst 34(7):1588–1630
58. Winkelman M (1994) Cultural shock and adaptation. J Couns Dev 73(2):121–126
59. Holt JB (1940) Holiness religion: cultural shock and social reorganization. Am Sociol Rev 5(5):740–747
60. Carver CS (1997) You want to measure coping but your protocol's too long: consider the brief COPE. Int J Behav Med 4(1):92–100
61. García FE, Barraza-Peña CG, Wlodarczyk A, Alvear-Carrasco M, Reyes-Reyes A (2018) Psychometric properties of the Brief-COPE for the evaluation of coping strategies in the Chilean population. Psicol: Reflexão e Crítica 31:1–22. https://doi.org/10.1186/s41155-018-0102-3
62. https://novopsych.com.au/assessments/formulation/brief-cope/. Last accessed 17 Aug 2021
63. Lewin K (1951) Field theory in social science. Harper Row, London
64. Baulcomb JS (2003) Management of change through force field analysis. J Nurs Manag 11(4):275–280
65. Paquin JP, Koplyay T (2007) Force field analysis and strategic management: a dynamic approach. Eng Manag J 19(1):28–37
66. Hlalele BM (2019) Application of the force-field technique to drought vulnerability analysis: a phenomenological approach. Jàmbá: J Disaster Risk Stud 11(1):1–6
67. Youssef AE, Mostafa AM (2019) Critical decision-making on cloud computing adoption in organizations based on augmented force field analysis. IEEE Access 7:167229–167239
68. Mak AH, Chang RC (2019) The driving and restraining forces for environmental strategy adoption in the hotel industry: a force field analysis approach. Tour Manage 73:48–60
69. Jokić Ž, Božanić D, Pamučar D (2021) Selection of fire position of mortar units using LBWA and Fuzzy MABAC model. Oper Res Eng Sci: Theory Appl 4(1):115–135
70. Hristov N, Pamucar D, Amine MSME (2021) Application of a D number based LBWA model and an interval MABAC model in selection of an automatic cannon for integration into combat vehicles. Def Sci J 71(1):34–45
71. Torkayesh AE, Pamucar D, Ecer F, Chatterjee P (2021) An integrated BWM-LBWA-CoCoSo framework for evaluation of healthcare sectors in Eastern Europe. Socio-Econ Plann Sci 78:101052
72. Yazdani M, Pamucar D, Chatterjee P, Torkayesh AE (2021) A multi-tier sustainable food supplier selection model under uncertainty. Oper Manag Res 1–30. https://doi.org/10.1007/s12063-021-00186-z
73. Chien LM, Tu KJ (2021) Establishing merger feasibility simulation model based on multiple-criteria decision-making method: case study of Taiwan's property management industry. Sustainability 13(5):2448
74. Ryńca R, Ziaeian Y (2021) Applying the goal programming in the management of the 7P marketing mix model at universities-case study. PLoS ONE 16(11):e0260067

75. Ecer F, Pamucar D, Mardani A, Alrasheedi M (2021) Assessment of renewable energy resources using new interval rough number extension of the level based weight assessment and combinative distance-based assessment. Renew Energy 170:1156–1177
76. Pinar A, Boran FE (2020) A q-rung orthopair fuzzy multi-criteria group decision making method for supplier selection based on a novel distance measure. Int J Mach Learn Cybern 11(8):1749–1780
77. González-Rodríguez MR, Martín-Samper RC, Köseoglu MA, Okumus F (2019) Hotels' corporate social responsibility practices, organizational culture, firm reputation, and performance. J Sustain Tour 27(3):398–419
78. Lorsch JW, McTague E (2016) Culture is not the culprit. Harv Bus Rev 94(4):21
79. Azeem M, Ahmed M, Haider S, Sajjad M (2021) Expanding competitive advantage through organizational culture, knowledge sharing and organizational innovation. Technol Soc 66:101635
80. Szczepańska-Woszczyna K (2014) The importance of organizational culture for innovation in the company. In: Forum scientiae oeconomia, September, vol 2, no 3, pp 27–39
81. Bashkin O, Davidovitch N, Asna N, Schwartz D, Dopelt K (2021) The organizational atmosphere in israeli hospital during COVID-19: concerns, perceptions, and burnout. Int J Environ Res Public Health 18(11):5544
82. Sapta I, Muafi M, Setini NM (2021) The role of technology, organizational culture, and job satisfaction in improving employee performance during the Covid-19 pandemic. J Asian Financ, Econ, Bus 8(1):495–505

Chapter 15
Personality Prediction System to Improve Employee Recruitment

Mihir Satra, Faisal Mungi, Jinit Punamiya, and Kavita Kelkar

Abstract Personality is an important factor for predicting whether an applicant would be a perfect fit for the company. The personality of a candidate can give the recruiters an insight about the candidate and hence it can improve candidate selection. The fate of an organization depends on its employees, which makes the selection of the best candidate a very crucial matter for an organization. In the current scenario, the applicant will be selected for the particular job by going through his/her Curriculum Vitae (CV). But shortlisting and going through thousands of applicant CVs is a tedious and hectic task. Besides, one may not get a good idea about the personality of the candidate from a CV. The cost of bad hires can be saved if the recruiter selects the right candidate for a particular job role. Personality Prediction is finding out and comprehending the personality of an applicant which can be used in the present recruiting system. The personality of the candidate will not only help the recruiters in the selection but also provide the candidate with a good job role based on his personality. In this chapter, we have come up with a method to evaluate the personality of a candidate using different strategies. The proposed system asks CV-related and personality-based questions to predict and analyze his/her personality with the help of Machine Learning and Natural Language Processing which helps the organization to shortlist candidates based on the job profile and company requirements. Various Machine Learning Models were tested from which Logistic Regression provided the highest accuracy of 85.71%. Bidirectional Encoder Representations from Transformers or BERT is implemented to extract the keywords

M. Satra (✉) · F. Mungi · J. Punamiya · K. Kelkar
Department of Computer Engineering, K. J. Somaiya College of Engineering, Vidyavihar, University of Mumbai, Mumbai, Maharashtra, India
e-mail: mihir.satra@somaiya.edu

F. Mungi
e-mail: faisal.mungi@somaiya.edu

J. Punamiya
e-mail: jinit.p@somaiya.edu

K. Kelkar
e-mail: kavitakelkar@somaiya.edu

S. Kautish et al. (eds.), *Computational Intelligence for Modern Business Systems*, Disruptive Technologies and Digital Transformations for Society 5.0,
https://doi.org/10.1007/978-981-99-5354-7_15

to provide recruiters an understanding about the candidate's personality. Thus, the system will help the human resource to select the right candidate for the desired job profile, which in turn will provide an expert workforce for the organization.

Keywords Personality · Personality prediction · Curriculum vitae · Machine learning · Natural language processing · Keyword extraction · Logistic regression · BERT

1 Introduction

A Personality Assessment provides the recruiter with useful insights regarding the candidate's behavior in a work context and can help in predicting the company fit and job performance. The recruiter can access the personality traits of the candidate based on the Big Five Model. The Big Five Model is a universal model that is used in the field of psychology to predict the personality of any individual. The "big five" consists of broad categories of personality traits. It is also called the OCEAN model based on the categories—Openness to Experience, Conscientiousness, Extraversion, Agreeableness, and Neuroticism. The big five personality traits are a set of broad, bipolar trait dimensions that constitute the most widely used model of personality structure. This helps us rely on this model as any individual is a mixture of multiple and diverse personality traits [1]. Hence, we implemented multiple methods to predict the diverse traits of an individual's personality.

The objective of the proposed chapter is to implement a system that will help identify the personality traits exhibited by a candidate so that the recruiter can have a better understanding about them and can make better decisions to hire the most appropriate candidate for their position and organizational culture. The system uses the power of Machine Learning (ML) and Keyword Extraction to predict the personality of an individual. Together ML and keyword extraction can help achieve better accuracy and thus predict better results. Machine Learning is used to predict the overall personality of the candidate by taking the input from the answers to personality questions answered by the candidate. Keyword Extraction to extract the keywords from the answers to the subjective Curriculum Vitae (CV) related questions given by the candidate. The relevant keywords are extracted and displayed as a Word cloud to help the recruiter gain insight into the personality of the candidate. Furthermore, we use Fuzzy Logic Interpretation from the scores obtained from the subjective questionnaire of the candidate and with the help of the Big Five Model, to predict and elaborate the personality of the candidate in detail. Thus, with the help of these three strategies, we have tried to predict the personality of the candidate in a diverse manner.

2 Literature Survey

Recent research in 2021 explained the prediction of a candidate's personality by parsing CV data to extract keywords and match it with the Big Five Personalities, to produce input in the Machine Learning Model—Random Forest to predict the personality of the candidate [2]. Research in May 2019 focused on making a system that prompts the candidate to create a CV. Keywords from the CV are then extracted and are used as input in TF-IDF which calculates the number of times a word appears in the CV. Based on the word frequency, a predicted personality is displayed as the result [3]. One of the research projects in January 2018 explained the importance of the Big Five personality traits in our daily lives. It helped us understand the meaning of each of the personality factors and how they impact our life outcomes [1]. Chandrashekhar Singh and Prabhakar Kumar analyzed the effect of personality on work performance. They concluded that conscientiousness and agreeableness have a big impact on work performance and candidates who score high in these personality factors are a great asset for an organization [4]. Additional research on the impact of personality type on job productivity by Janjua Najam-us-Sahar in 2016 explained the impact and relationship of each Big Five personality trait with employee productivity [5]. One of the research projects in June 2021 predicted the personality using CV and psychometric analysis. The proposed system parsed CV data for keywords and tried to match them with keywords required by the admin to get a probability of the predicted personality [6]. Another research in May 2021 explained the impact of the Big Five Model in the analysis of personality assessment using Machine Learning. The proposed system uses social media data to analyze the personality of the user using data vectorization and machine learning. The system is built on the Big Five Model and explains its importance over other personality models [7].

3 Proposed Methodology

The hiring process for companies has become a daunting task nowadays. With an ever-increasing number of applicants trying for vacancies, it has become very tedious for recruiters to peruse through all the applicants and their profiles and then shortlist them for further rounds. Thus, recruiters may hire candidates that are not suitable for the given job role and may not be fit according to the company's requirements.

A detailed literature survey revealed that when it comes to employee recruitment, prediction of personality has mostly been done from CVs which do not contain information that has relevance to predicting a candidate's personality. The work experience, education or skills shown in a CV do not help in predicting the personality of the candidate. Research shows that not much work has been in checking the relevance of the data. Previous systems are highly dependent on CV for data which will provide inaccurate information pertaining to the personality of the candidate [2]. Also, models such as Term Frequency-Inverse document frequency (TF-IDF) are used to extract keywords that depend on the number of times a word occurs in a document rather than giving importance to the relation of the word to the sentence [3]. Hence, we have focused our research on getting relevant information that can help predict the personality of an individual so that we can fulfill the research gaps in previous systems.

Also, social media data is apocryphal and cannot be depended upon as individuals are pretentious on social media platforms. Predicting personality from social media data can be displayed as an added perspective but not as a primary system [7]. Furthermore, previous systems have predicted only one dominant personality type, but with the Big Five Model we are able to show that an individual is a combination of multiple personality traits.

We have proposed a Personality Prediction System which can be used in the present recruiting system to filter out and automate this process. It will generate a comprehensive personality report of the candidate. The system will ask the candidate to answer questionnaires that are based on personality questions. Their responses to these questions will be used to determine the possible personality portrayed by the user. Firstly, the system will predict the overall personality exhibited by the user from the objective questionnaire answers with the help of machine learning. The system will analyze the scores achieved by the user in each of the five traits of the Big Five model and explain the significance of the scores.

The keywords from the answers submitted by the candidate through the subjective questionnaire will be extracted using Keyword Extraction Model—BERT. These keywords are filtered using lemmatization and further used to generate a word cloud to give a visual representation to the recruiter to help them understand the candidate's nature. These approaches keep us focused on asking the candidate relevant questions to predict the personality rather than extracting information from CVs which may not contain precise and relevant data. Lastly, using the Big Five Model one can understand the personality of the candidate on the basis of a score spectrum and the combination of the five personality factors. On the basis of the score, the personality of the candidate can be predicted for each of the personality factors in the Big Five Model [1, 7]. By using three strategies, we have diversified the personality of the candidate and not limited our research to a particular personality trait. Since the system uses multiple approaches to predict the personality, it helps us overcome the errors from data authenticity and improves the accuracy of the personality predictions.

Figure 1 depicts an overview of the proposed system design and how the candidate personality profile is built using all three strategies.

4 The Big Five Model

The Big Five Model also known as the OCEAN Model is a highly reputed model in the field of psychology and helps us predict personality in a broad spectrum rather than just deciding whether an individual is introverted or extroverted. An individual is a mixture of multiple personality traits and the Big Five Model helps in depicting

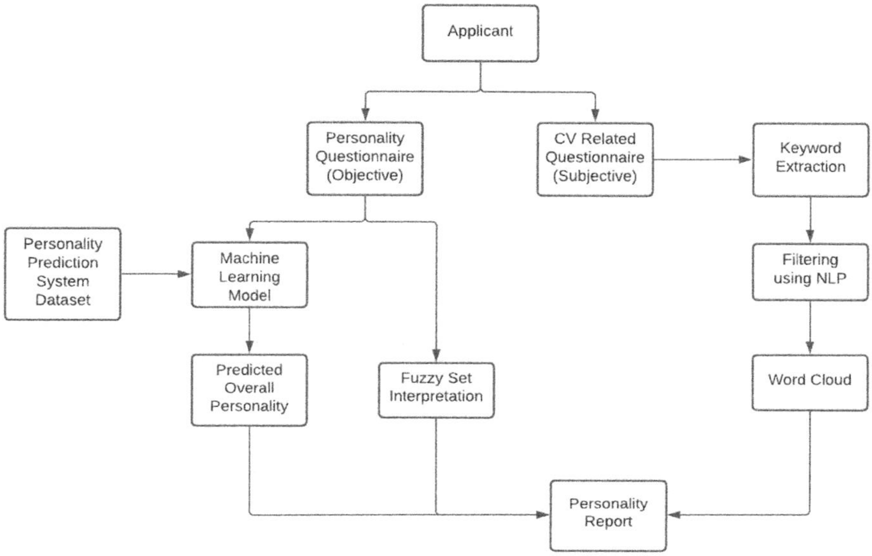

Fig. 1 Proposed system design

this diversity of personality traits. The five personality traits of the Big Five Model are—Openness to Experience, Conscientiousness, Agreeableness, Extraversion, and Neuroticism [1, 8].

Openness to Experience determines how innovative and curious one is. Those who are Open to Experience are intellectually curious and think of new experiences and ideas. They like to experience new things in life and are well suited for a company's design teams as they excel in creative roles.

Conscientiousness is one of the most important factors a recruiter will look into since it determines how much an individual will excel and experience success in his/her career. Such individuals are disciplined and follow a rigid plan to pursue and achieve their goals. These individuals are well suited for various company roles especially when it comes to management and delivering for the company.

Extraversion is the factor that determines the social skills of an individual. They love to meet new people and interact socially to gain experience from the world. These people thrive when it comes to sales and marketing jobs or in client-side interactions representing the company.

Agreeableness is a factor that measures the willingness of an individual to do a task. It is the ability that decides whether an individual will cooperate socially with his/her peers and colleagues. They are affable, helpful, and trustworthy and are required in every team for the smooth functioning of the team.

Neuroticism is a measure of the emotional stability of an individual. They can't control their feelings which affects their ability to make decisions and thoughts and hence have mood swings. A recruiter may not want to hire a person with neuroticism

Table 1 Personalities that come under each big five personality trait

Openness to experience	Conscientiousness	Agreeableness	Extraversion	Neuroticism
Imaginative	Leadership	Altruism	Sociableness	Pessimism
Curious	Planning	Modesty	Optimism	Jealousy
Insightful	Ambitious	Patience	Friendliness	Fear
Varied interests	Thoroughness	Politeness	Talkativeness	Anxiety
Originality	Self-disciplined	Loyalty	Outgoing nature	Lack of confidence
Daring	Persistence	Helpful	Assertiveness	Instability
Prefers variety	Reliability	Amiability	Teamwork	Oversensitivity
Cleverness	Consistency	Cheerful	Merriness	Insecurity
Creative	Hardworking	Unselfish	Energetic	Self-criticism
Perceptive	Resourceful	Kind	Ability to express	Nervousness
Intellect	Energetic	Tact	Affectionate	Testiness
Eager to learn	Perseverance	Patient	Socially confident	Moody

although neurotic individuals can excel at creative roles and hardworking roles if they show potential towards other Big Five personality factors.

The few personalities that come under these five personality factors are as follows (Table 1).

5 Predicting Overall Personality Using Machine Learning

Machine Learning is a method using which computer systems can be trained using pre-defined datasets and used to predict, identify or classify with or without human supervision. We have used the Personality Prediction System Dataset which is an open-source dataset available on Kaggle. This dataset uses the Big Five Model personality traits and classifies the overall personality of the candidate.

The system asks objective questions to the candidates and based on their answers ranging in intensity from Strongly Disagree to Strongly Agree we calculate the score for each personality attribute. The system asks 75 questions and the score for each personality trait is generated on the basis of the given answer. The system then predicts using the ML Model and predicts the overall personality of the candidate [3].

On the dataset, we applied various machine learning models to try and predict the personalities of the candidates. We have applied Random Forest, Decision Tree, Naïve Bayes-Gaussian, K-Nearest Neighbors, SVM, and Logistic Regression. After

Table 2 Accuracy of various machine learning models

Model	Accuracy (%)
Random forest	32.38
Decision tree	28.57
GaussianNB	49.52
KNN	26.98
SVM-linear	69.52
SVM-polynomial	57.46
SVM-RBF	50.48
Logistic regression	85.71

applying them, we compared the performances of all these algorithms against each other [9].

Experimental Results:

We compiled the performances of these algorithms to have a better understanding of which algorithm would be more suitable for our data. Comparing them, we found that Logistic Regression was the best-suited algorithm followed by SVM in our case. Multinomial Logistic Regression gives us the highest accuracy since the dependent variable is ordinal and the values are continuous in nature. The dependent variables are mutually exclusive and there is no multicollinearity among the independent variables which is suitable and hence increases the accuracy of the model (Table 2).

Based on the score obtained, the model classifies the overall personality into five data points: Dependable, Extraverted, Serious, Responsible, and Lively.

Dependable—A trustworthy person who fulfills their commitments by taking accountability of their work and action is a dependable person. A dependable person is reliable and constantly improves oneself to help others. They are reliable and manage their commitments. If they are in a managing and leading position in a team, then he holds his entire team accountable to fulfill their commitments.

Extraverted—These people enjoy engaging socially out in the world. They want to be the center of attraction among peers and groups. They look for excited and enthusiastic people with whom they can blend in and be productive. They try to seek passion in their job and stimulation from their team. Such people are outgoing and highly optimistic about things and willing to be bold in social situations or take risks.

Serious—Such people have a sober and mature attitude towards life and are not inclined to show emotion or reveal one's thoughts. They know their own limitations as well as capabilities and work accordingly. They are least likely to show self-pride and self-importance and have no pretensions. They are hardworking and are focused towards their goals.

Responsible—A responsible person is one who is willing to take accountability of his/her decisions and actions. They do not make excuses for their failures and plan ahead of time so that they can adhere to their duties and obligations. Such people are self-disciplined and do not procrastinate during work. They know how to control their feelings and admit their own mistakes.

Lively—People who are lively show active interest and enthusiasm to do things. They are energetic and full of life. They are highly responsive and always on alert. A lively person is fun to hang around with and is easy to talk with. Such people are great for improving the environment of the company.

6 Keyword Extraction Using Natural Language Processing

Keyword extraction was applied to the subjective answers submitted by the candidates. These questions are based from the Curriculum Vitae (CV) but framed from the aspect of personality [10, 11]. For example, what unique personality do you have that will benefit the company? From these answers, we had to extract important keywords which could be an indicator of personality traits. We used various keyword extraction algorithms such as TF-IDF, Rake, Spacy, pyTextRank, and BERT [2]. The output from each keyword extractor was analyzed and BERT extracted the majority of the keywords from the answers. Bidirectional Encoder Representations from Transformers or BERT is developed by Google for its SEO and it extracts keywords based on the meaning of the word in the sentence it is used, rather than processing each word individually. This gives us each personality mentioned by the user.

A survey of sample questions was shared using Google Forms and data was collected to test the output and accuracy of the Keyword Extraction Model so that the majority of the keywords get extracted. Figure 2 shows the dataset of questions asked in the survey and extracted keywords for each answer given by different individuals for that question. Using this showed that BERT gave the highest accuracy of extracted keywords.

After extracting the keywords, we had to extract the roots of the given keywords. We applied stemming and lemmatization to see which was more apt for retrieving the root word. In stemming, the common prefixes and suffixes were cut off and the remaining was taken to be the root word. This however would not always result in

	If you could change one thing about your personality, what would it be?	What are your hobbies?	In your previous group projects(college or in any organization), what role did you generally have (eg: team leader) and how did you accomplish your tasks efficiently?	What will your family and friends say about your personality? (eg: Ambitious, Caring, Kind , etc)	What is your career objective? What drives you in your professional life? (Eg: Passion, money, energy, knowledge)	What kind of person would your boss, colleagues or professors describe you as? (eg: studious, hardworking, shy)	What unique personality trait do you possess which can be helpful for the company?
0	gain confidence	cricket,singing	project manager ,keep a perfect schedule for t...	Quiet person	not yet decided,self motivation	Helpful and hardworking	Hardworking and enthusiastic
1	Anger and Laziness	Sports and music	Frontend developer	Ambitious, kind, passionate, aggressive	Learn new things and expand skillset	Helpful and creative	Willingness to learn and creativity
2	Short tempered feelings	Dancing, acting, music, sports	A documentation and coding role..it went great	Ambitious	To be successful and earn loads of money	Enthusiastic	I don't bitch
3	Speech	Music	Worked both ways, defined expectations and eff...	Mature	To be self employed, urge to learn new things	Easy-going	Curious and ability to see things from differe...
4	Self-discipline	Reading, watching shows, cooking	I've been both a team leader and a usual team ...	Practical, ambitious, caring, confident	I'm a need based person and eventually my need...	Confident, no nonsense, practical and hardworking	My flexibility and empathy towards my team mem...

Fig. 2 Sample dataset of questions and extracted keywords from their answers

Fig. 3 Word cloud generated from the extracted keywords

an actual word as the root. So, this could not be considered due to erroneous root word generation. Porter Stemmer, a popular stemming algorithm was used for this purpose. Lemmatization, however, works differently. It takes the context of the word into consideration while trying to retrieve the root word. It applies morphological analysis to the words. Applying lemmatization in our case, we could see a significant improvement in the retrieval of root words. We applied WordNet Lemmatizer from the NLTK toolkit to generate the root words. Wordnet is a large and open-source database aimed at establishing semantic relationships between words. Using Wordnet Lemmatizer helps us convert the majority of the words to its root form.

After filtering the keywords, we generate a word cloud out of the filtered words. The purpose of the word cloud generation is to give the recruiter a brief description of the candidate by highlighting certain words that have been used to describe them. Using this, the recruiter can have a general overview of the applicant. The extracted keywords from the survey data are filtered using Lemmatization and using these keywords, a sample Word Cloud is generated as shown in Fig. 3. The sample Word Cloud shows different personalities and important factors extracted from the CV-related questionnaire.

7 Fuzzy Logic Interpretation

The Fuzzy Logic Interpretation is based on the scores of the Big Five Model. The average scores that were used as the input to the ML Model is used as the score of the candidate in each personality trait. Based on the scores the personality of the candidate can be predicted on the basis of the Big Five Model. The recruiter can use these scores and identify the perfect candidate for the job role. Also, it gives an idea about the candidate's nature and how it will impact the job performance. Using this interpretation, the recruiter may also open up various job categories for the candidate so that it is easy for the candidate to fit in. For example, an introvert and serious candidate may not be good at client-side interaction and hence a core developer might suit the candidate. As the Big Five Model is a combination of five personality factors, the recruiter will get to know the score and the meaning of each personality factor [3].

The interpretation of scores of the Big Five Personality Traits to predict the candidate's personality [1].

7.1 *Openness to Experience*

Candidates who score high in Openness to Experience are open to trying and experiencing new things. They are highly creative and can be used in places where creative thinking is required. They have a good imagination and are more concerned with dealing with new challenges. They think in an abstract way and work to build on those thoughts and concepts. Such people have high job satisfaction and are easily adaptable in their workplace [12].

Candidates who score low tend to have very few or common interests. Such candidates do not like change and follow a rigid pattern. They are old-fashioned in their thinking and resist new ideas or experiences. Even though one might not have a high score, closed individuals excel in job performance in areas where rules and procedures are the most important since they choose to do things in a monotonous way.

7.2 *Conscientiousness*

Those who score high in conscientiousness are ambitious and hardworking individuals. Such people like keeping things in order and planning their commitments so that they are prepared to fulfill them. They are persistent and pay attention to every detail. They complete their tasks on time and enjoy having a fixed schedule. They are extremely good decision makers and avoid erratic decisions. They keep high standards for themselves and will work hard to achieve those standards. Such people have better job performance, are fit for leadership roles, and are less likely to leave their workplace [12, 13].

Individuals with low Conscientiousness are less organized and are often in trouble with their higher authorities. They don't have any specific schedule and finish things at the last minute. Moreover, they procrastinate on important tasks and end up failing to complete their tasks. The benefit of having a low score is that such people are impulsive and are needed in situations where action is needed rather than thought.

7.3 Extraversion

High Score on Extraversion means that the person is very sociable and enjoys being active with others. They seek adventure and meet new people. They like to start a conversation and thrive in social situations. Such people excel when it comes to client-side interactions or marketing the company. In their workplace, they have better job performance and show strong leadership skills [12, 13].

Introverts are those who score low on Extraversion and do not like to constantly engage in social interactions. They are not very positive and find social interactions to be highly overwhelming. They avoid large groups and find difficulty in starting a conversation. They take time to process their emotions and prefer to have fewer but more close connections. They prefer being alone and enjoying their own company.

7.4 Agreeableness

People who score high on agreeableness are understanding, caring, and honest in nature. They look out for others around them and are always ready to help out. They engage with people around them and believe in the best of others. They show unselfish concern for the welfare of others and enjoy helping and being a part of other people's happiness. Agreeable individuals are easy to talk to and are respected highly by their peers. They get along with others easily and display tact in dealing with others. Such people have great job performance and better behavior in their workplace [12].

Disagreeable people keep themselves and their interests above everything else. They are concerned with others' well-being and focus on their own goals and tasks. They are viewed as unforthcoming and unfriendly. They are stubborn, self-centered, and know how to entice others to get what they want. If viewed from a positive aspect, Disagreeable people focus on their goals and ambitions which make them perfect in many roles in an organization.

7.5 Neuroticism

A high score on Neuroticism means that the candidate is prone to emotional instability and is highly reactive to emotions. Those who are ranked high in neuroticism tend to be very anxious, angry, and depressed. They experience these feelings routinely and are more emotionally touched about negative things. Such people easily get stressed and struggle to cope up in various situations. They are often insecure and are likely to be upset. They are mercurial and struggle in controlling their emotions. This leads to lower job satisfaction and higher stress levels in their workplace [12].

Those who have a low score are more emotionally stable and keep calm and composure in difficult situations. They are optimistic and have a good control over their emotions, such people worry less and rarely get sad or depressed. They have better behavior and bonding in their workplace [13].

8 Result—Candidate Personality Profile

After all the above processing and analyzing, a report of the candidate is generated from the above three strategies. This is made available to the recruiter and shows the personality traits exhibited by the candidate. The overall personality generated using the ML Model and the word cloud which was generated using Natural Language Processing, helps the recruiter gain knowledge about the candidate's personality and ask relevant questions to the candidate. The recruiter can then determine which personality traits are more suitable to their job requirements. For example, a person who would be working where there is a lot of human interaction should have great social skills and thus should be extraverted.

The result also contains a small description of the different personality traits, in case the employer needs help in understanding what each personality type involves. After perusing through various articles and research work based on personality, we thought that it would be prudent that we display all the scores the candidate got in each personality type rather than just assigning their dominant personality type as an individual is a mixture of multiple personalities. Doing this, we could provide a better understanding of the candidate to the employer. Hence, we showed the different scores in each of the five types of OCEAN model, and using the help of research articles, we explained the meaning of high and low scores in each personality factor. Using all the information provided on the candidate, the employer can have a more in-depth idea about the candidate.

9 Conclusion and Future Work

In this project, we have successfully designed a model for predicting the personality of the candidate based on CV-related and personality questions. Various research papers have been studied and, we have tried to fulfil the research gaps by asking for relevant information to individuals rather than extracting personality from CV. We then analyzed the different theories to determine the personality of individuals. The Big Five Model seemed to have more weightage than other models since it is a diverse and universal model and is widely used to predict personalities in the field of psychology.

We selected Logistic Regression for the ML Model which gives us an accuracy of 85.71% and predicts the overall personality of the candidate. For Keyword Extraction we have selected BERT as the output is accurate and covers all the words from the text by relating the extracted word with the sentence it is used in. Lastly, The Big Five Model is a strong tool that helps us understand the diversity of personality traits and predicts in detail the multiple personalities of an individual. The recruiter needs to use these insights accurately to select the perfect candidate for any job role and one who will suit the demands of the organization's culture.

To begin with, identifying the job's responsibilities, does the job role involve great pressure? Would a team player be fit for the role or an individual who prefers working individually? Would there be a lot of social interaction? Does the role involve innovative and creative ideas or demands strictness to rules and procedures? How well-suited the candidate would be depending on the organizational culture? Once these the recruiter identifies the needed personality for the role using the system, he can easily assess whether the candidate fits perfectly for the position and the organization's culture. Along with the other recruitment tests for aptitude and problem-solving, a personality prediction system will greatly improve the candidate selection process. This system can be used by various business and government sectors also where there are a large number of people applying for various job openings.

The future work of this system includes improving the accuracy and efficiency of the ML Model and keyword extraction. We plan on predicting user behavior and personality from social media such as LinkedIn, Facebook, and Instagram [14, 14]. Through this, we can generalize the use of our system to various other fields by having improvement of employee recruitment as our primary focus. Also, a job portal can be developed to help candidates apply for jobs and multiple companies shortlist candidates using the personality prediction system [16, 17].

References

1. Soto C (2018) Big five personality traits
2. Kulkarni A, Shankarwar T, Thorat S (2021) Personality prediction via CV analysis using machine learning. Int J Eng Res Technol (IJERT) 10(09)
3. Rout J, Bagade S, Yede P, Patil N (2019) Personality evaluation and CV analysis using machine learning algorithm. Int J Comput Sci Eng 7:1852–1857. https://doi.org/10.26438/ijcse/v7i5.18521857
4. Singh C, Kumar P (2017) Affect of personality on work performance 1. IJAIEM 6:77–79
5. Janjua N-u-S (2016) Impact of personality type on job productivity. J Hotel Bus Manag 5. https://doi.org/10.4172/2169-0286.1000119
6. (2021) Personality prediction with CV and using psychometric analysis. Int J Emerg Technol Innov Res 8(6):a93–a95. www.jetir.org | UGC and ISSN Approved. ISSN:2349-5162
7. Aslam N, Masood Khan K, Nadeem A, Munir S, Nadeem J (2021) Analysis of personality assessment based on the five-factor model through machine learning. Int J Sci Technol Res 10(05)
8. Raad B, Mlacic B (2015) Big five factor model, theory and structure. https://doi.org/10.1016/B978-0-08-097086-8.25066-6
9. Kaur G, Maheshwari S (2019) Personality prediction through curriculum vitae analysis involving password encryption and prediction analysis. Int J Adv Sci Technol 28
10. Arora A, Arora NK (2020) Personality prediction system through CV analysis. In: Hu YC, Tiwari S, Trivedi M, Mishra K (eds) Ambient communications and computer systems. Advances in intelligent systems and computing, vol 1097. Springer, Singapore

11. Jayaratne M, Jayatilleke B (2020) Predicting personality using answers to open-ended interview questions. IEEE Access 8:115345–115355. https://doi.org/10.1109/ACCESS.2020.3004002
12. Awadh AM (2012) The impact of personality traits and employee work-related attitudes on employee performance with the moderating effect of organizational culture: the case of Saudi Arabia
13. Ahmad J, Razzaq M, Hussain M (2014) Impact of big five personality traits on job-performance (organizational commitment as mediator)
14. Ngatirin NR, Zainol Z, Chee Yoong TL (2016) A comparative study of different classifiers for automatic personality prediction. In: 2016 6th IEEE international conference on control system, computing and engineering (ICCSCE), pp 435–440. https://doi.org/10.1109/ICCSCE.2016.7893613
15. Hassanein M, Hussein W, Rady S, Gharib TF (2018) Predicting personality traits from social media using text semantics. In: 2018 13th international conference on computer engineering and systems (ICCES), pp 184–189. https://doi.org/10.1109/ICCES.2018.8639408
16. Robey A, Suhkla K, Agarwal K, Joshi K, Joshi S (2019) Personality prediction system through CV analysis. Int Res J Eng Technol 6
17. Satelkar NR, Sawant AA, Vengurlekar MV, Sawant VG (2019) Personality prediction system through CV analysis. Int J Sci Dev Res 4

Part III
Computational Intelligence for Operational Excellence, Supply Chain and Project Management

Chapter 16
Towards Operation Excellence in Automobile Assembly Analysis Using Hybrid Image Processing

E. Sandeep Kumar and Gohad Atul

Abstract The business of a company relies on the delivery of quality products and services with optimal resource usage and it has been a constant endeavor to form timely strategies for improvements in operational efficiency. Many a time product development companies can find opportunities to cut down on repetitive and labour-intensive business processing tasks. Automating the routine process can be a wise business strategy to improve operational efficiency. In this direction, the usage of artificial intelligence concepts like machine learning, deep learning, and reinforcement learning in software product development has been the top choice of many business firms. We believe in, developing value-added differentiators based on recent technological advances in large data management, image processing, and AI/ML algorithm technologies to speed up the drive towards operational excellence. These technologies are at an inflection point, have never been seen before, and definitely can aid in further advancement of business strategies. In this chapter, we discuss one such tool that uses advanced image processing and deep learning algorithms to segment the failure regions from disintegrated automobile assembly parts images. Image segmentation is an aspect of image processing that finds its vast applications in industries and with the advent of machine learning techniques, segmentation has become handier in terms of its computational efficiency. In our technical approach, we use fully convolutional neural networks to segment the region of interest (failure regions) from the image obtained after disintegrating the automobile part, specifically the engine DNox Supply module. One interesting aspect of this work was making segmentation achieve an accuracy of 87% for validation and 98% for training with the sparse dataset. The proposed methodology helped by bringing in intelligent automation instead of manual intensive activity for identifying the region of interest around the failures or abrasions seen in the assembly parts. The generated business reports

E. Sandeep Kumar (✉) · G. Atul
Department of ETM, Robert Bosch Engineering and Business Solutions, Bengaluru, India
e-mail: Erudiya.SandeepKumar@in.bosch.com

G. Atul
e-mail: AtulAnil.Gohad@in.bosch.com

S. Kautish et al. (eds.), *Computational Intelligence for Modern Business Systems*,
Disruptive Technologies and Digital Transformations for Society 5.0,
https://doi.org/10.1007/978-981-99-5354-7_16

are shared with the OEM (Original Equipment Manufacturers) for further improvements in the quality of the parts. As illustrated, we can bring in around 40–50% productivity gains along with upwards of 30% cost reduction.

Keywords Operational efficiency · Image segmentation · Machine learning · Fully convolutional networks · Sparse dataset

1 Introduction

Every business big or small wants to increase revenue by taking operationally efficient decisions. Operational efficiency is the ratio of a company's input and output, which determines the system's overall performance [1]. However, operational efficiency includes processes such as marketing, production, resource utilization, sales, supply chain, and inventory management. The goal here is to provide a quality delivery in a cost and time-effective manner. According to the latest survey of 1300 global CEOs [2], 77% say their focus on driving revenue growth is to create operational efficiencies. According to [3], Ford recently started using drones in its London engine factory to inspect hard-to-reach areas and diagnose machine statuses. The drones fly around the factory and use cameras to zoom in on areas that humans can't reach. Investing in drones not only allows the factory to operate more smoothly, but it saves money and allows the plant to continually operate instead of having to shut down periodically to perform the inspections by hand. This was an intelligence strategy taken by Ford towards operational efficiency. IBM Watson's chatbot [4] is another good example where chatbots take up the charge of providing effective customer service. Today's pandemic situation has demanded work from homes where more calls reach customer centres for numerous queries. There are always chances of human attendees getting burned out in responding to these calls. An AI-based automation tool can provide answers to simple queries improving operational efficiency. Many such examples can be quoted and it's quite evident today that many such companies are moving towards smarter ways of making operations efficient. In this chapter, the use of an AI tool for automation of the failure analysis process is discussed in detail and hence providing a flavour of an inter-disciplinary approach towards improving operational efficiency.

Image processing is predominantly being used in many industrial applications [5–7]. Few among them include: measuring the length and aspect ratio of a machine part, finding the corrosion based on shape deformation or colour degradation, measuring the position of an object, segmenting the required region of interest based on its unique patterns, and so on. Image segmentation is one of the challenging as well as widely adapted image processing methods for industrial applications [8]. Though the last few decades followed traditional methods of image segmentation, with the advent of machine learning recently, the community started to adapt learning methods to achieve the same. In the last few years, deep learning-based image segmentation has been extensively used in industries for various applications. These include encoder-decoder models, multiscale and pyramid-based architectures, visual attention-based and generative adversarial networks [9]. Each one of these models has been applied to a variety of industrial applications. In this chapter, we will see one such application of a class of machine learning algorithms called deep learning (DL) which used an encoder-decoder model to segment the region of interest from RGB images taken by disintegrating automobile parts belonging to the industry failure catalogue. DL algorithm was used to segment the failure region from the images. This output was further optimized by a wrap of image processing code to crop the region out, thereby the cropped images were further taken directly to the final report generation. Plentiful research and adaption of deep learning for industry applications are seen in the existing literature. In [10], the authors use a customized deep-learning architecture to detect the defects in the fabric. The architecture had two stages the segmentation and the detection stage. In [11], authors use adversarial deep learning networks to identify the cracks in the concrete. For this purpose, the authors used 1200 annotated images with 3000 unannotated images, and the neural networks were trained using the semi-supervised learning method. In [12], the authors proposed a hybrid encoder-decoder model that used the VGG16 network as the encoder and UNet at the decoder end. The method is exclusively trained to detect the corrosions on the surface. The network was trained with the Heidelberg dataset. In [13], the authors proposed a compact CNN for surface anomaly detection. For training and testing, they relied on the DAGM dataset. Therefore, on observing the current state-of-the-art works, it is quite evident that deep learning algorithms are prevalent in the industry and the architecture varies depending on the application, data availability, and accuracy requirements. Good accuracy is witnessed in these works which further motivated us to incorporate deep learning architectures in our work as well. The further of this chapter is organized as follows: Sect. 2 discusses the motivation, Sect. 3 on the dataset used, Sect. 4 with the proposed methodology in detail, and finally Sect. 5 with the conclusions.

2 Motivation

If an automobile part fitted on a vehicle undergoes failure in the field, they are returned to the supplier for analysing the failure so that appropriate design corrections can be realized to prevent similar failures in future products. Each of these field failure parts is subjected to "tear down" and images of all its subcomponents are taken and stored, which then are analysed to identify which sub-component failed and to identify how does the failure manifest. Having this information, it is then possible to identify the root cause of the failure so that appropriate design corrections can be arrived upon, and finally, a report is generated manually. However, since most images from the camera are not optimized to maximize the region of interest in the frame, it is necessary to crop the images to ensure the region of interest is maximized in the frame. We also elaborate on how to crop the images to maximize the region of interest in the frame, using CNN. The cropped images are then used for the final report generation. For this purpose, a simple GUI was created which provides an easy one-click setup for the user to generate the final report. This tool is one such example where AI-based automation can make industrial operations effective and efficient.

3 Dataset

The images were populated from various disintegrated parts of the engine DNOX module specifically the supply and dosing module. The parts like a filter, valve needles, electrical connectors, screws, etc., were considered for imaging. The images were not restricted to background and were taken at fields; however, care was taken that the images were taken in good lighting and with uniform distance between the camera and the object. Around 490 images were collected, out of which 80% images were used for training and the remaining 20% for validation. This accounted for 392 images for training and test for validation. The sample images are shown in Fig. 1.

It is to be noted that the segmentation can be the entire component available in the image like the image in the left-most upper corner where the entire module is supposed to be cropped. In other cases, the specific region in the whole image needs to be segmented like the second-row first image where only the circular disc shape is required. Hence, the chosen algorithm must consider all these variations. One notable aspect is the lack of a huge number of data samples. None of the companies maintains the failure images unless and until it is necessary, in many cases the failure samples are not available at all. The developed algorithm should digest and learn from this sparse data as well.

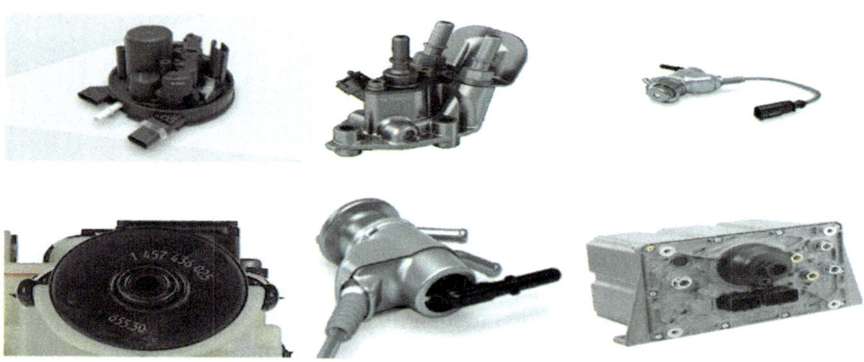

Fig. 1 Sample images used from the dataset

4 Proposed Methodology

This section has two parts—the first part discusses the image segmentation in detail and the second part with the GUI. The GUI also has some additional features along with segmentation which will be addressed at the end of this section.

4.1 Image Segmentation

It is quite evident that convolutional neural networks [14] are widely used in general for many computer vision applications industry-wide. In this context, a fully convolutional network (FCN8) [15] is used for the image segmentation problem we are dealing with. The architecture is shown in Fig. 2.

It's a well-established deep learning architecture for semantic segmentation [16]. Conventionally in a CNN, the image goes through a series of convolutional layers down-sampling the given image and then passing this to a dense layer leads to a label for the input image as a whole. This is a typical image classification task. But in FCNs, the image is down-sampled and up-sampled simultaneously. After every pooling layer, the image feature map is up-sampled to match the dimensions of the original image. This is done at any of the layer outputs and at the end, all the up-sampled maps are combined to obtain one heat map. This image will have the

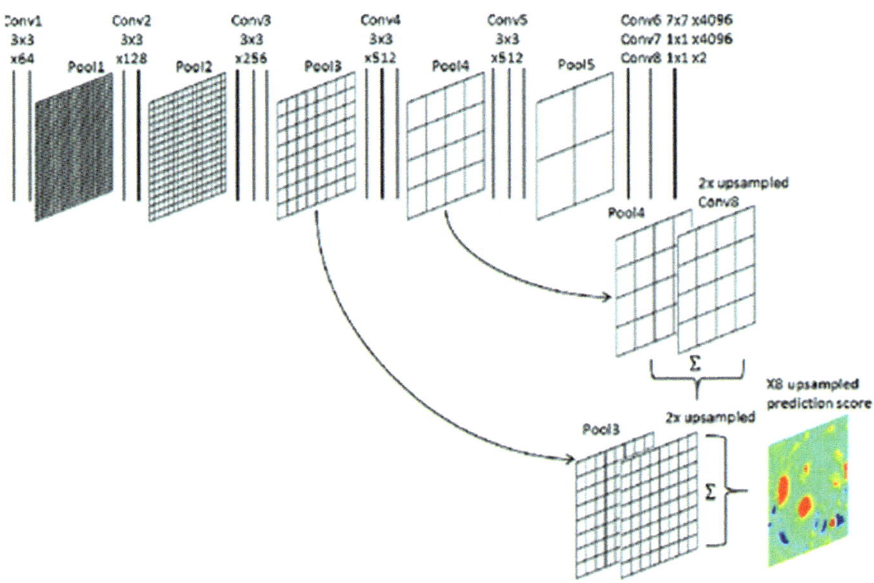

Fig. 2 Fully convolutional neural network

region of interest segmented (in turn all the pixels labelled). There are an ample number of FCN architectures and every method has a different way of up-sampling and combining. In our work, we employed FCN-8 architecture. FCN is trained with 392 images with their respective masks and a learning rate of the model was 0.001 with RMSProp as the optimizer. The image region of interest extraction falls into the category of semantic segmentation, where every pixel is labelled. For this purpose, the training process should include masks of all the training images. In the training process, 490 images and their corresponding masks were used. These masks are generated as part of the labelling or annotation process, which is currently carried out manually. This is a one-time activity that includes collecting all the spare part images from the database and manually creating masks using tools like MS-Paint or GIMP. In the current experiment, it took 17 h for creating a mask for 490 images (single manual effort). Though there are plentiful of tools available for annotating images [17, 18], the majority of them are available as online platforms. The images used in this project are governed by firm privacy issues and hence using any such kind of online platforms is strictly avoided in actual industrial practices for sensitive data. There are few stand-alone image segmentation tools available for annotation, however, they need the same amount of manual effort that is spent in tools like GIMP. The annotation tools come in extremely handy and useful in case of object recognition which has definitive shapes and features, however for the problem statement that is dealt in this chapter the image part has to be segmented. These parts appear in a variety of shapes, colour and sizes. Hence, it is mandatory to manually annotate these images using tools like GIMP.

The learning rate and other parameters of the neural network were adjusted after a series of trials and errors until a required accuracy was reached. The images were randomly split into 80–20 ratios in each iteration to achieve a robust model.

In Fig. 3, the central black region is the region of interest and hence in the mask, the region has to be explicitly brought out. The FCN was trained for 60 epochs and every time the training and validation accuracy was tabulated. The epoch number was fixed based on a series of trials and errors, and the best number was taken into account. The training accuracy was 98.9% and the validation accuracy was 87%. The trained model was saved as a.h5 file to be loaded later. The obtained training accuracy for different epochs is shown in Table 1 and it confirms that for 60 epochs the obtained accuracy is high.

The obtained result from the model is shown in Fig. 4. The validation accuracy tells how well the network performs when it is used on the fresh dataset.

The predicted image from the model is subjected to further post-processing using traditional image processing which is used to extract the perimeter of the predicted region of interest to draw a bounding box. The extracted rectangular region is pasted into the cell locations of an excel sheet. The cell locations for a spare part are hard-coded, and a specific cropped image will be pasted in a pre-defined cell location in excel (in this work, MS excel is used to generate the final report). The coding platform used to implement the deep learning algorithm was python with TensorFlow and all relevant packages using an HP Zbook 8th gen computer with an i7 processor.

Fig. 3 Image and its mask with the region of interest marked

Table 1 Epochs versus training accuracy

Epochs	Training accuracy (%)
60	98.92
50	98.72
40	98.10
30	97.90
20	97.52
10	97.14

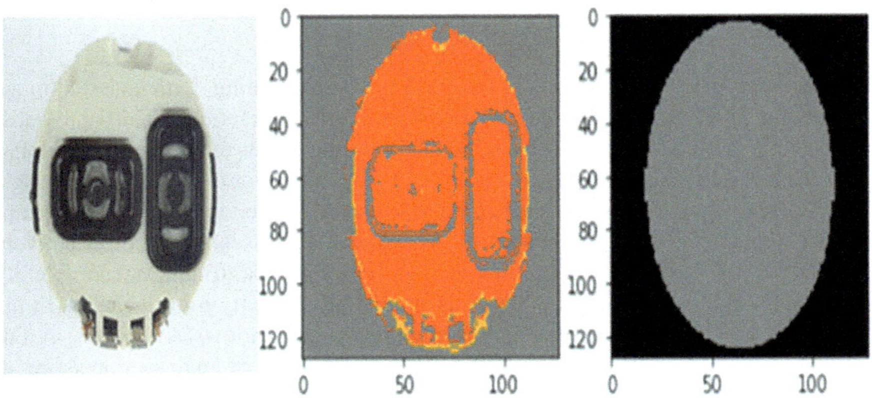

Fig. 4 **a** Input image **b** obtained segmented region prediction from the model **c** mask used to train that image sample

4.2 *Designed GUI for the Segmentation Task*

The UI design is carried out using the packages available for python like PysimpleGUI. The project was named "LEaAF: Learn Extract and Analyse Failures". The

GUI has a selection box to select project modules, based on the selected module, the images will be loaded for segmentation. This module differentiation is required for the images to be stored in a pre-defined location in the generated investigation report in the form of an excel sheet. Accordingly, different reports will be generated for different modules. The tool processes one image set at a time and hence generates one report at a time. The user can provide a folder path for images to be analysed. All the images are read from this location in one go for processing by the model. The moment the folder is read, the trained deep learning model file will be activated alongside the image processing algorithm to extract the suitable region of interest. The cropped images that are available as output from the model are persisted in the path provided for the same.

The user can also provide the folder paths for additional tasks of extracting and storing the measurements tables into the final report from other data files available as Excel sheets. In the background, our scripts perform tasks of excel data aggregation from different files placed in various paths, along with the cropped images to finally generate a semi-automated investigation report.

The LEaAF tool was validated for around 85 images of the supply module and about 55 images of the dosing module to obtain expected cropped images (extracted region of interests) that are used for generating specific reports (the algorithm as explained in the previous section was trained for all the variations of these).

5 Conclusions

This chapter discusses the use of combining deep learning with classic image processing algorithms for the optimal region of interest extraction as a drive towards the usage of automation tools for making business operations efficient. The work has been a successful implementation as a one-click process for a user in generating a ready-to-use report. The work has helped to substantially reduce the time consumption for report generation compared to the manual processes. The tool effectively handled the variations in the data concerning its lighting, background, and complexity with limited data. Our tool has shown significant productivity improvements in the overall process of analysing the failure regions for automotive assemblies. A fair estimate puts these gains at around 40–50% for each cycle of analysis consisting of a batch of around 80 assembly parts. Typically, in a year, around 5–6 such analysis cycles are expected, and this has shown promise to even be applied to other similar analyses originating from other departments. Also, when we factor in the dependency on skilled experts' saved time, the teams benefit from a reduction of around 30% in costs. The productivity and cost benefits of such an AI and Image processing-based analysis tool can only become better as more and more data samples are made available for our model's iterative learning. As discussed in the earlier parts of this chapter, the reduction in the manual efforts thereby makes business in a firm more effective and reliable.

The technique dealt with in this chapter can be used in any such application, industry-wide, where image analysis and segmentation are to be performed.

References

1. Blaney B (2021) What is operational efficiency and how can we improve it?. https://tipalti.com/operational-efficiency/. Accessed 23 Aug 2021, 8.13 PM
2. Annual Global CEO Survey (2021) CEO's curbed confidence spells caution. https://www.pwc.com/gx/en/ceo-survey/2019/report/pwc-22nd-annual-global-ceo-survey.pdf,internet. Accessed 24 Aug 2021, 11.18 AM
3. Ferris R (2021) Ford is using drones to keep an eye on its UK factories—and save money. https://www.cnbc.com/2018/08/31/ford-is-using-drones-to-keep-an-eye-on-its-factories.html. Accessed 24 Aug 2021, 11.21 AM
4. IBM AI customer service (2021) Improving the customer service, the landscape. https://www.ibm.com/watson/ai-customer-service-smartpaper/#landscape. Accessed 24 Aug 2021, 11.21 AM
5. Fondevik SK, Stahl A, Transeth AA, Knudsen OØ (2020) Image segmentation of corrosion damages in industrial inspections. In: IEEE 32nd international conference on tools with artificial intelligence (ICTAI), pp 787–792. https://doi.org/10.1109/ICTAI50040.2020.00125
6. Persoon E, Nijholt G, Maguire G, O'Brien J (1990) Industrial image processing by means of an image recognition integrated system. In: Proceedings of the 10th international conference on pattern recognition, vol 2, pp 402–407. https://doi.org/10.1109/ICPR.1990.119390
7. Fabijanska A, Kuzanski M, Sankowski D, Jackowska-Strumillo L (2008) Application of image processing and analysis in selected industrial computer vision systems. In: International conference on perspective technologies and methods in MEMS design, pp 27–31. https://doi.org/10.1109/MEMSTECH.2008.4558728
8. Tewari P, Surbhi P (2016) Evaluation of some recent image segmentation method's. In: 3rd international conference on computing for sustainable global development (INDIACom), pp 3741–3747
9. Minaee S, Boykov YY, Porikli F, Plaza AJ, Kehtarnavaz N, Terzopoulos D (2021) Image segmentation using deep learning: a survey. IEEE Trans Pattern Anal Mach Intell. https://doi.org/10.1109/TPAMI.2021.3059968
10. Huang Y, Jing J, Wang Z (2021) Fabric defect segmentation method based on deep learning. IEEE Trans Instrum Meas 70:5005715. https://doi.org/10.1109/TIM.2020.3047190
11. Shim S, Kim J, Cho G-C, Lee S-W (2020) Multiscale and adversarial learning-based semi-supervised semantic segmentation approach for crack detection in concrete structures. IEEE Access 8:170939–170950. https://doi.org/10.1109/ACCESS.2020.3022786
12. Mittal S, Chopra C, Trivedi A, Chanak P (2019) Defect segmentation in surfaces using deep learning. In: International conference on issues and challenges in intelligent computing techniques (ICICT), pp 1–6. https://doi.org/10.1109/ICICT46931.2019.8977696
13. Racki D, Tomazevic D, Skocaj D (2018) A compact convolutional neural network for textured surface anomaly detection. In: 2018 IEEE winter conference on applications of computer vision (WACV), pp 1331–1339. https://doi.org/10.1109/WACV.2018.00150
14. Albawi S, Mohammed TA, Al-Zawi S (2017) Understanding of a convolutional neural network. In: International conference on engineering and technology (ICET), pp 1–6. https://doi.org/10.1109/ICEngTechnol.2017.8308186
15. Long J, Shelhamer E, Darrell T (2015) Fully convolutional networks for semantic segmentation. In: IEEE conference on computer vision and pattern recognition (CVPR), pp 3431–3440. https://doi.org/10.1109/CVPR.2015.7298965
16. Jordan J (2021) Details of semantic segmentation. https://www.jeremyjordan.me/semantic-segmentation/. Accessed 15 June 2021, 6.45 PM

17. Image annotation tool. https://toloka.ai/ml/computer-vision?. Accessed 19 Oct 2021, 1.32 PM
18. Image annotation tools. https://medium.com/data-folks-indonesia/5-best-free-image-annotation-tools-80919a4e49a8. Accessed 19 Oct 2021, 1.33 PM

Chapter 17
Industry Revolution 4.0: From Industrial Automation to Industrial Autonomy

Pradeep Bedi, S. B. Goyal, Anand Singh Rajawat, Jugnesh Kumar, Shilpa Malik, and Lakshmi C. Radhakrishnan

Abstract Industrial 4.0 (4th industrial revolution) embodies rising technological advances withinside the improvement of clever manufacturing strategies. Industry 4.0 has great potential for many manufacturing companies to allow customization of products, provide flexibility to meet new needs in real-time and produce highly efficient jobs. The 4th Industrial Revolution and rising technologies—which include the Internet of Things, artificial intelligence, robots, and greater productions—affect the emergence of the latest manufacturing techniques and commercial enterprise fashions that transform fundamental manufacturing. The next generation of our industry is Industry 4.0—with the guarantee of improved production flexibility, as well as greater customization, improved productivity, and better quality. This allows companies to access additional products designed for each in a shorter period of time and better market standards. Intelligent manufacturing performs a critical position in Industry 4.0. Ordinary tools are transformed into intelligent objects that you will hear, handle, and perform in an intelligent environment. This makes industrial autonomy a reality where different countries work together to develop technology and start the next generation which is a 5.0 industrial revolution. Here in our work, we've mentioned the effect of Industry 4.0 on making the arena digital. Industry 4.0 is ubiquitous;

P. Bedi (✉)
Galgotias University, Greater Noida, Uttar Pradesh, India
e-mail: pradeep.kcu21@gmail.com

S. B. Goyal
City University, Petaling Jaya, Selangor 46100, Malaysia

A. S. Rajawat
School of Computer Science and Engineering, Sandip University, Nashik, India

J. Kumar
Echelon Institute of Technology, Faridabad, India

S. Malik
Aravali College of Engineering & Management, Faridabad, India

L. C. Radhakrishnan
Institute of Management and Technology (IMT Business School), Dubai, UAE

S. Kautish et al. (eds.), *Computational Intelligence for Modern Business Systems*, Disruptive Technologies and Digital Transformations for Society 5.0,
https://doi.org/10.1007/978-981-99-5354-7_17

however, we goal to take a more in-depth study of the cutting-edge enterprise imaginative and prescient and display the enterprise 4.0 destiny trends. Additionally, this chapter introduces generation to transport from Industry 4.0 to Society 5.0 and anticipates the destiny from Industry 4.0 to Industry 5.0.

Keywords Automotive industries · Digital technologies · IoT (internet of things) · Industry 4.0 · RAMI 4.0

1 Introduction

The term "Industry 4.0" become first utilized in 2011 by [1]. Used as a project title, part of the "Hightech strategy" of the German Government. This assertion is primarily based totally on the imaginative and prescient of the fourth business revolution to come.

a. Pushed by steam and water power, the first industrial revolution took place around 1750, allowing industrial systems to be mechanized. That was a revolution, although it contributed, among several other aspects, to the innovation of steam engines, the use of water, as well as heat energy. Then all types of additional devices will contribute to the industrial societal transition by employing trains, production mechanizations, and smog loads.
b. The II Industrialization began with electricity. Electrification enabled mass production in 1870.
c. New information and communication technologies made it possible for industrialization in the 1960s [2]. It was all about the development of computer technology, computer network systems (WAN, LAN, MAN), the growth of industrial robots, networks, and, no doubt, the proliferation of the Internet.
d. IV Industrialization should use CPS (Cyber-Physical Systems) to make industrial enterprises digital [2].

Industrial 4.0 demonstrates a progressive pattern towards mechanization and information marketing in the establishment and cycles within an integrated business, which includes developments such as IoT, advanced knowledge, AI, in-depth, realistic unpopularity of taxpayers we see, clever thinking, distributed computers, artificial plants, as well as creative arts, a modern web of things. This use of the computer creates an interconnected system in which plant machinery is expanded with remote sensory access to visualize and visualize the interaction of the universe and resolve independent decisions. This will provide faster response times, taking into account the ongoing communication between the structures.

This part of the chapter specializes in the vision of possibly the largest alternative to form our lifestyle choices: Industry 4.0. We will likely speak approximately the essentials, the improvement of robots and cycle enhancements, the best manner to show small and medium enterprises (SMEs) into Industry 4.0 [3], the monetary and academic component, the go-back of speculation, and more. forward.

Industry 4.0 'reflects the 4th industrialization of the emergence of modern, digital manufacturing technology. It is a combination of several significant digital technology developments. Such innovations comprise artificial intelligence and advanced robotics and; sensors; computing; IoT; data collection and analytics; digital manufacturing (3D printing). Applications-as-a-service as well as other emerging marketing modeling techniques; mobile phones, and other smart devices. Industry 4.0 leads to the evolution of machines that can use self-configuration, self-optimization, and Artificial Intelligence (AI) to accomplish complex tasks to achieve expense-efficiencies and services or products of a higher quality.

Industry 4.0 incorporates information, computing power, communication, intelligence and analytics, the interaction of individual machines, and digital conversion to physical transformation. The business is digitized, and processes are organized systematically, from manufacturing and purchasing to production, distribution, and operation. The parameters leading to the market growth are the emergence of cloud computing services, technological developments in the electronics sector, adoption by various countries of smart factory design and factory technological innovations, and government initiatives. We have shown Industry 1.0 to 2.0 to 3.0 to 4.0 snapshot in Fig. 1.

Industry 4.0's main drivers are digitization through the convergence of horizontal and vertical supply chains, product, and service delivery digitization, online marketing strategies and consumer holding, quicker, scalable, and reliable production, and growing smart manufacturing prominence. The analysis states that the significant challenge for Industry 4.0 market is the network security threat [4].

IR 4.0 provides a platform for developing and monitoring a process that controls the distribution process of different plant species. It can monitor the growth of large

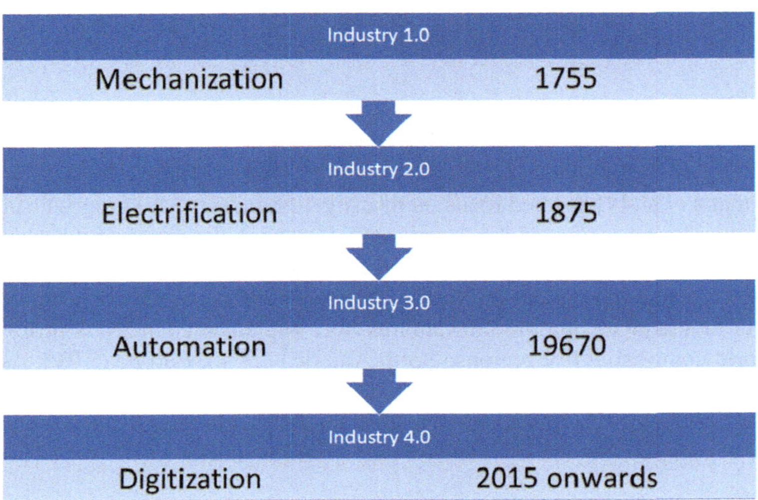

Fig. 1 Industry revolution

Table 1 Digitization stage with needs and benefits

Digitization stage	Needs and benefits
Semi-automated	To improve security, adequacy, and accessibility Stretch out to accomplish safe, more effective activities
Automated	To improve influence information for ongoing choices to improve effectiveness Stretch out to improve the usefulness
Semi-autonomous	To improve influence information for ongoing choices to improve effectiveness Reach out to improve efficiency
Autonomous orchestration	To coordinate various areas and capacities Some stretch out to catch new business openings with expanded adaptability
Autonomous operation	To Integrate ecosystem Some extend to optimize value chain with autonomous operations

numbers of industries. End users or customers are experiencing rapid technological advancement and an ever-changing business environment. IR 4.0-based technologies provide solutions and services that will facilitate digital transformation (DX) in their business and pave the way for the future when industries will move from Industrial Automation to Industrial Autonomy (IA2IA). Independence can be embedded in process plants in a variety of operational areas such as quality control and processes, strategies and preparation, supply chain management, field operations, maintenance, and engineering. We have developed the IR 4.0 technology range and their impact and industry problem-solving strategies and automation. In addition, we shared the impact of IR 4.0 and discussed industry expectations (of 2020). We have summarized the various stages of digitalization of their needs and benefits in Table 1.

1.1 Current Scenario of IR 4.0: Literature Survey

Kagermann et al. [4] Studied focus on investigation industry 4.0 proposals utilizing systematic digital review of the literature to determine the scope of their launch in various regions. Industry 4.0 programs in developing nations have created observations to be used as guidelines for the East African Population. Survey results showed that 117 industrial revolutions 4.0 ventures were implemented in 56 countries across the world comprising five regions: North America (28%), Europe (37%), Asia and Oceania (17%), Caribbean, as well as Latin America (10%), as well as Africa and Middle East (8%). ‘E was measured at 25% globally. It is reported that there is still a wide gap throughout the Industry 4.0 race across countries [5].

Research work analyses the existing state of the art of Energy 4.0 in Kazakhstan, their prospects, advantages, obstacles as well as government vision to bring additional discussions in the development of Industry 4.0's latest technical transition, which will be of a structural and fundamental character. The study examines the global manufacturing transformations and the subsequent global energy measures taken over the past ten years through demonstrating their NRI (Networked Readiness Index) schemes, which display Kazakhstan's locations based on various criteria defined by reputable international bodies. Besides, the paper discusses the ways of implementing the Energy 4.0 Approach in technologically developed countries. Throughout this report, we often propose the techniques and suggestions correlated with the extension of Industry 4.0 policy to the Kazakhstan energy sector that can be adopted based on the latest business growth strategy during the emerging global technological advancements [6].

This study examines current developments in digital technology, including their effect on industrial automation and, specifically, the 4.0 industry revolution. Communication technologies like Ethernet, cellular networks, and web-based applications are well-considered to have rendered distribution and information sharing increasingly efficient, scalable, and extensive across distributed industrial automation. In other terms, ICT (information and communication technology) offers (1) indigenous factory automation services, (2) control, oversight, and tracking benefits, and (3) high efficiency, better reliability, as well as more versatility. In this study, we discuss topologies of industrial interactions, and IoT (Internet of Things) enablers, which substantially reduce latency. They suggested that the current challenging priority on latency as well as protection on the wireless access points [7].

Industry 4.0 are structures formulated to change the way of the latest manufacturing operations by using some cutting-edge technologies, innovations sometimes used during the construction of Industrial IoT systems, data analytics, or robotics. One such technology is blockchain, which can bring trust, protection, and decentralization to various sectors of the industry. This study explores the advantages and drawbacks that occur while designing Industry 4.0 implementations leveraging blockchain-based smart networks. Moreover, this work comprehensively analyzes Industry 4.0 innovations focused on the most appropriate blockchain applications. The goal is to furnish a comprehensive show for future experiments in Industry 4.0 to decide whether blockchain will boost the next wave of cyber-secure industrial uses [8].

Throughout the scope of the automobile industry, this paper illustrates a significant difference between the specifications endorsed by the existing MES (manufacturing execution systems) as well as the criteria suggested by the ISA (International Society Automation) Industrial Norms, like ISA-88, ISA-95, on which Industry 4.0 is based. In this study, they fill the gap while adopting a model-based technique to project planning along with a method of gap assessment. The tasks are divided mainly into three stages: I phase of choice of industrial MES resources, (ii) phase of modeling characteristics, (iii) phase of the review of gaps dependent on the model specification. We utilized established, credible sources like MES model research reports and white papers, which provide in-depth and detailed data regarding different evaluation measures and tool supplier's database for the existing MES scenario during the recruitment process of the MES tool. And during the specification modeling process, using the general-purpose SysML (Systems Modeling Language) defined requirements generated from the purposes of ISA-95 as well as ISA-88 manufacturing specifications. During the gap research process, we notice the discrepancy among the following specification and the alignment with the current development tools [9].

This work provides the concept of a science fiction novel with a clip, "The Day Before the Remote for the Futuristic Control Panel 2025," designed to explore new possibilities for communication strategies or intelligent automation for potential Industrial 4.0 plants. The aim of the concept was to conduct experimental design experiments on the impact of specific UX objectives (user experience) on a modern production framework. All work initially describes how well the UX objectives are embedded in the science fiction frame with the proposed clip and, secondly, how well the concept is tested in the 2nd study configuration: as video streaming is included in web-based testing and discussions with process management staff. A critical contribution is to reveal how the technology fiction improvement technique can use video presentation in future-targeted UX evaluation and the way user-targeted tactics may be used to discover UX goals through the use of the technology fiction mode [10].

This study outlines the growing needs that must be met in the coming years, ranging from industry to 4.0 paradigm of production processes for years to come.s It is important to integrate existing technologies effectively with current models. The CPPS system (Cyber-Physical Production Systems) is the basis for the automated distribution of platforms and intelligent control. Fortunately, integrated systems need to be introduced into the global production system, which makes that possible. This analysis explored methods that present design and model-based technologies to bring the idea into reality [11].

In the 4th industrial revolution, Industrial 4.0 embodies future developments in market growth to achieve intelligent production processes such as counting on Cyber-Physical Systems (CPS), CPPS architecture, and sensible manufacturing facility deployment and operations. With regard to strategic planning, key innovations, opportunities, and threats, all research reflects the important concept of Industrial 4.0. The strategic management process involves developing a CPS system, addressing two key topics focused on a smart production system or an advanced industry, managing three applications (horizontal installation, direct installation, and end-to-end installation), and implementing eight system integration proposals. specification, productive activities, etc. Finally, it has contributed to the transformation of the producing enterprise to construct Industry 4.0 in China [12]. We have made Table 2 of the summary of book reviews.

Table 2 Summarization of the literature review

Sr. no	Author	Research	Aspect/keywords
1	Bongomin et al. [5]	The author's research focuses on industry research recommendations 4.0 that use systematic digital reviews of texts to determine the scope of their presentation in various regions	4.0 industry revolution, East African Population, systematic digital
2	Kada et al. [7]	Discussing topologies of industrial interactions, IoT (Internet of Things) enablers, which substantially reduce latency	4.0 industry revolution,
3	Alimkhan et al. [6]	Research work analyses the existing state of the art of energy 4.0 in Kazakhstan; this brings additional discussions in the development of industry 4.0's latest technical transition	4.0 industry revolution, Networked Readiness Index
4	Fernandez-Carames and Fraga-Lamas [8]	The study explores the advantages and drawbacks that occur while designing industry 4.0 implementations leveraging blockchain-based smart networks	Industry 4.0, blockchain-based smart networks, IoT systems
5	Kannan et al. [9]	Research illustrates a significant difference between the specifications endorsed by the existing MES (manufacturing execution systems) as well as the criteria suggested by the ISA (International Society Automation) Industrial Norms, like ISA-88, ISA-95, on which Industry 4.0 is based	Industry 4.0, production execution systems, Industrial Norms
6	Schoepf [10]	This work provides a clip-illustrated science fantasy concept, "The day before the Remote Supervisor in a Futuristic control panel in 2025", designed to explore prospects for new techniques of communication or smart automation for potential Industry 4.0 plants	UX analysis, science fiction model, Industry 4.0
7	Pérez et al. [11]	The analysis examined the criteria that introduce a design and technology based on models to bring the idea a reality	Industry 4.0
8	Zhou et al. [12]	The fourth industrial revolution, Industrial 4.0 embodies future developments in market growth to achieve sophisticated production processes such as relying on CPS	Industry fourth system, CPS, Strategic management

2 Structure of Industry 4.0

2.1 Industrial Automation and Monitoring Principles and Objectives

Industrial Automation and Control (IAC) have performed a significant task in modernizing industries as well as organizations since its incorporation into the manufacturing processes through (the 1960s). Also, on the other side, it improved the efficiency, versatility, sustainability, and protection of industrial processes. On the other side, this reduced the operating and maintenance time and costs. To minimize cost, connected with human controllers. The IAC categorization depends on the convergence of AI-related technologies (such as robots and computers), information systems, and interaction (speech recognition) within a unified model. The goal of the AI-based IAC is to incorporate cognitive technology to enable smart manufacturing and operations. Learning, preparation, thinking, and understanding are the key features of AI-based systems to conduct challenging industrial processes [13]. These functionalities are developed and constructed using complex mathematics, computer technology, philosophy, and other methods and strategies. Among several methods are implementation, computational quest, statistical methods, reasoning, likelihood, and game theory, such as in Fig. 2 [14].

It is very well recognized that the advent of smart hardware technologies linked with actuation, sensing, data gathering, data processing, and interaction demonstrates the IAC processes are more refined. Utilizing high-level techniques as well as embedded software methodology [15], IAC networks are becoming smarter and more self-operating.

Automation is typically accomplished via a four-level structural IAC configuration from high to low stages like communication or business, coordination, and monitoring output, power, and ground level. Figure 3 shows the IAC framework configuration. As per consumer needs or market research, development schedules and demands are prepared at the highest level (I / E tier). There's more commercial aspect to this level than the technological aspect. The level of SPC (Supervising and Production Control) is one of the most critical groups in IAC system architecture since it provides the process factor supervision, determining production goals, data gathering, indexing, and care. The degree of monitoring and production monitoring interacts with management, manufacturing, maintenance, efficiency, etc. The level of control involves personal and collective measures that incorporate industrial automation technologies from PLC to obtain high-level system output [16].

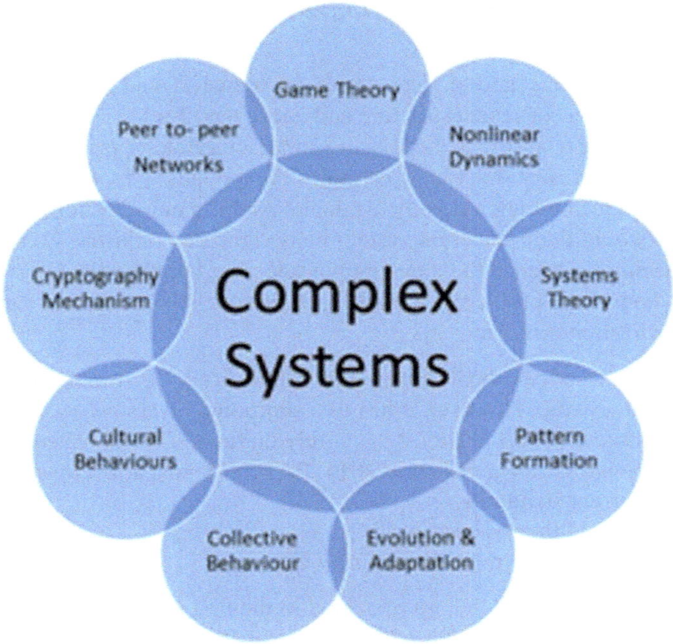

Fig. 2 Industrial-automation sciences and innovations

Fig. 3 Hierarchy of industrial automation and control systems

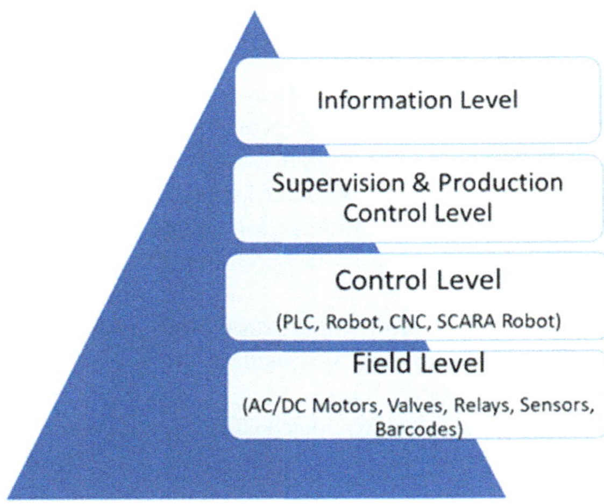

2.2 Reference-Architecture-Model-Industry 4.0 (RAMI 4.0)

The various programs and projects have encountered and described the same problem; a lack of a consistent structure and vocabulary for I4.0. This Leading to the evolution of separate abstracts models and architectural descriptions for I4.0 [13]

The idea of Industry 4.0 would certainly be applied by most businesses using equipment and technologies already available. And if a new development system is introduced, would the production system be designed as Industry 4.0 system from the very beginning. Hence the challenge of adapting current principles into a new concept is one of the problems. A RAMI 4.0, was developed in Germany to solve this standardization problem [17].

- Metamodel describes the features of the Industrial 4.0 development program that play a key role. However, it has two additional layers below, which include different structures of Industry 4.0 and rely on the globe-renowned architectural model of Smart Grids launched in 2014. RAMI4.0 must allow three parties:
- Defining current standards,
- Identifying and filling vacancies,
- Acquisition of the fragmentation of new values.

Two parts, type, and model are protected in the first place. If an idea, design, or product remains a proposal that is not easily achieved, it is called a genre. Phase 2 of the model is local and operational from development to the associated world (as the final phase of Industrial 4.0 growth for all firms, clients, and associated providers). The 3rd dimension of RAMI 4.0 is structured as follows in the active layers [16].

- The inventory contains materials such as machines, conveyor belts, PLCs, documents, archives, software, and ideas.
- The incorporation layer provides asset information on how it could be digitally accessed. It integrates components associated to IT, such as sensors, HMI combination, and computer-assisted technological process management.
- Adjustment of communication using a common data set-up and predefined procedures is a feature of the communication layer. It also offers integration layer services.
- A layer of information processes that incorporates data in useful information.
- Formal descriptions of functions are given in a functional layer. This layer also contains ERP features.
- A business layer provides business model mapping and links to multiple business processes.

As a first collection of the basic technical elements of Industry 4.0, the RAMI4.0 is recorded as DIN SPEC 91,345 in Germany. The definition of Industry 4.0 is viewed as the precondition for its implementation and as a model for international acceptance [18].

2.3 Industry 4.0 Advantages and Disadvantages

2.3.1 Advantages

Industry 4.0's fundamental aim is to make production—including related businesses like logistics—quicker, more effective, and much more customer-centric, but at the same time moving across optimizing and automotive techniques as well as finding new market opportunities and structures. Almost all Industry 4.0 major advantages are connected to advantages of digital industrial innovation through the use of IoT in production, functional or enterprise operational efficiency, data-powered value networks, overall technological innovation, smart factories, and several other aspects on our site. Yet describes below some of Industry 4.0's essential advantages [16].

1. Improved automation and efficiency have been one of the first goals of the Industrial 4.0 systems. In many other terms: total cost, improved stability, waste minimization, automatic avoidance of delays and errors, faster acceleration to do more in real-time, and performance of the entire production chain, because speed is important for all, digital paper-based processes, ability to quickly interrupt problems.
2. Real-time information for the real-time value chain in the Current Economy: Industry 4.0 is for the whole product life cycle naturally, production doesn't stand alone. Multiple stakeholders are involved if you look at the overall value chain and the environment in which industry operations exist. All those are customers, and consumers want to improve productivity, no matter where the distribution chains are located. Unless the end consumer requires better products quickly and it has raised expectations about consumer experience, service, quality, and goods that are produced at the specific moment they need, this will affect the entire value chain, all the way up to production and even beyond. In a progressively real-time economy, speed is not only a competitive edge and consumer requirement, and it is a factor of synchronization, value, and cost development. Moreover, customers always expect it [17].
3. Superior industry sustainability by automated servicing and controlling opportunities: If an industrial asset breaks, it must be repaired. Supporting people and engineers is often moving a lot, which requires time and money. Whenever a main industrial resource, like an industrial robot in an automotive production plant, relinquishes, it is not only the machine that is disabled. Production is impaired and costs tons of money and angry buyers, although the output may also be completely interrupted. It is the hardest nightmare of all, as the sustainability of enterprise is an incredibly high issue.
4. Higher quality goods: real-time tracking, improved efficiency facilitated by IoT, as well as robots (collaborative robots).
5. Favorable working standards and sustainable growth: Developing a working state in the field or factory depends upon real-time temperature, humidity, other statistics, improved security, and fast detection, and the list continues.

6. Enhanced agility: In the manufacturing industries, we require the same usability or skill they anticipate by promoting IT businesses and innovations, like cloud computing. This is partially due to the former customization concept but is primarily about using technology, Big Data, AI, robotics as well as cyber-physical systems to forecast and meet seasonal demand, output fluctuations, upscale or downscale possibilities; in all other words: all variations that are often almost predictable can be rendered unpredictable or predictable. And it can be done from a time and scale viewpoint, resulting in increased transparency, flexibility, and the prospect of exploiting resources as per optimum processing demand [19].

2.3.2 Disadvantages

- Negative Effect of Facts Sharing in an Aggressive Environment

Using Industry (non-analog) digital communications 4.0 using horizontal, vertical, and vertical integration allows data exchange in the communication platforms in a competitive sense, leading to a transparent business environment. This means that two concerns (a) From the high level of transparency will contribute to risks such as industrial intelligence, cyber-attacks, and additional issues like data rights and access. (b) The organization that determines stadium needs can discourage and push other companies to start a business with a well-defined, one-stop shopping [20].

- Complete Implementation of Industrial 4.0 is Required for Success

For this program to be successful, the full implementation of Industrial 4.0 is essential. This is a challenge as the launch of Industrial 4.0 has to be planned for many production structures or sizes. Approval of financing is risky because some companies use it to reinvent remaining production and transportation processes due to monetary constraints. Fourth Industry 4.0 cannot be used in one way, and sustainable results cannot be achieved. Aligning and connecting to existing infrastructure and production processes can lead to more complex and complex costs for organizations such as SMEs.

- Managing Grievances of Workers and Unions

The impact of Industrial 4.0 on workers seems to be significant. It is predicted that independent robots or low-level CPS work will be completed. New operational and maintenance skills for Business 4.0 systems are required for employees. In addition, employees must embrace emerging technologies and be prepared to adapt. The user's concern for implementing the technology project is very high at the time of the launch of any new technology. Employers are critical to the success of Enterprise 4.0 using technology. The adoption by Industrial 4.0 technologies is also an important issue. Trade union unions may oppose Industry 4.0 proposals because they involve significant changes in the way normal operations are performed. Union opposition was caused by a lack of adequate dissemination of information about Industrial 4.0 or a lack of confidence in unions at the beginning of Industrial 4.0.

- Cybersecurity

Industrial 4.0 ensures that people, products, and equipment are a digital network. Besides, digital additional value services, intelligent data collection, and business-processes work closely with transparency. CPS transparency and digital access to CPS indicate a number of online security issues, including process management, cyber blockade, cyber security, product-protection, data-protection, and additional penetrating data-protection. Cybersecurity would be important, and the nonexistence of proper policy will interfere with the company's non-public performance. Therefore, drastic action is required in Industrial 4.0 programs.

- High Initial Costs

Industry 4.0 can lead to horizontal, vertical, and vertical integration. In terms of business needs to be planned and implemented, the formation requires a significant initial investment in time and cost. The price of capital is huge, and the price of Industry 4.0 should be increased. However, the initial costs required to integrate the 4.0 industry are high in the long run and can even lead to profits. Next, these costs will decrease. Another point to consider is that it often applies to high starting costs at the beginning of high-skilled jobs. This is self-contained and includes all the major loads required to build software operating systems [20].

3 Building Block Technologies of IR 4.0 Overview

3.1 Artificial Intelligence (AI)

AI describes computers as humans who can "think"—recognize complex patterns, process knowledge, draw conclusions, and make recommendations. AI is used in many ways, from finding patterns in large stacks of unstructured information to moving the phone automatically.

AI can contribute to robot-teams collaborating and cooperating to accomplish such tasks specified aimed at a particular reason. It seems that the dominant field of research will be artificial intelligence, and field technologies will extend to any sector that requires human intelligence. Systems and devices based on AI can be employed in the management functions of companies (i.e. decision-making robots). Intelligent decision-makers will find Decision Support Structures. Human work profiles will change, and emerging professionals, like knowledge engineering and data scientists, seem to be more common than traditional professions. Society would now be more able to assimilate improvements and accept society's transformation. Further, AI extends to deep learning for sensible in the context of cognitive radio [21].

3.2 *Blockchain*

Blockchain is a secure and decentralized way to collect and share data with third-party intermediaries without relying on them. The most established blockchain technology is the digital Bitcoin currency. The technology can be used differently, including traceability of supply chains, the anonymity of sensitive medical data, and the combat against voter fraud.

3.3 *Big Data/Statistics*

Big data is excessive-quality, high volume, or very special facts that call for new processing for you to make advanced decisions, understanding, and system improvement. With the development of IoT and intelligent/related devices, extra information is generated and is to be had on the PLC and SCADA device levels. Unfortunately, those structures aren't able to store massive quantities of various information, whilst this information is crucial to the relevance, forecasts, and performance of business processes. That's why Big Data is the generation that had to shop ground facts. Big Data isn't a lead to itself however encourages information for later evaluation, referred to as analytics. In mathematics, different mathematical and mathematical technology together with sensible wisdom, system learning, and information mining are incorporated to system massive information. The cause of Analytics is to decide the correlation among system parameters and configurations primarily based totally at the real records of what came about on the shop ground contained in Big Data. Then improvement or AI strategies are used to enhance system-making plans or manufacturing parameters. Finally, consequences are displayed in dashboards for evaluation and decision-making.

3.4 *Cloud*

Cloud computing (CC) "is the idea of permitting customers get entry to a number of customizable laptop assets inside an all-inclusive, easy, and much-wished network (e.g. networks, servers, statistics, applications, and offerings), that may be brought and without difficulty deployed with confined administrative or communique effort. and carrier providers". One of the important thing blessings of CC is the cap potential to get entry to an extensive variety of offerings remotely from anywhere (demanded statistics at any time). In order to expand and enforce a cloud computing/manufacturing solution, different key necessities should be met: (i) service-centric, (ii) news, service quality, (iii) collaboration, (iv) tolerance, (v) load balancing, and (vi) material management. Cloud Production is a manufacturing model of cloud computing.

3.5 IoT

IoT explains various concepts to link to the network and be recognizable by other devices from medical protective equipment that controls the users' physical condition to cars and tracking devices placed in parcels. A tremendous plus for businesses is that consumer feedback can be obtained from continuously connected products, helping them to accurately determine how goods and modification campaigns are used accordingly by customers. Many industrial applications also exist, like farmers who use IoT sensors to monitor soil attributes and inform decisions like when to fertilize.

3.6 Cyber Security

Given the volume and scope of data or knowledge posted or shared on network (NW), growing demands for higher protection are being made as more and more companies go online. Industrial systems are progressively vulnerable to threats. Cybersecurity refers to necessary steps taken to ensure information security over the web or offline information that can be accessed from the web. With the development of the internet, cybersecurity has become one of the most common objects of concern for government, financial organizations, and business entities. Cybersecurity as a whole involves each step taken to save and defend data from all sorts of unauthorized access via the internet.

3.7 Virtual Reality

The technology of VR can open up a vast array of great probabilities for industries such as automobiles, automobiles, and the manufacturing of composite kinds of equipment. From investigating properties in 360 °C to finding situational knowledge, several customs VR technology may make things easier for industries. It can help save time and cost of production. Briefly, makers will be able to come up with more elegant and precise solutions. It can facilitate generating share shared virtual workspace connecting several people around similar projects. Users from several sites will perceive another, visualize and work on similar virtual models. It can enhance communication b/w co-workers to follow validation procedures lacking a physical meeting.

3.8 Robotics and Automation

The design, manufacture, and use of robots for personal and commercial applications is the topic of robotics. Though robotic assistants in all homes are yet to be seen, technical advancement has made robotics progressively sophisticated and dynamic. They are used in such diverse fields as manufacturing, security, and human resources.

3.9 3D Printing

It permits developed enterprises to print their parts, with less tooling, at a lower cost and faster than by traditional procedures. Plus, designs can be modified to ensure a great fit.

3.10 Simulation

Simulation modeling is the approach used to help explain or simulate the behavior of an assembled structure or process using real or imagined models. A physical, mathematical, or other model is constructed as an analog representation. Simulation outcomes for the evaluation of the efficacy and deadlock prevention methods of the negotiation mechanism proposed. For generating feedback, it has introduced multi-agent systems. Some parameters have been observed, which have a major influence on the system's efficiency. Simulation technology for the use of CPSs is also used for the production of Industry 4.0. The simulation also offers simulated system control and monitoring before full development and physical operation [22].

3.11 System Integration

Linkages are vital for every business to flourish, and Industry 4.0 seeks to evolve the idea of fully integrated IT systems. Imagine a manufacturing industry that is closely interlinked with engineering, production, marketing, and after-sales [23]. In the same way, businesses are often more advanced, resulting in completely distributed data integration networks and supply chains. The collaboration between businesses and industries would be facilitated [24].

4 IR 4.0: BIM (Building Information Modeling)

BIM is a technology that describes Engineering projects that contain intellectual property and its data and thought instructions, thus representing each object and its natural ingredients and features. BIM has five key features: "visualization, integration, simulation, efficiency, and editing ability".

The great advantage of BIM is that it also mimics digital projects, which were created at the beginning of the first project. Thus, BIM reduces planning errors, accelerated ratings, measures additional costs, and change indicators [25]. Therefore, BIM will greatly increase the performance of the construction industry through its many digital tools and processes. IR 4.0 is a major step forward that puts a new trend in what digital data can do. By building a central storage facility to capture digital information about a project or asset, BIM will increase progress.

By using complete and reliable data, time efficiency and cost can be increased and errors are greatly reduced in all segments of the project cycle, varying from design to creation and routine care [25]. The benefits of digital data can be achieved by installing BIM in the first phase of the project. For industrial construction, BIM technology is used. Based on three-dimensional visualization, BIM offers a 6D model that helps allow the selection and use of integrated components and forms time, cost, and stability. However, this not only guarantees quality and time but also reduces waste and increases the efficiency of the project as a whole. BIM Technologies and AR (Augmented Reality) will work together in the construction process to provide practical information on body parts. The integrated and compatible use of BIM and digital production further enhances structural design, advanced monitoring, and reliable information distribution. BIM provides especially useful information to pre-built component providers to improve the design, integration, integration, and maintenance processes. Data on BIM simulations converts the use of pre-built objects into more efficient ones in the design process. The digital documents provided by the building information will growth the data size of each component. In this regard, constructive participants must be able to control the resulting extra data/information. The Process and using this data will allow consumers to get a full view of this development, which will lead to new businesses in a digital environment.

At the top of the BIM cloud, all participants have full access to information about the installed project. 3D scanners make digital versions of existing systems for project integration. Other traditional features can be acquired very quickly in the development process, and flexibility control can be achieved with non-functional aerial vehicles and embedded sensors. All information collected is now transferred to the cloud in real-time, and all stakeholders can access the data as needed to correct the steps [26, 27].

5 IR 4.0: Smart Factories

The main focus of IR 4.0 in this section

Industry 4.0 is the most recent phase of the manufacturing sector due to IoT and data access. Industries called "Smart Factories" have become very important, especially in Europe.

The terms "Smart Manufacturing," "Smart Factory," "Factory of the Future," and "Intelligent Factory" all express the idea of what industrial production will look like in the future.

In this vision, the Smart Factory will be much smarter, more flexible, and more dynamic.

Production processes are organized differently by all production chains—from manufacturers to ethics and to control the product life cycle—closely linked to all areas within the business boundary.

Seamless connection of independent production steps. Affected processes include:

- Factory planning and production.
- Product development.
- Logistics.
- Enterprise resource planning (ERP).
- System operating systems (MES).
- Technology control.
- Individual sensors and actuators in the field.

Equipment and equipment in a factory have the potential to improve performance by making your own decisions and making your own. This is in stark contrast to the activities of the established system, as it is today.

Automation was already, to some extent, part of the factory, and thus high levels of change were nothing new. The word "automation" therefore means a person, observant task or process to be performed. In the past, automated and consecutive situations in which computers had made "decisions," which included lifting a valve or opening and closing a pump driven by a set of specific rules. With the use of AI and the growing complexity of cyber-physical systems [28] that can integrate virtual machines with business processes, automation is increasingly requiring complex human improvement decisions that are often made by humans. Lastly, and perhaps most importantly, the term "smart factory" recommends a combination of store floor decisions and details with the entire IT or OT landscape supply chain and the wider business. It may ultimately change production processes to improve customer–supplier relationships.

This definition shows that smart factories go beyond simple automation. A smart factory is a measurable tool, capable of adapting to NW-wide, flexible, and learning in new contexts in real-time or near real-time, or independent production processes7. Smart factories can operate within four factory walls and connect to a global network of networks such as a digital distribution network.

A smart factory is a system that is self-organizing, improves network performance, adapts to new conditions, learns from them in real-time or near real-time, and runs complete production processes independently.

However, it is important to note that due to the rapid technological advances [29], the intelligent industries identified and presented in this paper should not be regarded as "the end point." Rather than a "single and customized" approach to modernizing the industry, it reflects a continuous transformation, a continuous process of developing and sustaining a fast learning process [10].

The true power of a smart factory lies in its ability to grow and develop across the ever-changing needs of the company—whether it changes consumer demand, grows into new markets, creates new goods or services, makes operation and maintenance optimized predictable, and responsive, using new processes. or technology and progressive development in the near real-time. With mass accounting and simulation skills—as well as large intellectual property networks—smart industries can make companies respond to trends in ways that have been challenging or unpredictable in the past.

Assets—identified as plant structures such as material handling systems, tools, valves, and pumps—should be able to communicate independently with the central control system in order to operate intelligent plants. These control systems may take the form of a development system or a digital network delivery stack. The latter is an integrated, structured hub, which serves as a central data entry point for all smart factories, and a comprehensive digital delivery system, which integrates and integrates information into decision-making.34 organizing resources, IoT platforms, and analytics. Connecting goods and installations, data logging, and digital business operations, you will need to use the various digital and portable technologies that are part of Industrial 4.0—including mathematics, further development, automation, HDT, AI, and intelligence technology, advanced materials, and a growing reality [30]. We have described the many business categories and specific business processes in Table 3.

5.1 Default and Robots

Robots are the key to the successful launch of Industrial 4.0 in production environments. Industrial Changes 4.0'are critical features of independent production processes used by intelligent robots that can complete safety, flexibility, flexibility, and reliable operations. Without the need to divide the workplace of robots, integration into human workplaces becomes more economical and efficient and more open to multiple applications. In the stream of activity only in confined spaces, intelligent robots can take the place of humans.

Industry 4.0 integrates large automation and data exchange into production environments, including areas such as cyber-physical systems, IoT and CC, among

Table 3 Business process within the smart factory

S.no	Business sector/process	Major factor
1	Customers (IoT-based)	Responsive manufacturing
2	Suppliers (IoT-based)	Smart supply network
3	Manufacturing (workshop floor-oriented using IoT)	Intelligent products
		Self-driving vehicles
		Smart maintenance
		Intelligent sensors
		Mobile workforce workers
		Cyber-physical systems
4	Manufacturing (software-oriented using IoT)	Cloud storage accessing processing
		Data analytics
		Cyber security against cyber threats using encryption and protection mechanism
		Next-gen manufacturing: automated and smart decisions scheduling
5	Manufacturing (workshop floor-oriented using non-IoT-based)	3D printing
		Robotics
		Advanced materials
6	Manufacturing operations (IoT-based)	Digital twin Additive manufacturing Advanced planning and scheduling Cognitive bots and autonomous robots
7	Warehouse operations (IoT- and non-IoT-based)	Augmented reality Autonomous robots
8	Inventory tracking (IoT-based)	Sensors Analytics
9	Quality (IoT-based)	Optical-based analytics Real-time equipment monitoring
10	Maintenance (IoT-based)	Augmented reality Sensors
11	Environmental, health and safety, public places (IoT- and non-IoT-based)	Sensors to monitor environmental conditions, lack of moment, and handle dangerous equipment

others. With Industrial 4.0, manufacturers can use "smart" industries where individual consumer products can be exchanged quickly. Industrial 4.0 is an independent manufacturing method inspired by the concept known as "Io," the idea that robots can interact within themselves through a mesh of connected objects, devices, and computers. And independent robots are an excellent example in many fields, including manufacturing. Robotic functions can be more integrated and automatic than ever by connecting to a central server or website. Cleverly and without human input you can complete tasks. Consumables can be transported to the factory using autonomous mobile robots (AMRs), blocking obstacles, connecting ship partners, and determining real-time departure and departure [31].

5.2 Monitoring and Control

Monitoring, as well as control of production processes, is already a factor in the development and industrialization of industries. Monitor system is the management of sensory measurements to determine the state of processes (e.g., power, vision, temperature). The tool user performs repetitive monitoring tasks; for example, it detects incomplete, defective tools and conversations in the sound they produce. Filtered sensor measurements are used in unmanned monitoring algorithms to assess process status and user input. Detailed signal analysis of sensory measurements follows complex process conditions. The process of controlling the dynamics of dynamic processes (e.g., feed, speed, cutting depth) to control processes [31]. Users of machine tools control the online and offline process by changing the feed and speed to clear the conversation, initiate emergency breaks to respond to failure, and rewrite the component system to increase cutting depth to reduce burr formation, and more.

5.3 Imitation

3D simulation in intelligent industries describes how real-world process or system behavior mimics or is shown over time, such as the performance of dynamic stock structures, processes, and production activities predicted, tested, and verified. The rapid development of computational resources has made extensive use of industrial modeling techniques to refine the structure and function of buildings. An important point of imitation tools in the future industry is its ability, due to the precise representation of the structure and the flexibility of the factory, to provide the basic concepts of true industry intelligence. This allows for better conduct and control of production facilities to be achieved and lower costs such as direct and indirect employee development to be established [32].

5.4 Artificial Intelligence

AI is the so-called computer instructional model or human behavior that mimics, such as comprehension, learning, perception, decision-making, etc. [33]. It involves developing computer programs that can perform complex tasks. The concept of AI represents and is the driving force behind the digital mind of Industry 4.0. It requires the use of a learning machine that allows the machines to predict its future performance as well as an independent learning experience. The goal of multi-agent systems also facilitates the interaction of b / w machines with virtual connectors via distributed AI [34]. AI allows the power consumption of each machine to be captured, the analysis cycle adjustment is also adjusted to the next level, due to the sensors embedded in the production systems. It may also determine whether the performance data is incorrect. The system improves its efficiency and makes better predictions as the amount of data grows. Intelligent systems from Industry 4.0, for example, can identify, automatically configure, and differentiate product defects in the accuracy and level of objects within the transport chains by visualization, using AI technology [33].

6 IR 4.0: Digitalization of Industries

There is still a lot of discussion about the concept of digital transformation, especially in any industry. Depending on the sector it is used, the term also implies something new for business. How one industry adopts modern technologies will vary widely.

Here's what digital transformation presently implies for nine top industries [35].

- Manufacturing

The whole manufacturing industry has been dominated by digital transformation. Several factories, such as artificial intelligence, have been turned into "smart factories" to simplify their activities with emerging technology. Robots operate on the assembly line inside these "smart factories". IoT systems gather machine learning technologies to process data and data. The manufacturing process is enhanced, and the performance is increased.

- Transportation

Data is an important aspect of the transportation industry's digital transformation. Truckers use data to assess the most effective shipping routes, and trucking firms use data to track potential security threats. Shipping firms rely on 3D scanners to follow path packages. In the next decade, the transport industry will be oblivious to the development of autonomous vehicles and drone deliveries.

- Hospitality/Tourism

The influx of next-generation technology will never make a vacation the same again. Especially hotels have undergone a massive digital transformation. Visitors can use mobile devices to check into hotels; AI is utilized to satisfy demands for room service, as well as IoT devices capture user data to enhance their consumer experience. If you don't like travel, a virtual reality (VR) vacation is now available to take you from your home's comfort.

- Health Care

Healthcare is one of the main information processing industries. New smart hospitals and medical facilities provide AI for diagnosis and surgery. Some hospitals integrate blockchain technologies to enhance access to and protection of medical files. The way patients receive care changes wearable devices with major healthcare applications and the extension of telehealth remote services.

- Financial Services

Mobile payments are the icy point when it comes to digital transformation of financial services. AI is used to provide customer service, manage assets, and detect fraud before it happens. For added financial security, some financial institutions use Blockchain technology. The growth of the FinTech sector is a testament to technological excellence.

- Education

The performance gap in the education system can be closed by emerging technology. Data from assessments and other instruments might assist teachers in accommodating the specific lesson plans of each student. Mobile Bluetooth features can be used to speed up school attendance. Any educational institutions use Augmented Reality (AR) and VR for lessons and virtualized trips to students.

- Energy

Energy firms are also called upon to comply with novel regulations and remain competitive; thus, digital transformation is necessary. Automated machinery and utilization of sensors give useful data to power providers on possible failures of equipment. New Smartphone features allow field operators to operate faster. This results in improved customer service, which reduces unintended costs.

- Media

New technologies, particularly at a time of digital transformation, have often been picked up rapidly by the media industry. The only beginning is an expansion of content obtainable on mobile devices. Media enterprises gather data and use machine learning to suggest personalized content for consumers. This involves advertising products aimed at individuals. As future AI news anchors increase, media is evolving quickly.

- Retail

The emergence of modern retail technologies has the shopping experience revolutionized. New safety features of AI can stop theft before it happens. Smart shelving enables workers to recover products efficiently in-store [35]. The majority of retail outlets gather more information from their customers to improve customer experience. That consistent association between industrial facility work processes—from plan to manufacture, post-handling, assessment, and gathering [36].

7 IR 4.0: Automation to Autonomy

Automation means a collection of man-made robot functions.

Autonomy is a state in which a private robot or part of a device operates without specific human commands.

Automation can be done by creating static environments where robots are designed 24/7 for months or years to perform the same function. Automation benefits highly controlled production environments, which produce a large volume of the same product as automotive automotive production automotive standard models. The epitome of automation may be the automated factory for the removal of lights when robots operate without human intervention.

On the other hand, autonomy allows the robot to respond to and correct its behavior without anyone directing those changes. Autonomy has an incredible ability to mix high, low volume conditions or process product size. But for such a change to occur in the aircraft, the robot must be able to operate without explicit instructions in unfamiliar conditions. Robots must be able to use and rearrange.

Road to private robots:

Advances in industrial robots and AI give power to the manufacturing industry to overcome these problems. In particular, the following methods are important for building production independence:

- Using Sensory Data Intelligently

The robot can handle unstable situations because it can sense its location. Compulsory sensors/torches and sensors are the most widely used sensors to control the movement of robots.

- Digital Geometry Editing

Any offline applications allow users to build toolkits from digital CAD data directly. This is a very good idea, but most of these methods require a significant human response during the formation and fine-tuning of the individual, which makes them human hours more expensive in the transformation of the small mass. For example, suppose a robot arranges various geometric shapes. In that case, offline simulation is a valuable strategy, but this approach is not possible in large quantities as small as 3D printing due to the required engineering function.

- Holding System Information

When programmers of standard robots quickly create a series of stationary robots, all information about the process is lost. In processes that are at risk of variability, it is a challenge to create a system that works consistently. This is not a means of communication between a robot and a working clip.

- Learning Experience

As the sensory data, models, and processes become more complex, it becomes more difficult to build reliable models and controls from the very beginning. Recent advances in improving learning show encouraging results in addressing these difficulties and in the robot model "learning". A robot can benefit from its successes and failures by performing millions of steps without prior knowledge of the process [36].

8 IR 4.0: Implementing Total Production Maintenance (TPM)

TPM is a manufacturing method that aims to maintain the production and longevity of machinery by eliminating six losses. Six losses are the most important cause of product loss from a mechanical point of view. They are classified as random stops, fixed stops, small stops, slow stops, slow cycles, production rejection, and start-up rejection.

There are seven basic steps to implementing TPM, but TPM programs can be tailored to the specific needs and circumstances of the company, as outlined below:

1. Announce TPM Implementation Plans

Without encouragement from senior management, the successful TPM program will not be completed. In some cases, managers have to deal with operators and resist other employees in the transformation and doubts about the suitability of TPM. Publicly announcing business plans and occasionally managing the benefits of TPM implementation will help break down those barriers.

2. Identify the Location of the TPM Pilot Program

You could focus on one piece of equipment that is easily upgraded or that seems difficult and that needs to be adjusted for profits to improve quickly. Find the SMART target in this first phase of the TPM process: Measurable, Clear, Realistic, Reachable and Timed.

3. Focus on Restoring Targeted Equipment to the Primary Operating System

Just report the base of the current operation once the test location is established. Set up TPM implementation and define it. Using the 5S method (Set, Sort, Shine, Save, and Set) method. First, focus on implementing an independent repair program and provide training to machine operators when appropriate.

4. Start Measuring Total Performance (OEE)

However, random stop times should be reported, and clarification is provided in all cases if you decide to monitor OEE. Specify an "unknown cause" or "unstructured stop time" in cases where the operator is unaware of the causes of the stop.

5. Look for Major Causes of Loss

At this stage, a co-operative team of about 4 to 6 employees should be provided to evaluate key causes of unemployment. This team will develop a strategy to remove the root of the failure and will set a time limit for this program.

6. Use Effective Care

At this stage, you focus on preventing collapse or stopping time. Identify key pain points that may lead to mechanical failure and build a common, predictable prevention. Lastly, launch a feedback system to track the progress of your system on an ongoing basis. This response system may include repair logs, monthly or monthly surveys, as well as data collection thermographic analysis or vibration.

6. Use Specific TPM Concepts as Needed

For example, when designing and/or building new equipment in a manufacturing facility, the management of previous equipment may be compromised. Perhaps an increase in usability, easier cleaning or lubrication processes, or the introduction of protective measures could be used to accelerate the construction of new equipment. If there are problems with production or problems in contract management, then this management problem should be resolved as soon as possible [37].

9 IR 4.0: Smart Factory Analytics SaaS Implementation

Using the Smart Factory Analytics solution can be a risky proposition. Top leadership continues to work to develop Industrial 4.0 technology to achieve strategic profitability.

You may or may not be surprised to learn that as an advocate for all the benefits that can be provided by the Smart Factory Analytics solution [38]:

- Only 9% of organizations recognize the full benefits of their software.
- 57% of start-up projects exceed the budget.
- Only 43% of projects are left in the budget
- 80% of organizations offer nothing or no development when using software.

To reduce corporate SaaS deployments, the Strategic Roadmap for Industry 4.0. Success. Researchers have identified 5 Critical Success Factors (CSFs) teachings that are critical to success:

- Set Realistic Goals.
- Go Before Running.
- Select the Test Program field.
- Find a Champion and Your Team.
- Forget IoT Key Key to Success—Process.

10 IR 4.0 Adoption Strategy

In 2011, the term Industry 4.0 has been mostly associated with the production and implementation of new, dizzying-speed technology. The three main factors underpinning the idea of Industry 4.0 are broad numbers, IoTs, and additive technologies [39].

Seven key recommendations for businesses to adopt an Industry 4.0 approach and boost their chances of thriving in this modern situation were highlighted in the research findings.

1. Design the business size and match the priorities with the overall strategy of the company. You are defining your objectives and prioritizing your actions: increasing the organizational performance of your company, optimizing the value chain, and evaluating new business models.
2. Enable an interactive working environment for learning, change, and experimentation. Facilitate a productive environment.
3. Define the skills to be developed, whether internally or with external providers.
4. Recruiting and managing talent, giving preference to multidisciplinary teams, and making data analyses a powerful advantage.
5. To start to develop an ideal partner network, select a team of providers with tested technologies.
6. Take an environment perspective and build on network management abilities.
7. Start pilot projects, verify outcomes, and systematize processes for learning.

11 Recommendation for HR for a Successful Transition to Industry 4.0

There will be a lot of technology in the industry and there will be a lot of interaction between people and machines. Changes should be made to manufacturers using the 4.0 switch. Six winning practices deal with employee development in the 4.0 industrial era [40].

1. Train and Develop Talents

To train talent within the organization itself, it will be necessary for effective skills development programs. To ensure change, new jobs will be transferred to staff and training should be provided. Interesting ways to view employees with more experience, using real extended extensions or online learning.

2. Bringing the Staff Transformation of the Industry for 4.0 Years

Involving data scientists, research and development experts, or software programmers, businesses compete with many technology companies. They are recognized for providing their employees with the most comfortable work environment. Therefore, the most common methods in human resources are less important. Technology is needed, and flexible work schedules are similar to normal working hours.

3. Post Decisions

Classification and decision-making processes should be viewed in the same way as the previous issue. If we enrich our employees to do more difficult tasks, they will need more independence to make decisions. For example, engineers are forced to wait for machine operators to approve mechanical repairs. Or, if the manager only receives a real-time warning, the production coordinator does not wait for the manager's approval. The best interest of the organization is to eliminate hierarchical structures.

4. Make Employees Responsible for Innovation

Employees must be responsible for progress and results if they are properly trained and independent in their work. The employee should understand and take ownership of the results rather than just a small part of the process. In the meantime, management should encourage this work.

5. Gather Based on Strength

Traditional employment does not respond well to the requirements of Industrial 4.0 based on experience and diploma. Manufacturing firms are expected to have diverse staff capable of learning and working simultaneously on multiple projects. Conversations should be given preferences based on the organization's community.

6. Strategically Plan the Needs of the Workforce Related to the Industry 4.0

Listing all the important facts about employees is also important: possible parental leave, planned retirement, part-time jobs, subcontractors, etc. With a realistic picture of our staff, it is important to accurately predict and plan for changes in training and recruitment requirements.

12 IR 4.0 Global Trades and Investment

a. Trade Openness

- Trade: Total imports and exports of goods and services are measured as a share of GDP. Average unit: % GDP.

b. Trade Facilitation and Market Access

- Trade tariffs: Average rate of trade-weighted tariffs. An applicable tariff is a tariff imposed on imports of goods. The weighted average of all tariffs imposed, including the preferential rates applied to the rest of the world by a country, is measured as this indicator. The weights are the trade patterns of the comparison group in the importing country. Measure unit: % duty

- Prevalence of Trade Barriers: Executive opinion survey: "In your region, about how far are the ability of manufactured products to compete in their domestic markets limited by non-tariff barriers (for example, health and product standards, technical and labeling criteria, etc.)? Unit of measure: 1–7 (best) (1 = strong limit, 7 = not constrained at all)".
- Logistics Performance: Average score of 5 components from International Logistics Performance Index:

i. Cultures: cultural efficiency and border control permit
ii. Easy to schedule a post: easy to schedule a competitive price post
iii. Quality logistics services: efficiency and quality of logistics-truck services, custom brokerage, and referral
iv. Tracking and tracing: the ability to track and trace assets
v. Duration: Frequency of goods shipped during the expected or limited delivery Unit unit: 1–5 (preferably) standard delivery

c. Investment and Financing

- Greenfield Investment: A 5-year average for each of Greenfield FDI's announced projects for US $ (millions). Greenfield investment is a type of direct investment in other countries where the parent company is legally involved in a foreign country. Measurement unit: US $ millions.
- FDI inflow: 5-year average FDI country residual or economic flow. FDI inflows and outflows include payments made to foreign direct investors in the FDI business (either directly or through additional related businesses) or earnings from FDI business by direct foreign direct investors. FDI flow data is provided on a residual basis (expenditure makes lower debit credits between direct investors and their external agents). Measurement unit: US $ millions.
- Internal Credit to the Private Sector: Financial services provided by financial firms to private companies as a percentage of GDP. Financial services are loans, purchases of unequal securities and commercial credit, and additional accounts receivable generate payment claims. Measurement unit: % GDP [41].

13 Conclusion and Future Work

13.1 Discussion

We conducted a study of IR 4.0 acceptance in various countries based on reports [37, 42–45]. The global transformation of production systems is difficult, and the future of production in a two-tier world could be polarized. The test focuses on the 100 leading and most economically developed countries and economies to benefit from the transformation of production in only 25 countries from Europe, North America, and East Asia. More than 75% of the global production volume (MVA) has now been produced by those 25 countries and is well positioned for future growth. By comparison, sales of robots in the Republic of Korea, China, Germany, Japan, and the United States cost about 70%. Germany, Japan, and the United States hold the ranks of high-tech industrial robots, while China is the fastest-growing sector. Small, Medium, and Medium Enterprises (MSMEs) are on their way to all countries as they have a wide range of knowledge, understanding, and investment potential. Bringing countries to different levels of development and MSMEs and reducing the journey of change will require global solutions and significant investment [46].

As countries change production processes, a number of alternatives may emerge. Not all countries will try to achieve advanced development in the future. Any country that is a low-cost labor destination may wish to take advantage of general development opportunities within the existing paradigm in the short term. Developed countries should focus more on developed development. Some sectors of the global economy may be ahead of international production. It will be necessary for any country to differentiate itself and build on competitive advantages by developing its own strategic plan and a broader future economically productive economy.

There is room for change in all countries. At the beginning of this period of transformation, no nation has achieved the frontier of readiness, not to mention exploited the maximum potential in its production of the Fourth Industrial Revolution. Even the most modern and complicated countries in every part of their country are not strong since different sub-regions differ in readiness.

Since the new technological paradigm is creating a cluster of new industries, there is potential for leapfrogging, but only a few countries can capitalize. New paradigms of technology are a way to allow lagging nations to catch up so they would be able to enter new industries at a later stage without the expense of being locked into existing technologies. Emerging technologies linked to the Fourth Industrial Revolution provide the potential. However, are countries ready? Although short cuts are attractive, a minimum level of capacity is required to leapfrog. The best way to step into a new model is likely to be for the high-potential countries and economies such as Australia and the UAE and the nascent or leading countries near the high-potential border archetype. These countries do not have a strong current production base yet likely have the capital and other capabilities right to take advantage of the opportunities to leapfrog into a new model of production.

The Fourth Industrial Revolution will cause targeted reshoring and nearshoring in the global supply chains and other structural changes. Inertia is one of the greatest barriers to the change in global value chains, as manufacturing moving from one place to another is too expensive for most industries. The introduction of advanced technology would shift the economic-benefit calculus for increasing production processes and, eventually, the attractiveness of the location. If a value chain is newly established, countries may take their place or lose their share. At any point in the global value chain, the future of development needs such skills and capabilities. The preparation and growth of specific capabilities are crucial for countries, which render them an attractive production destination in the world's value chains. If countries cannot rapidly develop capability and connectivity, they will rapidly lose out.

Readiness for the future of production calls for global and regional strategies, not just federal governments. Countries that plan to grow in production in the future need to invest in coordinators and create a policy to leverage prospects and to create or improve the production base. Emerging technology regulations and legal frameworks, which are established at a sector-specific level—versus independent of each country—will lead to massive efficiencies and increase business satisfaction across global value chains. Regional collaboration will also help countries leverage their combined strength for global competitiveness.

To accelerate transformation, modern and creative public–private cooperation approaches are required. Any country faces issues that the private sector or the public sector alone can not solve. Legacy and developing economies in particular, by making more aggressive use of the private sector to solve macro-level problems, will speed up the readiness and transformation. However, innovative public–private cooperation approaches to replace existing models will help policymakers collaborate with industry, academia, and society successfully and rapidly in unlocking new value.

13.2 Comparison

The fourth industrial revolution in Industrial 4.0 is based on its promise and encompasses a wide range of technological advances in the value chain. Industrial 4.0 technology automation or robotics, Artificial Intelligence (AI), Internet of Things (IoT), additional production-transforming industrial production methods. With the advent of digital technology, the boundary between the real and the real world is slowly declining, and so-called cyber-physical production systems are developing. We have demonstrated comparisons of traditional production with Industrial 4.0 production in Table 4.

The table above lists the planned changes in Industry 4.0 presented by business models. The paradigm shift in Industry 4.0 ranges from consistent, efficient, and manual production to flexible and automated production. It shows the transition

Table 4 Comparison between industry 4.0 and traditional manufacturing

What is changing	Traditional manufacturing	Industry 4.0 manufacturing
Product	Standardized	Personalized and customized
Process	Stable and manual	Agile and automatic
Supply chain	Stock-based planning	Dynamic and predictive
Factories scale	Large industries in central areas	Small industries in isolated areas
Client relationship in the community	Low and indirect	Up and direct
Success metric	Low cost, high efficiency	High ROCE (Return On Capital Employed)

from mass production to mass customization that variable production and shorter lead times can be made possible. Also, the transition from large manufacturing industries to smart industries with high-tech tools that can build more goods at lower cost will take place. The ability to remotely use technology such as a magnified reality will also be demonstrated.

Business success metrics will also change from low-cost production to high return on investment. Industry 4.0 promotes sustainability by promoting more customization, reducing labor costs, and reducing complex costs. At the same time, it helps to reduce resources by making the app more flexible and efficient. This suggests that there are three distinct paths—(1) the use of outdated technology, (2) the automation of implementation, and (3) the adoption of Industry 4.0. Industrial Route 4.0 promotes sustainability as a reduction in labor costs and a growing level of material consumption, compensated by increased technological decline. This improves performance. Jinns have also improved due to increased profitability of goods due to increased quality and flexibility. Industrial 4.0 also uses buildings efficiently by reducing changing time, machine downtime, installation and repair times, etc.

13.3 Concluding Remarks

Industries of any nation are the backbone. The national economy has a positive impact on industrial development. Increased per capita income, overall inflation, availability of affordable goods, and all aspects of success reflect industrial growth. National prosperity is directly related to industrial development.

Industrial construction leads to the development of the country. The emergence of large-scale industries contributes to the national economy through mass production and the use of the latest technology. Industrialization is a major source of social, technological, and financial growth.

The current research is based on a research perspective on the growth and development of Industry 4.0. This is because Industry 4.0 technology is used in different areas between countries or industries.

Industrial 4.0 is a new twenty-first century production system that allows businesses to establish "intelligent" services and products by reducing costs, as well as improved productivity, where the human parameter is important for implementation and research, with a strong focus on literature available in the field.

13.4 Future Work

Proponents of Industrial 4.0 refer to this concept as a smart home—a "smart factory" network, in a figurative sense. In a smart home, automated technologies such as scanning, sensors, voice, systems, and facial recognition enhance modern home security and luxury features—electrical appliances, lights, clocks, and alarms.

The implementation of Industrial 4.0, which may change gradually as young people adopt and move forward in the pocket and existing firms struggle and compete to stay relevant, will help producers of all sizes and ages. Over the next decade, Industry 4.0 could make new industry leaders for unknown companies.

Upcoming research and forecasting will be the next change—Industry 5.0. If the current transformation emphasizes the transformation of factories into intelligent IoT-enabled applications using cognitive computing and interconnect by cloud servers, Industry 5.0 will focus on the return of human hands and minds to the industrial framework.

Industry 5.0 is a transformation that integrates people and machines and finds ways to work together to improve productive and efficient resources. Ironically enough, among companies now with Industrial 4.0 concepts, the fifth transformation is likely to continue. Even businesses that use state-of-the-art equipment do not automatically burn a large number of their employees and become computerized.

Other concerns a few manufacturers have shared about the current change can be expressed in the concept of Industry 5.0. Cognitive computing and cyber machines, in particular, will eliminate the need to treat people and take millions out of work. Instead, Industrial 4.0 may eventually reorganize human activities in the manufacturing sector to support workers. Humans can work comfortably when machinery is tight.

Industrial 5.0's major forecasts include the interaction between computer intelligence and human intelligence. Integrated personal and computer equipment can reach new levels of speed and perfection in development. Version 5 could be better for the environment, as companies develop renewable energy and waste management systems.

Overall, the growth of Industrial 5.0 will reflect its full realization, which Industry 4.0 designers dreamed of in early 2010. Initially, with advanced artificial intelligence and human power in factory robots, inter-computer communication, and robots, and people eventually have a purpose and educate each other. And what could be better than a good working relationship in an industrial environment?

Industry 4.0 Today, Industry 5.0 Tomorrow: As technological advances progress rapidly, changes can occur rapidly in the next decade and beyond. It is only natural that talking about the 5th revolution will soon coincide with the fourth in the next work, considering the pace of these developments.

Overall, Industry 4.0 allows for modern digital transformation. This will allow autonomous to autonomous frameworks that can help out one another. The innovation will help take care of issues and track measures while additionally expanding usefulness and productivity.

References

1. Kagermann H. Lukas WD, Wahlster W (2011) Industrie 4.0: Mit dem Internet der Dinge auf dem Weg zur 4. industriellen Revolution. VDI nachrichten
2. Bauernhansl T, Hompel MT, Vogel-Heuser B (2014) Industrie 4.0 in Produktion. Automatisierung und Logistik: Anwendung Technologien Migration. Springer, Wiesbaden
3. Rajawat AS, Pradeep Bedi SB, Goyal PK, Shukla AZ, Jain A, Khan MM (2021) Reformist framework for improving human security for mobile robots in industry 4.0. Mobile Info Syst 2021:4–10. https://doi.org/10.1155/2021/4744220
4. Kagermann H, Wahlster W, Helbig J (2013) Umsetzungsempfehlung für das Zukunftsprojekt Industrie 4.0. Tech. rep. acatech
5. Bongomin O, Nganyi EO, Abswaidi MR, Hitiyise E, Tumusiime G (2020) Sustainable and dynamic competitiveness towards technological leadership of industry 4.0: Implications for East African community. J Eng. https://doi.org/10.1155/2020/8545281
6. Alimkhan A, Makhambayev A, Ukaegbu IA (2019) The Fourth Industrial Revolution: Towards Energy 4.0 in Kazakhstan. In: International Conference on Advanced Communication Technology, pp 527–532. https://doi.org/10.23919/ICACT.2019.8701979
7. Kada B, Alzubairi A, Tameem A (2019) Industrial communication networks and the future of industrial automation. Ind Syst Eng Confer. https://doi.org/10.1109/IASEC.2019.8686664
8. Fernandez-Carames TM, Fraga-Lamas P (2019) A review on the application of blockchain to the next generation of cybersecure industry 4.0 smart factories. IEEE Access 7:45201–45218. https://doi.org/10.1109/ACCESS.2019.2908780
9. Kannan SM, Suri K, Cadavid J, Barosan I, Brand MVD, Alferez M, Gerard S (2017) Towards industry 4.0: gap analysis between current automotive MES and industry standards using model-based requirement engineering. In: IEEE International conference on software architecture workshops, ICSAW 2017: Side Track Proceedings, pp 29–35. https://doi.org/10.1109/ICSAW.2017.53
10. Schoepf T (2016) Plug-and-Produce is Key for the Smart Factory of the Future—Part 3. https://www.belden.com/blogs/plug-and-produce-is-key-for-the-smart-factory-of-the-future-part-3
11. Pérez F, Irisarri E, Orive D, Marcos M, Estevez E (2015) A CPPS Architecture approach for Industry 4.0. In: IEEE International Conference on Emerging Technologies and Factory Automation, ETFA, 2015-October. https://doi.org/10.1109/ETFA.2015.7301606
12. Zhou K, Liu T, Zhou L (2016) Industry 4.0: towards future industrial opportunities and challenges. In: International Conference on Fuzzy Systems and Knowledge Discovery, pp 2147–2152. https://doi.org/10.1109/FSKD.2015.7382284
13. Hayden E, Assante M, Conway T (2014) An abbreviated history of automation & industrial controls systems and cybersecurity. SANS Institute
14. Stouffer K, Pillitteri V, Lightman S, Abrams ML, Hahn A (2015) Guide to industrial control systems (ICS) security. National Institute of Standards and Technology Special Publication, pp 800–82
15. Zhang, P (2008) 4 - Digital Controllers for Industrial Control BT - Industrial Control Technology. 429–568. http://www.sciencedirect.com/science/article/pii/B9780815515715500050
16. Wollschlaeger M, Sauter T, Jasperneite J (2017) The future of industrial communication: automation networks in the era of the internet of things and industry 4.0. IEEE Ind Electron Mag 11(1):17–27. https://doi.org/10.1109/MIE.2017.2649104
17. Berger R (2014) Industry 4.0: ?e New industrial revolution how Europe will succeed. Roland Berger. Munich, Germany
18. CGI Global (2017) Industry 4.0: making your business more competitive. CGI Global. Montreal. Canada
19. Industry 4.0 - The next industrial revolution? https://www.eeweb.com/blog/renesas/industry-4.0-the-next-industrial-revolution accessed on 10th Sept 2023
20. Sony M (2020) Pros and cons of implementing Industry 4.0 for the organizations: a review and synthesis of evidence. Prod Manuf Res 8(1):244–272. https://doi.org/10.1080/21693277.2020.1781705

21. Goyal SB, Bedi P, Kumar J et al (2021) Deep learning application for sensing available spectrum for cognitive radio: an ECRNN approach. Peer-to-Peer Netw Appl 14:3235–3249. https://doi.org/10.1007/s12083-021-01169-4
22. Alkoc E, Erbatur F (2010) Productivity improvement in concreting operations through simulation models. Build Res Inf 25(2):82–91. https://doi.org/10.1080/096132197370462
23. Balina S, Baumgarte D, Salna E (2017) Cloud based cross-system integration for small and medium-sized enterprises. Procedia Comput Sci 104:127–132. https://doi.org/10.1016/J.PROCS.2017.01.084
24. Oztemel E, Gursev S (2018) Literature review of Industry 4.0 and related technologies. J Intell Manuf 31(1):127–182. https://doi.org/10.1007/S10845-018-1433-8
25. Alaloul WS, Liew MS, Zawawi NAWA, Mohammed BS (2018) Industry revolution IR 4.0: future opportunities and challenges in construction industry. MATEC Web Conf 203:02010. https://doi.org/10.1051/MATECCONF/201820302010
26. Khosrowshahi F, Arayici Y (2012) Roadmap for implementation of BIM in the UK construction industry. Eng Constr Archit Manag 19(6):610–635. https://doi.org/10.1108/09699981211277531/FULL/XML
27. Kymalainen T, Kaasinen E, Aikala M, Hakulinen J, Heimonen T, Paunonen H, Ruotsalainen J, Lehtikunnas L, Mannonen P (2016) Evaluating future automation work in process plants with an experience-driven science fiction prototype. In: Proceedings - 12th International Conference on Intelligent Environments, pp 54–61. https://doi.org/10.1109/IE.2016.17
28. Uppal M, Gupta D, Juneja S, Dhiman G, Kautish S (2021) Cloud-based fault prediction using IoT in office automation for improvisation of health of employees. J Healthc Eng. https://doi.org/10.1155/2021/8106467
29. Kautish S, Reyana A, Vidyarthi A (2022) SDMTA: attack detection and mitigation mechanism for DDoS vulnerabilities in hybrid cloud environment. IEEE Trans Ind Info. https://doi.org/10.1109/TII.2022.3146290
30. Burke R et al The Smart Factory. https://www2.deloitte.com/content/dam/insights/us/articles/4051_The-smart-factory/DUP_The-smart-factory.pdf
31. Gaur K, Virdi GS (2018) Engineering and technology a high impact factor. Int J Innov Res Sci. https://doi.org/10.15680/IJIRSET.2018.0703082
32. Schlaepfer RC, Koc M, Deloitte AG (2015) Industry 4.0: challenges and solutions for the digital transformation and use of exponential technologies. audit, tax. consulting. corporate finance
33. Lee J, Davari H, Singh J, Pandhare V (2018) Industrial artificial intelligence for industry 4.0-based manufacturing systems. Manuf Lett 18:20–23. https://doi.org/10.1016/J.MFGLET.2018.09.002 accessed on 10th Sept 2023
34. Xu J, Huang E, Hsieh L, Lee LH, Jia QS, Chen CH (2017) Simulation optimization in the era of Industrial 4.0 and the Industrial Internet. J Simul 10(4):310–320. https://doi.org/10.1057/S41273-016-0037-6 accessed on 10th Sept 2023
35. Lerner S (2019) Digital transformation: 9 industries leading the way. https://www.enterprisedigi.com/artificial-intelligence-machine-learning/articles/industries-digital-transformation accessed on 10th Sept 2023
36. Oqton (2018) From automation to autonomy - how the new era of robots will revolutionize manufacturing. https://www.oqton.com/from-automation-to-autonomy/ accessed on 10th Sept 2023
37. Lachance EA (2019) Implementing Total Production Maintenance (TPM) with IIoT and smart factory analytics technologies. https://www.worximity.com/en/blog/implementing-total-production-maintenance-tpm-with-iiot-and-smart-factory-analytics-technologies accessed on 10th Sept 2023
38. Lachance EA (2019) Critical success factors to a successful smart factory analytics saas implementation. http://www.worximity.com/en/blog/critical-success-factors-smart-factory-analytics-saas-implementation accessed on 10th Sept 2023
39. Roig C (2020) How to implement an industry 4.0 strategy. https://dobetter.esade.edu/en/industry-4.0-strategy-implementation?_wrapper_format=html

40. Consolante E (2019) 6 HR tips for a successful transition to industry 4.0. https://www.worximity.com/en/blog/6-hr-tips-for-a-successful-transition-to-industry-4.0
41. Kearney AT (2019) Readiness for the future of production report. World economic forum's system initiative on shaping the future of production. https://www3.weforum.org/docs/WEF_White_Paper_Readiness_Future_Production_.pdf accessed on 10th Sept 2023
42. Borg A (2016) The world economic forum COVID action platform. https://www.weforum.org/agenda/2020/03/world-economic-forum-launches-covid-action-platform/ accessed on 10th Sept 2023
43. Pawar A (2020) COVID-19 action agenda leaders on the front line: why social entrepreneurs are needed now more than ever. http://www3.weforum.org/docs/COVID19_SocEnt_Alliance_Report_2020.pdf accessed on 10th Sept 2023
44. Szabo M (2020) Leaders on the front line: COVID social enterprise action agenda. https://www.weforum.org/platforms/covid-action-platform/live-updates/week-ending-18-september#uplink-selects-15-innovative-solutions-to-covid-19-challenges
45. Bagrodia T (2020) Cultural requirements in M&As. https://www.peoplematters.in/article/culture/cultural-requirements-in-mas-16839
46. Yang F, Gu S (2021) Industry 4.0, a revolution that requires technology and national strategies. Complex Intell Syst 7(3):1311–1325. https://doi.org/10.1007/S40747-020-00267-9

Chapter 18
Artificial Intelligence and Automation for Industry 4.0

Amrita Chaurasia, Bhakti Parashar, and Sandeep Kautish

Abstract The key premise of smart factories and enterprise 4.0 is the application of AI by employing robots to perform hard activities, lower fees, and improve the high-quality of products and solutions. Artificial intelligence (AI) is infiltrating the industrial sector with the help of cyber-bodily systems, fusing the physical and digital worlds. Artificial intelligence (AI) makes manufacturing smarter and more capable of coping with modern difficulties like customizable needs, faster time to market, and an expanding spectrum of sensors in equipment. The usage of bendy robots combined with artificial intelligence facilitates the production of a wide range of products. AI technologies can be used to analyse massive volumes of real-time data collected from a variety of sensors (such as data mining). AI is ushering in a new industrial revolution with intelligent automation, massive data, and networking. Time or place, data integration universally with networks evolves and allows completely automated supply chains, Industry 4.0 will bring the integration of horizontal and vertical systems with businesses, departments, features, and talents will become much more cohesive. Extra systems will be enhanced with embedded computers as the Internet of Things becomes more industrialized, and they will be connected using standard technologies. This allows machines to communicate and interact with one another, and a more centralized machine controller becomes increasingly vital. As cross-company, universal data-integration networks expand and enable totally automated value chains in Industry 4.0, horizontal and vertical system integration among firms, departments, functions, and capacities will become much more cohesive. Industrial auto solutions and the Internet of Things will also add embedded computing to more objects and connect them using standard standards.

A. Chaurasia
Christ University, Gaziabad, India

B. Parashar (✉)
Vellore Institute of Technology University, Bhopal, India
e-mail: sharma.bhakti@gmail.com

S. Kautish
LBEF Campus, Kathmandu, Nepal
e-mail: dean.acd@coeruniversity.ac.in

S. Kautish et al. (eds.), *Computational Intelligence for Modern Business Systems*, Disruptive Technologies and Digital Transformations for Society 5.0,
https://doi.org/10.1007/978-981-99-5354-7_18

Keywords Industry 4.0 · Artificial intelligence · Intelligent automation · Internet on things · Business · Big data and connectivity

1 Introduction

Industry 4.0 is a new field that brings the communication of machines in place of humans. All work is turned and performed with the help of computerization in the industry as well as at a personal level [1]. It advises that a group of representatives from business, politics, and academia promote the concept as a way to improve manufacturing enterprises' competitiveness [2]. "A device's capacity to properly receive external data, research from such facts, and apply such learnings to accomplish set objectives and tasks through flexible adaptation" is how artificial intelligence (AI) is described [3]. At one point, AI was split up into many software domains. Natural language processing, automated programming, robotics, computer imagination and prescient, automated theorem proving, intelligent information retrieval, and so on were just a few of them.

Artificial intelligence has resulted in a shift in the way businesses operate, owing to a new type of human–machine interaction. Intelligent factories [2]. are now at the heart of Industry 4.0, which are defined by cloud-based interactions between humans and cyber-physical systems. Intelligent factories have automated structures and include digital enablers that allow machines to communicate with one another and with the factory's overall systems through an Internet of Things configuration. These abilities are in high demand by industries across the board, as they seek to maintain the competitiveness of their manufacturing units in an increasingly technology environment [1].

Industrial automation receives a lot of attention when it comes to artificial intelligence in manufacturing, but it's only one component of the smart factory revolution, which is a natural next step in the search of [3]. Artificial intelligence also has the potential to open up entirely new commercial opportunities in the industrial industry. One of Artificial Intelligence's benefits is that it allows robots to learn from the data they collect from factory activities, allowing them to improve their skills in every engagement [4]. After all, this is one of Industry 4.0's essential pillars, and it will help factories become more autonomous and productive.

We've heard a lot about Industry 4.0, but it usually revolves around automation. Here, we are looking at advanced applications of Artificial Intelligence (AI) techniques, from using deep learning algorithms to generate predictions to implementing a full-fledged Digital system engagement [5]. After years of being a distant ideal for most firms, Industrial AI technologies are now actual and readily available. Of course, the most crucial concept for extracting genuine value from an Industrial AI solution is determining which one is appropriate for a certain business problem [6].

Generative design uses AI and automation algorithms to create several design solutions that are all viable for the same goal at the same time, robotics involves the use of robotic and collaborative machines to assist operators in completing methodical and/or extremely exact activities, [7]. Until human capital is educated to deal with processing, programming, and data systems will this sort of functionality be useful. Companies must spend not just in technology capital but also in the education and training of competent employees [8].

2 Application of AI and Automation in Business

The shift from traditional automation based on autonomous industrial robots to networked "hyperphysical systems" has transformed manufacturing operations and established new market standards. As a result, both producers and consumers have benefited from this [9]. Initially, ordinary work needing low- to medium-level certifications and skills were the target of automation or substitution. Two common examples are check-in agents at airport counters and accountants [10]. Just-in-time manufacturing: Adapted real-time production methods have reached a new degree of efficiency. These AI-driven manufacturing systems can respond to requests and produce parts in a timely manner. Sensors track components and arrange them according to demand patterns and algorithms to decrease lead times [11]. There are three foundations that have emerged from AI approaches [12]. (Fig. 1).

2.1 Product Optimization

In its broadest meaning, the term "product optimization" refers to a systematic method of product creation in which the investigator changes the formula and processing conditions over time. Because it reduces time, cost, and risk in the development process, product optimization has become a more useful tool in recent years

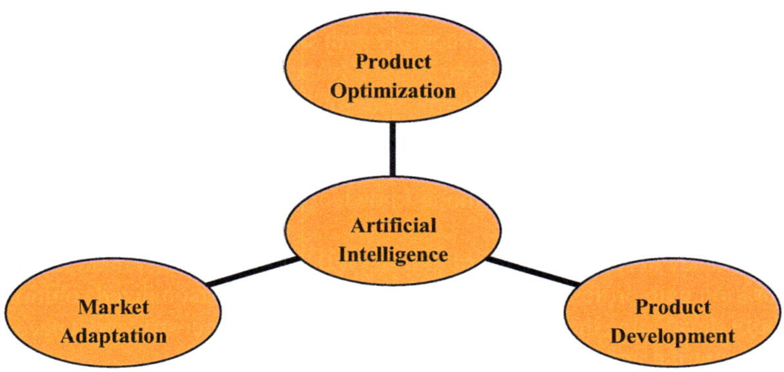

Fig. 1 General application of artificial intelligence in business

[13]. It is now viable to gather and measure intelligence from various elements of the manufacturing environment advent of the Internet of Things, artificial intelligence and low-cost sensors. This creates a lot of data when coupled with standard data collection systems like SCADA and DCS. This data may be utilized to make informed decisions and conduct efforts that will improve production efficiency.

2.2 Market Adaptation

Artificial intelligence (AI) is transforming the marketing environment and will entirely transform it in the near future. Although marketing is one of the most common corporate applications of AI today, early adopters are aiming to make money from it [14]. Despite the fact that AI is used in the majority of businesses today, many companies still lack high-level implementation. Several marketers have expressed an interest in implementing AI in the near future, with nearly all of them intending to do so fully. Only 20% of marketers [14] said they used one or more artificial intelligence (AI) technologies in their companies.

2.3 Product Development

To enhance the quality and long-term viability of manufacturing processes while decreasing costs, smart manufacturing systems require innovative concepts. In this context, artificial intelligence (AI)-driven technologies based on I4.0 Key Enabling Technologies (e.g., Internet of Things, advanced embedded systems, cloud computing, big data, cognitive systems, virtual and augmented reality) are poised to establish new industrial paradigms [15]. Artificial intelligence (AI) can assist in the resolution of important challenges in sustainable manufacturing, such as energy resource optimization, logistics, supply chain management, and waste management, to name a few. In this context, smart manufacturing is increasingly using AI into green manufacturing processes in order to comply with stricter environmental requirements [14]. In the era of growing digitization and advanced big data analytics, embracing excellent data for creating and delivering state-of-the-art services will enable creative business models and management approaches [16, 17] and have a range of implications. Among other things, digitization and big data analytics disrupt business models and effect employment among knowledge workers, just as automation did for industrial workers.

The intricacy of AI application in Industry 4.0 [18] on the other hand, necessitates collaboration with professionals to develop relevant and personalized solutions. The expense of developing the requisite technology is extremely costly, and it necessitates extensive internal and technical knowledge (Fig. 2).

Fig. 2 Business transformation with the help of artificial intelligence

- **New product introduction:** Production lines are turned into information systems that help people make crucial decisions about things like product lines. This makes the transfer from raw materials entering the manufacturing process to the completed product easier, facilitating demand adaptability [19].
- **Shift of consumerization:** Between assets and consumers, one of the most important transformations in mindset has occurred. Consumers are linked to the industry via information networks, and they want higher-quality products and more personalized experiences. Manufacturers, on the other hand, can produce customized goods while maintaining efficiency owing to digital designs and clever technology [20].
- **Advancement in labour market:** While AI approaches have a long learning curve, they are progressing at an irreversible rate. As a result, job titles like "Data Scientist" are becoming increasingly popular [19]. Governments must invest in education, which is becoming an increasingly essential component in decreasing unemployment, as the labour market shifts toward more analytical and skilled applicants in the field of technology [20].
- **Business intelligence:** Artificial intelligence sharpens this advantage, which benefits the global economy greatly. 82% of Spanish firms are already experimenting with AI, according to IBM's "The Global Race for AI". AI enables companies to scale up their production models while maintaining process quality, which is critical given the market's increasing competitiveness [2].
- **Automation and learning techniques:** are AI methodologies such as deep learning (DL) and machine learning (ML) that, when used appropriately, have a major positive impact on a company's ROI. By incorporating predictive maintenance systems into manufacturing processes and replacing visual inspections with robots or cobots that perform quality controls tenfold more precisely and effectively [21], automatic learning increases product quality. Complex

algorithms are also created via machine learning, allowing for “Manufacturing”, in which data collected during manufacturing is analysed and changes are made automatically. Deep Learning, a subtype of Machine Learning, creates its own neural networks for unsupervised learning, significantly increasing the approaches’ autonomy [22].

3 Application of Artificial Intelligence and Automation in Other Fields

Artificial intelligence (AI) is a combination of technologies that allows software and machines to sense, comprehend, act, and learn on their own or in collaboration with humans [18]. Artificial Intelligence can boost industrial manufacturing efficiency when compared to human labour. Additionally, AI can enable robots to undertake tasks that a human would be unable to, such as managing hazardous raw materials or microscopic components [9]. Similar observations have been made by social observers. As IT and automation evolved, [23] predicted that labour expenses will become a less and smaller part of manufacturers’ cost structures. It is a branch of computer science that uses machines, software, and computer platforms to try to replicate human intelligence and [1] it takes advantage of digital technology that has improved our intelligence and productivity while also changing the way we communicate, learn, shop, and play [24].

It’s critical to understand that many of these industrial robots aren’t very sophisticated right now. They can accomplish a variety of skilled activities in many circumstances, although they are designed in a limited fashion [25]. To reconfigure it if you require more. The fact that this technology is improving with Industry 4.0 is driving ongoing expansion in the area. Artificial Intelligence improves each time, and its costs fall. Implementing more complicated AI algorithms has also enabled businesses to assess the acquisition of new technologies that enable them to solve problems and make more complex and secure decisions [26].

3.1 Artificial Intelligence and Technology

Novelty, flexibility, and enchantment are the hallmarks of artificial intelligence. Radical innovations and destabilizing systems are innovative components that create new markets or disrupt existing markets or products [27]. AI refers to a machine’s capacity to simulate human skills such as thinking, learning, planning, and creativity. It allows technological systems to sense their surroundings, respond to what they see, solve issues, and act in order to achieve a specified objective. The computer receives data that has already been prepared or collected by its own sensors, such as a camera, analyses it, and replies. By analysing the impact of past acts and operating autonomously, AI systems are capable of adjusting their behaviour to a degree.

- **Speech recognition**

Automatic speech recognition (ASR) is one of natural language technology's most rapidly expanding and economically promising applications. Speech is the most natural conversational channel for humans in many situations, including dictation, querying database or information-retrieval systems, or simply delivering orders to a computer or other device [28]. Over the last several decades, technology has steadily progressed to the point where properly built systems for appropriately restricted applications are now a reality. Commercial solutions are now available for a wide range of activities and parameters, such as.

First a database retrieval job, and large-vocabulary dictation, voice control of medical equipment, and large-vocabulary dictation, and large-vocabulary dictation, and large-vocabulary dictation, and large-vocabulary dictation, and large-vocabulary dictation, and large-vocabulary dictation, and large-vocabulary dictation, and large-vocabulary dictation (several 10,000 s of words, as in a general dictation task).

Second the speaking style, which comprises read speech, prepared speech, and spontaneous communication.

Third the language domain is another thing to think about: task-oriented and confined (e.g., database query) vs open and uncontrolled (e.g., programming) (nontask-oriented, human-to-human conversation). Less restricted languages have a larger vocabulary and allow for more spontaneous speech.

Fourth criterion to consider is speaker specificity, which may be divided into two categories: speaker-dependent (system trained on test speaker) and speaker independent. (Native vs. nonnative speech is another issue to consider.)

Fifth factor is channel quality, which comprises bandwidth (the frequency range of human speech) and distortion (for example, telephone).

Sixth acoustic environment is also one of the component, which includes the volume and type of background noise (such as static, nonhuman sounds vs background speech and crosstalk from other speakers) [28].

- **Natural language processing**

The effectiveness of neural networks in processing unstructured data like natural language and photos raises the question of whether we may relax the fundamental assumption of database administration, which is that the data we process is represented as fields of a pre-defined schema. What if data and inquiries could instead be expressed as short natural language sentences, with queries being answered using these sentences? What if relevant image data could be smoothly merged with text and structured data in neural databases, with preliminary empirical evidence of its potential? For decades, database systems have struggled to provide various benefits that neural databases provide.

The first and most fundamental advantage is that there is no pre-defined schema in a neural database. As a result, the database's scope does not have to be determined ahead of time, and any data that becomes important as the application is used can be saved and searched.

The second advantage is that changes and queries may be expressed in a variety of natural language formats, making it easy for any user to utilize. A standard database query, on the other hand, must be based on the database structure. Even when the data is structured using a more flexible formalism like RDF, each relation has a unique name that must be used in updates and queries. Third, thanks to recent developments in machine translation, the language of queries and responses may differ from the language of data in the neural database.

Finally, the neural database is built on a pre-trained language model that already includes a large amount of knowledge that can be used to create better replies to a wider range of queries [28].

Foundations of Design

It allows producers to look at the possibility of converting to Industry 4.0 technology. The following are the design principles based on the components listed below:

Integration: The Internet of Things and the Internet of People require objects, machines, and people to be able to interact with one another. This is the most important concept that genuinely distinguishes a smart factory.

Cyber-physical system: CPSs must be able to recreate and generate a virtual duplicate of the actual environment. CPSs must also be able to monitor items in the immediate vicinity. Simply, everything must be virtualized.

Cloud computing: Refers to a CPS's capacity to function autonomously. This allows for the creation of bespoke goods and the resolution of problems. This also makes the manufacturing environment more adaptable. The problem is outsourced to a higher level in situations of failure or conflicting aims. However, even with such technology in place, quality assurance is still required throughout the process.

Real-time capability: A digital economy must be able to gather data in real-time, store or analyse it, and make decisions based on fresh information. This includes not just market research but also internal operations such as a machine breakdown on a production line. Smart objects must be able to recognize the problem and reassign duties to other devices. This also helps significantly to production flexibility and optimization.

Customer: Focused Manufacturing: Production must be centred on the customer. In order to create products that suit the needs of customers, people and smart objects/ devices must be able to connect efficiently via the Internet of Services. At this moment, the Internet of Services becomes crucial.

Adaptability: In a volatile industry, the capacity of a Smart Factory to adapt to a new market is critical. In a normal circumstances, an average business would probably take a week to analyse the market and modify its output accordingly. Smart factories, on the other hand, must be able to react to seasonal changes and market trends swiftly and smoothly.

3.2 Artificial Intelligence and Agriculture

Apart from the risks, AI technology has a lot of advantages for mankind. For example, on cell phones, there is an AI application called "Climate Basic" that assists farmers in increasing their land yields. "To decide how to optimise yields for each plot, the programme uses local temperature and erosion records, predicted precipitation, soil condition, and other agricultural data." The application of AI, according to the Department of Agriculture, has resulted in the largest crops in the country's history. There are various autonomous machine prototypes that can harvest without human assistance, as well as automatic systems for diagnosing illnesses or pests in particular plantations [29] According to Leontief, agriculture in the United States has been increasing. Mechanized, and the manufacturing sector will follow suit. Technical progress favours skilled over unskilled labour by boosting its relative productivity and demand [30–33], also digitization has resulted in large losses of middle-class jobs.

- **Transportation:** Smart technology can help with route estimations or provide suggestions for more efficient modes of transportation, such as alternative routes, traffic sign detection, self-driving vehicles, and sensors. Automation in agriculture is a major source of concern and a contentious issue throughout every nation. Global population is rapidly rising, and as a result, the demand for food also rapidly increasing. Farmers' traditional methods are insufficient to meet rising demand; therefore, they must wreak havoc on the land by applying toxic pesticides in greater quantities. This has a significant impact on agricultural practises, and as a result, the land stays barren and devoid of fertility [29].
- **Agricultural management:** Crop diseases, lack of storage management, pesticide control, weed management, lack of irrigation, and water management are some of the issues that plague the agriculture industry, and all of these issues may be handled utilizing the strategies listed above. Today, it is critical to decipher concerns such as the use of hazardous pesticides, regulated irrigation, pollution management, and environmental consequences in agricultural operations.
- **Agriculture mechanization:** Farming techniques that are automated have been shown to promote soil gain and improve soil fertility [29]. IoT was used to suggest a method for flower and leaf identification and watering that might be applied on a botanical farm.2.3.

3.3 Artificial Intelligence and Manufacturing

Industrial automation receives a lot of attention when it comes to artificial intelligence in manufacturing, but it's only one component of the smart factory revolution, which is a natural next step in the search of efficiency. Artificial intelligence also has the potential to open entirely new commercial opportunities in the industrial industry [34].

Automation has risen as one of the most transformative technologies in the business sector, having the ability to transform management and business operations. In the 4.0 industry, its principal uses are: predictive repair and maintenance are used to improve OEE. Quality 4.0 is achieved by operational excellence, which enhances product quality on a constant basis.

3.4 Artificial Intelligence and Social Justice

Fundamental socioeconomic shifts should not be neglected. As digitized, big data-based systems are considered as more cost-effective and error-prone than humans, machines are rapidly replacing people in cognitive activities [35]. By supporting equitable growth, sustainable development, and well-being, AI should help both people and the environment. To maintain a fair and just society, AI systems should be created in a way that respects the rule of law, human rights, democratic ideals, and diversity, and they should incorporate appropriate protections, such as allowing for human intervention when necessary. To guarantee that everyone is on the same page. There should be transparency and ethical disclosure around AI systems so that people can criticize AI-based outcomes. Throughout their entire cycles, AI systems must be strong, secure, and safe, and potential risks must be continuously reviewed and handled. AI system developers, deplorers, and operators should be held accountable for their appropriate operation in accordance with the aforementioned criteria [11].

3.5 Artificial Intelligence and Ethics

The number of public and commercial groups releasing ethical guidelines to guide the development and use of AI continues to rise, as does the number of AI risks. Indeed, many people believe that this is the most effective proactive risk mitigation option [36]. Fairness. For the ethical management of trustworthy AI, there are complementing values-based principles such as:

- Accountability.
- Human agency.
- Transparency.

- Privacy.
- Respecting human rights.

Establishing ethical principles can assist organizations in safeguarding human rights and liberties while also enhancing wellness and the common good. These ideas can be translated into norms and practises by organizations, which can then be regulated [36].

3.6 *Artificial Intelligence and Government*

A growing number of governmental and private organizations, ranging from tech corporations to religious institutions, have issued ethical standards to guide the development and use of AI, with some even pushing for laws based on science fiction to be expanded. Although the ethical AI principles field is wide, there are some commonalities. We distilled over 90 sets of ethical principles, totalling over 200 principles, into nine key ethical AI principles.

We can see and capture the worries about AI that are expressed by tracking the principles by company, kind of organization, sector, and territory by tracking the principles by company, type of organization, sector, and geography. These can then be transformed into norms and practises, which can subsequently be managed.

These essential ethical AI concepts are based on universally recognized fundamental human rights, international declarations, agreements, and treaties, as well as a review of current codes of conduct and ethical standards from a variety of organizations, enterprises, and initiatives.

The circumstances of knowledge that enable organizations to decide whether an AI system is consistent with an ethical principle are represented by epistemic principles, which are the precondition for an inquiry into AI ethicality. They include interpretability and reliability considerations.

Meanwhile, general ethical AI principles describe behavioural concepts that are applicable across a wide range of cultural and geographical contexts, and they propose how AI solutions should behave when faced with moral decisions or difficulties in a certain field of application. Accountability, data privacy, and human agency are among the principles. Distinct industries and types of organizations gravitate toward different principles, including government agencies, private-sector enterprises, academics, think tanks, associations, and consortiums. While all companies value fairness, sectors that deal with physical assets are more likely to prioritize safety over industries that deal with information assets. Lawfulness and compliance, on the other hand, are most usually found in guidelines issued by government agencies, associations, and consortiums, and while all organizations are required to respect the law, few regard it as an ethical standard.

4 Artificial Intelligence and Industry 4.0

Industry 4.0 enables the automation of manufacturing processes by delivering customized and adaptive mass production technology. As a result, traditional manufacturing methods are being replaced by new ones and smarter manufacturing is being implemented, with machines exhibiting qualities such as self-awareness, self-prediction, self-comparison, self-configuration, self-maintenance, self-organization, and resilience. This means that machines will be self-contained and capable of [37] making judgments and improving themselves through a continuous learning process on their own. Artificial intelligence (AI) and other innovative technologies will speed up inspection detection procedures and increase the capacity to detect new types of flaws. Many project limits and technological constraints exist in the implementation of machine vision applications, which must be taken into account during the selection, implementation, and validation phases for industrial solutions, otherwise, the system may fail to perform as planned. Despite the fact that contemporary machine vision systems with AI applied to IQC exist, they tend to be situation-specific and lack a standard KPI to assess their success or possibility for replication in other industrial contexts. There is presently no way to determine which type of AI approach is most appropriate and how to integrate it with existing or future machine vision application systems [38]. An Industry 4.0 system consists of several elements/phases that must be designed to meet the needs of the manufacturer such as:

- Data collecting from the past.
- Sensors are used to capture real-time data.
- Combination of data.
- Communication protocols, routing, and gateway devices all contribute to connectivity.
- Connections to PLCs.
- Monitoring and analysis dashboards.
- Machine learning and other approaches are used in AI applications.

By employing approaches and procedures, Artificial Intelligence (AI) improves the precision with which automated actions are carried out. Industry 4.0 delves into a set of technologies and sensors that allow for even more advancement in AI processes and applications for business operations, resulting in improved performance and new prospects. Contributions to Industry 4.0, as well as research and comparisons of proprietary and open-source technological features. AI used to be a concept with several application areas.

AI is a subset of Industry 4.0, and it is concerned with how to give computers the complexity to respond intelligently in a wider range of situations. It fully embraces computer science's enthusiasm for abstraction, programming, and logical formalisms, as well as detail programming for algorithms over behavioural data, synthesis over analysis, and engineering (how to do) over science [39]. AI is unfriendly to psychology; psychology is inextricably linked to A I [40].

5 Ideologies of Artificial Intelligence

5.1 Automation

Another RPA-focused service is Automation [41] which has the added benefit of providing information about AI approaches and algorithms. As an RPA tool for ERP setups, it is especially prone to be linked or integrated with SAP and Oracle ERPs, as well as other ERPs from other organizations. It's particularly prone to being linked or integrated with SAP and Oracle ERPs, as well as other ERPs from other companies, and it covers a wide range of applications, including human resources, customer relationship management, and supply chain management. "Digital Workers" is a term that describes a process that is both automated and intelligent [37]. Atomizer creates a system for automating processes that incorporates a cognitive automation module as well as analytical data processing capabilities.

RPA may automate rules-based processes requiring routine operations, structured data, and deterministic outcomes, such as transferring data from several input sources like email and spreadsheets to ERP and CRM systems. The bulk of RPA programmes has been utilized to automate jobs in service business operations such as confirming insurance premium sales, issuing utility bills, paying health care insurance claims, and managing personnel records [42].

It provides a set of data that enables the configuration, operation, and execution of automated processes as a multi-functional application. Internally, the Automation Anywhere tool uses its Bot tool [43] to execute various AI methods and algorithms, such as fuzzy logic, Artificial Neural Networks, and natural language processing, to extract information from documents and enhance the efficiency of document validation. In this sense, it appears that the IQ Bot platform has made several AI approaches or algorithms available through the Automation Anywhere intelligent word processing tool [43].

5.2 Optimization

RPA-as-a-Service: automation software that executes and instigates processes and workflows throughout the business and is simple to access and use for anybody with an internet connection, on any device. Automation self-service is also a feature of cloud-based automation, with easy drag-and-drop actions and graphic flow charts allowing non-technical teams to automate on their own. Cloud RPA allows users to automate any process using a web-based interface that can be accessed from any browser [43].

6 Suggestions

5.1 The way those ideas should be viewed is heavily influenced by cultural differences. As a result, the principles must first be contextualized to represent the community's beliefs, social norms, and behaviours. These "local behavioural drivers" are divided into two categories: compliance ethics, which is concerned with the rules and regulations that apply in a particular jurisdiction, and beyond compliance ethics, which is concerned with social and cultural norms. Investigators must aim to identify stakeholders, their values, and any tensions or conflicts that may occur for them as a result of IT during the contextualization process.

5.2 There has been a lot of debate over how to measure fairness in terms of an individual, a specific decision, and a specific setting. Simply declaring that systems must be "fair" does not provide guidance on who, what, where, or how that fairness should be applied, and different authorities have varied perspectives on fairness. Contextualization would necessitate important stakeholders defining fairness in their own terms.

5.3 To limit regulatory uncertainty, it is critical to link ethical concepts to specific human rights. Furthermore, including human rights principles into AI activities aids in the establishment of moral and legal accountability, as well as the development of human-centric AI for the greater good.

5.4 In order to clearly transform important ideas into particular standards that impact concrete design and governance to shape the creation and usage of AI systems, they must also be aligned with organizational values, existing business ethics practises, and commercial objectives. AI ethical frameworks that are actionable, with explicit accountability and real approaches, should be developed by organizations.

5.5 To foster trustworthy AI innovation, the government must make public and private investment in research and development easier. Through digital infrastructure, technology, and data and knowledge exchange channels, the government must encourage open AI ecosystems.

7 Conclusion

Industry 4.0, the present revolution, includes the Internet of Things, intelligent automation, intelligent devices and processes, and cyber-physical systems.

When all of these ideas and technologies are merged, they provide a dramatic shift in industrial processes, altering the workflow of digital operations across the organization.

In order to optimize these activities, they are increasingly embracing the automation of some steps utilizing robots (RPA). In addition, many RPA technologies now contain intelligent techniques and algorithms (AI), allowing for increased levels of intelligence in the automation of corporate processes. For a substantial number of businesses throughout the world, Industry 4.0 is now a reality. Still, it is worth noting that all the changes required to fully participate in this industrial revolution and reap the benefits that it may bring will not occur overnight.

Because of the complexities of applying artificial intelligence in industrial automation, firms must work with experts to develop bespoke solutions. Attempting to construct the requisite technology is expensive, and most manufacturers lack the essential skills and experience.

References

1. Renjen P (2021) Deloitte in India recognised as a leading inclusiveworkplace.ClinMedNetPrints. https://www2.deloitte.com/content/dam/insights/us/articles/us32959-industry-4-0/DI_Industry4.0. Accessed 23 Aug 2021
2. Hohnen P (2007) Corporate social responsibility an implementation guide for business. NetPrints. https://www.iisd.org. Accessed 24 Aug 2021
3. Zhou J, Zhou Y, Wang B, Zang J (2019) Human–cyber–physical systems (HCPSs). In the context of new-generation intelligent manufacturing. Engineering 5(4):624–636. https://doi.org/10.1016/j.eng.2019.07.015
4. Guo Q et al (2009) A novel approach for multi-agent-based Intelligent manufacturing system. Inf Sci 179(18):307–390. https://doi.org/10.1016/j.ins.2009.05.009
5. Lee J, Bagheri B, Kao HA (2015) A cyber-physical systems architecture for industry 4.0-based manufacturing systems and systems architect. Manuf Lett 3:18–23. https://doi.org/10.1016/j.mfglet.2014.12.001
6. Shi J, Wan J, Yan H, Suo H (2011) A survey of cyber-physical systems. In: Wireless communications and signal processing (WCSP). 2011 International conference on. IEEE, pp 1–6. https://doi.org/10.1109/WCSP.2011.6096958
7. Tuptuk N, Hailes S (2018) Security of smart manufacturing systems. J Manuf Syst 47:93–106. https://doi.org/10.1016/j.jmsy.2018.04.007
8. Bahrin MAK, Othman MF, Azli NHN, Talib MF (2016) Industry 4.0: a review on industrial automation and robotic. Jurnal Teknologi 78:137–143. https://doi.org/10.11113/jt.v78.9285
9. Ben-Daya MH (2017) Internet of things and supply chain management: a literature review. Int J Prod Res. https://doi.org/10.1080/00207543.2017.1402140
10. Autor D, Dorn D (2013) The growth of low-skill service jobs and the polarization of the US labor market. Am Eco Rev 103(5):1553–1597
11. Branger J (2015) From automated home to sustainable healthy and manufacturing home a new story enabled by the Internet-of-Things and Industry 4.0. J Manag Anal 2(4):314–332. https://doi.org/10.1080/23270012.2015.1115379
12. Baxter NE (1989) Research guidance: not giving it your best product testing with consumers.american society for testing and materials, pp 10–22. https://doi.org/10.1520/STP19490S
13. Bughin J et al (2007) Artificial intelligence. The next digital frontier. Mc building change the manufacturing landscape: an Industry 4.0 perspective. Int J Mech 2–24. https://doi.org/10.5281/zenodo.1336426. Accessed on 28 Oct 2021
14. Carvalho TP et al (2019) A systematic literature review of machine learning methods applied to predictive maintenance. Comput Ind Eng. https://doi.org/10.1016/j.cie.2019.106024
15. Gupta NA (2017) Literature survey on artificial intelligence. ClinMed Net Prints. https://www.ijert.org/research/a-literature-survey-on-artificial-intelligence-IJERTCONV5IS19015.pdf. Accessed on 18 Sept 2021
16. Dallasega PR (2018) Industry 4.0 as an enabler of proximity for construction supply chains: a systematic literature review. Comput Ind 9:205–225. https://doi.org/10.1016/j.compind.2018.03.039

17. Boyd D, Crawford K (2012) Critical questions for big data. Inf Commun Soc 15(5):662–679
18. Brynjolfsson E, McAfee A (2014) The second machine age: work, progress, and prosperity in a time of brilliant technologies. W.W Norton, New York, NY
19. Brettel MF (2014) How virtualization decentralization and network building change the manufacturing landscape: an Industry 40 perspective. Int J Mech. https://doi.org/10.5281/zenodo.1336426
20. Brennan LF (2015) Manufacturing in the world: Where next? Int J Oper Prod Manag 35(9):1253–1274. https://doi.org/10.1108/IJOPM-03-2015-0135
21. Thoben KD et al (2017) Industrie 4.0 and smart manufacturing—a review of research issues and application examples. Int J Autom Technol 11(1):4–16
22. Lee J, Ardakani HD, Yang S, Bagheri B (2015) Industrial big data analytics and cyber- physical systems for future maintenance & service innovation. Procedia CIRP. https://doi.org/10.1016/J.PROCIR.2015.08.026
23. Buer SV, Strandhagen JO, Chan FTS (2018) The link between Industry 40 and lean manufacturing: mapping current research and establishing a research agenda. Int J Prod 56(8):2924–2940. https://doi.org/10.1080/00207543.2018.1442945
24. Ford M (2009) The lights in the tunnel. automation, accelerating technology and the economy of the future. Acculant Publishing, Wayne, PA
25. Intel gov (2011) Principles of artificial intelligence ethic for the intelligence community clinmednetprint. https://www.intelligence.gov/images/AI/Principles_of_.AI_Ethics_for_the_Intelligence_Community.pdf
26. Tao F, Cheng Y, Da Li X, Zhang L, Li BH (2014) Cloud computing and Internet of things-based cloud manufacturing service system. IEEE Trans Ind Inform 10(2):1435–1442. https://doi.org/10.1109/TII.2014.2306383 Accessed 23 Aug 2021
27. Kirtan Jha AD (2019) A comprehensive review on automation in agriculture using artificial intelligence. Artif Intell Agric 2:112–114. https://doi.org/10.1016/j.aiia.2019.05.004
28. Christensen CM (1997) The innovator's dilemma: when new technologies cause great firms to fail. Harvard Business School Press, Boston, Mass
29. Stolcke A (1997) Linguistic knowledge and empirical methods in speech recognition. AI Mag. https://doi.org/10.1609/aimag.v18i4.1319
30. Judit Nagy JO (2018) The role and impact of industry 4.0 and the internet of things on the business strategy of the value chain—the case of hungary. Sustainability 10:1–24. https://doi.org/10.3390/su10103491
31. Bekman E, Bound J, Machin S (1998) Implications of skill-biased technological change: international evidence. Quart J Econ 113(4):1245–1279
32. Heathcote J, Storesletten K, Violante G (2010) The macroeconomic implications of rising wage inequality in the United States. J Polit Econ 118(4):681–722
33. Autor D (2014) Polanyi's paradox and the shape of employment growth. Paper presented at the Kansas City's economic policy symposium on "Re- Evaluating Labor Market Dynamics", Jackson Hole, WY
34. Levy F, Murnane R (2013) Dancing with robots: human skills for computerized work, Third Way Next. Fresh Thinking
35. Eleanor Bird JFS (2020) The ethics of artificial intelligence: issues and initiatives. ClinMed-NetPrints. https://www.europarl.europa.eu/thinktank/en/document
36. Acemoglu D, Autor D (2013) Skills, tasks and technologies: implications for employment and earnings. In: Card D, Ashenfelter O (eds) Handbook of labor economics, vol 4B. Elsevier, Amsterdam, pp 1043–1171. Accessed 09 Oct 2021
37. Eleanor B et al (2020) Future of science and technology (STOA).STOA | Panel for the future of science and technology. pp 4–124: NetPrint. https://www.europarl.europa.eu/RegData/etudes/STUD/2020/634452/EPRS_STU(2020)634452_EN.pdf
38. Ribeiro J et al (2021) Robotic process automation and artificial intelligence in industry 4.0—a literature review.51–58. Accessed 28 Aug 2021
39. Silva RL, Rudek M, Szejka AL, Junior OC (2018) Machine vision systems for industrial quality control inspections. In: Product lifecycle management to support industry 4.0: 15th IFIP WG

5.1 international conference, PLM 2018, Turin, Italy, July 2–4, 2018, Proceedings 15. Springer International Publishing, pp 631–641
40. Nilsson N (1980) Principles of artificial inteligence. Tioga Press, Menlo Park CA, p 366
41. Newell A (1980) A textbook that points the way review of Nils Nilsson. In: A. Newell. California: Tioga Press
42. Unknown (2021) Automation anywhere robotic process automation to ERP automation anywhere. NetPrints. https://www.automationanywhere.com/lp/automate-any-erp-process-with-rpa Accessed 25 Aug 2021
43. Lacity M, Willcocks LP (2016) A new approach for automating services. MIT Sloan Manag Rev 58:40–49

Chapter 19
Process of Combined Thinking for Long-Term Sourcing

Chiranjib Bhowmik, Sudipta Ghosh, Sumit Das Lala, and Amitava Ray

Abstract The purpose of this research is to develop a trustworthy decision-making tool that integrates the thinking process, the concept of six sigma, and lean thinking in a long-term supply chain. Supplier quality was recognized as a supply chain barrier in this study because ecologically friendly materials result in a sustainable supply chain. The purpose of this study is to fill a vacuum in the existing literature by emphasizing the connections between theory of constraints, six sigma, and lean thinking for continuous improvement. This study combines the notion of constraints, six sigma, and lean thinking into one evaluation model. Before selecting the best provider with the least amount of environmental impact, the programme looks at a variety of qualitative and quantitative parameters. Supplier 1 appears to be the best, followed by the others. The sensitivity analysis of the model assumes that "higher is better," and the model's output identifies the best supplier with the least environmental impact.

Keywords Theory of constraints · Six sigma · Lean-thinking · Supplier selection · Sustainability · Sensitivity analysis

C. Bhowmik (✉)
Department of Mechanical Engineering, Techno India University, Salt Lake, Kolkata, West Bengal 700091, India
e-mail: chiranjibbhowmik18@gmail.com

S. Ghosh
Department of Mechanical Engineering, Durgapur Institute of Advanced Technology and Management, Durgapur, West Bengal 713214, India

S. Das Lala
Department of Mechanical Engineering, Parul Institute of Engineering and Technology, Parul University, Vadodara 391760, India

A. Ray
Department of Mechanical Engineering, Jalpaiguri Government Engineering College, Jalpaiguri, West Bengal 735102, India

S. Kautish et al. (eds.), *Computational Intelligence for Modern Business Systems*, Disruptive Technologies and Digital Transformations for Society 5.0,
https://doi.org/10.1007/978-981-99-5354-7_19

1 Introduction

With rising public awareness and government laws on sustainability policies, it is becoming more vital for businesses to focus on strategic sourcing, as successful supplier selection is critical to an industry's success. Any organization's economic development is dependent on sustainable sourcing. Now a days, natural resources are decreasing at an alarming rate, thus sustainable and green thinking can help to avert resource depletion and adverse environmental effects. Many businesses have gained a competitive edge by focusing on environmental impact reduction. As a result, firms have focused on sustainable policies based on traditional methodologies. Due to globalization in business, competitive market situations, and the nature of customer needs, selecting the best supplier is no longer enough [1, 2]. Global supply chain management difficulties have developed in decision-making frameworks that provide a place to track the concerns of numerous components across the supply chain. Significant effort has gone into developing decision models for supply chain concerns during the last few decades [3, 4]. Appropriate collective supply chain methods are established based on market distinctness in order to fulfill advanced levels of consumer expectations at a lower cost [5]. Surprisingly, selecting the right supplier also plays a crucial role because it saves acquisition expenses [6]. A typical supply chain management system is made up of a number of interconnected essential mechanisms [7]. Any organizations sustainability also depends on its supplier. Therefore, businesses can become more sustainable with better policy and product management, as well as lower prices and timely delivery [3, 4]. The characterization of the buyer's degree of agreement is a new component in today's supply chain management [8, 9]. Supply chain (SC) resilience literature [10, 11] focuses on supply chain management (SCM), risk, vulnerability, logistics, and teamwork. SC resilience does not mean that everything can be recovered after a disruption, but it is an important component of SCM, which includes protectiveness, structured and integrated capability exploration, and measured reactions to uncertainty [12]. As a result of globalization or optimal supplier selection, the supply chain's complexities are compromised in the current situation, putting the environment at risk. The future of SCM cannot be predicted accurately, despite modern forecasting methodologies and a variety of strategies. As a result, enhanced flexibility and the ability to react swiftly to global changes in market conditions, as well as forecasting skills, are becoming increasingly vital. Decision makers handle purchasing, production, distribution, planning, and other logistic responsibilities in traditional supply chains separately from supply chain managers. So, therefore keeping in mind the current scenario, this research has tried to connect three continuous improvement techniques simultaneously such as; theory of constraints (TOC), six sigma (SS), and lean thinking (LT) concept in an integrated manager to select the optimum supplier for sustainability. The integrated management philosophy is highlighted in this study as a strategy to address global supply chain risk, with many target functions such as customer requirements, technical requirements, low production/inventory costs, and so on. The main reason behind choosing the strategy of continuous improvement in a combined manner is that, utilizing its existing resources without incurring new investment in the process. The business world has challenges in managing a changing and uncertain future, necessitating flexibility, or an organization's ability to endure, adapt, and persist in a chaotic global business environment. The combined thinking process represents a new paradigm in quality control procedures, allowing for new decision-making processes that can optimize industrial activities. Because of rising competition and marketing demands for environmental sustainability, businesses have begun to recognize the need to take initiatives to reduce their environmental effect [13].

Enterprises need to apply methods for decreasing their environmental effect as a result of increasing modern competitive and marketing pressures for environmental sustainability [14]. As society becomes more concerned about environmental issues, manufacturing organizations are forced to embrace green techniques in order to meet their objectives in a significant and cost-effective manner. Both economic and environmental performance can be improved by implementing internal and external sustainable methods [15]. According to Seuring et al. [16], the organization must link supply chains with long-term development. The concepts of green SCM emerged from the realization that isolated implementations of environmental practices by companies are not as effective as collective actions that can make the entire supply chain greener. Ageron et al. [17] developed a framework from which the concepts of green SCM emerged. For measuring environmental performance, Dey et al. [18] utilized analytical hierarchy process (AHP) to assess the green supply chain performance of the selected firms in the United Kingdom (UK) as well as benchmark those firms.

In this research, authors utilized three management philosophies in an integrated manner for the selection of optimal supplier for sustainability. The term TOC refers to a management concept introduced by Goldratt (1984) in his book The Goal [13], which consists of five steps with a focus on process improvement. Thereafter, another strategy, i.e., six sigma is utilized to overcome the uncertainty that exists inside any organization which uses a series of arithmetical strategies [19]. Simultaneously, another five-step approach is incorporated into this research which was introduced by Kalashnikov et al. [20]. This method helps to optimize the situation based on the advantages that reduce the investment cost. According to Aldowaisan et al. [21] low sigma levels always affect normal distribution process, where the degree of variation reduction grows exponentially. Research highlights that the sigma level of any product is a good measure of its quality [21]. Aguilar-Escobar [22] investigated how TOC behaved in the logistics of hospital medical records. As a result of their efforts, the level of service and employee productivity grew dramatically. Mitchell and Kovach [23] studied the marine transportation supply chain strategy using six sigma. In order to establish a more sustainable supply chain, Bastas and Liyanage [24] reported how quality management systems affect society behavior. For enhancing manufacturing quality and throughput, Bhowmik et al. [25] proposed an integrated model based on TOC and six sigma. Result showed the integrated methodology helps to identify the optimal supplier in terms of quality. Dubey et al. [26] proposed an integrated paradigm based on benefits, opportunities, costs, and risks for assessing supplier relationships in a risky supply chain context. Despite the fact that previous studies have largely used evolutionary algorithms [27], but this study helps to identify the optimum suppliers using an integrated strategy. The novelty of this work is that the study helps to identify the optimum supplier using the combined methodology which always suggests using the existing resources without future investment. Unlike previous studies, this study gives a comprehensive methodology for identifying the best supplier for long-term product quality.

The study's design is detailed in the following section. Section 3 emphasizes the effectiveness of the proposed study framework as well as the results. The sensitivity analysis of supplier selection is explained in Sect. 4, and the study is wrapped up in Sect. 5.

2 Literature Review

In these challenging times of enormous market competition, when environmental degradation is threatening economic and social sustainability, it is very much essential that industry professionals, academic practitioners, and policy makers share their knowledge in a concurrent manner to design environmentally conscious solutions that move organizations toward real sustainable development. One of the globally appreciated solution is strategic sourcing, a notion that aims to ameliorate financial prosperity, minimizes environmental burdens, and prevent unnecessary and excess consumption of natural resources. During the last decade, the importance of sourcing practices has grown endogenously in industry and researchers' community because it promotes sustainable development by creating value in the economy. Sourcing has been one of the most crucial decision for industry practitioners on account of ever-increasing consumption levels and for sustainable development as well. While organizations in developed nations have already gained competitive advantage by implementing effective sourcing practices, it is still in its infancy stage due to presence of various hurdles such as lack of investment, lack of awareness among stakeholders, lack of infrastructure, lack of legislation, and commitment from top management. As far as environmental degradation is concerned, corporate executives are supposed to collaborate and cooperate with suppliers for green innovation and operational excellence. Various sourcing frameworks can be found in previous literature which is listed in Table 1.

Table 1 Prior art on sourcing

Authors	Focus area	Industrial domain	Criteria considered	Methods used
Ghosh et al. [28]	Environmentally-conscious sourcing	Automobile manufacturing sector	Investment in corporate social responsibility (CSR), investment in research and development (R&D), % of energy used from renewable source, total energy consumption, total carbon-di-oxide (CO2) emissions, total waste generation	Entropy, technique for order preference by similarity to the ideal solution (TOPSIS)
Ghosh et al. [29]	Strategic sourcing for green supply chain management (GSCM)	Automobile manufacturing sector	Investment in CSR, investment in R&D, utilization of renewable energy, total energy consumption, total CO2 emissions, total waste generation	Entropy, complex proportional assessment (COPRAS) and the Grey relational analysis (GRA) methods
Ghosh et al. [30]	GSCM sourcing framework	Manufacturing, service, and process organizations	Investment towards environmental protection, renewable energy utilization (% of total energy), total scrap material generation, total energy consumption, top management commitment for GSCM, system for effective recycling	Entropy, TOPSIS, COPRAS, GRA
Ghosh et al. [31]	Supplier's performance evaluation framework under uncertain environment	Leather, pharmaceutical, and steel industries	Design for proper utilization of resources, optimization of process parameters to increase quality and reduce scrap and wastes, environmental impact of harmful emissions released from production, use of eco-friendly packaging material, utilization of renewable energy	Fuzzy-analytical hierarchy process (AHP), fuzzy

(continued)

Table 1 (continued)

Authors	Focus area	Industrial domain	Criteria considered	Methods used
Ghosh et al. [32]	Green supplier selection	Steel industry, car manufacturers, and food processing company	Use of energy-efficient technologies, total energy consumption, effective waste management, frequency of environmental accidents, quality, reuse of material, design for proper utilization of resources, total CO2 emission, reduce scrap material, internal investment recovery, generation of hazardous waste, use of eco-friendly packaging materials, top management commitment, collaboration with suppliers for green purchasing, investment in CSR	Item analysis, principal component analysis, simple additive weighting
Ghosh et al. [33]	Green supplier selection	Pipe fitting industry, automobile sector, IT sector	Process parameters, design criteria, manufacturing criteria, process optimization, environmental planning, reuse of materials, design for proper utilization of resources, quality, reduce scrap, robust environmental policy, green procurement policy, cooperation with supplier, cooperation with customer for green initiatives	Item analysis, principal component analysis, scoring model
Ghosh et al. [34]	Key performance parameters of GSCM for sustainability	Tea processing firms	Decrease in consumption of toxic materials, reduction of usage of harmful materials, reduction of air emission, recycling, robust environmental policy, employee motivation, waste elimination, reduction of accidents, solid waste reduction, wastewater reduction, quality improvement, scrap rate reduction, minimum inventory, on-time delivery, optimum design, energy consumption cost	Pareto analysis
Bhowmik et al. [35]	Sustainable sourcing	Energy firm		Synthesized AHP
Bhowmik et al. [36]	Sustainable sourcing	Energy firm		BOCR-AHP

3 Research Design

This research employs TOC, Six Sigma, and Lean Thinking process simultaneously. The individual methodologies are described in the following sub-sections.

3.1 Theory of Constraints (TOC)

It is one of the continuous improvement philosophies which was developed by Goldratt and Cox [13]. TOC is a logic-driven methodology for identifying the weakest link or the limiting factor (constraint) that stands as a resistance in the way of achieving goal and then improving the constraints in a hierarchical manner until it is no longer the constraint. Thus, TOC helps in increasing organizations' throughput. Constraints is often referred as 'Bottleneck' in various operations. TOC primarily considers that any system must contain at least one constraint which resists the system from achieving the desired goal and also considers that the presence of constraints in the system implies that opportunity is available for further improvement [37]. The five focusing steps of TOC are shown in Fig. 1 and explained below:

Step 1: *Identify the system constraint (What to change?):* It is the first and foremost step where system constraints or bottlenecks are identified. Identifying constraints is crucial as the performance of the entire supply chain depends on it. There are so many types of constraints such as raw material, production capacity, distribution, cost, market, policy, infrastructure, etc. The constraints may be physical (man, material,

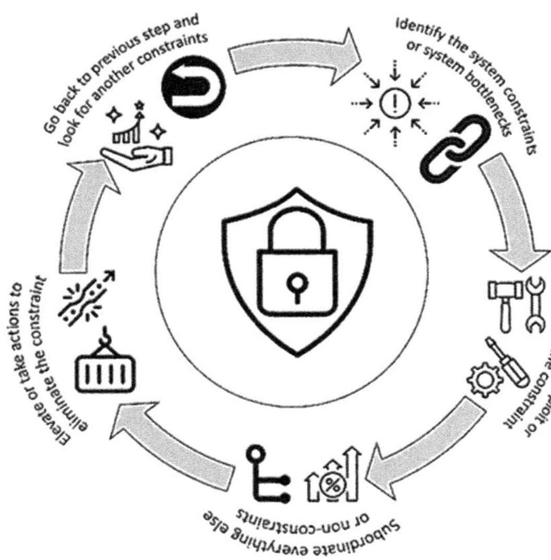

Fig. 1 Five focusing steps of TOC

machine, etc.) or managerial (methods, policies, etc.). It is also necessary to prioritize the constraints according to their impact on the performance of the organization.

Step 2: *Exploit the system constraint (What to change to?):* In this step, decision maker has to decide how to do everything possible to utilize the constraint to a greater extent or to its maximum capacity. It is necessary to obtain as much capability as possible from the constraints undergoing no expensive modification or significant changes in system or without hampering other components. If the constraint is physical, then it has to be reduced or eliminated as effective as possible. Managerial constraints cannot be exploited but can be replaced with a new policy or method which will help in increasing the throughput.

Step 3: *Subordinate everything else (How to cause the change?):* This step implies that all other components except the non-constraints, must be adjusted to a setting that enables the constraint to perform at its maximum or peak effectiveness. Once this task is accomplished, then the overall performance is evaluated to determine whether the constraint has shifted to other components. If the constraint gets eliminated in this step, then directly jump to the last step otherwise go for the next step.

Step 4: *Elevate the constraint (How to sustain the change?):* After exploitation and subordination, if the constraint still remains the most critical factor in the system, then rigorous improvement measures should be applied on the system to eliminate the constraint. Major changes are incorporated in the system, in this step. As the constraint's performance increases eventually, the system will encounter a new constraint.

System 5: *Go back to previous step and repeat the cycle:* If the constraint is broken in any of the above steps, then go back to initial step and identify new constraints. Do not stop and don't let inertia become the constraint because TOC is a continuous process and no policy or method will yield satisfactory result or perform appropriately for all the time in all circumstances.

3.2 Six Sigma (SS)

It is a statistical data-driven approach for eliminating defects within product, process or service. It is also a continuous improvement methodology. It was first introduced by Bill Smith in 1980 (Motorola Company). Sigma implies standard deviation, which is basically a measure of variation in a data set. Six sigma has a process mean which is six standard deviations from the nearest specification limit. The accuracy level obtained by six sigma processes in 3.4 defects per million products. DMAIC is one of the two most popular methodologies (DMAIC & DMADV). DMAIC is the acronym for five phases, namely, Define, Measure, Analyze, Improve & Control. The DMAIC method is shown in Fig. 2.

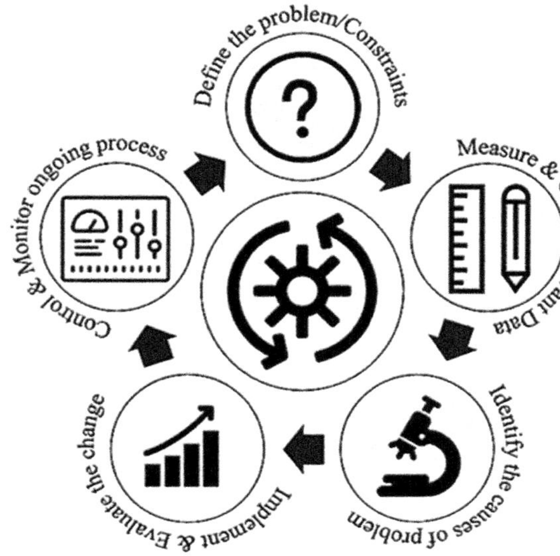

Fig. 2 DMAIC (six sigma) methodology

3.3 *Lean Thinking (LT)*

Lean thinking is, in essence, a business methodology. It has its roots in the history of how Japanese manufacturing methods have been adopted by businesses and organizations all over the world. As the name implies, lean thinking is also a mindset. It is a philosophy that treats work in a lean manner, emphasizing value to the client above everything else. Lean thinking is altering your perspective of your company and how you see how business operations work, not just using particular technologies or adjusting a few steps in a business process. It's a new style of thinking that you adopt to alter how you approach your business.

3.4 *Integrated TOC-Six Sigma-Lean Methodology*

In this study, combined thinking process is proposed to find out the effectiveness in identifying supply chain limitations and removing them from the existing system. Firstly, the system's weakest parameter, which prevents the system from attaining its aim, is identified based on outlining the six sigma and lean thinking. Secondly, the TOC philosophy helps to limit and subordinate the processes that is being studied. In this subordination process research utilized Shannon's entropy method. In this case, six sigma helps to quantify the value and analyzing it sequentially. As a result, the value stream of a Lean process can be mapped. Six sigma uses statistical tools to increase the limitations in the next stage (improvement step). Control step in six sigma is used to ensure that all of the previously mentioned preventive actions are

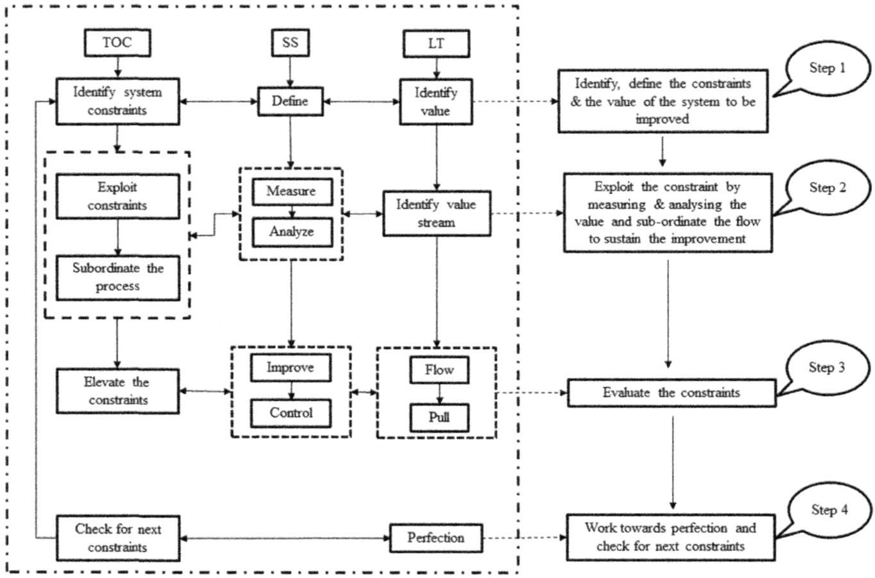

Fig. 3 Proposed research framework

successfully applied to the system. The flow and pull method aids in the removal of functional barriers and the improvement of the system. Finally, the integrated system adheres to Kaizen and the practice of continual improvement. The optimal supplier ranking is done using grey relation analysis (GRA) method. To avoid inertia in the entire system, the search of precision should be pursued by resuming the initial step of TOC to uncover new constraints. As a result, the lean system can be treated as a measure of perfection. The proposed research methodology is depicted in Fig. 3 and is tested on three different suppliers.

4 Application, Result, and Discussion

The distinction in the characteristics of their sustainable supply chain regulations was the primary rationale for selecting those suppliers. Furthermore, the selected supplier groups are industry pioneers in their respective sectors. The following is the research framework adopted in this study to select the optimum supplier. The steps are delineated below.

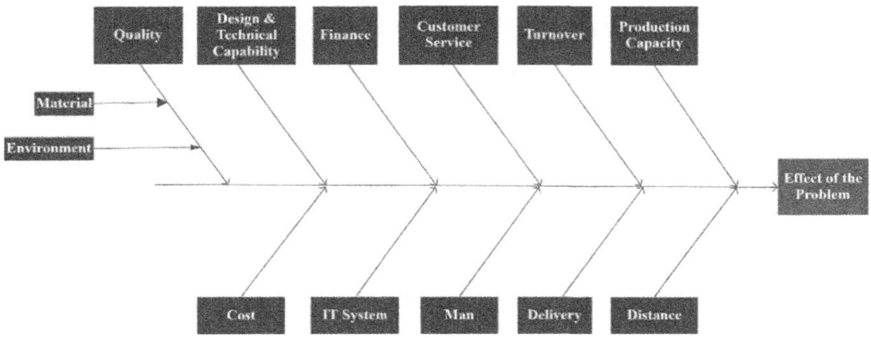

Fig. 4 Ishikawa diagram to identify the root cause

4.1 Determine the Value of the Constraint and the Process that Has to Be Improved

According to the integrated thinking process, the first phase was to maintain the improving by identifying the constraints that stopped the system to reach its goal [37–40]. The authors noticed that the project team made a number of advancements depending on the options they uncovered while choosing the study location. Study identified that the quality of the suppliers is the main effect rather than other factors using an Ishikawa diagram, which is a team brainstorming method used to find potential root causes of the problem as shown in Fig. 4.

4.2 Exploit the Constraint by Measuring and Analyzing the Value

In this section study decided to exploit the constraints to measure the improvement. Each supplier's qualitative and quantitative value is represented in the supplier matrix from published literature [38, 40]. Once the value-added and non-value-added events have been determined, the activity flow will be focused toward improving the situation [40, 41]. Result of Shannon's entropy probabilistic technique shows that quality has been given the most priority. The supplier process parameters and their corresponding priority weights are; quality (0.228), finance (0.019), customer service (0.038), production capacity (0.091), design & technical capability (0.095), IT system (0.081), turnover (0.021), distance (0.030), delivery (0.190) and cost (0.208). As demonstrated in Fig. 2, the most influencing parameters on quality are material and environment. As a result, the constraint can be detected, and the system's bottleneck can be specified. Integrated methodology ensures that product and service movement is uninterrupted throughout the process. As a result, new processes can be implemented to ensure that progress is maintained.

4.3 Increase the Limitations' Severity

The final ranking of the suppliers was determined using grey relation analysis (GRA) and the ranks are (0.708, 0.570, 0.650) for supplier 1, 2, and 3 respectively. The supplier grades have normalized values of 0.37, 0.30, and 0.34, respectively. Because the first two phases of this study were successful in recovering the supplier selection problem to a large extent, there was no need to extend the limitations or improve the existing parameters. The decision is acceptable and should be implemented if the vendor's high quality improves production throughput.

4.4 Work Toward Perfection and Keep an Eye Out for the Next Set of Constraints

The integrated methodology repeats the process monitoring to eliminate any non-value processes, keep the flow going, and fulfill customer expectations. Once the best supplier has been chosen, the system's environmental impact will be lessened, and it will no longer be a bottleneck. If any potential new restrictions emerge, the management team should return to step 1 and resume the loop, examining the supplier's performance and customer needs.

5 Sensitivity Analysis

A mathematical model is constructed to combine the priority weights of the supplier with reference to quality of the service provided by the supplier. The equation of the model is as follows [39];

$$SI_i = \alpha \times PWM_i + (1 - \alpha) \times QWM_i \tag{1}$$

where,

$$QWM_i = \left[OFQM_i \times \sum_{i=1}^{n} OFQM_i^{-1} \right]^{-1} \tag{2}$$

QWM_i = Quality weights measure, PWM_i is the priority weights measure of suppliers. SI_i is the Selection Index; α is the attitude of the decision maker varies from $0 \le \alpha \le 1$.. The priority weights of the suppliers (PWM_i) are 0.708, 0.570 & 0.650 respectively [38]. *OFQM*s are designed in such a manner, which indicates non-dimensional value of each supplier.

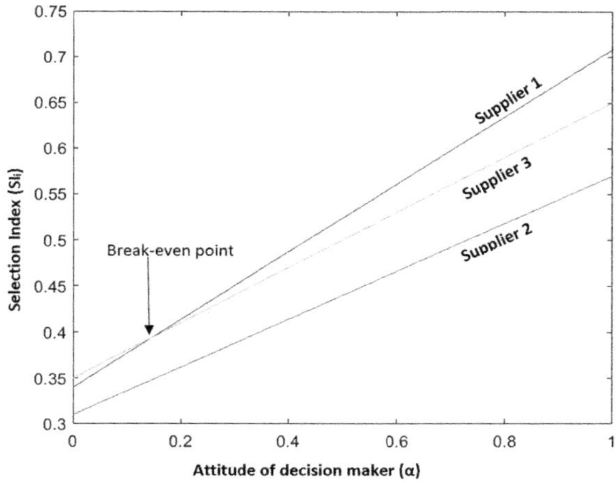

Fig. 5 Sensitivity plot

The selection of α is a major issue. The value of α depends upon the decision maker's preference regarding the importance of quantitative and qualitative factor measures. Hence, a sensitivity plot to analyze the effect of α on sustainable supplier selection problem is strongly recommended. If the value of α is 0.26, either supplier-1 or supplier-3 will be selected. As the attitude of the decision maker inclines towards a sustainable environment, supplier-1 is the optimum supplier as our proposition is "higher the better" as shown in Fig. 5.

6 Conclusion

This research focuses on an integrated technique for evaluating and selecting sustainable suppliers. The proposed integrated methodology helps to move the organization closer to encouraging enslaved members to control the constraint that prevents it from participating in global supply chain management system. The combined thinking process of supplier selection method currently focuses on forwarding SCM. The outcomes of each criteria performance are statistically analyzed to determine the organization's potential benefits. Instead of adopting the traditional methodology, any organization can use this integrated methodology for sustainable supplier evaluation to uncover and priorities opportunities for improving their sustainability performance from a holistic perspective. The approach considers a variety of qualitative and quantitative criteria to determine the best supplier for minimizing environmental damage. The results suggest that supplier 1 is the best supplier, followed by the rest. This research has provided a prospective framework for long-term supplier selection, which can be expanded further by completing a survey to compare the overall story of the organization's supply chain problems to those who do not. Other soft-computing methodologies can be combined with the suggested method, and the results can be compared.

References

1. Amindoust A, Ahmed S, Saghafinia A, Bahreininejad A (2012) Sustainable supplier selection: a ranking model based on fuzzy inference system. Appl Soft Comput 12(6):1668–1677
2. Talluri S, Narasimhan R (2004) A methodology for strategic sourcing. Eur J Oper Res 154(1):236–250
3. Cox A, Chicksand D (2006) Aligning brand and supply chain strategies. CPO Agenda, autumn, pp 46–50
4. Zarei M, Fakhrzad MB, Paghaleh MJ (2011) Food supply chain leanness using a developed QFD model. J Food Eng 102(1):25–33
5. Christopher M, Towill DR (2002) Developing market specific supply chain strategies. Int J Logist Manag 13(1):1–14
6. Ghodsypour SH, O'brien C (2001) The total cost of logistics in supplier selection, under conditions of multiple sourcing, multiple criteria and capacity constraint. Int J Prod Econ 73(1):15–27
7. Truong TH, Azadivar F (2005) Optimal design methodologies for configuration of supply chains. Int J Prod Res 43(11):2217–2236
8. Hsu CC, Kannan VR, Keong-Leong G, Tan KC (2006) Supplier selection construct: instrument development and validation. Int J Logist Manag 17(2):213–239
9. Huang SH, Keskar H (2007) Comprehensive and configurable metrics for supplier selection. Int J Prod Econ 105(2):510–523
10. Norrman A, Jansson U (2004) Ericsson's proactive supply chain risk management approach after a serious sub-supplier accident. Int J Phys Distrib Logist Manag 34(5):434–456
11. Peck H (2006) Reconciling supply chain vulnerability, risk and supply chain management. Int J Log Res Appl 9(2):127–142
12. Lin RH (2009) Potential use of FP-growth algorithm for identifying competitive suppliers in SCM. J Operat Res Soc 60(8):1135–1141
13. Goldratt EM, Cox J (1984) The goal: excellence in manufacturing. North River Press
14. Lazuras L, Panagiotis H, Ketikidis A, Bofinger B (2010) Promoting green supply chain management: the role of the human factor. In: 15th Panhellenic logistics conference and 1st Southeast European congress on supply chain management
15. Sarkis J, Zhu Q, Lai K (2011) An organizational theoretic review of green supply chain management literature. Int J Prod Econ 130:1–15
16. Seuring S, Gold S (2007) Conducting content analysis-based literature reviews in supply chain management. Supply Chain Manag Int J 17(5):544–555
17. Ageron B, Gunasekaran A, Spalanzani A (2012) Sustainable supply chain management: an empirical study. Int J Prod Econ 140(1):168–182
18. Dey PK, Cheffi W (2013) Green supply chain performance measurement using the analytic hierarchy process: a comparative analysis of manufacturing organizations. Prod Plan Control 24(8–9):702–720
19. Chen C, Lin Y (2012) Integration of kano model into TOPSIS method for effective product assessment. Appl Mech Mater 145:475–479
20. Kalashnikov V, Benita F, López-Ramos F, Hernández-Luna A (2017) Bi-objective project portfolio selection in Lean Six Sigma. Int J Prod Econ 186:81–88
21. Aldowaisan T, Nourelfath M, Hassan J (2015) Six Sigma performance for non-normal processes. Eur J Oper Res 247(3):968–977
22. Aguilar-Escobar VG, Garrido-Vega P (2016) Applying the theory of constraints to the logistics service of medical records of a hospital. Eur Res Manag Bus Econ 22(3):139–146

23. Mitchell EM, Kovach JV (2016) Improving supply chain information sharing using design for six sigma. Eur Res Manag Bus Econ 22(3):147–154
24. Bastas A, Liyanage K (2019) Integrated quality and supply chain management business diagnostics for organizational sustainability improvement. Sustain Prod Consump 17:11–30
25. Bhowmik C, Gangwar S, Ray A (2018) Integrating six-sigma and theory of constraint for manufacturing process: a case study. In: Soft computing: theories and applications. Springer, Singapore, pp 607–617
26. Dubey VK, Chavas JP, Veeramani D (2018) Analytical framework for sustainable supply-chain contract management. Int J Prod Econ 200:240–261
27. Rajpurohit J, Sharma TK, Abraham A, Vaishali A (2017) Glossary of metaheuristic algorithms. Int J Comp Inform Syst Indus Manag Appl 9:181–205
28. Ghosh S, Mandal MC, Ray A (2021) Selection of environmental-conscious sourcing: an empirical investigation. Benchmarking 28(6):2130–2155
29. Ghosh S, Mandal MC, Ray A (2022) Strategic sourcing model for green supply chain management: an insight into automobile manufacturing units in India. Benchmarking 29(10):3097–3132
30. Ghosh S, Mandal MC, Ray A (2022) Green supply chain management framework for supplier selection: an integrated multi-criteria decision-making approach. Int J Manag Sci Eng Manag 17(3):205–219
31. Ghosh S, Mandal MC, Ray A (2022) A PDCA based approach to evaluate green supply chain management performance under fuzzy environment. Int J Manag Sci Eng Manag https://doi.org/10.1080/17509653.2022.2027292
32. Ghosh S, Bhowmik C, Mandal MC, Ray A (2021) Green supplier selection: an empirical investigation. In: Modeling, simulation and optimization. Springer, Singapore, pp 723–735. https://doi.org/10.1007/978-981-15-9829-6_57
33. Ghosh S, Mandal MC, Ray A (2021) Green supplier selection using statistical method." In: Lecture notes on multidisciplinary industrial engineering. Singapore: Springer Singapore, pp 397–413
34. Ghosh S, Mandal MC, Ray A (2022) Investigating the key performance parameters of green supply chain management for sustainability in tea processing firms using Pareto analysis. J Inst Eng (India) Ser C https://doi.org/10.1007/s40032-022-00888-8.
35. Bhowmik C, Bhowmik S, Ray A (2020) Optimal green energy source selection: an eclectic decision. Energy Environ 31(5):842–859
36. Bhowmik C, Bhowmik S, Ray A (2022) Green energy sources selection for sustainable planning: a case study. IEEE Trans Eng Manage 69(4):1322–1334
37. Bhowmik C, Ray A (2015) The application of theory of constraints in industry: a case study. Int J Extens Res 2:21–29
38. Yang CC, Chen BS (2006) Supplier selection using combined analytical hierarchy process and grey relational analysis. J Manuf Technol Manag 17(7):926–941
39. Ray A, Sarkar B, Sanyal S (2009) The TOC-based algorithm for solving multiple constraint resources. IEEE Trans Eng Manage 57(2):301–309
40. Bhowmik C, Zindani D, Bhowmik S, Ray A (2020) Sustainable supplier selection using combined thinking process. In: Soft computing: theories and applications. Springer, Singapore, pp 281–289. https://doi.org/10.1007/978-981-15-0751-9_26.
41. Ghosh S, Mandal MC, Ray A (2022) Exploring the influence of critical parameters on green supply chain management performance of small and medium-sized enterprise: an integrated multivariate analysis-robust design approach. Clean Logist Supply Chain 4(100057):100057

Chapter 20
Technological Reforms of Global Projects Using Artificial Intelligence

Medhavi Yadav, Siddharth Shahi, Himanshu Ahuja, and Mridula Batra

Abstract Artificial Intelligence or machine intelligence is a combination of multiple types of technologies which work in tandem to help machines to sense their surroundings, understand complex problems, provide solutions and learn new things mirroring a human-like level of intelligence. With the increasing capabilities and sophistication of AI systems, it is now being used in multiple ranges of sectors including pharmaceuticals, finance, transport, energy, public services, cybersecurity, and automotive. In the last few years Artificial Intelligence is said to have self-evolved, software has been created by researchers that borrow concepts from Darwinian evolution, and its usage is resulting in better efficiency, lesser time to complete tasks, and higher accuracy in workloads. The latest 2020 global projects by Artificial Intelligence include it in the health sector to help diagnose and treat COVID-19, to spot brain tumors, etc.; in Technology and Cyber Security to spot critical Microsoft security bug, to distinguish between bots and humans, quantum information processing, etc. AI in automation to reduce traffic congestion and fuel consumption is an example of AI not only helping in making our life easier and fast but also helping the environment. The upcoming stage of Artificial Intelligence is the era of Augmented Intelligence, which is going to link humans and machines seamlessly together.

Keywords Artificial Intelligence · Technological reforms · Global projects

---

M. Yadav (✉) · S. Shahi · H. Ahuja · M. Batra
Department of Computer Applications, Manav Rachna International Institute of Research and Studies, Haryana, India
e-mail: medhavi_yadav2019@manavrachna.net

S. Shahi
e-mail: siddharth_shahi2019@manavrachna.net

H. Ahuja
e-mail: himanshu_ahuja2019@manavrachna.net

M. Batra
e-mail: mridula.fca@mriu.edu.in

S. Kautish et al. (eds.), *Computational Intelligence for Modern Business Systems*,
Disruptive Technologies and Digital Transformations for Society 5.0,
https://doi.org/10.1007/978-981-99-5354-7_20

1 Introduction

"Artificial Intelligence" is a vast field with an obvious definition that induces machines to perform the tasks that are usually done with human intelligence. AI is defined as a set of skills, which does not require any specific technical means to achieve those skills. This field has worked on the idea that human intelligence can be described so accurately that a machine can be made to mimic the same.

AI has further two main distinctions as shown in Fig. 1.

- **Narrow AI:** It is also called "Weak AI," this type of artificial intelligence works for a limited range and is a way of imitating some human intelligence. Weak AI tends to focus on doing a very good job and while these machines may seem smart and intelligent, they work with more limitations and limitations than most basic human intelligence.

  Machine learning takes the data of a computer and helps it "learn" how to get better at tasks progressively by using statistical techniques, removing the requirement for a huge amount of specially written code as it is done without being specifically programmed for the task. Machine learning has the depths of both supervised learning as well as unsupervised Learning.

  Deep learning is a form of machine learning, which works based on the construction of a human-inspired neural network. These networks have many deep and hidden layers where all the data is processed to promote the machine's share of depth in its reading, linking, and input focus for excellent results.

- **Artificial General Intelligence (AGI):** It is also known as "Strong AI". AGI is a machine with intelligence that is generalized and applies that intelligence to find a solution for any problem just like a human being [1].

In the twenty-first century, AI types and techniques have experienced an improvement following exponential improvement in computer processing power, theoretical understanding, and massive amounts of information; and AI techniques became a necessary part of the technology business, serving to unravel several difficult issues in engineering, package engineering and research and additionally in further life aspects.

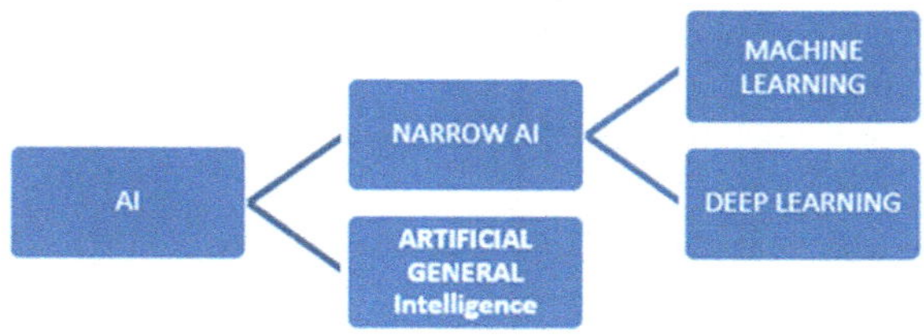

Fig. 1 Various distinctions in AI

1.1 "Artificial Intelligence is Self-evolving"

Artificial Intelligence (AI) is making progress and improving. Researchers have built package that follows the ideas from "Darwinian evolution.", as well as "only the fittest will survival" to make AI programs that adapt and improve without human involvement.

A program called machine ML-Zero has been developed which will help develop AI programs with effectively no human input. It operates by creating a population of one hundred candidate algorithms via willy-nilly combination of mathematical operations. Next, it performs tests on them by giving them an easy task, like an object or image recognition task where it has to identify whether the given picture shows a car or an apple. With every cycle, this program compares the algorithms' performance with other previously self-designed algorithms. Copies of the best-performing algo's are mutated/evolved by willy-nilly replacement, removal or edits on a code to create new variations of the best algorithms. These newborns reach out to people, and the old programs are removed. And then this cycle keeps repeating itself [2].

2 Artificial Intelligence Emerging Projects

2.1 AI in the Health Sector

2.1.1 AI to Help Diagnose, Treat COVID-19

In order to effectively diagnose and treat COVID-19, a data science center has been developed which is helping in developing new applications utilizing AI to do the same. Drawing on resources from the three thousand eight hundred bed networks of eight big apple town hospitals and a graduate school, the Mount Sinai COVID Science Center will make use of technologies like Big Data and Machine Learning to help diagnose and treat the novel coronavirus and many of its symptoms.

Another key objective of the center is to help hospital networks to make more efficient use of its bed capacity, intensive-care facilities, and other resources.

Several machine-learning algorithms were developed to assist doctors and others to create assessments such as finding out which COVID-19 patients are possible to need a ventilator and which of them are recovering and are getting ready to be discharged. These algorithms generate predictive scores, which are entered into a patient's EMR file. Displayed as a color-coded entry on their screens, the scores can then be used by doctors, case managers, social workers, and others to flag critical issues and optimize the use of hospital resources.

The Mount Sinai center is also readying a new diagnostic tool for COVID-19. Based on a very simple machine-learning algorithm, the diagnostic system compares computed tomography (CT) scans of the chest in tandem with patient characteristics including gender, age, symptoms, blood type, and possible contact with a known carrier of the virus. The algorithm is currently being trained against the actual clinical findings of Mount Sinai's doctors.

The Mount Sinai physicians are able to correctly diagnose the disease 75% of the time, while recent tests in which the machine-learning algorithm analyzed the same lung scans and clinical data for more than 900 patients correctly diagnosed the coronavirus in 84% of the cases [3].

2.1.2 AI to Spot Brain Tumors

Artificial Intelligence models are to be trained that can identify brain tumors. The group plans on training robust models using the largest brain tumor dataset to date. By employing the privacy-preserving technique of federated learning, the organizations will be able to contribute to that dataset without actually sharing their patient data.

It's a learning paradigm based on decentralized data. Rather than relying on data pooled together in a single location, an algorithmic model is trained in multiple iterations at different sites. In the healthcare sector, this offers a degree of privacy for hospitals and other organizations that want to pool their resources to train a deep learning model without actually sharing their data or letting it out of their possession.

The initiative will be sponsored for $1.2 million, within the NIH (National Institutes of Health) [4].

2.1.3 AI-Powered Apps to Diagnose Mental Health Issues

A few new speech-based app has been developed by researchers also including those from the University of Colorado at Boulder, in the United States, that utilizes artificial intelligence to diagnose the mental health status of its users.

The current diagnosis of mental problems by professionals relies on listening to the patient's talk and conversation. Shifts in tone or pace of speech may hint at mania or depression.

The mobile app asks patients a series of questions that can last from 5 to 10 min. The users are asked how they are doing mentally and emotionally, or to tell a short story or to listen to one and repeat it. It assesses the speech samples submitted, compares them to the previous samples by the same user and the users of the overall broader population, and based upon those results, rates the user's mental state.

Upon comparing these Artificial Intelligence backed app-based results to the ones by clinicians as given in Table 1, it was found that these AI models can be at least as accurate as the clinicians [5].

Researchers at IBM square measure victimization transcripts and audio inputs from psychiatric interviews, including machine learning techniques to seek out patterns in speech. This methodology can eventually facilitate clinicians to accurately predict and monitor schizophrenia, psychosis, mania, and depression. To lessen such problems and support the psychological health of patients, numerous organizations

Table 1 Top AI-based mental health apps

Top AI-based mental health apps (2020)	Main features
BioBase	Trace, measure, and manage stress
Woebotv	Master skills to scale back stress and live happier
Youper	Emotional health assistant, science-based approaches
Wysa	Meditation and mindfulness audios
Ginger emotional support	Confidential, emotional support and guided self-care
Depression CBT self-help guide	Understanding depression, interaction in self-care behaviors
Replika	Chatbot companion powered by AI
Unmind	Neuroscience, positive psychology, mindfulness
Tess	Emotional wellness coping strategies

are developing AI-based applications for smartphones so as to enhance user welfare [6].

2.1.4 Machine Learning to Analyze and Identify Haematological Malignancies

Myelodysplastic syndrome (MDS) is a disease that disturbs the maturing and differentiation of blood cells caused by defects in the stem cells in the bone marrow. Global incidence of Myelodysplastic syndrome is 4 cases per 1 lac people a year. To diagnose MDS, a bone marrow test is expected to likewise examine hereditary changes in bone marrow cells. The condition is characterized by gatherings to decide the idea of the issue in more detail.

Image analysis using neural networks and machine learning can help identify the hard to observe by human-eye details in tissue. Research led at the University of Helsinki showed that the procedure makes it conceivable to precisely decide hereditary changes in the disease cells of patients experiencing myelodysplastic condition, a dangerous blood disorder.

In the review led at the University of Helsinki, infinitesimal pictures of MDS patients' bone marrow tests were analyzed using an image analysis technique dependent on AI. The examples were stained with haematoxylin and eosin (H&E staining), a methodology that is used for routine diagnostics for the infection. The slides were digitized and broke down with the assistance of computational profound learning models.

By employing machine learning, the digital image dataset could be analyzed to precisely recognize the most well-known hereditary changes influencing the movement of the disorder, like mutations and chromosomal aberrations. The higher the quantity of variant cells in the examples, the higher the unwavering quality of the outcomes produced by the prognostic models [7].

2.2 *Technology Reforms and Cyber Security*

2.2.1 Artificial Intelligence to Enhance Complex Systems

An algorithm that can determine what data needs to be fed into a fiber optic network in order to get the desired result at the other end has been invented. In any system, you need some kind of input and output, with an action taking place in between. But when that action is particularly complex or requires large amounts of synchronized data, to know what input is needed to get the right output.

Investigators have developed a system of guessing a picture to show their process. In a maze-like network of lasers, light beams move from one magnifying glass to the next and from one series to another, capturing coded information. On the other hand, the information is displayed on a small screen, where a series of green images appear.

The algorithm can make it easy to turn on the light or start an effect or action remotely or create 3D images and holograms. The invention draws to the goal of creative networks in practice. They are the basis of artificial intelligence and allow the systems used to engage in machine learning [8].

2.2.2 New AI Chips to Take the Processing to Next Level

A new generation of chips designed especially for artificial intelligence-based workloads might find a series of new application areas for the technology. A number of chips are designed based on a model that imitates functions of a human brain, after which likely answers are obtained through patterns and relations among obtainable information.

Many of the newly designed AI chips are comprised of interconnected circuits that simultaneously conduct a lot of simple calculations to get answers and the accuracy of the deduced answers improves as the AI model learns and adapts. New chips have neural networks that calculate probabilities so that robots and drones achieve a better contextual understanding of their environment and make smarter decisions [9].

2.2.3 AI—to Differentiate Between Humans and Bots Based on Twitter Activity

Artificial intelligence (AI) is helping to identify differences between human user and fake accounts on Twitter.

The manually verified data set consisted of eight point four million tweets from 3500 human accounts and three-point four million from five thousand bots. The researchers also noticed that human users replied 4 to 5 times extra than the bots. The real users became more interactive with time.

The length of the tweets via human users together shrunk as the experiment unfolded "The amount of knowledge that is exchanged diminishes," says Ferrara. According to him, this change is because of the psychological depletion over time, within which folks reduce the effort and are less likely to expand mental effort.

On the other hand, Bots show no change in their interactivity or the amount of knowledge they tweeted over time.

The group of researchers then combined all the results to coach an existing bot detection algorithmic program, known as Botometer, with the help of the new results the Botometer was significantly more accurate in noticing and detecting bot accounts on Twitter compared to before when it didn't take into consideration of the temporal arrangement of the posts. [10]

2.2.4 Researcher Incorporate New Computer Vision, Uncertainty into AI for Robotic Prosthetics

Researchers have designed a new computer program that will be collaborated with existing hardware to allow people with prosthetics or exoskeletons to help duplicate movements like walking in a much safer and a lot more natural way on different types of pieces of land so they can easily overcome any obstacle. The new framework incorporates pc vision into prosthetic leg management with the help of strong AI-based algorithms that will modify the computer code to take more account of uncertainty and provide more movement to the user [11].

2.2.5 AI to Auto-Update Websites When Facts Change

One major truth about the web is that it has heaps of obsolete data. Many organizations pour millions of dollars into content moderation and curbing fake news. But there are tons of old information that change over time. Simply contemplate the numerous news stories written in the early long stretches of the COVID-19 pandemic, before we find out about how the infection was communicated. That data is still out there, and the most we can do to limit its effect is to cover it in indexed lists or proposition admonitions that the substance is old.

Scientists from MIT depict apparatuses to handle these issues. They intend to lessen the measure of off-base or obsolete data on the web and furthermore make profound learning models that powerfully acclimate to ongoing changes.

Author Tal Schuster, a PhD student in MIT's Computer Science and Artificial Intelligence Laboratory (CSAIL) claims the model to be monitoring article updates, identify meaningful changes, and suggest edits to other similar articles. The sensitive automatic fact verification models will Significantly verify such edits and update the predictions accordingly. Empowering likewise people to have an adaptable outlook and update their convictions within the sight of new proof was past the degree here. However, helping the altering system of old articles can as of now basically decrease the measure of old data on the web [12].

2.2.6 AI-Generated "Smart Clothes" to Measure Your Movements

Lately, there have been huge technological advancements in wearable devices as now, smartwatches have been developed that can monitor our blood oxygen levels and breathing rate. But there are more possibilities to smart wearables like tracking physical activities and sports which can monitor and report how to even improve your technique.

A dress has been developed by MIT recently that utilizes some rather uncommon filaments in order to monitor an individual's development through touch. Apart from other things, the analysts observed that these "smart garments" can also detect if the person wearing them is walking, running, sleeping or carrying out some other activities.

The scientists from the research group at MIT'S Computer Science and Artificial Intelligence Lab (CSAIL) have observed that their garments can also be used for athletic preparation and restoration. If they have the consent of the patient, these garments can even be used to monitor the well-being of the concerned patients and detect emergency cases like one in which they have fallen or lost consciousness.

Multiple garments (models) have been developed ranging from gloves, socks to even full vests. The "material hardware" used in these "smart" clothes makes use of a blend of materials—common material strands used for making clothes to some uniquely designed practical filaments which possess the capability to sense movements and any sort of interaction between the individual and the cloth [13].

2.2.7 Artificial Intelligence in the Metaverse

The potent AI, parsing huge volumes of data at lightning speed to generate insights and drive action when applied to augmented and virtual reality unlocks a whole new degree of verisimilitude, building a smarter immersive new universe, called Metaverse.

The metaverse is an extensive virtual space that uses augmented and virtual reality in combination with artificial intelligence and blockchain to create scalable and accurate virtual worlds. Users can interact with 3-dimensional digital objects and 3-dimensional virtual avatars of people in a complex manner that mimics the real world.

Facebook, now known as Meta, is well recognized for its work in artificial intelligence and sophisticated AI algorithms. The company's AI research spans diverse areas like self-supervised speech processing, content analysis, computer vision, robotic interactions, whole-body pose estimation, etc.

Digital humans and accurate avatar, built entirely using AI tech, are 3D versions of chatbots that exist in the metaverse and are essential to the landscape of the metaverse. The use of Multilingual accessibility of AI for breaking down natural languages, converting it into a machine-readable format, performing analysis, arriving at a response, converting the results back into a readable format, and sending it to the user in just a fraction of a second.

AI assisting human–computer interactions, called Intuitive interfacing and other factors proves that ultimately, without AI, it will be difficult to create an engaging, authentic, and scalable metaverse experience [14].

One of the best examples of this technology set is NVIDIA's Omniverse. The collection of components that this platform provides to create digital worlds and mimic the real world is amazing. From massive landscapes and simulations to virtual spaces to test autonomous robots / AIs to AI voice technology, NVIDIA technology is a great example of how AI will be key to creating digital spaces where social interaction will take place in Metaverse [15].

2.3 Automotive AI

This graph of Fig. 2 represents the deployments of Artificial Intelligence in the automation industry in various countries across the world. The US is leading the world in AI deployment.

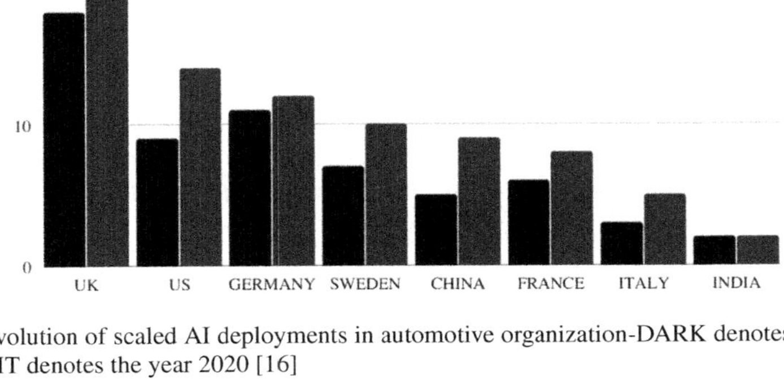

Fig. 2 Evolution of scaled AI deployments in automotive organization-DARK denotes mid-2017 and LIGHT denotes the year 2020 [16]

2.3.1 AI Enables Autonomous Cars to Predict Movement of Traffic with Impressive Accuracy

The GAN of the coauthor creates trajectories that follow the existing scenes in the squares, given access to high-definition scene maps and detection and tracking systems and a tracking system informed by LiDAR, radar, and camera sensors in the vehicle. The GAN emits a self-closing motor indicator, and the base is centered and the x- and axes are defined by the car headings and left sides, respectively.

For individual vehicles with future GAN predictions, scene context data and map issues are included with an RGB image that can be interpreted by a matrix object. The images capture ten meters behind cars and thirty meters in any part of them, similarly ten meters behind.

In testing, the AI system and a few other foundations were forced into the Google Tensor Flow machine learning framework and acquired a wide range of information, a real-world set of information (ATG4D) containing 240 h of driving experience in a variety of traffic situations. Every car in 0.1 s created a single point of data and 0.4 s ago the excavations obtained, acceleration, headers, and total conversion rates of 7.8 million data points.

Motion prediction is one of the most important parts of self-driving technology, modeling future performance and uncertainty of the characters being followed in self-driving environments. An in-depth analysis of quality and measurement has shown that the strategy succeeds in the current GAN-based moving movement prediction of surrounding players, creating many relevant and accurate clues [17].

2.3.2 AI to Reduce Traffic Congestion and Improve Fuel Average

Researchers at the Oak Ridge National Laboratory (ORNL) have created a computerized vision system that helps keep cars moving faster at intersections and has the added benefit of reducing fuel consumption by using artificial intelligence and machine learning.

The team used Gridsmart's 3 smart stoplight cameras, and to collect real-world information on car photos as they walked through lab intersections, they hired a car management services company. They have taught cameras how to identify vehicle types and their calculated fuel mileage using Artificial Intelligence and Machine Learning and send that information to a traffic light. Sensors that are powerful and easy-to-install are hung at a crossroads and they help in telling if a car stays on track or not [18].

2.4 AI in Scientific Searches

The capacity of researchers in various fields of data production and storage has increased to the point where scientists face a flood of data. This made it difficult for the researchers to analyze huge amounts of data for calculating patterns and for insights. With AI learning strategies, it can play an important role in reducing investigator work by circulating data analysis. According to various scientists, "Once the system is trained with sufficient examples, it demonstrates the brilliant power of predicting the outcome of a straightforward model. This interprets some AI potency that might convert hours or days of work into seconds.

2.4.1 Artificial Intelligence to Predict Which Planetary Systems Will Survive

Instead of simulating a given configuration for a billion orbits—the normal brute-force approach, which might take around ten hours to calculate—Tamayo's AI model did the same simulation for 10,000 orbits, which took only a fraction of a second. From this brief summary, the astrophysicist calculated ten abbreviated metrics that took the power of system resources and then trained the machine learning algorithm to predict from these ten factors which systems would remain stable.

The AI model known as Benjamin Spock—Stability of Planetary Orbital Configurations Klassifier—determines whether or not the systems can "live long and prosper." Benjamin Spock determines the semipermanent stability of planetary configurations around 100,000 times quicker than the previous approach, breaking the machine bottleneck. Benjamin Spock will also determine quick instabilities in compact systems, making an attempt to try and do stability-strained characterization [19].

2.4.2 AI Can Help Historians in Solving Ancient Puzzles

Digital humanities are battling for funding against a lot of future-focused applications of AI. The AI model learned the way to reverse the erosion method and predict what the original fragments seemed like. Researchers then explained how the model ought to take a look at whether or not fragments fit along [20].

2.5 *AI in Marketing*

The graph of Fig. 3 is a projection of how much revenue has been generated so far and will be generated in the future by making use of Artificial Intelligence.

2.5.1 The Value of AI in Marketing Field

Artificial Intelligence (AI) is inflicting multiple new ideas in marketing and rising as a game changer altogether in many areas of the marketing field.

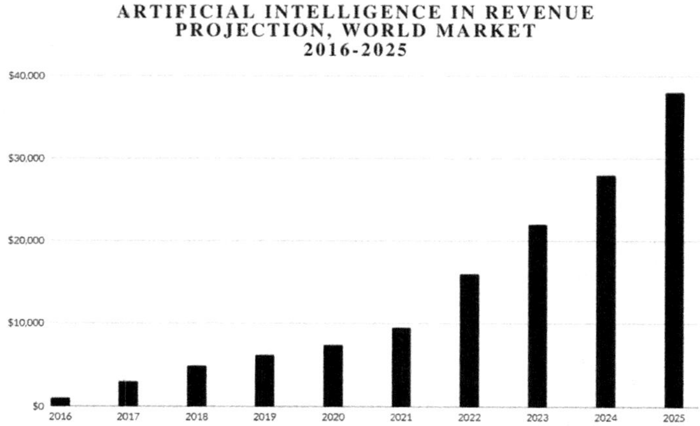

Fig. 3 Artificial intelligence in revenue projection from 2016 to 2025 [21]

Artificial Intelligence (AI) is acting as a major help as well as competition for nearly all industries. AI can produce a large impact on the marketing ways among multiple organizations and provide new methods to take decisions based on analytics and big data.

Within the field of marketing AI has additionally shown its utility and its forthcoming advantages in the future The American Marketing Association defines marketing as "a service, institutional set, and process of building, communication, delivery, and exchange of some donations with the values of customers, partners, and the community at huge". AI has created an excellent change within the sales units and marketing companies, inflicting revolutionary changes within the field. For instance, AI will facilitate phasing the customers by aiding to gather their information, managing and analyzing data in real-time, and providing personalized and personalized service and sales based on customer segments. Many AI tools already in existence also possess the ability to improvise promotion and advertising processes by offering social media ads, flyers, and custom emails automatically without the involvement of the user. AI is also able to improve corporate governance and market performance through its advanced algorithms and extensive data analysis and complex business and financial data to assist business selection and predict inflation. In addition, AI will contribute to delivery and payment processes by offering advanced router programs that learn traffic items in real-time systems and 24/7 viewing systems to stop fraud [22].

2.5.2 Artificial Intelligence in Shopping

Shopping is an important part of daily life and Artificial Intelligence is actually changing the way that we shop online. The development of AI as an in-depth learning and technology dependence combined with the increase in sales is now being used by e- commerce-based retailers like Amazon to improve the lives of consumers that are online.

Both are software that learns to perform complex tasks without the need for human control. Traditional search engines such as Google have been using algorithms that select keywords to show the results, they think are closest to what users want. While this method works for those people who want newsletters and the like, it has not been the best thing for people who shop online. With the advancement of AI, online search has become much clearer with image search to find reasons with vendors.

Marketing companies use a combination of math, machine learning, in-depth learning, computer visualization, and face-to-face competition in online business. The latest, for example, is the same technology that drives self-driving cars. Amazon is already using most of this technology in its new Amazon Go stores. For now, this is included in one store in Seattle, but an extension will follow.

Currently, four horse riders in the retail business are fighting for our wallets. Alexa by Amazon, Google's Now, Siri by Apple, and Tmall Genie by Alibaba are at the forefront of the battle to become a voice trading platform, supported by AI.

We can already see a dramatic change in spending and expending. On average, consumers using Amazon Echo spend more than the average Amazon Prime customers: US $ 1,700 compared to US $ 1,300. AI makes sales cheaper, faster, and more efficient, covering everything from customer service to product delivery. It also redefines the purchase information, makes it smoother and simpler, and if we allow AI to buy for us—ultimately invisible [23].

2.5.3 Usage of Artificial Intelligence in Flipkart

Flipkart uses artificial intelligence as a tool to interpret multiple petabytes of data which they can use to determine the likes and dislikes of customers. This data is useful for Flipkart to improve the online shopping experience of the customers and helps them in deciding which product to purchase and which endeavors to pursue.

Flipkart has been making use of AI for carrying out multiple types of tasks like image speech, discovery, intent modeling, logistics, consumer support, text processing, address understanding, and contriving separate items or private labels through AI and ML models.

NLP (Natural Language Processing) is used in Chatbots, which gains knowledge through every interaction between customers and also helps the platform in avoiding any potential negative reviews [24].

The strategy is helping the e-commerce giant get a ringside view of these innovations and even access them. These range from startups that understand the emotions of shoppers using artificial intelligence, companies that provide mobile robots that sort packages and improve personnel productivity, to firms that capture data using on-farm sensors [25].

3 Conclusion

Artificial intelligence architecture is basically a collection of algorithms developed to imitate human intelligence in ways. Machine learning is one of them, and in-depth learning is one of those ways. The intended use of AI can be simple or complex. Objectives can be clearly defined or done. It is a system of comprehension and interpretation of information to be communicated from external information and applying what is being learned to achieve specific goals and activities in a flexible manner.

Artificial intelligence has played a great role in human life by not only making tasks easier but also more effective and unusually fast. The advancements and reforms in the current tech, from diagnosing diseases to fixing bugs in software, from finding the age of fossils to conserving fossil fuels, all these would have not been possible without AI.

AI is a rising tech field and research in it is growing which would lead to success in making the Healthcare, Automotive, Finance and economics, Cybersecurity, Military, Hospitality, Audit, Art, etc. sectors more advanced and tech-driven.

References

1. Artificial-intelligence; B&B Tech
2. Gent E (2020) Artificial intelligence is evolving all by itself. In: Advance replicates decades of AI research in days
3. Kass EM (2020) Mount Sinai center developing ai to help diagnose, treat COVID-19. Retrieved 16 July 2020
4. Stephanie Condon for between the lines "Intel, UPenn partner with 29 health organizations to train AI to spot brain tumors". Retrieved 11 May 2020
5. University of Massachusetts Amherst (2020) UMass Amherst Computer Scientists Tested 'FluSense' in Campus Clinic Waiting Rooms. Retrieved 19 March 2020
6. Top AI-based mental health apps of 2020. From cmbclinicaltrials. Accessed 8 April 2020
7. University of Helsinki (2021) Image analysis based on machine learning reliably identifies haematological malignancies challenging for the human eye. helsinki.fi. Accessed 22 March 2021
8. Rahmani B, Loterie D, Kakkava E, Borhani N, Psaltis D, Moser C (2020) Actor neural networks for the robust control of partially measured nonlinear systems showcased for image propagation through diffuse media. Retrieved 15 July 2020 from Nature Machine Intelligence
9. Shah A (2020) New Computer Chips Could Power AI to Next Level. Wall Street J Art Retrieved 2 June 2020
10. Pozzana I, Ferrara E (2020) Measuring bot and human behavioral dynamics. Front Phys. Retrieved 22 April 2020
11. Zhong B, da Silva RL, Lobaton E, Li M, Huang H (2020) North Carolina State University, the Joint Department of Biomedical Engineering at North Carolina State University and the University of North Carolina at Chapel Hill. IEEE Trans Autom Sci Eng
12. CSAIL/TransPerfect (2021) MIT Computer Science & Artificial Intelligence Lab. Auto updating websites when facts change. Retrieved 29 March 2021
13. Conner-Simons A (2021) MIT CSAIL; MIT Computer Science & Artificial Intelligence Lab; "Smart clothes" that can measure your movements. Retrieved 24 March 2021

14. Greener R (2021) Artificial Intelligence in the metaverse: bridging the virtual and real. xr today. Retreived 9 Dec 2021
15. Pereira D (2021) How AI will shape the Metaverse. Retrieved 21 Dec 2021
16. Capgemini Research Institute (2019) AI in automotive executive survey. December 2018–January 2019.
17. Wang E, Cui H, Yalamanchi S, Moorthy M, Chou F-C, Djuric N (2020) Uber advanced technologies group "Improving movement predictions of traffic actors in bird's-eye view models using GANs and differentiable trajectory rasterization". Publisjed on 12 June 2020
18. Kanowitz S (2020) How AI can reduce traffic congestion and fuel consumption. GCN Retrieved 25 March 2020
19. Artificial Intelligence predicts which planetary systems will survive. The Times of India Article. Retrieved 14 July 2020
20. Minsky C (2020) How AI helps historians solve ancient puzzles. Financial Times Article. Retrieved 1 July 2020
21. DATA STORY: Next investment opportunity? Artificial Intelligence market projected to grow 25-fold in eight years from moneycontrol. Retrieved 16 October 2017
22. "What is AI marketing" from Marketing Evolution
23. "How Artificial Intelligence is transforming the E-commerce Industry" Countants. Retrieved 10 May 2019
24. Rangaiah M (2021) How Flipkart uses Artificial Intelligence (AI) ?. Retrieved 14 Jan 2021
25. Abrar P (2021) Robotic packers, AI that reads buyer emotions: Flipkart nurtures startups. Retrieved 19 January 2021

Chapter 21
Choosing the Optimal Route for a Delivery Vehicle in X Express Company Using Clarke and Wright Algorithm

Željko Stević and Mladen Gavranović

Abstract Choosing the optimal route in transport is a daily challenge for all companies in the world that deal with the transport of goods. Routing is the selection of the optimal path in the network of roads on which vehicles should move, i.e., defining the most appropriate and fastest mode of transport from the starting point to the endpoint. X express company in Bosnia and Herzegovina, with 16 distribution centers, currently has over 350 delivery vehicles equipped with GPS devices. Drivers of vehicles are faced daily with the selection of the optimal route so that all shipments are delivered as soon as possible, with minimal transport costs and so that all customer requirements are met. The most commonly used algorithm for solving vehicle routing problems is the Savings Algorithm or the Clark and Wright algorithm. Along with the use of the Clark-Wright algorithm, this chapter presents the routing of a delivery vehicle with predefined locations for the shipments in the delivery area of the distribution center Banja Luka. For all 27 delivery locations, the distance from the central warehouse and possible savings were defined, after which the connection was made and the final optimal route was determined. The obtained results were compared with the actual mileage of the delivery vehicle, where the functionality and significance of the Clark-Wright algorithm were proven.

Keywords Routing · Optimization · Delivery · Transport · Algorithm

Ž. Stević (✉) · M. Gavranović
Faculty of Transport and Traffic Engineering, University of East Sarajevo, Doboj, Bosnia and Herzegovina
e-mail: zeljkostevic88@yahoo.com

S. Kautish et al. (eds.), *Computational Intelligence for Modern Business Systems*, Disruptive Technologies and Digital Transformations for Society 5.0,
https://doi.org/10.1007/978-981-99-5354-7_21

1 Introduction

The problem of vehicle routing is the process of choosing the optimal way in the network of roads on which vehicles should move. Whether it is business logistics or some other transport/transportation, it is desirable to find the most suitable and fastest way from point A to point B. Various routing methods can be applied in different areas, mostly in logistics and transport networks. One route represents the complete route that one vehicle travels from the warehouse to each selected location and back to the warehouse. Routing should allow all user requirements to be met, as well as all restrictions to be met. Each user is served exactly once, and the solution to the problem must be such that the carrying capacity of the vehicle is not exceeded at any time [1]. Transport management is a central problem in a company's logistics, and the selection of optimal routes is one of the key functions in the process. The solution of VRP is a group of optimal routes; those routes start and end in the hub, satisfying demands by serving customers. The process requires route selection, estimating alternate routes, and predicting driving time based on criteria variables, objective functions, and constraints [2]. If the creation of routes for the delivery of goods to user locations is left to the computer, multiple benefits can be achieved–the obtained solution is more efficient, and at the same time less time is spent on planning the transport of goods. The data show that by using computer optimization, companies can achieve savings on transportation up to 5 percent [3]. A large number of authors throughout history were interested in solving such problems, and this has led to a large number of methodologies and ways of solving such problems. An excellent overview of the methodologies applied so far can be found in the works of Nagy and Salhi [4], Drexl and Schneider [5] as well as Prodhon and Prins [6]. Nambiar and others [7] considered the case of locating distribution centers with the aim of minimizing the distance traveled. The authors defined clusters or better to say groups of clusters based on the distance between them and the capacitive capabilities of the vehicles that should serve them. The goal was that all users could be served from the depot. Liu and Lee [8] discussed stochastic customer demand and the cost of investing in distribution centers. The aim of this chapter is to optimize routing in order to achieve greater effects in the company, increase efficiency and quality of service, reduce transport time, and reduce delays in shipment delivery. In this way, we reduce the number of kilometers traveled, which affects fuel consumption and environmental protection. The efficacy of the logistics system depends on vehicle scheduling and routing optimization, which influences profitability and customer satisfaction [9]. In this chapter, the main focus is on the routing of the X express delivery vehicle to pre-designed locations for the delivery of shipments. The total number of locations is 27, and the total mass of all shipments does not exceed the carrying capacity of the vehicle. Routing was performed using the Savings Algorithm method or the Clarke and Wright algorithm.

2 About X Express and Used Algorithm

X express company was founded in 2016 with the aim of setting new standards in the field of express delivery services. In 2016, X express started the business with a few dozen vehicles and 6 distribution centers. In the first five years of its existence, the company X express has expanded its business and through 16 distribution centers, with more than 350 vehicles, it delivers shipments throughout Bosnia and Herzegovina. By using modern technological solutions, the company has improved the security and efficiency of the distribution network, and in favor of that delivery quality reaches 98% with more than ten thousand users. X express offers fast delivery services, and basic services are included the transport of documentation, packages, and pallets. In addition to the delivery of shipments in the territory of Bosnia and Herzegovina, service users have the opportunity to send shipments internationally to Serbia. In addition to basic services, users can choose from a wide range of additional services such as [10]:

- delivery of tender documentation,
- return of documentation,
- delivery of shipments on Saturdays,
- cash on delivery,
- return of the replacement consignment,
- opening of the consignment,
- shipment insurance,
- SMS notification,
- Web client service,
- retention and storage of consignments,
- urgent delivery of shipments.

Within a network of 16 distribution centers, three parallel central warehouses have been formed. The central warehouse in the distribution network represents the hub of night lines where the distribution and routing of shipments to the distribution centers is performed. All distribution centers schedule shipments according to delivery lines within a defined time. When distributing shipments and determining delivery couriers, the problem with routing and delivery methods appears. Further in this chapter, a real example and the optimal solution for one delivery line are presented. The Clarke and Wright algorithm is one of the most popular and commonly used algorithms in solving vehicle routing problems. Back in 1964, Clarke and Wright proposed this algorithm based on the principle of gluttony. The algorithm calculates the savings S_{ij} between users *i* and *j*, assuming that c_{i0} represents the cost of travel from source to user *i*, and whose cost of travel from user *i* to user *j*. The savings algorithm is written as follows [11]:

1. Calculate the savings $S_{ij} = C_{i0} + C_{0j} - C_{ij}$, Za $I,j = 1,\ldots,n$, I $I \neq j$. Rank S_{ij} savings and organize them from largest to smallest;
2. Create a list of savings, starting with the first entry in the list (maximum S_{ij}). For savings under consideration (S_{ij}), it is necessary to add arcs (i, j) of a certain

route, only if adding (*i, j*) will not violate the route restrictions. The following three cases are taken into account:

(a) if *i* and *j* have not yet been assigned to a particular route, a new route is initiated that includes both;
(b) if exactly one of the two points *i* or *j* is part of an already existing route, and that point is not internal to that route, then the arc (*i, j*) is added to that route. If the point is internal and does not violate the capacity, then (*i, j*) is added to that route, if it violates the capacity, it is necessary to create a new route with the point (user);
(c) If *i* and *j* are already included in two different pre-existing routes, and no point is internal to the route, then the two routes are merged by merging *i* and *j*. In case the points are internal, unification cannot occur;

3. If the S_{ij} savings list is not fully utilized, you need to go back to Step 2. Otherwise, it ends.

3 Using the Clark and Wright Algorithm on a Practical Example

The courier at X express has 27 different addresses for delivery in one day. The vehicle has a load capacity of 1400 kg, while the total weight of all shipments is 151 kg. Below will be shown what is the optimal route with the minimum distance traveled in order for each shipment to be delivered.

3.1 Constructing Elementary Routes and Defining Their Lengths

Below are the routes and their lengths:

(0-1-0), F = 8,9 + 8,9 = 17,8

(0-2-0), F = 7,7 + 7,7 = 15,4

(0-3-0), F = 6,6 + 6,6 = 13,2

(0-4-0), F = 5,1 + 5,1 = 10,2

(0-5-0), F = 8,2 + 8,2 = 16,4

(0-6-0), F = 7,4 + 7,4 = 14,8

(0-7-0), F = 7,6 + 7,6 = 15,2

(0-8-0), F = 6,6 + 6,6 = 13,2

(0-9-0), F = 16,4 + 16,4 = 32,8

(0-10-0), F = 16,3 + 16,3 = 32,6

(0-11-0), F = 11,5 + 11,5 = 23

(0-12-0), F = 7,8 + 7,8 = 15,6

(0-13-0), F = 8,5 + 8,5 = 17

(0-14-0), F = 9,5 + 9,5 = 19

(0-15-0), F = 9,5 + 9,5 = 19

(0-16-0), F = 7,7 + 7,7 = 15,4

(0-17-0), F = 10,4 + 10,4 = 20,8

(0-18-0), F = 10,7 + 10,7 = 21,4

(0-19-0), F = 17,8 + 17,8 = 35,6

(0-20-0), F = 17,8 + 17,8 = 35,6

(0-21-0), F = 14,2 + 14,2 = 28,4

(0-22-0), F = 17,5 + 17,5 = 35

(0-23-0), F = 18 + 18 = 36

(0-24-0), F = 8,8 + 8,8 = 17,6

(0-25-0), F = 6,4 + 6,4 = 12,8

(0-26-0), F = 7,8 + 7,8 = 15,6

(0-27-0), F = 10,6 + 10,6 = 21,2

3.2 Possible Savings

After defining the route lengths, we determine the possible savings for all connected nodes:

$S_{12} = C_{10} + C_{02} - C_{12} = 8{,}9 + 7{,}7 - 4{,}6 = 12$

$S_{13} = C_{10} + C_{03} - C_{13} = 8{,}9 + 6{,}6 - 6{,}1 = 9{,}4$

$S_{1\text{-}27} = C_{10} + C_{0\text{-}27} - C_{1\text{-}27} = 8{,}9 + 10{,}6 - 8{,}2 = 11{,}3$

$S_{23} = C_{20} + C_{03} - C_{23} = 7{,}7 + 6{,}6 - 2{,}8 = 11{,}5$

$S_{34} = C_{30} + C_{04} - C_{34} = 6{,}6 + 5{,}1 - 7{,}1 = 4{,}6$

$S_{45} = C_{40} + C_{05} - C_{45} = 5{,}1 + 8{,}2 - 4{,}4 = 8{,}9$

$S_{46} = C_{40} + C_{06} - C_{46} = 5{,}1 + 7{,}4 - 9 = 3{,}5$

$S_{56} = C_{50} + C_{06} - C_{56} = 8{,}2 + 7{,}4 - 2{,}3 = 13{,}3$

$S_{67} = C_{60} + C_{07} - C_{67} = 7{,}4 + 7{,}6 - 1{,}5 = 13{,}5$

$S_{68} = C_{60} + C_{08} - C_{68} = 7{,}4 + 6{,}6 - 5{,}4 = 8{,}6$

$S_{78} = C_{70} + C_{08} - C_{78} = 7{,}6 + 6{,}6 - 1{,}7 = 12{,}5$

$S_{89} = C_{80} + C_{09} - C_{89} = 6{,}6 + 16{,}4 - 6{,}8 = 16{,}2$

$S_{8-11} = C_{80} + C_{0-11} - C_{8-11} = 6{,}6 + 11{,}5 - 12 = 6{,}1$

$S_{9-10} = C_{90} + C_{0-10} - C_{9-10} = 16{,}4 + 16{,}3 - 0{,}7 = 32$

$S_{10-11} = C_{10-0} + C_{0-11} - C_{10-11} = 16{,}3 + 11{,}5 - 3{,}3 = 24{,}5$

$S_{11-12} = C_{11-0} + C_{0-12} - C_{11-12} = 11{,}5 + 7{,}8 - 2{,}7 = 16{,}6$

$S_{11-13} = C_{11-0} + C_{0-13} - C_{11-13} = 11{,}5 + 8{,}5 - 3{,}1 = 16{,}9$

$S_{12-13} = C_{12-0} + C_{0-13} - C_{12-13} = 7{,}8 + 8{,}5 - 1{,}1 = 15{,}2$

$S_{13-14} = C_{13-0} + C_{0-14} - C_{13-14} = 8{,}5 + 9{,}5 - 1{,}7 = 16{,}3$

$S_{13-15} = C_{13-0} + C_{0-15} - C_{13-15} = 8{,}5 + 9{,}5 - 4 = 14$

$S_{13-16} = C_{13-0} + C_{0-16} - C_{13-16} = 8{,}5 + 7{,}7 - 9{,}5 = 6{,}7$

$S_{14-15} = C_{14-0} + C_{0-15} - C_{14-15} = 9{,}5 + 9{,}5 - 0{,}1 = 18{,}9$

$S_{14-16} = C_{14-0} + C_{0-16} - C_{14-16} = 9{,}5 + 7{,}7 - 4{,}2 = 13$

$S_{15-16} = C_{15-0} + C_{0-16} - C_{15-16} = 9{,}5 + 7{,}7 - 2{,}9 = 14{,}3$

$S_{15-17} = C_{15-0} + C_{0-17} - C_{15-17} = 9{,}5 + 10{,}4 - 7{,}4 = 12{,}5$

$S_{16-17} = C_{16-0} + C_{0-17} - C_{16-17} = 7{,}7 + 10{,}4 - 3{,}5 = 14{,}6$

$S_{17-18} = C_{17-0} + C_{0-18} - C_{17-18} = 10{,}4 + 10{,}7 - 0{,}2 = 20{,}9$

$S_{18-19} = C_{18-0} + C_{0-19} - C_{18-19} = 10{,}7 + 17{,}8 - 2{,}3 = 26{,}2$

$S_{18-20} = C_{18-0} + C_{0-20} - C_{18-20} = 10{,}7 + 17{,}8 - 4{,}5 = 24$

$S_{18-21} = C_{18-0} + C_{0-21} - C_{18-21} = 10{,}7 + 14{,}2 - 6{,}4 = 18{,}5$

$S_{19-20} = C_{19-0} + C_{0-20} - C_{19-20} = 17{,}8 + 17{,}8 - 0{,}1 = 35{,}5$

$S_{19-21} = C_{19-0} + C_{0-21} - C_{19-21} = 17{,}8 + 14{,}2 - 3{,}1 = 28{,}9$

$S_{20-21} = C_{20-0} + C_{0-21} - C_{20-21} = 17{,}8 + 14{,}2 - 1{,}9 = 30{,}1$

$S_{21-22} = C_{21-0} + C_{0-22} - C_{21-22} = 14{,}2 + 17{,}5 - 1{,}4 = 30{,}3$

$S_{22-23} = C_{22-0} + C_{0-23} - C_{22-23} = 17{,}5 + 18 - 1 = 34{,}5$

$S_{22-24} = C_{22-0} + C_{0-24} - C_{22-24} = 17{,}5 + 8{,}8 - 5{,}8 = 20{,}5$

$S_{23-24} = C_{23-0} + C_{0-24} - C_{23-24} = 18 + 8{,}8 - 5{,}3 = 21{,}5$

$S_{23-25} = C_{23-0} + C_{0-25} - C_{23-25} = 18 + 6{,}4 - 10{,}3 = 14{,}1$

$S_{24-25} = C_{24-0} + C_{0-25} - C_{24-25} = 8{,}8 + 6{,}4 - 4{,}7 = 10{,}5$

$S_{24-26} = C_{24-0} + C_{0-26} - C_{24-26} = 8{,}8 + 7{,}8 - 7{,}9 = 8{,}7$

$S_{25-26} = C_{25-0} + C_{0-26} - C_{25-26} = 6{,}4 + 7{,}8 - 2{,}8 = 11{,}4$

$S_{26-27} = C_{26-0} + C_{0-27} - C_{26-27} = 7{,}8 + 10{,}6 - 3{,}4 = 15$

$S_{26-1} = C_{26-0} + C_{01} - C_{26-1} = 7{,}8 + 8{,}9 - 14{,}5 = 2{,}2$

$S_{27-1} = C_{27-0} + C_{01} - C_{27-1} = 10{,}6 + 8{,}9 - 19{,}2 = 0{,}3$

3.3 Merging Routes and the Final Solution

According to the non-growing array of each savings, it is necessary to merge the routes. It should be noted that the total mass of all shipments does not exceed the total load capacity of the vehicle:

$S_{19-20} = 35{,}5$; node 19 is first in route 0-19-0, node 20 is first in route 0-20-0 and then route 0-19-20 is formed;

$S_{22-23} = 34{,}5$; node 20 is first in route 0-22-0, node 23 is first in route 0-23-0 and then route 0-22-23-0 is formed;

$S_{9-10} = 32$; node 9 is first in route 0-9-0, node 10 is first in route 0-10-0 and then route 0-9-10-0 is formed;

$S_{21-22} = 30{,}3$; node 21 is first in route 0-21-0, node 22 is first in route 0-22-23-0 and then route 0-21-22-23-0 is formed;

$S_{20-21} = 30{,}1$; node 20 is first in route 0-20-0, node 21 is also first in route 0-21-22-23-0 and then route 0-20-21-22-23-0 is formed;

S19-21 = 28.9; node 19 is first in route 0-19-0, node 21 is not the first, nor the last in its route so that this route cannot be formed;

$S_{18-19} = 26{,}2$; node 18 is first in route 0-18-0, node 19 is first in route 0-19-20-21-22-23-0 and then route 0-18-19-20-21-22-23-0 is formed;

$S_{10-11} = 24{,}5$; node 10 is last in route 0-9-10-0, node 11 is first in route 0-11-0 and then route 0-9-10-11-0 is formed;

$S_{18-20} = 24$; node 18 is first in route 0-18-19-20-21-22-23-0, node 20 is not the first, nor the last in its route so that this route cannot be formed;

$S_{23-24} = 21{,}5$; node 23 is first in route 0-18-19-20-21-22-23-0, node 24 is first in route 0-24-0 and then route 0-18-19-20-21-22-23-24-0 is formed;

$S_{17-18} = 20{,}9$; node 17 is first in route 0-17-0, node 18 is first in route 0-18-19-20-21-22-23-24-0 and then route 0-17-18-19-20-21-22-23-24-0 is formed;

$S_{22-24} = 20{,}5$; node 22 is neither the first nor the last in route 0-17-18-19-20-21-22-23-24-0 and this route cannot be formed;

$S_{14-15} = 18{,}9$; node 14 is first in route 0-14-0, node 15 is first in route 0-15-0 and then route 0-14-15-0 is formed;

$S_{18-21} = 18{,}5$; node 18 is neither the first nor the last in route 0-17-18-19-20-21-22-23-24-0 and this route cannot be formed;

$S_{11-13} = 16{,}9$; node 11 is last in route 0-9-10-11-0, node 13 is first in route 0-13-0, and then route 0-9-10-11-13-0 is formed;

$S_{11-12} = 16{,}6$; node 11 is neither the first nor the last in route 0-9-10-11-13-0, so a new route cannot be formed;

$S_{13-14} = 16{,}3$; node 13 is last in route 0-9-10-11-13-0, node 14 is first in route 0-14-15-0 and then route 0-9-10-11-13-14-15-0 is formed;

$S_{89} = 16{,}2$; node 8 is first in route 0-8-0, node 9 is first in route 0-9-10-11-13-14-15-0 and then route 0-8-9-10-11-13-14-15-0 is formed;

$S_{12-13} = 15{,}2$; node 12 is the first in route 0-12-0, and node 13 is neither the first nor the last in route 0-9-10-11-13-14-15-0, so this route cannot be formed;

$S_{26-27} = 15$; node 26 is first in route 0-26-0, node 27 is first in route 0-27-0 and then route 0-26-27-0 is formed;

$S_{16-17} = 14{,}6$; node 16 is first in route 0-16-0, node 17 is first in route 0-17-18-19-20-21-22-23-24-0 and then route 0-16-17-18-19-20-21-22-23-24-0 is formed;

$S_{15-16} = 14{,}3$; node 15 is last in route 0-8-9-10-11-13-14-15-0, node 16 is first in route 0-16-17-18-19-20-21-22-23-24-0 and then route 0-8-9-10-11-13-14-15-16-17-18-19-20-21-22-23-24-0 is formed;

$S_{23-25} = 14{,}1$; node 23 is neither the first nor the last so this route cannot be formed;

$S_{13-15} = 14$; node 13 is neither the first nor the last so this route cannot be formed;

$S_{67} = 13{,}5$; node 6 is first in route 0-6-0, node 7 is first in route 0-7-0 and then route 0-6-7-0 is formed;

$S_{56} = 13{,}3$; node 5 is first in route 0-5-0, node 6 is first in route 0-6-7-0 and then route 0-5-6-7-0 is formed;

$S_{14-16} = 13$; node 14 is neither the first nor the last so this route cannot be formed;

$S_{15-17} = 12{,}5$; node 15 is neither the first nor the last so this route cannot be formed;

$S_{78} = 12{,}5$; node 7 is last in route 0-5-6-7-0, node 8 is first in route 0-8-9-10-11-13-14-15-16-17-18-19-20-21-22-23-24-0 and then route 0-5-6-7-8-9-10-11-13-14-15-16-17-18-19-20-21-22-23-24-0 is formed;

S_{12} = 12; node 1 is first in route 0-1-0, node 2 is first in route 0-2-0 and then route 0-1-2–0 is formed;

S_{23} = 11,5; node 2 is last in route 0-1-2-0, node 3 is first in route 0-3-0 and then route 0-1-2-3-0 is formed;

$S_{25\text{–}26}$ = 11,4; node 25 is first in route 0-25-0, node 26 is first in route 0-26-27-0 and then route 0-25-26-27-0 is formed;

$S_{1\text{–}27}$ = 11,3; node 1 is first in route 0-1-2-3-0, node 27 is last in route 0-25-26-27-0 and then route 0-25-26-27-1-2-3-0 is formed;

$S_{24\text{–}25}$ = 10,5; node 24 is last in route 0-5-6-7-8-9-10-11-13-14-15-16-17-18-19-20-21-22-23-24-0, node 25 is first in route 0-25-26-27-1-2-3-0 and then route 0-5-6-7-8-9-10-11-13-14-15-16-17-18-19-20-21-22-23-24-25-26-27-1-2-3-0 is formed;

S_{13} = 9,4; node 1 is neither the first nor the last so this route cannot be formed;

S_{45} = 8,9; node 4 is first in route 0-4-0, node 5 is first in route 0-5-6-7-8-9-10-11-13-14-15-16-17-18-19-20-21-22-23-24-25-26-27-1-2-3-0 and then route 0-4-5-6-7-8-9-10-11-13-14-15-16-17-18-19-20-21-22-23-24-25-26-27-1-2-3-0 is formed;

$S_{24\text{–}26}$ = 8,7; nodes 24 and 26 are neither the first nor the last so that this route cannot be formed;

S_{68} = 8,6; nodes 6 and 8 are neither the first nor the last so that this route cannot be formed;

$S_{13\text{–}16}$ = 6,7; nodes 13 and 16 are neither the first nor the last so that this route cannot be formed;

$S_{8\text{–}11}$ = 6,1; nodes 8 and 11 are neither the first nor the last so that this route cannot be formed;

S_{34} = 4,6; node 3 is last in route, node 4 is first in same route then this route cannot be formed;

S_{46} = 3,5; node 6 is neither the first nor the last so this route cannot be formed;

$S_{26\text{–}1}$ = 2,2; nodes 26 and 1 are neither the first nor the last so that this route cannot be formed;

$S_{27\text{–}1}$ = 0,3; nodes 27 and 1 are neither the first nor the last so that this route cannot be formed;

In addition to the final solution (Fig. 1), it's necessary to calculate the length of the optimal route:

f* = 570,6–472,6 = 98 km.

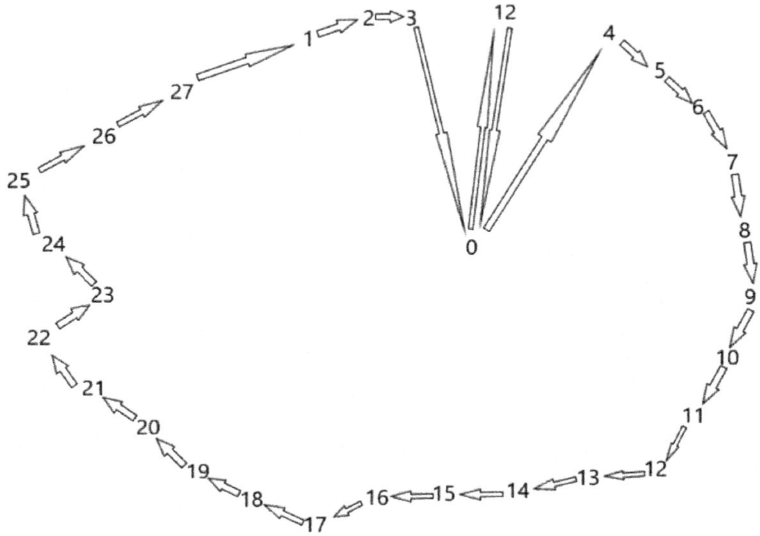

Fig. 1 The final solution of the optimal route

4 Conclusion and Discussion

According to the obtained results, a significant difference can be seen between the actual mileage and the optimal solution. Figure 2 shows the trajectory of the vehicle taken by the courier to deliver all 27 shipments.

According to the GPS report [12], the actual mileage is 166.58 km, and 19.99 l of fuel is consumed. The difference between the optimal route and the actual mileage is 68.58 km. Fuel consumption for the optimal route would be 13.33 l. The savings of 68.58 km and 6.66 l of fuel represent a significant difference if we take into account that the fleet of the company X express has over 350 vehicles (Fig. 3).

The savings algorithm is a simple method for obtaining the optimal route, but the fact is that nowadays there are advanced routing systems. In addition to assigning loads to each means of transport, calculating the optimal route, sequence of site visits, minimum empty runs, advanced software systems take into account the real-time traffic situation, thus avoiding congestion, calculating less stops on the road, etc. All of the above points to significant savings that can be achieved by properly routing vehicles.

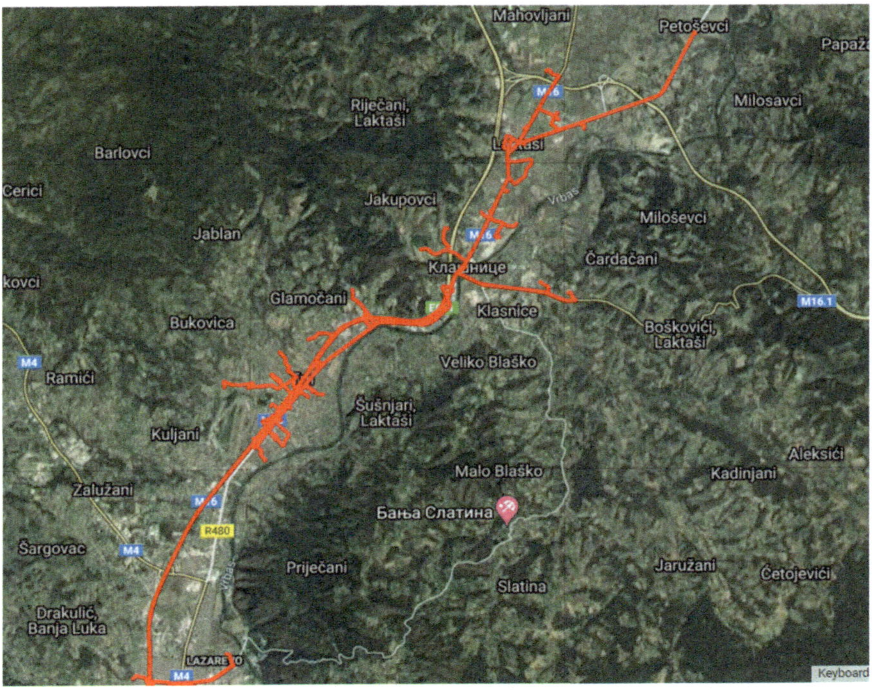

Fig. 2 The final solution of the optimal route shown on the map

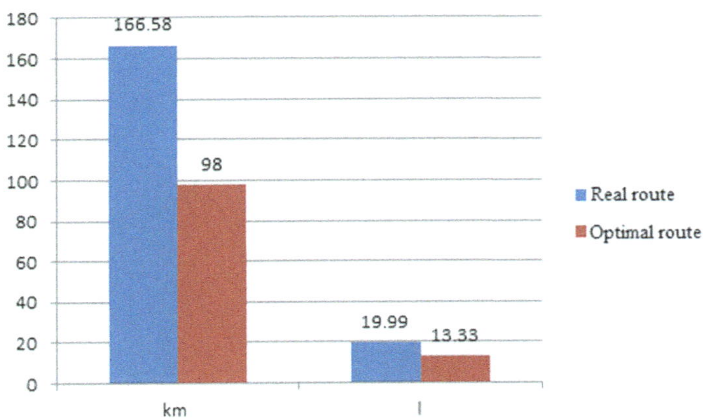

Fig. 3 Comparison of real and optimal routes

References

1. Chen W (2009) N, Application of queuing theory to dynamic vehicle routing problem. Glob J Bus Res 3(2):85–91
2. Abosuliman SS, Almagrabi AO (2021) Routing and scheduling of intelligent autonomous vehicles in industrial logistics systems. Soft Comput 25:11975–11988
3. Hasle G, Lie KA, Quak E (2007) Geometric modelling, numerical simulation, and optimization. Springer, Berlin
4. Nagy G, Salhi S (2007) Location-routing: Issues, models and methods. Eur J Oper Res 177(2):649–672
5. Drexl M, Schneider M (2013) A survey of location-routing problems. Technical Report LM-2013–03
6. Prodhon C, Prins C (2014) A survey of recent research on location-routing problems. Eur J Oper Res 238(1):1–17
7. Nambiar JM, Gelders LF, Van Wassenhove LN (1981) A large scale location-allocation
8. Liu S, Lee S (2003) A two-phase heuristic method for the multi-depot location routing problem taking inventory control decisions into consideration. Int J Adv
9. Rahman MA, Hossain A-A, Debnath B, Zefat ZM, Morshed MS, Adnan ZH (2021) Intelligent vehicle scheduling and routing for a Chain of retail stores: a case study of Dhaka. Bangladesh. Logistics 5:63
10. https://www.x-express.ba/usluge/dodatne-usluge
11. Caccetta L, Alameen M, Abdul-Niby M (2013) An improved clarke and wright algorithm to solve the capacitated vehicle routing problem. ETASR—Eng Technol Appl Sci Res 2(2):413–415
12. https://www.global-gps.ba/

Chapter 22
Diet and Food Restaurant in the Covid-19 Time by Machine Learning Approaches

Md. Babul Islam, Swarna Hasibunnahar, Piyush Kumar Shukla, Prashant Kumar Shukla, and Paresh Rawat

Abstract Covid-19 is a curse to the people of this century and there is no such thing as a plague. If someone is infected with Covid-19, we must follow the doctor's instructions as to which foods should be eaten more frequently at that time. We attempt to discuss this in details. The objective of this chapter is to see how a country's food affects its Covid-19 mortality rate. With so many diverse eating cultures throughout the world, it'd be fascinating to examine which food groups can best predict a country's death rate. We used a machine learning model (Linear Regression and Random Forest/Regression Tree) to estimate the proportion of fatalities caused by the coronavirus pandemic, taking into consideration statistical data on the population's eating habits (food types: animal, eggs, fish, beer, etc.). We may deduce which sorts of meals have a greater influence on the ultimate outcome based on the model's predictions. Furthermore, throughout the Covid-19 period, the economic environment has changed considerably. In this chapter, we examined restaurant meals in a variety of ways.

Keywords Covid-19 · Food habit · Restaurant food · Random forest · Regression tree · Linear regression · Machine learning

---

Md. Babul Islam (✉)
School of Information Engineering, Huzhou University, Huzhou, China
e-mail: babulcseian@gmail.com

S. Hasibunnahar
School of Life Science, Huzhou University, Huzhou, China
e-mail: swarnamic@gmail.com

P. K. Shukla
Department of CSE, UIT RGPV, Bhopal, MP, India
e-mail: pphdwss@gmail.com

P. K. Shukla
Engineering School and Technology, Jagran Lakecity University, Bhopal, India
e-mail: prashantshukla2005@gmail.com

P. Rawat
S. N. Technology, Bhopal, MP, India

S. Kautish et al. (eds.), *Computational Intelligence for Modern Business Systems*, Disruptive Technologies and Digital Transformations for Society 5.0,
https://doi.org/10.1007/978-981-99-5354-7_22

1 Introduction

The World Health Organization (WHO) designated the coronavirus (Covid-19) outbreak a worldwide pandemic in 2020 [1]. Covid-19's transmission, mechanism, and therapy are all unknown at this time. Covid-19 is thought to be transmitted by person-to-person contact and respiratory droplets, therefore social distance, hand-washing, and wearing facial covers are essential for limiting the virus's transmission [2].

Many nations have limited public access to Market or Supermarket, stores, and recreation facilities because there is no vaccine to prevent illness and no assured cure if sick. However, the situation has changed. The severity of the limitations varies by country and is determined by the rate of infection. Despite the fact that governments throughout the world have implemented lockdown measures, the number of new instances continues to rise [3]. Although the purpose of such limitations is to prevent interpersonal interaction and therefore transmission, many experts are concerned about the long-term effects that shutdown may have on people's physically and mentally health [4]. The amount of hours allowed for exterior physical exercise has been reduced, as has people's ability to good food. A previous study has shown that heightened stress, which is expected during a worldwide pandemic, may have a significant influence on a person's lifestyle patterns [5]. Anxiety and stress have been linked to an increase in the consumption of alcohol and sweet foods, as well as an energy imbalance due to lower energy expenditure during lockdown [5]. Positive lifestyle patterns, such as more time for cookery and fewer fast-food meals, may have evolved as a result of the epidemic. Many countries have since lifted their embargoes. Despite the growing and considerable interest in the pandemic's influence on numerous health outcomes, no worldwide assessment has examined the link between Covid-19 and eating habits. The purpose of this review is to examine the research on the influence of lockdown on dietary changes in different populations. The goal will be to analyse eating patterns and themes in order to better understand the nutritional and behavioural changes that occurred throughout the initial shutdown.

During the first Covid-19 lockdown phase, it will be determined how such changes may affect health & wellbeing in the near term, and this information will be used to feed research into the pandemic's long-term implications for diet modifications and health outcomes.

In this chapter, machine learning algorithms were used to predict which food is the best for Covid-19 from huge data. Also, we analysed the food restaurants for the customers' eating habits in different ways. We have used more than two data sets (Fig. 1).

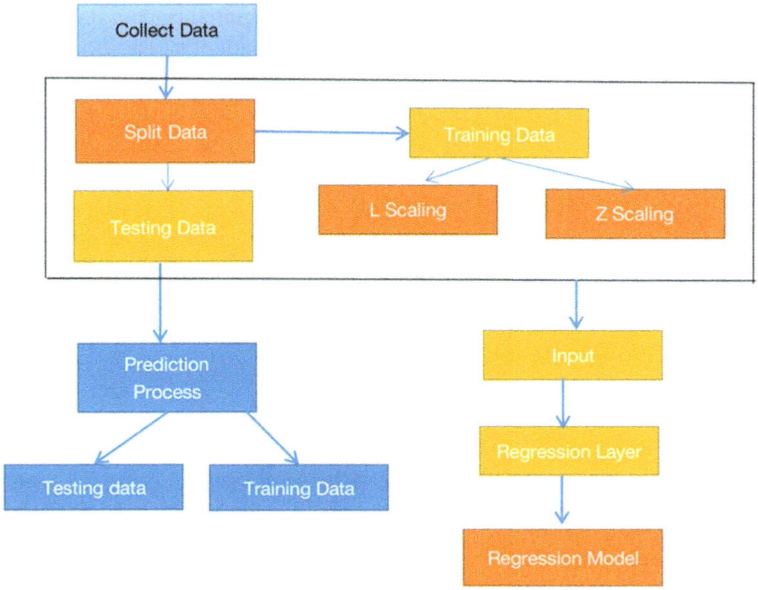

Fig. 1 General diagram of regression model

2 Background Study

The availability of fresh food has been hindered since the COVID-19 epidemic began, and people are spending a plethora of time indoors and limiting their physical activity. Some beneficial habits, such as cooking, may have developed as a result of spending more time at home. Dietary modifications during the first lockdown were the focus of this investigation. Themes and patterns were analysed, as well as correlations with other lifestyle variables.

The Google Scholar (GC), PubMed, and Science Direct (SC) databases were searched between June and July 2020, and the outcome was selected for eligibility based on title, abstract, and full text. Studies that looked at how COVID-19 lockdown affects diet were included in this search, as were publications published in English and published in 2020 (or in pre-print). The following were the exclusion criteria: scholarly, systematic, and narrative studies evaluating prior research.

Numerous factors, such as the patient's behaviour, habits, and nutrition, contribute to the slowing of the Coronavirus pandemic. These initiatives include government-implemented mitigation strategies that emphasize the potential of the political-economic, food, and climatic sectors. All steps are appropriate to encourage egalitarian, healthy eating, and active living, but they must also target other epidemics and reduce the long-term demand for health care [6, 7].

During the COVID-19 pandemic, the [8] suggested a correlation matrix (CM) for food habits and general characteristics of the sample. They show that an increase in vegetable, sweet food, and snack intake has a substantial impact on patients with diabetes, and that physical is not active worsens the patient's condition.

In [9], a research was given that included perceptions of health hazards caused by food in the early aftermath of a dietary crisis. Their research looked at the differences in health conditions depending on risk perceptions and attitudes toward fresh produce and game meat, and it depended on data obtained from 1,008 participants in January 2020. A good diet may fend off any viral attack by supplying a particular amount of vitamin. As a result, individuals' physical and mental health is maintained through good eating habits [10].

3 Methodologies

In terms of data sets, we used kaggle and chose two different data sets. One of the data sets focuses on food-related public assistance programs across the world, such as SNAP and WIC. The Food Security Indicators for the World 2016–2020 is a good place to start if you're looking for a more global picture of food security.

Another one is that the data may be utilized to convey the tale of what cafes were like that in 2020, what was trendy, what would become more prominent, and how large and small businesses differed. I'm interested to see what kind of insight this data may provide! We'll use the Covid-19 health dataset to train and assess the model. Because the number of labelled samples is limited, we'll use a traditional machine-learning model.

In the chapter of Data Exploratory and Analysis, we evaluated the food quality using 8 rows and 29 columns, namely, Alcoholic Beverages, Animal Fats, Animal Products, Eggs, Fish Seafood, and so on.

Import relevant modules, another module, the next hidden code cell imports the packages we'll need to explore, process, and assess the data, as well as construct, execute, and evaluate the model. We will also apply it to data acquisition, exploration, and processing. In the data set, we will apply the Z score as well as Linear regression, Random Forest machine learning algorithms to analyses and predict the result.

The data that has been trained is categorized into one of the specified classes. One of these methods' drawbacks is that they can't handle incomplete data. But thankfully, existing data can be used to fill in the gaps. To analyse COVID-19 trends, a variety of categorization techniques are utilized [11].

The regression approach is a tool to determine characteristic that converts a data item into a meaningful predictive variable. Regression is a statistical technique for establishing a link between the goal and predictor variables. We may also use regression to model the relationship between a dependent and independent variable. Variables such as age, temperature, pay, and so on may be easily predicted when a given quantity of data has been trained. Few of them are listed below.

Linear Regression (LR)—for forecasting, linear regression is a straightforward technique. It depicts the direct and indirect variables' linear connection. If there is just one input, the model is referred to as simple linear regression (SLR), and if

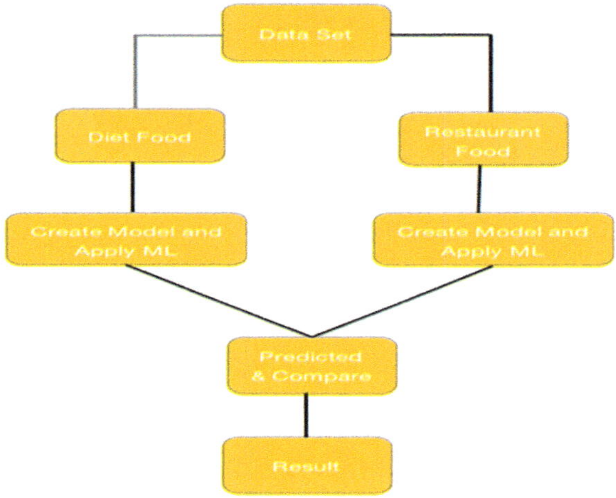

Fig. 2 Work-flow diagram

there are many inputs, the model is referred to as multiple linear regression. Coronavirus illness has resurfaced in new cases (COVID-19), which were predicted using a heterogeneous linear regression model [12].

When the data is extensive, classification using a decision tree might lead to overfitting. Random forests can help us overcome these constraints. To increase accuracy, not similar decision trees are categorized and then combined together. This is referred to as "bagging". The RF algorithm was used to predict COVID-19 health in [13] reference. The spatial–temporal distribution (S-T-D) of COVID-19 daily occurrences throughout the earth was estimated using a random forest machine learning technique [14, 15].

Below have a look at our workflow diagram (Fig. 2).

4 Result and Analysis

4.1 Processing of Data

In this part, we'll look at how to normalize our data so that we can get a normal distribution, as well as how to deal with records that have missing values. Releasing records, we'll discard such records since the label has missing data and because of the context of the problem. Starting with the first data set, **visualizes** data distribution features will assist us in deciding whether or not we need to normalize our data. We'll use pandas to plot the histogram for each feature (Fig. 3).

We may deduce the following by viewing the histograms: animal_products, obesity and Vegetal Products have a distribution that is about typical. We'll presumably just use the z-score technique to scale their values. Cereals–Excluding Beer, on

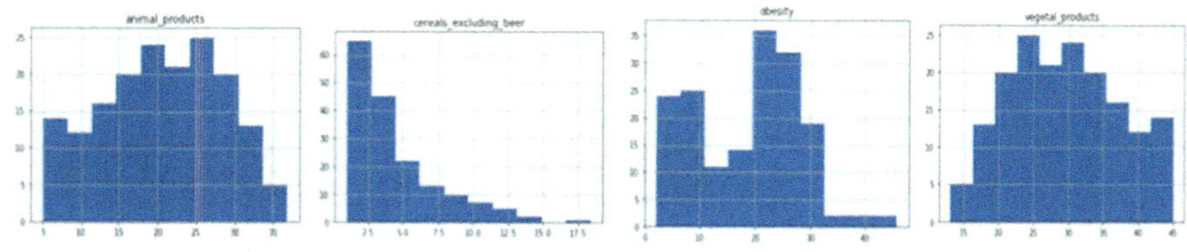

Fig. 3 Data visualization with features distribution

the other hand, exhibit a skewed distribution to the right. Perhaps a log scalling will assist us in obtaining a normal distribution for those two characteristics.

4.2 Data Normalization (DN)

In the previous section, we created a histogram for all the characteristics and discovered that the values of Animal fats and Cereals—with the exception of beer, distribution is not evenly spread. We'll look at z-score and log scalling in this part. Keep in mind that, we'll execute the scalling on a duplicate of diet_data simply to see how the results change. The real scalling will occur throughout the model-building process (Fig. 4).

We appear to have gotten a much more normal distribution for all of our features after applying z-score and log scaling. So, when it comes to model building, we'll absolutely adhere to this strategy.

4.3 Data Noise Label Normalization (D-N-L-N)

The final step before building the model is to remove the noise from the label values and put the label within the same range as the features. We'll just maintain the first four numbers following the floating point to make the computations simple. We

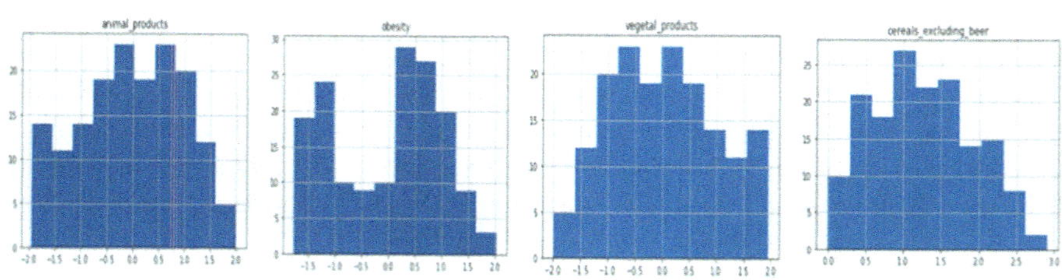

Fig. 4 Data normalization

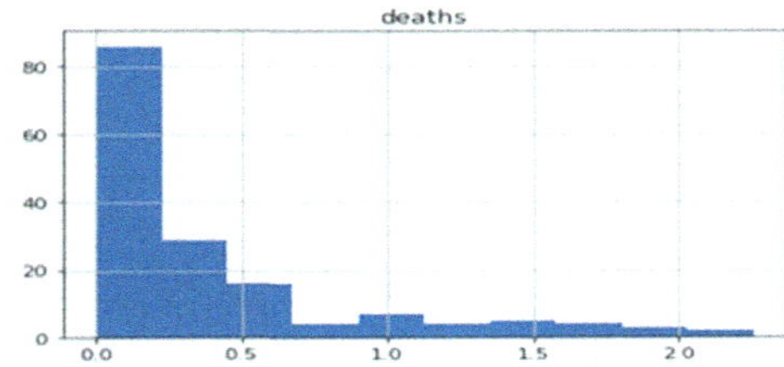

Fig. 5 Show death rate

include + 1 at logging to prevent logging 0 values, which causes -inf outcomes (Fig. 5).

4.4 Creating ML Model

The methods of building, training, and assessing the model are covered in the sections below. We separated the features from the label and divided the data-set into training and testing data. We'll build the input layer that our model will need. The normalization methods that we covered in the Data normalization step will be considered when defining the columns. Finally, we'll generate a function for creating and compiling a basic linear regression model (Fig. 6).

Despite the fact that the predictions are not particularly precise owing to the small number of instances used for training, after analysing the model, we can still see that when the percentage of characteristics with a negative correlation increases and the number of features with a positive connection decrease, the percentage of fatalities decreases. So, by adjusting the proportions of fat income types by a little bit, we can make a difference in the long run of the day.

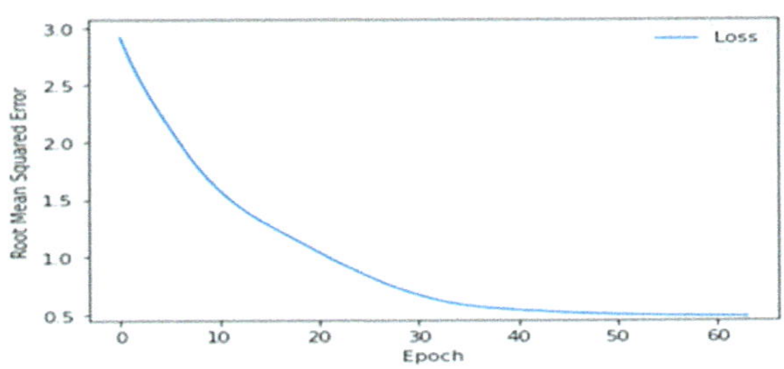

Fig. 6 Regression model

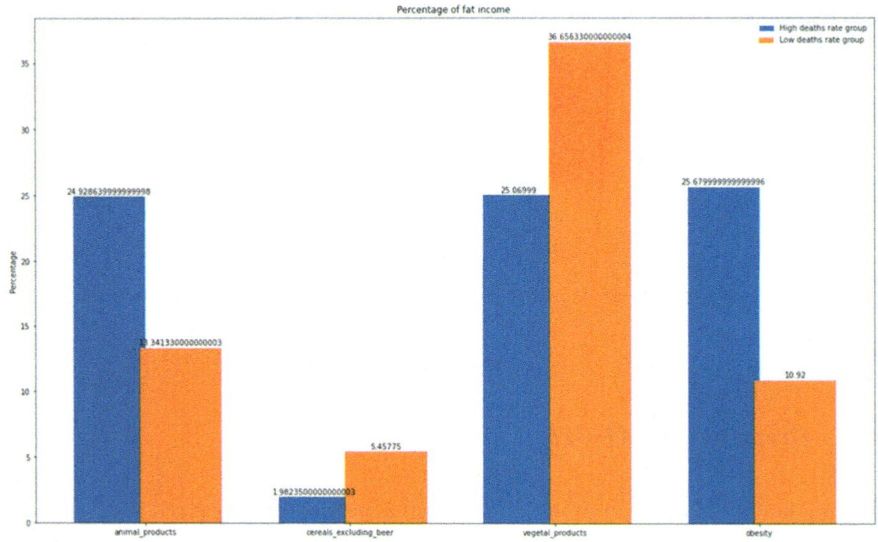

Fig. 7 Diet versus Covid percentage

4.5 Compare Covid-19 Versus Diet

This section covers the steps involved in evaluating the data and confirming our idea. We'll begin by looking at the dataset's statistics. Following monitoring and evaluating such facts, we believe that sorting the records by the percentage of deaths and then picking 10 rows, five of which represent records with the greatest percentage of deaths and the other five representing records with the lowest, would be a decent strategy. Last but not least, we'll take the average of each set of records' values and compare them using pie charts (Fig. 7).

As we could see from the bar chart, a population that eats a healthy diet rich in vegetables and grains have a lower mortality rate than a population that is more obese and consumes more animal products. Finally, we can affirm that a population with good food and lifestyle has a low risk of mortality due to the coronavirus pandemic based on this data.

4.6 Supervised Approaches

We used supervised methods to train models and make predictions, such as Linear Regression and Random Forest/Regression Tree. Some errors were found, such as mean squared error and coefficient of determination. As a result, we must need to enhance our models. Finally, here are the results (Fig. 8).

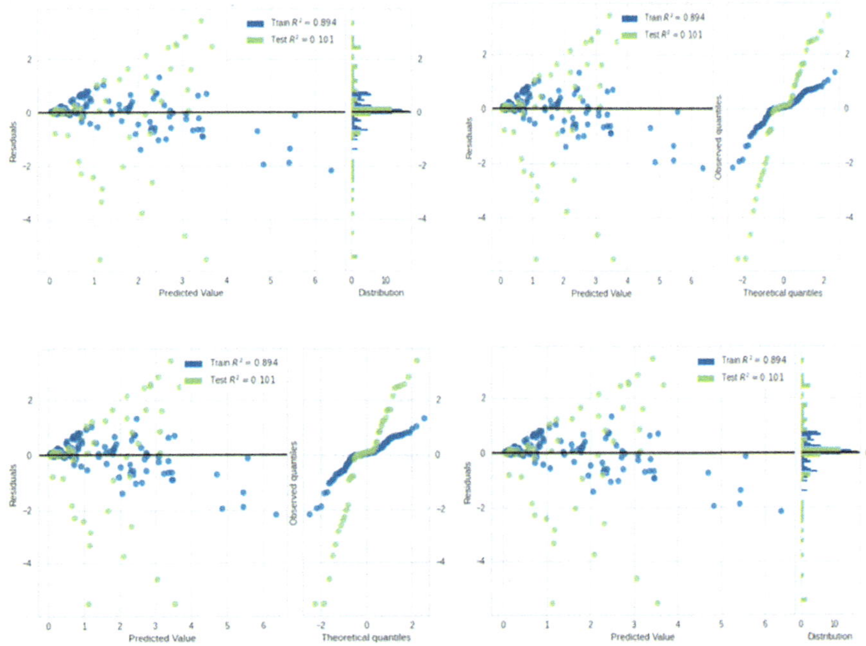

Fig. 8 Residuals for random forest regressor with QQ plot

4.7 Principal Component Analysis (PCA)

In the sector, we have discussed about the Principal Component Analysis (PCA). Below is the result with cluster analysis (Figs. 9 and 10).

In summary, a country's COVID-19 confirmed and active cases can somehow be explained relatively well by food categories such as the calorie contents of oil crops, and the protein content in infant food and miscellaneous food. On the other hand, the

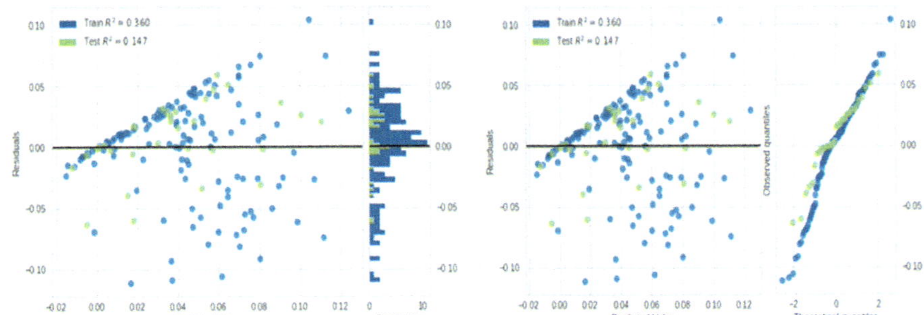

Fig. 9 Residuals for linear regression model with QQ plot

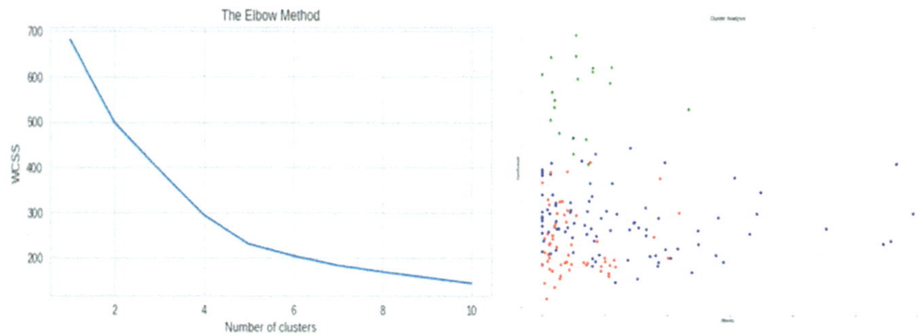

Fig. 10 K Means and cluster analysis

same cannot be said about the death and recovered cases. This could be due to the fact that these models do not satisfy the necessary model assumptions of having equal variance and normally distributed residuals. However, it is also important to note that mortality has not had an outcome, and hence the first model should only be taken as a grain of salt. However, recall that this model only talks about the correlation between food categories and the rate of deaths.

There is no evidence to suggest that a country's diet has an effect on the spread of COVID-19. Additionally, there are also many other factors causing the spread of COVID-19 that are totally uncorrelated with diet, e.g., how active the general public is, the preventive measures implemented by the countries, density of population, etc.

4.8 Restaurant Business Analysis

We'll use these statistics to see how restaurants behave during Covid-19, what factors influence sales during Covid, how customers eat during this period, and how this impacts future restaurants. Using the data set of the top 250 restaurants, attempt to find out what happened to top restaurant sales in 2020 and extract useful information. Follow the Feature Engineering process, which includes testing for null values, manipulating YOY statistics, and creating category features. The EDA Data Visualization is demonstrated (Fig. 11).

4.9 YOY Sales Indicator

Despite the fact that the restaurants were on the top list for the whole year, roughly 35% of them had negative indications.

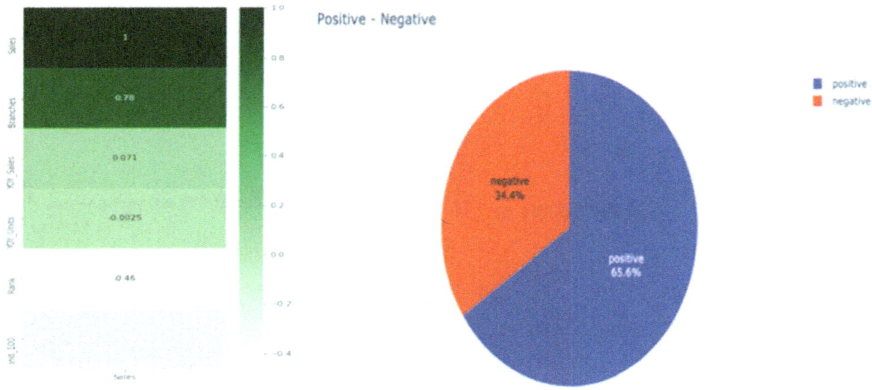

Fig. 11 Distribution and correlation with sales, show the effect rate

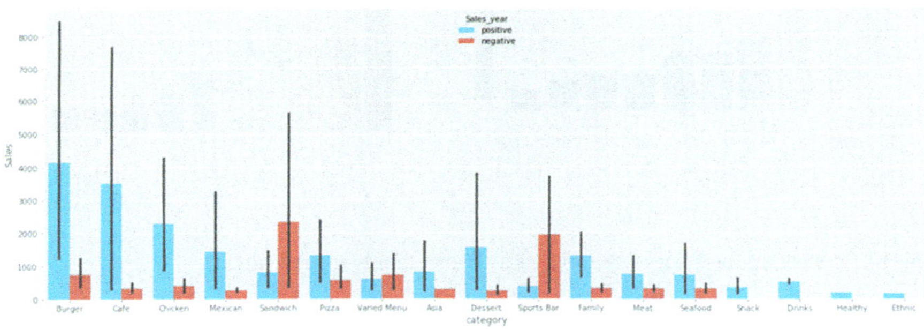

Fig. 12 Sales indicator

4.10 Category Sales Indicator

The varied cuisine, sandwiches, and sports bar all garnered more negative than positive marks (Fig. 12).

4.11 Top Restaurant in Best Category

We have analysis some category ways and we conclude some statement. A number of branches have the strongest correlation with sales, Top Restaurants are negative review of 35% YOY, Burger Restaurants will have the highest sales in 2020, Quick Service is the best way to obtain high sales, and Sports Bar and Sandwich Restaurants will lose a lot of money (Figs. 13, 14, 15, 16 and 17).

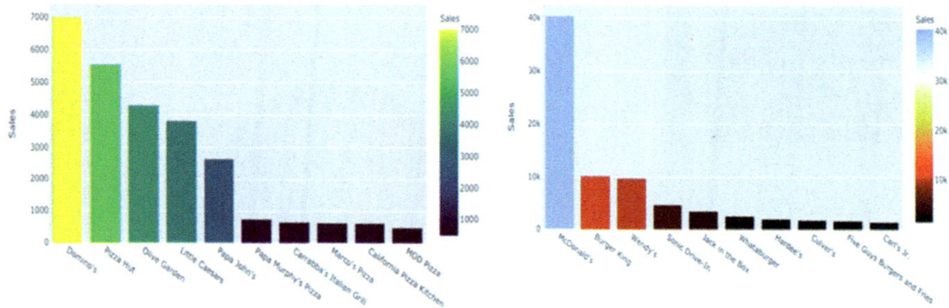

Fig. 13 Restaurant food sales category (pizza, burger)

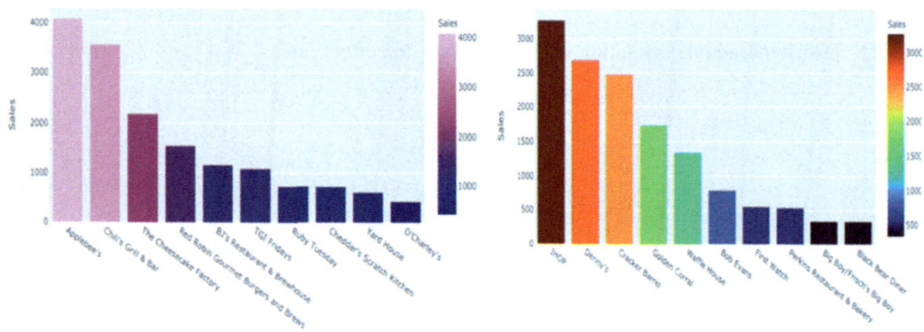

Fig. 14 Restaurant food sales category (varity menu, family top restaurant)

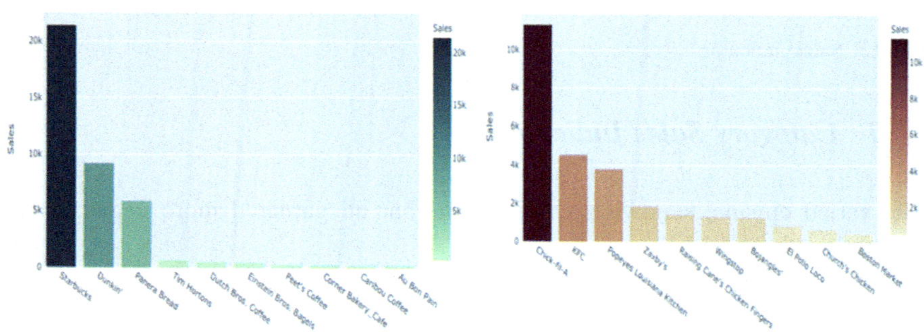

Fig. 15 Top restaurant food Covid-19 time (café top, chicken)

4.12 Is Corona Infection Linked with Diet?

We're now looking at if there's a link between certain food products and high Coronavirus infection rates. Below correlation.

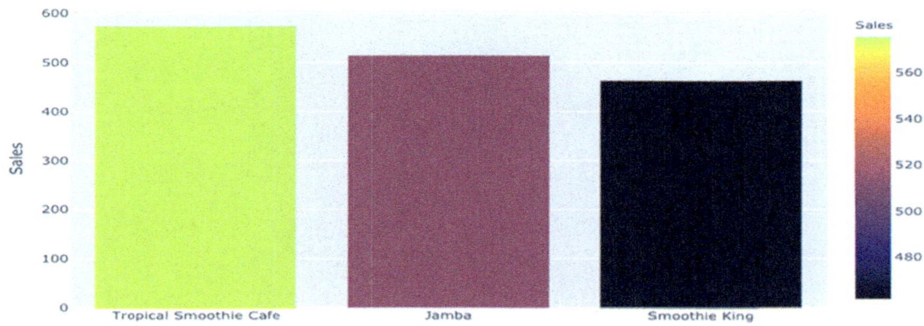

Fig. 16 Top restaurant food Covid-19 time (drinks top)

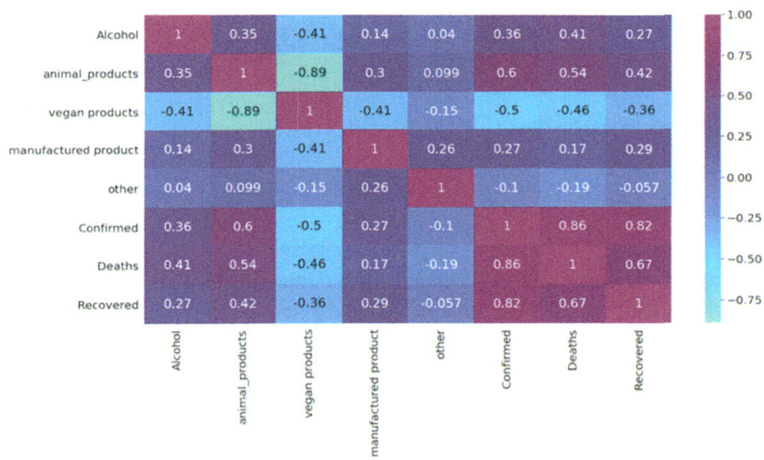

Fig. 17 Food versus Covid

In contrast to animal products, which have a positive connection with the number of confirmed cases (+6), vegan products have a negative link with the number of confirmed cases (−5). This is a plausible explanation for why sales at Evergreens Restaurant increased the greatest during the Corona epidemic (Fig. 18).

In the summary from the restaurant food analysis, we could say that the majority of people ate organic food during the Corona epidemic, Restaurants that depend on crowds have lost a lot of money, The number of branches makes a significant impact in sales and aids in providing quick services, Because of its tremendous potential for producing big sales, franchising is worth considering, COVID-19 Severity can be reduced by eating Vegan/Vegetables.

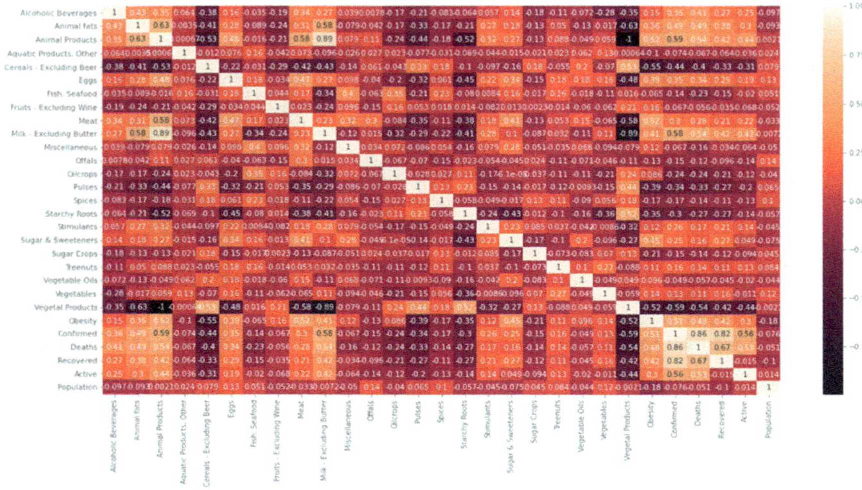

Fig. 18 Food compared with Covid-19

5 Conclusion

In this chapter, we have analysis restaurant business in the epidemic time (Covid-19) and which food is the best for human body and strongly fights with Covid-19 Virus. We utilized two data sets and applied separate algorithms to determine that those who ate vegetables had less confirmed case/death rate than those who ate animals. The same things happened in a restaurant business analysis. Those who sell animal-related food items, drop in sales, their business falls in; while those who sell vegetable-related food products, are upright in sales and have a stable business. Also, Linear Regression (LR) and Random Forest/Regression Tree, machine learning model have been applied for data analysis.

References

1. Listings of WHO's Response to COVID-19. https://www.who.int/news-room/detail/29-06-2020-covidtimeline
2. NHS. Social distancing: what you need to do–Coronavirus (COVID-19). https://www.nhs.uk/conditions/coronaviruscovid19/social-distancing/what-you-need-to-do/. Accessed 25 Sept 2020
3. WHO/Europe. Coronavirus Disease (COVID-19) Outbreak
4. Cheval B, Sivaramakrishnan H, Maltagliati S, Fessler L, Forestier C, Sarrazin P (2020) Relationships between changes in self-reported physical activity and sedentary behaviours and health during the coronavirus (COVID-19) pandemic in France and Switzerland. J Sports Sci

5. Mattioli AV, Sciomer S, Cocchi C, Maffei S, Gallina S (2020) Quarantine during COVID-19 outbreak: changes in diet and physical activity increase the risk of cardiovascular disease. Nutr Metab Cardiovasc Dis
6. Oni T, Micklesfield LK, Wadende P, Obonyo CO, Woodcock J, Mogo ERI, Odunitan-Wayas FA et al (2020) Implications of COVID-19 control measures for diet and physical activity, and lessons for addressing other pandemics facing rapidly urbanising countries. Glob Health Action 13(1):1810415
7. Détang-Dessendre C, Guyomard H, Réquillart V, Soler L-G (2020) Changing agricultural systems and food diets to prevent and mitigate global health shocks. Sustainability 12(16):6462
8. Ruiz-Roso MB, Knott-Torcal C, Matilla-Escalante DC, Garcimartín A, Sampedro-Nuñez MA, Dávalos A, Marazuela M (2020) COVID-19 lockdown and changes of the dietary pattern and physical activity habits in a cohort of patients with type 2 diabetes mellitus. Nutrients 12(8):2327
9. Xie X, Huang L, Li J, Zhu H (2020) Generational differences in perceptions of food health/risk and attitudes toward organic food and game meat: the case of the COVID-19 crisis in China. Int J Environ Res Public Health 17(9):3148
10. Islam MB et al (2022a) Twitter opinion mining on COVID-19 vaccinations by machine learning presence. In: Proceedings of Third Doctoral Symposium on Computational Intelligence: DoSCI 2022. Springer Nature, Singapore
11. Kantardzic M (2020) Data mining: concepts, models, methods, and algorithms
12. Rath S, Tripathy A, Tripathy AR (2020) Prediction of new active cases of coronavirus disease (COVID-19) pandemic using multiple linear regression model. Diabetes Metab Syndrome Clin Res Rev
13. Iwendi C et al (2020) COVID-19 patient health prediction using boosted random forest algorithm. Front Public Health 8:357. https://doi.org/10.3389/fpubh.2020.00357
14. Yeşilkanat CM (2020) Spatio-temporal estimation of the daily cases of COVID-19 in worldwide using random forest machine learning algorithm. Chaos Solit Fract 140:110210. https://doi.org/10.1016/j.chaos.2020.110210
15. Islam MB et al (2022b) Pandemic outbreak time: evaluation of public tweet opinion by machine learning. In: IEEE International Conference on Current Development in Engineering and Technology (CCET)

Chapter 23
Crowd Counting via De-background Multicolumn Dynamic Convolutional Neural Network

Santosh Kumar Tripathy, Naman Kaushik, Subodh Srivastava, and Rajeev Srivastava

Abstract The current state-of-the-art density map-based crowd counting methods have focused on designing convolution neural network (CNN)-based models to exploit multiscale features to handle crowd shape change due to perspective distortion. However, the significant concerns with such approaches are using static kernels and being not adaptive to input data. Again, the multiscale features should be more attentive towards background minimization. Hence, this chapter proposes a de-background multicolumn dynamic CNN for crowd counting to address the issues. The proposed model can handle crowd shape change due to perspective distortion and learn to minimize the background influence while doing crowd counting. Two benchmark crowd counting datasets, Mall and UCSD, are used to show the model's effectiveness. In addition to this, a separate ablation study has been conducted to show the effect of individual modules of the proposed model.

Keywords Multicolumn CNN · Dynamic CNN · Multiscale features · De-background multiscale features

---

S. K. Tripathy (✉) · R. Srivastava
Computing and Vision Lab, Department of Computer Science and Engineering, Indian Institute of Technology (BHU), Varanasi 221005, Uttar Pradesh, India
e-mail: santoshktripathy.rs.cse18@iitbhu.ac.in

R. Srivastava
e-mail: rajeev.cse@iitbhu.ac.in

N. Kaushik
Department of Mechanical Engineering, Indian Institute of Technology (BHU), Varanasi 221005, Uttar Pradesh, India
e-mail: namank.cd.mec17@iitbhu.ac.in

S. Srivastava
Department of Electronics and Communication Engineering, National Institute of Technology, Patna, Bihar 800005, India
e-mail: subodh@nitp.ac.in

S. Kautish et al. (eds.), *Computational Intelligence for Modern Business Systems*, Disruptive Technologies and Digital Transformations for Society 5.0,
https://doi.org/10.1007/978-981-99-5354-7_23

1 Introduction

Several research communities have vastly explored vision-based crowd counting solutions in recent years. One of the primary motivations behind vision-based crowd counting is its effectiveness in controlling crowd disasters. Nevertheless, the vision-based crowd counting system is affected mainly by two challenging issues like crowd head or body shape change due to perspective distortion and influence of crowd backgrounds. The literature addresses these challenges differently, mainly using deep learning concepts. For example, multicolumn CNN [1], single-column scale adaptive CNN [2], multiscale generative adversarial networks (GANs) [3] have been developed to address the variation of crowd shape by extracting multiscale spatial features from the crowd scene. Head attention mechanism [4] has been developed to minimize the cluttered background effect and focus on the head regions of the crowd scene for crowd counting.

However, the state-of-the-art approaches [1, 3, 5–7] utilize CNN with a fixed convolution kernel for crowd counting, and thus, the kernels are not adaptive to different input sequences. The multiscale models have been developed to handle perspective distortion, but these features must be free from background influences. In addition to this, there is a lack of use of dynamic CNN for crowd counting. Hence, based on these studies, the following are the probable research gaps that have to be addressed,

- Lack of use of dynamic CNN for crowd counting.
- Instead of static convolution kernel, the impact of dynamic convolution kernels for crowd counting has to be studied.
- There is a lack of applying head attention mechanism in multiscale features to make the multiscale features more attentive towards crowd head regions to minimize the influence of crowd backgrounds.

This chapter proposes a crowd counting solution via de-background multicolumn dynamic convolutional neural networks to fulfill the above research gaps. The following are the main contributions or objectives of the proposed approach.

- A multicolumn CNN with dynamic kernels has been designed to extract multiscale spatial features for crowd counting.
- Before multiscale feature fusion, each column of the proposed model is attentive towards crowd head regions to minimize the crowd scene's background influence.
- The fused multiscale de-background spatial features are used for crowd density estimation.
- To show the effectiveness of the proposed approach, extensive experiment and ablation studies have been conducted on two publicly available datasets: Mall and UCSD.

The rest of the chapter is structured as a brief literature review on vision-based crowd counting is discussed in Sect. 2, the proposed model and its optimization process are discussed in Sect. 3, Datasets and the performance measures are explained in Sect. 4, Experimental analysis has been done in Sects. 5, 6 and 7 discuss ablation study and conclusion, respectively.

2 Literature Review

This section starts with a brief discussion of the taxonomy of vision-based crowd counting approaches followed by recent developments on CNN-based crowd counting approaches.

2.1 Taxonomy of Vision-Based Crowd Counting Approaches

The available crowd counting solutions can be categorized broadly into single image-based and video-based approaches. The following Fig. 1 shows a brief taxonomy of vision-based crowd counting approaches. The vision-based crowd counting approaches could be categorized into four different ways,

- Taxonomy based on dataset modality.
- Taxonomy based on counting process.
- Taxonomy based on learning trends.
- Taxonomy based on different AI-learning paradigm.

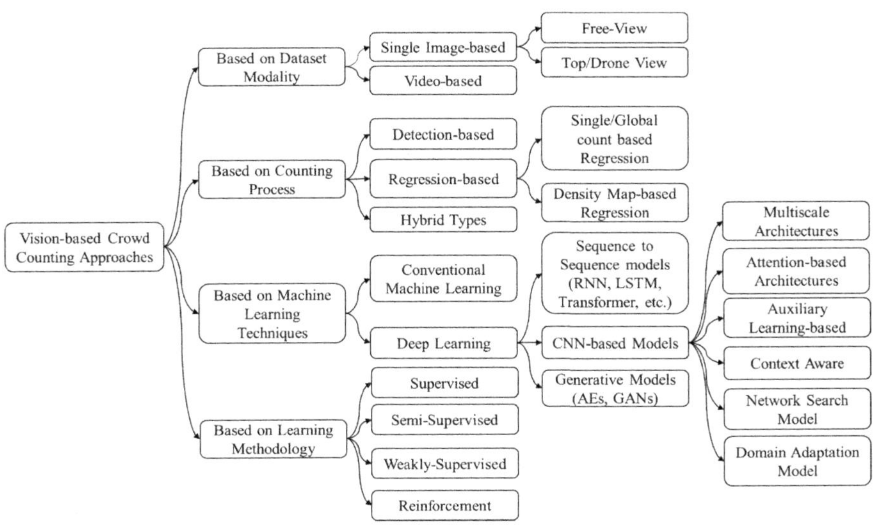

Fig. 1 Taxonomy of vision-based crowd counting approaches

Based on the available dataset modality, the crowd counting approaches can be categorized into single image-based [8, 9] and video-based approaches [10–12]. The single image-based approaches could be categorized into free view [8] and top view or drone view [9] based counting approaches. The single image-based approaches [8] rely on spatial features for crowd counting. On the other hand, the video-based approaches rely on spatial–temporal features for crowd counting.

However, the crowd counting approaches could be categorized based on the counting process or methods adopted. Based on this intuition, the crowd counting approaches are categorized into detection-based [13], regression-based [14–20], and hybrid approaches. The detection-based counting approaches take the help of body part detectors for counting. Nevertheless, the performance is degraded in the dense and low-resolution crowd scenes. The regression-based approaches apply several regression techniques to map the crowd features onto the ground-truth crowd count values. However, based on the available ground-truth crowd count annotations, the regression approaches can be categorized into two types: single or global count-based and density map-based regression techniques. The former regression technique has the ground-truth crowd counts as global count value, and the available approaches map the crowd scene features onto the global crowd count values of the crowd scenes. The state-of-the-art global count-based regression techniques utilize conventional and deep learning (formally CNN) techniques, but such approaches act as ridge regressors and do not consider the spatial distribution of crowds in the scene. The latter approach overcomes such drawbacks. The density map-based regression approaches utilize the ground-truth annotated crowd heads to generate crowd density maps. Techniques like Gaussian distribution, geometric adaptive Gaussian distribution are used to obtain crowd density maps. Recently, different hybrid approaches have also been developed for crowd counting. Such approaches take advantage of both detection and regression techniques for crowd counting.

The crowd counting approach could be categorized into two types based on the learning trends that evolved over the last two decades, i.e., conventional machine learning [14–16, 21] and deep learning-based crowd counting approaches [8–12]. The former approaches extract texture and or motion features from the crowd scene followed by regression approaches for crowd counting. Regression approaches like Gaussian kernel-based regressor, ridge regressor have been used. But the later approaches utilize deep learning concepts for crowd counting. Deep models like CNN, sequential, and generative models have been vastly explored in this area. Sequential models like LSTM, Bidirectional LSTM, Conv-LSTM, and vision-transformers have been utilized. Generative models like autoencoders, generative adversarial networks have also been used for crowd counting. Nevertheless, the different architectures using CNN and its variants have been vastly utilized in this area.

The multiscale architectures using CNN have been proposed to extract multiscale features from the crowd scene to handle perspective distortion issues in the crowd scene. Different attention-based CNN structures [4, 11] have also been proposed for crowd counting. Different multitask learning-based approaches using CNN are also proposed. Such approaches show crowd counting as the main task and other tasks like density classification as an auxiliary task. The crowd scene context-aware CNN architecture model has also been proposed in the literature. Different dataset domain adaptation has been an issue in any vision-related research field. To address such issues, domain adaptation-based CNN has also been proposed.

Another way in which the vision-based crowd counting approaches have been categorized into four types is based on the learning approaches used. Crowd counting models can be categorized as supervised [8–12], semi-supervised [22], weakly-supervised [23], and reinforcement learning-based approaches [24] have also been used for crowd counting.

2.2 *A Brief Review of CNN-Based Crowd Counting Approaches*

The current research trends utilize CNN architectures and proposed several models for crowd counting by addressing challenges in the crowd scene like crowd shape change, the influence of crowd backgrounds, domain adaptability. The crowd shape change issue has been handled in the literature by proposing several multiscale CNN models. At first, Zhang et al. [1] combined different columns of multilayers of CNN with varying kernel shapes to extract multiscale features for crowd density estimation. But, Cheng et al. [25] improved the multicolumn learning capability by exhibiting a mutual learning mechanism between different columns of the multicolumn CNN, thereby improving multiscale features' capacity. Zou et al. [5] identified that the multicolumn CNN(MCNN) [1] is limited in handling crowd scenes with varying crowd densities and proposed an adaptive capacity multiscale CNN (ACM-CNN) for varying crowd densities. With the help of coarse, refined, and smooth networks, the ACM-CNN can learn varying crowd densities and generate high-quality crowd density maps.

Zhang et al. [6] proposed a dilated multicolumn CNN (DMCNN), a lightweight structure compared to MCNN, and extracts scale-invariant features for crowd counting. Deb et al. [26] performed perspective-free crowd counting via multicolumn dilated CNN (AMDCNN) aggregation. Li et al. [27] proposed a congested-scene recognition-network (CSR-Net) for crowd counting. The authors used dilated CNN kernels in the CNN structure and produced density maps of high quality. Onoro-Rubio et' al. [28] proposed a scale-aware counting CNN network (CCNN) for the cross-domain (different types of entities) counting. Authors achieved good performance for crowd and vehicle counting. Kumagai et al. [7] proposed a mixture of CNNs (MoCNN) for crowd counting. The authors proposed an adaptive integration of CNNs for crowd counting. The CNNs are specialized to specific scenes for crowd counting. Miao et al. [10] utilized spatial and temporal features for crowd density estimation in video datasets. Tripathy et al. [11] proposed an attentive multi-stream CNN (AMS-CNN) for video crowd counting. The AMS-CNN [11] does not simply fuse the features of each stream for density estimation; rather, each stream's feature maps are attentive towards crowd density estimation followed by multi-density map fusion for crowd counting. Again Tripathy et' al. [12] adopted a transfer learning mechanism to extract multiscale features from different video cues for crowd density estimation. Wang et al. [29] proposed a multi-density map fusion network for crowd counting. The model is built on VGG-16 and can extract scale-invariant features. Sindagi et' al. [4] proposed a hierarchical attention mechanism in CNN architecture (HA-CNN) for crowd counting. The authors also proposed a novel idea of removing background interference from the crowd scene and focusing on the head regions while generating crowd density maps. Wang et al. [30] performed a pixel-wise attention mechanism in a single column of CNN layers for crowd density estimation. Li et al. [31] proposed a crowd counting model via a cross-level parallel network built on a VGG network. Recently, Wei et al. [32] proposed a scene adaptive attention network (SAA-Net) for crowd counting. The SAA-Net is built on a deformable transformer and can handle crowd shape variation and perform in complex scenes. Sajid et al. [33] proposed a plug-and-play patch rescaling module (PRM) for effective crowd counting. The authors proposed a multi-resolution fusion strategy and attention mechanism and can handle scene perspective distortion, varying crowd densities, and clutter background effects in the crowd scenes.

3 Proposed Model

The primary motivation behind the design of the proposed model is to enhance the counting capacity of the existing MCNN model [1]. The MCNN-based models use static convolution kernels for multiscale feature extraction, which are fixed and not adaptive to different types of images. Also, such architectures do not take any measures to suppress the effect of backgrounds. Hence, to overcome such issues, the proposed model introduces the utilization of dynamic convolution kernels [34] instead of static kernels in the multicolumn architecture. Instead of simply fusing the multicolumn features, each column features are first attentive towards head regions followed by fusion. The idea of mapping feature maps towards head regions is adopted from HA-CCN [4]. Figure 2 shows the overall architecture of the proposed model. The following subsections elaborate on the proposed model in detail. According to Fig. 2, the proposed model constitutes of three things,

- Multi-column wise head attention modules.
- De-background Multiscale Feature Fusion.
- Density Estimation Module (DME).

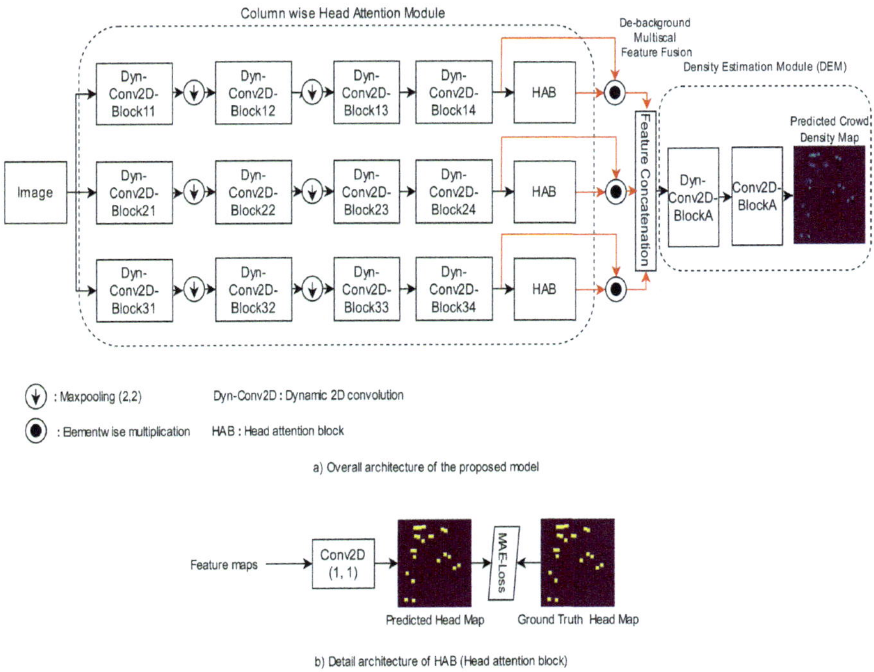

Fig. 2 Detail architecture of the proposed model

3.1 Architecture Details

The proposed model utilizes a three-column architecture to fulfill the objective function. The proposed model follows the MCNN as far as three-column design and the number of kernels are concerned. However, the proposed model replaces the 2D-CNN with the dynamic 2D-CNN during the column design. The details of the layers of the proposed model are illustrated in Table 1. The proposed model is built on three parallel columns of dynamic 2D convolution blocks. Each column has four dynamic convolution blocks. Each dynamic convolution block constitutes a dynamic 2D CNN layer followed by ReLU activation, followed by a batch normalization (BN) layer. The number of output feature maps and kernel size information is mentioned in Table 1. Two Max-Pooling layers are used, each after three columns' first and second dynamic convolution blocks. The Max-Pooling layers down-sample the feature maps into their half size. Three head attention block (HAB) takes the feature maps of three columns and produces head maps for each column. The architecture of HAB is shown in Fig. 2a. The predicted head map of each column is elementwise multiplied with the feature maps of the fourth dynamic convolution block. All three columns' head attentive feature maps are fused and fed into dynamic convolution Block A for further feature modeling. The Dyn-Conv2D-BlockA is followed by a merged layer

Table 1 Layer details of the proposed model

Block name	Layer name	No. of kernels	Kernel shape	Block name	Layer name	No. of kernels	Kernel Shape
Dyn-Conv2D-Block11	Dynamic2D Convolution	16	(9, 9)	Dyn-Conv2D-Block23	Dynamic2D Convolution	20	(5, 5)
	ReLU	NA			ReLU	NA	
	BN				BN		
Dyn-Conv2D-Block12	Dynamic2D Convolution	32	(7, 7)	Dyn-Conv2D-Block24	Dynamic2D Convolution	10	(5, 5)
	ReLU	NA			ReLU	NA	
	BN				BN		
Dyn-Conv2D-Block13	Dynamic2D Convolution	16	(7, 7)	Dyn-Conv2D-Block31	Dynamic2D Convolution	24	(5, 5)
	ReLU	NA			ReLU	NA	
	BN				BN		
Dyn-Conv2D-Block14	Dynamic2D Convolution	8	(7, 7)	Dyn-Conv2D-Block32	Dynamic2D Convolution	48	(3, 3)
	ReLU	NA			ReLU	NA	
	BN				BN		
Dyn-Conv2D-Block21	Dynamic2D Convolution	20	(7, 7)	Dyn-Conv2D-Block33	Dynamic2D Convolution	24	(3, 3)
	ReLU	NA			ReLU	NA	
	BN				BN		
Dyn-Conv2D-Block22	Dynamic2D Convolution	40	(5, 5)	Dyn-Conv2D-Block34	Dynamic 2D Convolution	12	(3, 3)
	ReLU	NA			ReLU	NA	

(continued)

Table 1 (continued)

Block name	Layer name	No. of kernels	Kernel shape	Block name	Layer name	No. of kernels	Kernel Shape
	BN				BN		
Dyn-Conv2D-BlockA	Dynamic2D Convolution	30	(3, 3)	Conv2D-BlockA	2D CNN	1	(1, 1)
	ReLU	NA			ReLU	NA	
	BN						

(Conv2D-BlockA) made of Conv2D layer with a kernel size of (1, 1) and activation of ReLU. The Conv2D-BlockA will be the output layer used to generate crowd density maps. The proposed model is trained in a two-stage manner. During the first stage, the multi-column-wise head attention module is trained. On completion of training, the de-background multiscale features are extracted. During the second stage, the DME is trained.

3.2 De-background Multicolumn Feature Extraction

The background influences have to be minimized from the multiscale features. In the proposed model, the HAB will act as the head attention layer and produce each column's head map. In this regard, the model is trained end-to-end as far as the generation of head maps for three columns is concerned. We must have ground-truth head maps to obtain predicted crowd maps from three HAB layers. We follow the process adopted in HA-CCN to obtain ground-truth head maps where this chapter used the value of sigma to 0.5.

The input images are rescaled into [158 × 238], and the shape of the predicted head map is [39 × 59]. Hence the ground-truth head maps are obtained for the image of size [39 × 59], and their corresponding ground-truth head points are also mapped to the shape [39 × 59]. Let the ground-truth head maps be denoted by a set $H = \{h_1, h_2 \ldots h_T\}$ and let the rescales images be represented by the set $I = \{i_1, i_2 \ldots i_T\}$. Here, T is the total number of sequences. Three losses were obtained from three HAB modules for each column. Let the predicted head maps obtained from three HABs be denoted by three different sets, PH^1, PH^2, PH^3. Here, $PH^1 = \left\{ph_1^1, ph_2^1, \ldots, ph_T^1\right\}$, $PH^2 = \left\{ph_1^2, ph_2^2, \ldots, ph_T^2\right\}$, $PH^3 = \left\{ph_1^3, ph_2^3, \ldots, ph_T^3\right\}$. Let, the three losses of HABs for three columns are denoted as $loss_1, loss_2, and\ loss_3$. All three losses are obtained using mean absolute error (MAE), which are defined as

$$loss_1 = \frac{1}{T}\sum_{i=1}^{T}\left|\left(h_i - ph_i^1\right)\right| \tag{1}$$

$$loss_2 = \frac{1}{T}\sum_{i=1}^{T}\left|\left(h_i - ph_i^2\right)\right| \tag{2}$$

$$loss_3 = \frac{1}{T}\sum_{i=1}^{T}\left|\left(h_i - ph_i^3\right)\right| \tag{3}$$

Let θ denotes all the learnable parameters of the network used in the first stage of training. The first stage model is trained by minimizing the combined losses of three HABs. Let $loss_{1st} = loss_1 + loss_2 + loss_3$ be the total loss of the 1st stage network. The final loss has to be minimized, which can be denoted as,

$$\underset{\theta}{argmin}\, loss_{1st} \tag{4}$$

This chapter used the Adam optimizer to optimize the loss as mentioned in Eq. 4.

3.3 Multiscale De-background Feature Fusion and Crowd Density Estimation

After training the multicolumn architecture with HABs, the predicted head maps were obtained from three columns. Then, the de-background column features were obtained by performing elementwise multiplication between the feature maps of the fourth dynamic 2D convolution blocks and the predicted head maps. After obtaining, column-wise de-background features, all these features are fused by concatenating them. For crowd density map prediction, these multiscale de-background features are fed into the density estimation block (DEB). Let $\varnothing$ denote all the learnable parameters of the DEB block. In this work, the ground-truth crowd density maps are obtained by applying the geometric adaptive kernel approach [1]. Let, the ground-truth crowd density maps for T number of frames are denoted as $G = \{g_1, g_2 \dots g_T\}$. Let the predicted density map of the DEB be denoted as $P = \{p_1, p_2 \dots p_T\}$. The loss between the ground truth and the predicted density maps captured by using MAE, which is represented as

$$L = \frac{1}{T}\sum_{i=1}^{T}|(g_i - p_i)| \tag{5}$$

The loss L has to be minimized, which can be denoted as,

$$\underset{\varnothing}{argmin} L \tag{6}$$

This chapter used Adam optimizer to optimize the loss as mentioned in Eq. 6.

4 Dataset and Performance Metrics

Two publicly available datasets: UCSD [21] and Mall [19] are used for the experiment. The UCSD dataset contains 2000 frames captured using a still camera. Among 2000 frames, sequences from 601 to 1400 are used for training, and the rest are for testing. The resolution of the frames is [158 × 238], and all the frames are grayscale. We have used the same grayscale images with the resolution of [158 × 238] as input to the proposed model.

The Mall dataset also contains 2000 frames captured from a mall using a standalone camera. The first 800 frames are used for training, and the rest are used for testing. All the frames are rescaled into [158 × 238] grayscale—the following Figs. 3 and 4 show some samples of Mall and UCSD. The experiments are conducted without using the ROI.

The performance evaluation is conducted using two widely used metrics in crowd counting approaches, i.e., MAE and RMSE, whose formulas are given below.

$$MAE = \frac{1}{T} \times \sum_{k=1}^{T} |g_k - p_k| \tag{7}$$

Fig. 3 A sample of the mall dataset

Fig. 4 A sample of the UCSD dataset

$$RMSE = \sqrt{\frac{1}{T} \times \sum_{k=1}^{T} |g_k - p_k|^2} \tag{8}$$

Here g_k *and* p_k is the ground truth and predicted headcounts. T is the total number of training sequences.

5 Experiment Setup and Results Analysis

The code of the proposed model is developed in Python using PyTorch. The code is executed in different computing nodes of the ParamShivay (Supercomputer). The learning rate and batch size for all the datasets are set to 0.001 and 64, respectively. The momentum for the batch normalization is set to 0.95. To avoid overfitting, the proposed model adopted dropout and early stopping. The dropout layer of 25% dropping rate is used in all the proposed model's dynamic convolution layers (except Dyn-Conv2D-BlockA).

5.1 Results Analysis on the UCSD Dataset

The proposed model gets MAE and RMSE of 1.46 and 2.26, respectively, on the UCSD dataset. The model's performance is compared with recent state-of-the-art methods and is also shown in Table 2. The proposed model performs similarly to the recent model, i.e., AMS-CNN [11]. The MCNN [1] is reimplemented on the UCSD without providing a region of interest (ROI) and achieves MAE = 2.4 and RMSE = 3.13. The model performs better than the state-of-the-art models like MCNN [1], Cross-Counting CNN [35], Switch-CNN [36], AMDCNN [26], FCN-rLSTM [37], MS-GAN [3] and Density map + MESA [38]. The ACM-CNN [5] obtains 1.01 and 1.29 of MAE and RMSE, respectively, which is better than the proposed approach. However, the results obtained by ACM-CNN [5] are based on ROI. Figure 5 shows the line graph plotting between ground-truth versus predicted crowd counts of the proposed model.

Table 2 Results analysis on UCSD

Approaches	MAE	RMSE
MCNN (without ROI) [1]	2.4	3.13
Cross-scene counting [35]	1.60	3.31
Switching CNN [36]	1.62	2.10
ACM-CNN (with ROI) [5]	1.01	1.29
AMDCNN [26]	1.74	
CCNN [28]	1.51	–
Count forest [21]	1.60	4.40
FCN-rLSTM [37]	1.54	3.02
MS-GAN [3]	1.78	3.03
Density map + MESA [38]	1.70	–
AMS-CNN [11]	1.46	1.82
Proposed model	1.46	2.26

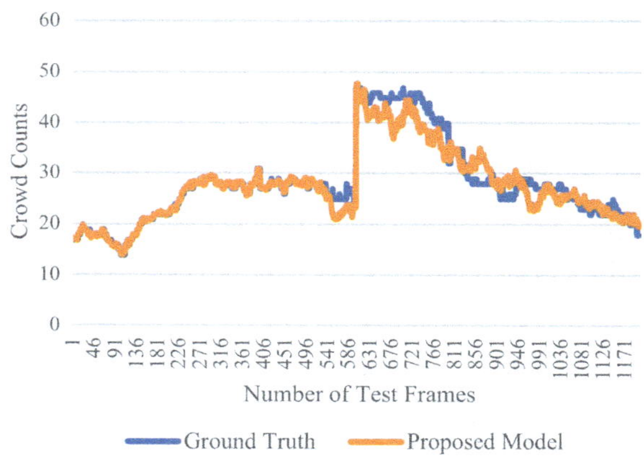

Fig. 5 Predicted versus ground-truth crowd counts of the proposed model on the UCSD dataset

5.2 *Results Analysis on the Mall Dataset*

The MAE and RMSE of the proposed model are 2.88 and 3.46, respectively, on the Mall dataset. The difference between MAE and RMSE is also very less as compared to MCNN [1], ST-CNN [10], DAL-SVR [39], Count Forest [21], CCNN [28], Bidirectional ConvLSTM [40] and MoCNN [7]. The model performs better as compared with recent state-of-the-art approaches like ST-CNN [10], DAL-SVR [39], CCNN [28], Bidirectional ConvLSTM [40], and MoCNN [7] as far as RMSE is concerned. For MAE, the proposed model performs better than state-of-the-art approaches like DMCNN [6], MCNN [1], ST-CNN [10], CNN-MRF [41], Count Forest [21], and CCNN [28]. However, Bidirectional ConvLSTM [40] achieves an MAE of 2.1 but acquires a high difference between MAE and RMSE. The proposed model achieves

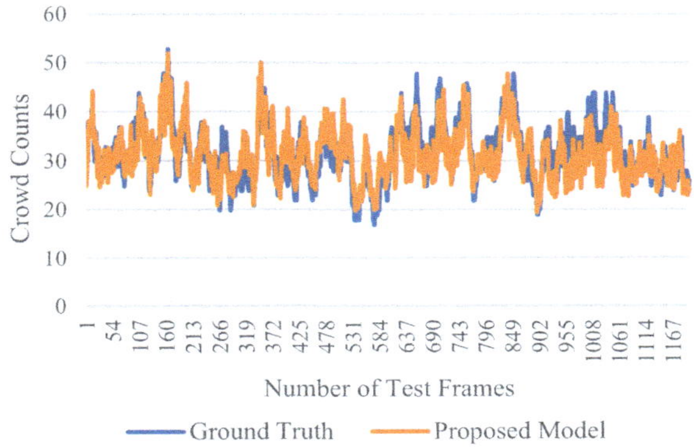

Fig. 6 Predicted versus ground-truth crowd counts of the proposed model on the mall dataset

quite comparable performance with the recently published paper AMS-CNN [11]. So, the proposed model can be a good choice for crowd counting. Figure 6 shows the line graph plotting between ground-truth versus predicted crowd counts of the proposed model.

6 Ablation Study

To show the effectiveness of the proposed model, an ablation study has also been conducted. During the ablation study, the proposed model is divided into the following models based on HAB and dynamic convolution layers.

- Proposed architecture without HAB.
- Proposed architecture without dynamic convolution layers.

The results of the ablation study are shown in Table 3. The proposed architecture without HAB obtains <MAE and RMSE> of <3.43 and 4.34> and <3.06 and 4.12> respectively on the Mall and UCSD dataset. But without dynamic convolution layers, the proposed architecture obtaine <MAE and RMSE> of <4.74 and 8.64> and <2.40 and 3.13 > on the Mall and UCSD dataset, respectively. However, the proposed model with dynamic convolution layers and HAB obtains better results. Hence it can be concluded that both dynamic convolution and head attention mechanisms are required to improve the performance of the proposed counting model.

Table 3 Results analysis on mall

Approaches	MAE	RMSE
DMCNN [6]	3.16	3.68
ACM-CNN [5] (ROI in testcases)	2.3	3.1
MCNN [1]	4.74	8.64
AMS-CNN [11]	2.47	3.08
ST-CNN [10]	4.03	5.87
CNN-MRF [41]	4.66	9.01
DAL-SVR [39]	2.40	9.57
Count forest [21]	5.75	10.88
CCNN [28]	5.36	9.34
Bidirectional ConvLSTM [40]	**2.10**	7.6
MoCNN [7]	2.75	13.4
Proposed model	2.88	3.46

7 Conclusion

In this chapter, a de-background multicolumn dynamic CNN has been proposed to address issues like crowd head scale variation and the effect of crowd scene backgrounds. The proposed model adopts the multicolumn architecture to extract multiscale features. To minimize the effect of backgrounds, column-wise head attention blocks (HAB) have been introduced. Finally, the head attentive multicolumn features have been fused to yield de-background multiscale features. This feature set has been given to the density estimation module (DME) to learn crowd density maps. The model's performance is evaluated on two datasets: Mall and UCSD. The proposed model obtains <MAE, RMSE> of <2.88, 3.87> and <1.46, 2.26> on the Mall and the UCSD datasets respectively. The obtained results were compared with recent state-of-the-art approaches, and from Tables 2 and 3, it can be observed that the proposed model provides promising results. Apart from this, in this chapter ablation study has also been performed. The main focus of the ablation study was to show the effects of HAB and also, dynamic convolution in the proposed model. The obtained results (Table 4) of the ablation study show that the proposed model performs better with HABs and dynamic convolutions. Although the current study addresses scale issues and minimizes background details, developing a cross-domain deep model for crowd counting is also essential. The cross-domain architecture will focus on minimizing the domain gaps between different datasets. Hence, future work will develop a more sophisticated cross-domain, perspective-free, and de-background deep model for crowd counting.

Table 4 Results of ablation study

Name of the models	The mall dataset		The UCSD dataset	
	MAE	RMSE	MAE	RMSE
Proposed model without HAB	3.43	4.34	3.06	4.12
Proposed model without dynamic convolution layers	4.74	8.64	2.40	3.13
Proposed model	2.88	3.46	1.46	2.26

Acknowledgements The support and the resources provided by "PARAM Shivay Facility" under the National Supercomputing Mission, Government of India at the Indian Institute of Technology, Varanasi, are gratefully acknowledged.

References

1. Zhang Y, Zhou D, Chen S, Gao S, Ma Y (2016) Single-image crowd counting via multi-column convolutional neural network. Proceedings IEEE conference computer vision pattern recognition, pp 589–597
2. Zhang L, Shi M, Chen Q (2018) Crowd counting via scale-adaptive convolutional neural network. Proceedings 2018 IEEE Winter conference application computer vision, WACV 2018, vol. 2018-Janua, no. 1, pp 1113–1121
3. Zhou Y, Yang J, Li H, Cao T, Kung S-Y (2020) Adversarial learning for multiscale crowd counting under complex scenes. IEEE Trans Cybern, pp 1–10
4. Sindagi VA, Patel VM (2020) HA-CCN: hierarchical attention-based crowd counting network. IEEE Trans Image Process 29(8):323–335
5. Zou Z, Cheng Y, Qu X, Ji S, Guo X, Zhou P (2019) Attend to count: crowd counting with adaptive capacity multiscale CNNs. Neurocomputing 367:75–83
6. Zhang Y, Lei J, Wang T, Wang L, Li G (2019) Crowd counting using DMCNN. ACM international conference proceeding series, vol Part F1481, pp 138–144
7. Kumagai S, Hotta K, Kurita T (2017) Mixture of counting CNNs: adaptive integration of CNNs specialized to specific appearance for crowd counting, pp 1–8
8. Yingying Zhang YM, Zhou S, Chen S , Gao S (2016) Single-image crowd counting via multicolumn convolutional neural network. CVPR, vol 2, no 35, pp 11431–11437
9. Liu W, Lis K, Salzmann M, Fua P (2020) Geometric and physical constraints for drone-based head plane crowd density estimation, pp 244–249
10. Miao Y, Han J, Gao Y, Zhang B (2019) ST-CNN: spatial-temporal convolutional neural network for crowd counting in videos. Pattern Recognit Lett 125:113–118
11. Tripathy SK, Srivastava R (2021) AMS-CNN: attentive multi-stream CNN for video-based crowd counting. Int J Multimed Inf Retr
12. Tripathy SK, Srivastava R (2021) A transfer learning-based multi-cues multi-scale spatial–temporal modeling for effective video-based crowd counting and density estimation using a single-column 2D-Atrous Net, pp 179–194
13. Liu Y, Shi M, Zhao Q, Wang X (2019) Point in, box out: beyond counting persons in crowds. Proceedings IEEE computer social conference computer vision pattern recognition, vol 2019-June, pp 6462–6471
14. Chen K, Loy CC, Gong S, Xiang T (2012) Feature mining for localised crowd counting. BMVC 1(2):1–11
15. WL, Senjian An SV (2007) Face recognition using kernel ridge regression. CVPR'07. IEEE Conference on, IEEE, pp 1–7

16. Chan AB, Liang ZSJ, Vasconcelos N (2008) Privacy preserving crowd monitoring: Counting people without people models or tracking. 26th IEEE conference computer vision pattern recognition, CVPR
17. Chen K, Gong S, Xiang T, Loy CC (2013) Cumulative attribute space for age and crowd density estimation. Proceedings IEEE computer social conference computer vision pattern recognition, no 2, pp 2467–2474
18. Wang C, Zhang H, Yang L, Liu S, Cao X (2015) Deep people counting in extremely dense crowds. In: MM 2015—proceedings of the 2015 ACM multimedia conference, pp 1299–1302
19. Shang C, Ai H, Bai B (2016) End-to-end crowd counting via joint learning local and global count. Proceedings international conference image processing. ICIP, vol 2016-Augus, pp 1215–1219
20. Hu Y, Chang H, Nian F, Wang Y, Li T (2016) Dense crowd counting from still images with convolutional neural networks. J Vis Commun Image Represent 38:530–539
21. Pham VQ, Kozakaya T, Yamaguchi O, Okada R (2015) COUNT forest: Co-voting uncertain number of targets using random forest for crowd density estimation. Proceedings IEEE international conference computer vision, vol. 2015 Inter, pp 3253–3261
22. Olmschenk G, Chen J, Tang H, Zhu Z, Dense crowd counting convolutional neural networks with minimal data using semi-supervised dual-goal generative adversarial networks
23. Sam DB, Sajjan NN, Maurya H, Babu RV (2019) Almost unsupervised learning for dense crowd counting. Proc AAAI Conf Artif Intell 33:8868–8875
24. Mehta K, Valloli VK (2019) W-Net: reinforced U-net for density map estimation
25. Cheng ZQ, Li JX, Dai Q, Wu X, He JY, Hauptmann AG (2019) Improving the learning of multicolumn convolutional neural network for crowd counting. In: MM 2019—Proceedings of the 27th ACM international conference on multimedia, pp 1897–1906
26. Deb D, Ventura J (2018) An aggregated multicolumn dilated convolution network for perspective-free counting. IEEE computer social conference computer vision pattern recognition Workshop, vol 2018-June, pp 308–317
27. Li Y, Zhang X, Chen D (2018) CSRNet: dilated convolutional neural networks for understanding the highly congested scenes. Proceedings IEEE computer social conference computer vision pattern recognition, pp 1091–1100
28. RJL-S Onoro-Rubio, Daniel, Towards perspective-free object counting with deep learning. Europe conference computer vision. Springer, Cham, pp 615–629
29. Wang Y, Zhang W, Liu Y, Zhu J (2020) Multi-density map fusion network for crowd counting. Neurocomputing
30. Wang B, Cao G, Shang Y, Zhou L, Zhang Y, Li X (2018) Single-column CNN for crowd counting with pixel-wise attention mechanism. Neural Comput Appl 9(2)
31. Li J, Xue Y, Wang W, Ouyang G (2019) Cross-level parallel network for crowd counting. IEEE Trans Ind Informatics PP(c):1–1
32. Wei X et al. (2021) Scene-adaptive attention network for crowd counting, vol 1, pp 1–12
33. Sajid U, Wang G (2021) Towards more effective prm-based crowd counting via a multi-resolution fusion and attention network. Neurocomputing 474:13–24
34. Chen Y, Dai X, Liu M, Chen D, Yuan L, Liu Z (2020) Dynamic convolution: attention over convolution kernels. Proceedings IEEE computer social conference computer vision pattern recognition, pp 11027–11036
35. Zhang C, Li H, Wang X, Yang X (2015) Cross-scene crowd counting via deep convolutional neural networks. Proceedings IEEE computer social conference computer vision pattern recognition, vol 07–12-June, pp 833–841
36. Sam DB, Surya S, Babu RV (2017) Switching convolutional neural network for crowd counting. Proceedings 30th IEEE conference computer vision pattern recognition, CVPR 2017, vol. 2017-Janua, pp 4031–4039
37. Zhang S, Wu G (2017) FCN-rLSTM : Deep Spatio-temporal neural networks for. Iccv, pp 3687–3696
38. Lempitsky V, Zisserman A, Learning to count objects in images victor. Adv Neural Inf Process Syst 3(3):1–5

39. Wei X, Du J, Liang M, Ye L (2019) Boosting deep attribute learning via support vector regression for fast moving crowd counting. Pattern Recognit Lett 119:12–23
40. Xiong F, Shi X, Yeung DY (2017) Spatiotemporal Modeling for crowd counting in videos. Proceedings IEEE international conference computer vision, vol 2017-Octob, pp 5161–5169
41. Han K, Wan W, Yao H, Hou L, Image crowd counting using convolutional neural network and Markov random field, pp 1–6

Chapter 24
Critical Factors and Their Relationship Affecting Bundling Practices in Indian Retail Industries: An AHP Approach

Rohan Pal, Kshitij Anand, Sushanta Tripathy, and Deepak Singhal

Abstract The advancement in technology has enabled sellers to discriminate based on customer-revealed purchasing intention. Sellers can now track the things purchased by buyers using various new technologies like sensors and RFID tags, and with this, there are new challenges in the implementation of bundling. To use this data, see what this new perceivability means for evaluating and market results. A detailed critical review was carried out on product bundling and their practices across the world in different markets. On the basis of our review of literature, it can be concluded that there are certain factors that are significantly more important than others. In this study, observations are made on how different factors compare to each other, and which should be prioritized. Information quality for the relevant bundling as per the market requirements is the most critical factor and transportation, while significant, is the least important.

Keywords Bundling · Retail sectors · India · AHP

1 Introduction

Bundling is becoming popular and is being used as a marketing strategy across a variety of industries [1]. Bundling is a prominent selling method in India, particularly in the food, software, retail, software, and tourism industries, and is used in various other industries [2]. Bundling, in both its pure and blended forms, has been proven to improve sales and/or profit by researchers. Researchers have also utilized bundling as a strategy for price discrimination, allowing retailers to split the market according to reservation prices. Given the multiple benefits of bundling, its power should be leveraged to generate profits for both a company and its clients. However, a detailed

---

R. Pal · K. Anand · S. Tripathy (✉) · D. Singhal
School of Mechanical Engineering, Kalinga Institute of Industrial Technology, Bhubaneswar, India
e-mail: sushant.tripathy@gmail.com

S. Kautish et al. (eds.), *Computational Intelligence for Modern Business Systems*, Disruptive Technologies and Digital Transformations for Society 5.0,
https://doi.org/10.1007/978-981-99-5354-7_24

grasp of customers' perceptions of bundled goods is essential for efficient utilization of bundling as a method for improving dealing and profitability.

However, there is not enough research into customer attitudes toward bundling [3]. How do customers prefer to purchase bundled products and services? Which attributes should be given more significance when choosing bundles? There are still queries, particularly associated with India. One of the main reasons and purpose of this research is to find solutions to these issues. This article aims to provide marketers with an understanding of the various aspects of bundling and how they can be targeted to these customers. The article starts with a review of the literature and its goals, then goes over the findings and their implications for future research.

With the growth of technology, retailers in both online and offline locations can now track customer behavior and utilize it to price things strategically. As a result, when a customer enters into a store to purchase new clothing, the prices are already on the shelf. So, if a buyer chooses a shirt and then proceeds to purchase a pair of pants, sellers are making a pricing selection at the same time.

As a result, vendors have found out a way to profit from the information by selling a bundle of things; however, prices are still determined ex-ante. Instead of selling pants and shirts, the department shop might sell an outfit. However, with the advancements in information technology, pricing decisions are now made in stages rather than all at once. If a seller so desires, he can now determine the shirt a buyer chooses before determining the pants costs the shopper will pay attention to. Because the buyer's preferences are revealed by the product selection, sellers can anticipate the buyer's willingness, and, as a result, the seller can boost the price of the pant that matches the specific shirt.

The method involved with changing rates for specific customers which is impacted by their will to pay is known as designated estimating. At the point when an extra item or administration is promoted to a previous purchaser [4] it is known as strategically pitching. This exploration checks out the designated evaluating choices accessible to a mechanically progressed all-seeing retailer having perceivability into a client's underlying item determinations and the capacity to strategically pitch dependent on that information.

With the growth of technology, retailers in both online and offline locations can now track customer behavior and utilize it to price products strategically. As a result, when a customer enters a store to buy brand-new clothing, the price is already on the shelf. So, if a buyer chooses a shirt and then proceeds to purchase a pair of pants, sellers are making a pricing selection at the same time. By using bundling that is by selling products into bundle sellers, have discovered a way to profit from the information on the; yet, prices are still fixed ex ante. Instead of selling pants and shirts, the department shop might sell an outfit. However, with the advancements in information technology, pricing decisions are now made in stages rather than all at once. If a seller so desires, he can now determine the shirt a buyer chooses before determining the pants costs the shopper will pay attention to. Because the buyer's preferences are revealed by the product selection, sellers can predict the buyer's willingness, allowing the seller to boost the price of the pant that matches the specific shirt.

This chapter aims to find a relationship between different factors affecting the adoption of bundling practices in the Indian retail market and establish an understanding of their importance.

2 Literature Review

The literature published in respectable journals indexed in the 'Web of Science' and 'Scopus' databases was examined. The terms "supply chain," "bundling," "important factors," "retailer," "mixed," "pure," and "Indian markets" were used. This literature review is divided into two subsections. Firstly, the literature related to bundling as a strategy in the Indian markets was studied, and secondly, there was a discussion on the literature that focuses on the factors that influence the success and failure of bundling strategies.

Bundling is the act of selling at least two unmistakable wares as well as administrations as a solitary unit or bundle (Stremersch and Tellis, 2002). Packaging can be isolated into two kinds: item packaging and estimating packaging [5]. Price bundling is the method involved with consolidating at least two items into a solitary unit and selling the total bundle at a rebate. Item packaging, then again, centers around explicit things.

According to Harris and Blair [6], lowering search costs through integrated bundling increases customers' willingness to spend. Although there are numerous benefits of bundling, firms can go too far with their bundling efforts bundling efforts and are deemed anticompetitive, which results in a kind of antitrust developed against the bundling of software services as shown by both Microsoft and IBM, which in turn leads to lengthy litigations.

Further review of the literature on bundling shows that several studies have examined the various advantages of bundling both from the point of view of the business and the customer. Many such studies suggest that bundling could provide a competitive advantage to the company and its customers leading to greater efficiency in terms of production, increased customer retention, etc.

Bundling can aid in the development of competitive advantage by raising entry barriers and extending monopolistic power, hence limiting competition. Bundling is a strategy that manufacturers may use to compete for retail shelf space. Another study claimed that bundling lowers both average and fixed costs. Average prices have lowered as a result of increased demand and the pooling of fixed costs across the various bundled items.

Harris and Blair [6] mentioned an extraordinary objective fact. They fostered an item proposal model dependent on the possibility that client inclination closeness from earlier buy conduct is basic in foreseeing current buy conduct. As indicated by these investigations, traders' clients favored great fixed valuing for future acquisition of that product.

The analytical hierarchy process (AHP) has been employed by many scholars within the framework of the Basic Supply Chain; Chan [7] provides an example of the AHP which has been employed to evaluate the performance of the hospitals. The examination of small and medium-sized firms' supply chains is described by Bhagwat and Sharma [8]. Varma et al. [9] have also used these methodologies to evaluate the chain's petroleum supply performance.

For example, Jovanovic and Krivokapic [10] employ AHP to establish BSC perspectives' key performance metrics. Bentes et al. [11] investigated the relative effectiveness of AHP mainly in the three departments of a Brazilian telecoms company's financial department. Hey used AHP in the fraud, collection, and finally revenue assurance. Najmi and Makui [12] present a framework that incorporates the principles of the BSC and SCOR model for comparing the benchmark and ideal supply chain using the combination of both the AHP and DEMATEL methodology.

Dey and Cheffi [13] propose a methodology for assessing and comparing how effective the green supply chain is using the AHP method. They have targeted mainly three industries automotive, cement, and carpet manufacturing. Leung, Lam, and Cao use the AHP and its variant, the analytic network process (ANP), to aid in the deployment of the BSC [14]. Wang et al. [15] used the supply chain operations reference (SCOR) to help and identify suppliers. Also, AHP has also been used by them. For optimizing the overall performance of the system, Bhagwat and Sharma [16] designed an integrated AHPPGP paradigm to help the suppliers. Proposes Agarwal et al. [17]

In several studies [18] have adopted an innovative approach to the management of the supply chain effectiveness by developing a prediction model using the rules of the fuzzy logic. Tsai and Hung [19] utilize a fuzzy goal programming technique with activity-based costing and a value-chain structure to improve the network which is currently used in the green supply chain.

Chan and Qi [20] suggest calculating an aggregated supply chain performance index using a combination of AHP and fuzzy sets. A few researchers have utilized an incorporated methodology utilizing fluffy AHP and fluffy TOPSIS to assess execution in the Turkish financial area; for example, Seçme, Bayrakdarolu, and Kahraman [21] have used the method to check the performance of various computer of the ODM companies the laptop model used was notebook, and Sun [22] similarly he has also used the same technique to help the industry practitioners.

So there have been numerous analytical models for the supply chain performance assessment challenge have been created in the literature throughout the years. However, no author has tried to create an AHP model for assessment and decision-making in the supply chain.

The current study intends to uncover the essential aspects that may influence revenue creation in the Indian retail business through bundling techniques, taking into account the range of benefits connected with bundling. A link between the recommended factors was also established. Companies will be able to improve and leverage the potential of bundling as a marketing tactic as a result of this knowledge (Table 1).

From the literature review, six specific factors were shortlisted as listed below (Table 2).

Table 1 Literature review

Name, Author, Year	Country	Key focus	Key findings	Research gaps
Bhatt et al. [23]	India	Studying the factors that affect the adoption of Enterprise resource planning (ERP) system in India	Cost of deployment was found to be the most important criteria. "Vendor credibility" was the least significant factor Criterion "User friendliness and Security" and "Need fulfillment" were second and third in the order of importance	DEMATEL and ANP methods can be used to find the relationship among the identified factors
Zhou et al. [24]	China	Study and optimization of the supply marketing order allocation problem (SMOAP) using time consistency and bundle discounts	A tabu search is proposed along with three novel strategies, namely. reduction & decomposition, score-based heuristic, and two-level perturbation Small scale trading has optimal solutions and large scale ones give acceptable solutions	Other algorithmic approaches need to be considered
Wei and Chen [25]	China	Study the influence of pricing strategy and choice of channels for evaluating an offline to online system. Further, focus is on the profits gained under varying market-dominant powers	It was observed that the wholesale price or sale price is not affected by the dominating player For bundled and independent sales, the overall profits of the O2O supply chain are equal, thus enabling a fair competitive market	Information in this study is considered to be symmetrical. Future studies should study the influence of market-dominant powers under information asymmetry

(continued)

Table 1 (continued)

Name, Author, Year	Country	Key focus	Key findings	Research gaps
Caiatia et al. [26]	USA	In this paper, MAAS method has been used to calculate a choice model that is based on a Netherlands-based survey	A binary mixed model was used to calculate and understand the outcomes of quality of services, social influence, social demographics, and transportation-related variables on the decision to subscribe	A stated choice experiment may show various flaws when used to evaluate the demand for a novel and unfamiliar service
Zhang et al. [27]	China	A company's software bundling commitment in conjunction with its piracy deterrence strategy is studied by the author in this book	Effects in mixed bundling, pure bundling (PB) was found to be the best strategy (MB) due to amalgamation of rivalry and cannibalization	Considerations of assiduity competition, similar to competition in a duopoly terrain, could give redundant useful information
Dolla and Laishram [28]	India	Facilities management, organizational theory, privatization	The analysis finds that invention, full growth, standard specificality, compass, rivalry, data imbalance, and sale characteristics all have a major impact on the execution and progress PPP systems, grounded on sale cost economics, agency, and transaction propositions	This methodology can be used to develop municipal infrastructure solutions by predicting substitute stock supply network potentiality and company-level organizational strategies. Future study might use this contribution to evaluate the framework in order to improve information of bundling theory and how to arrange network infrastructure PPP initiatives

(continued)

Table 1 (continued)

Name, Author, Year	Country	Key focus	Key findings	Research gaps
Cao et al. [29]	China	The connection between a retailer's order and a manufacturer's wholesale decision, as well as a retailer's bundle decision. How does the bundle affect the profit sharing between manufacturers and retailers?	Bundles help stores reduce the negative effects of fluctuations in demand for primary products and price discrimination between buyers of secondary products The bundle can lower the retailer's wholesale pricing elasticity, divide the retailer's purchasing choice into bundled and non-bundled categories, and drastically reduce the retailer's optimal order quantity between the two	Impact of competition between retailers to understand the bundling decisions and how this might affect the retailer Ordering both primary and secondary products as a bundle from the manufacturer
Dominique-Ferreira et al. [30]	USA	To investigate the impact of the bundling method on consumer price index, identify consumer price index for 3- and 5-star hotels	The optimal pricing and the extent of allowable prices for three- and five-star hotels are shown by the PSM results. Bundle strategy surveys show that Fivestar customers are less likely to have a mix of leader bundles	Respondents only responded based on each hotel's star rating, location, and available facilities, and it was difficult to collect clear answers. Due to the limited period of sending the questionnaire, only a few foreign tourists participated

(continued)

Table 1 (continued)

Name, Author, Year	Country	Key focus	Key findings	Research gaps
Shao and Li [31]	China	In a two-stage supply chain with channel rivalry, bundling, and product strategy, as well as the influence of bundling strategy on supply chain members' behavior Propose game models for various competition conditions that include a bundling strategy and a product-quality strategy to help supply chain members make the best judgments	Bundle strategies encourage external competitors (that is, new suppliers of complementary components) to provide average quality components, while bundle strategies influence the product strategies of channel competitors in the supply chain. not	Only ideal complimentary components are taken into account; however, the analysis should be expanded to include flawed complementary products as well More complicated channel battles with multiple competitors
Liao and Tasi [32]	China	The application of big data analytics to BPM in retail is a crucial topic	User profiles, product and brand classifications, along with sales associations are examples of big data analysis/data mining outcomes used to suggest management development and uncover alternatives to bundles of store layouts and sales business processes	Though there are numerous enterprises in the retail industry, only a handful of them are truly successful
Honhon and Pan [33]	China	The optimum strategy for bundling according to two or more vertically differentiated products Considers customer's basket shopping behavior	When consumers benefit from collaborative consumption of components, the product forms an optimally nested set	Most research is focused on horizontal bundling; this paper is one rare work on vertically different products being bundled

(continued)

Table 1 (continued)

Name, Author, Year	Country	Key focus	Key findings	Research gaps
Xu et al. [34]	USA	Three situations of no bundling, pure bundling, and mixed bundling are observed, and the firms' ideal pricing and quality policies are analyzed Under what circumstances will one bundling approach outperform the others?	Regardless of the degree of complementarity, it is better to use a bundled or mixed bundled strategy than to sell a single product individually. In unbundled and mixed unbundled scenarios, a high degree of complementarity leads to less profits. The conditions for a pure bundle strategy are reversed	On one side, this study does not decide profit participation, so additional studies of equally weighted profit participation are possible. However, marketing initiatives (such as advertising and promotions) can be included in the survey
Wan et al. [35]	China	App bundling approach using ambivalence theories	The paper identifies certain properties of an app bundle influenced by investigational research This research addresses the task of investigating bundles using ambivalence theory This study demonstrates the coexistence of positive and negative attitudes	Data was collected using two techniques, and it's probable that the two-part collection resulted in some inaccuracies Our research model did not include actual purchasing behavior, which is one of the drawbacks of all cross-sectional analysis

(continued)

Table 1 (continued)

Name, Author, Year	Country	Key focus	Key findings	Research gaps
Chakravarty et al. [36]	USA	This paper is about the impact of the degree of cooperation between the parties in the supply chain on increasing profits by bundling as a price-discriminatory service. Use of pure bundles for price discrimination	High-margin items with similar ratings and minimal correlation are good candidates for bundling into an integrated business and supply chain When bundling the supply chain, the retailer's margin is equal to the margin of each independent supplier, and when the suppliers are pooled, it is equal to the total margin	The scenario of mixed bundling was not investigated This study only used one retailer; however, other merchants can be used Going further upstream than just the provider is a good idea
Setia et al. [37]	USA	Forming a digital business strategy targeted toward localized customer service units and understanding its efficiency	Looking at how information quality is used in capability-building and how it is dependent on the characteristics of different local regions	Focused only on localized dynamics, but it is important to simultaneously look into local and central dynamics
Chang and Yang. [1]	Taiwan	Social marketing, social behavior, theory of planned behavior	Actual control elements (external objective factors) and social behavior incentives could direct influence social behavior	Only two social issues, namely bag-taking and blood donation, were used in this study
Gomes and Sonia Dahab [38]	Portugal	How bundling and unbundling processes are being maintained by inter-firm cooperation	Maintaining the coordination between different firms in the supply chain by either modularizing the process or creating standard procedures	A general framework is provided, may differ for specific regions

(continued)

Table 1 (continued)

Name, Author, Year	Country	Key focus	Key findings	Research gaps
Schoenherr et al. [39]	USA	Creating a conceptual model and putting it to the test using a large-scale poll of purchasing professionals who are familiar with bundling in B2B online auctions	1-The complexity of item specification has no direct impact on package performance, either adversely or positively, demonstrating that highly complicated and difficult-to-specify products can be auctioned successfully	Is it possible to test the model using data from other industries? The relationship between bundling and other supply management methods, such as expenditure analysis, strategic sourcing, or supplier base rationalization, is an essential topic to investigate and one that merits future research
Putten et al. [40]	Netherlands	Interaction between several logistic providers, automated transportation systems	The total gains that can be obtained from bundling partial loads are not very high With greater flexibility in timings, the potential savings can increase significantly	Assumes that only same day orders can be matched
Deck and Wilson [41]	USA	The market effect of corporations monitoring consumers and charging preferential rates depending on search history is being investigated	If a merchant is able to trace customers and differentiate prices, consumers will face the same price as if the seller offered a single price to all buyers	How will seller tracking affect customer search patterns? The development of client search behaviors may have an impact on the extent to which merchants use tracking technologies

(continued)

Table 1 (continued)

Name, Author, Year	Country	Key focus	Key findings	Research gaps
Geng et al. [42]	USA	If the consumer assessment of raw materials can be additively separated, this article considers a simple criterion for the approximate optimality of a bundle	Bundling is essentially ideal if consumer values do not fall too quickly, according to the report. The paper uses the example of bundling to argue that if the value of following commodities declines quickly, bundling will be poor	No research gaps could be found
Choudhary et al. [43]	USA	The competitive impact of personalized pricing (PP). Companies charge distinct prices to different consumers based on their eagerness to pay	When the PP firm's quality is high, both firms' qualities rise in comparison to the uniform price situation While it is ideal for a company that adopts PP to increase product difference, a company that does not adopt PP tries to diminish differentiation by coming closer in the quality domain While PP broadens market penetration, it also intensifies pricing competition among enterprises. The type of cost function influences whether companies win or lose by applying such PP policies because it entails a change in equilibrium attributes	Only take into account each firm's single product offering, despite the fact that in practice, organizations frequently offer many items Presume that a company with the capability to customize prices is aware of each customer's willingness to pay for each feasible level of quality

(continued)

Table 1 (continued)

Name, Author, Year	Country	Key focus	Key findings	Research gaps
Hitt and Chen [44]	USA	The paper focuses on customized bundling, a pricing technique that allows customers to select up to a quantity M of goods for a fixed price from a pool of N verities of goods	The right bundle size for customized bundling reduces both the diversity of customer preferences and the marginal costs of production	Because sometimes physical products or services have the key qualities of information commodities, such as low marginal costs, customized bundling does not have to be confined to information goods
Iyer et al. [45]	USA	All product changes, bond type changes that make the product more attractive to the company's loyal customers, and conquest type changes that allow the company to make the product more attractive to the loyal customers of its competitors. An example of changing the conquest type	Retention changes have the potential to make a competitor's behavior less aggressive, lowering the degree of competition in the market. Conquesting changes typically increase competition, which is harmful to all downstream businesses	The cost of implementing the changes is not taken into account

Table 2 Factor n

Notations	Factors
Factor 1 (F1)	Information quality
Factor 2 (F2)	Labor constraints
Factor 3 (F3)	Buyer availability
Factor 4 (F4)	Product diversification
Factor 5 (F5)	Transportation constraints
Factor 6 (F6)	Inter-firm cooperation

3 Methodology

To learn more about the elements that influence the retail industry's decision to adopt bundling, keywords like bundling and retail were used to search electronic resources like Google Scholar, Scopus, and Science Direct. In addition, there was a thorough study of the journals where articles about supply chain bundling are anticipated to emerge often.

Following a thorough review of the required journals, research was conducted in India to speak with retail establishments and warehouse companies about their bundling procedures and the issues that need confronting. The experts were requested to conduct a pairwise analysis of the criteria in order to determine the most critical element for AHP implementation. The AHP ranking method is based on a ratio scale, with the proportion of values indicating the intensity of desire.

Satty Preference Scale [46] tells the different linguistic preferences as follows. The numerical values of linguistic preferences such as Equally preferred, Moderately preferred, Strongly preferred, very Strongly preferred, and Extremely preferred as denoted as 1,3,5,7and 9 respectively. Similarly, Equally to Moderately preferred, Moderately to Strongly preferred, Strongly to Very Strongly preferred and Very Strongly to Extremely preferred are valued as 2,4,6, and 8, respectively.

4 Results Analysis

F1 is the factor that is preferred over all other factors. F2 is moderately preferred over F3 and F5, while F2 is strongly preferred over F4. F3 is strongly preferred over F5. F6 is moderately preferred over F3 and F5. F5 however, is the least preferred among these factors (Table 3).

After this data was collected, the sum of each column was calculated. This was followed by dividing each value by the column sum of the respective columns. The resulting matrix formed is the normalization matrix. Further, taking the row average of each row in the normalization matrix provides the priorities for each factor (Table 4).

Table 3 Pairwise comparison matrix of the main factors with respect to goals

	F1	F2	F3	F4	F5	F6
F1	1	2	4	3	4	5
F2	0.5	1	3	2	3	0.5
F3	0.25	0.33	1	0.33	2	0.33
F4	0.33	1	3	2	3	2
F5	0.25	0.33	0.5	0.33	1	033
F6	0.2	2	3	0.5	3	1

Table 4 Normalized matrix and priorities

	F1	F2	F3	F4	F5	F6	Priorities
F1	0.4	0.32	0.28	0.42	0.25	0.55	0.37
F2	0.2	0.16	0.21	0.28	0.19	0.05	0.18
F3	0.1	0.05	0.07	0.05	0.13	0.04	0.07
F4	0.13	0.08	0.21	0.14	0.19	0.22	0.16
F5	0.1	0.05	0.03	0.05	0.06	0.04	0.06
F6	0.08	0.32	0.21	0.07	0.19	0.11	0.16

Based on the obtained priorities, it can be observed that factor F1 is the most preferred among the six factors and factor F5 is the least preferred. However, before the results can be concluded, there is a need to perform a consistency check to ensure that the values obtained from the experts are not random or biased.

5 Calculation of Inconsistency Ratio

There are three necessary steps for the calculation of the consistency ratio:

1. The measure of consistency for each criterion is calculated.
2. After finding λmax by taking the average of different consistency measures, the consistency index (CI) can be calculated.
3. The consistency ratio (CR = CI / RI, where RI is the Random Index). For six factors. RI = 1.24

$$\begin{pmatrix} 0.37 \\ 0.18 \\ 0.07 \\ 0.16 \\ 0.06 \\ 0.16 \end{pmatrix} = \begin{pmatrix} 1 & 2 & 4 & 3 & 4 & 5 \\ 0.5 & 1 & 3 & 2 & 3 & 0.5 \\ 0.25 & 0.33 & 1 & 0.33 & 2 & 0.33 \\ 0.33 & 0.5 & 3 & 1 & 3 & 2 \\ 0.25 & 0.33 & 0.5 & 0.33 & 1 & 0.33 \\ 0.2 & 2 & 3 & 0.5 & 3 & 1 \end{pmatrix} \begin{pmatrix} 0.37 \\ 0.18 \\ 0.07 \\ 0.16 \\ 0.06 \\ 0.16 \end{pmatrix}$$

The maximum value of λ is calculated by taking the average of all 6 values obtained from the above calculation, which was found to be 6.49. The consistency ratio was found to be 0.08 and the random index was taken as 1.24 for six factors as given by the random index table. Since, in our work, the consistency ratio can be observed to be less than 0.10, our work can be considered consistent.

6 Discussion

Information quality has always been considered a key factor in the bundling procedure. Several different types of information is required during this process including what is the perceived value of individual products or services by the customer. If this perceived value is incorrectly measured, the bundle might not succeed in the market. For instance, if the bundle is priced higher, customers will not buy the bundle and if priced lower, the company might lose on the full revenue potential. Considering the fact that Indian customers are highly cost-conscious and younger generations are becoming more purpose-driven, information quality becomes furthermore critical. For good information quality, there is a need of skilled labor who are not only involved in bundling but also have skills to perform good user research and collect valuable information.

Inter-firm cooperation is slightly preferred over product diversification. For the bundling processes to happen hand-in-hand, different firms in the supply chain should have good agreement and products should be packaged based on the information collected initially.

7 Conclusion and Future Work

Implementing bundling practices in the Indian market has always been a complex issue with a wide variety of factors. With new technological advancements, this becomes further difficult to manage considering the adoption acceptance by various firms involved in the supply chain.

From our findings, it is evident the quality of information is the most critical among all other factors to ensure bundling practices can be successfully adopted in India. Not only is the availability of information important, but also the accessibility and clarity of the new technology. Transportation is the least important among our factors and is however a significant matter for the Indian market.

In the future, this study can be further extended by taking more factors into account like revenue, number of products, etc. Additionally, it might be interesting to see how adoption is different in different geographical locations across India.

References

1. Chang KF, Yang HW (2015) How to adopt social behavior to achieve efficient social marketing. In: Robinson L (eds) Marketing dynamism & sustainability: things change, things stay the same. Developments in marketing science: proceedings of the academy of marketing science. Springer, Cham. https://doi.org/10.1007/978-3-319-10912-1_17
2. Kumar SR (2007) Consumer behaviour and branding: concepts, readings and cases—the Indian context. Chennai, India: Pearson Education India

3. Mazumdar T, Jun SY (1993) Consumer evaluations of multiple versus single price change. J Consumer Res 20(3):441–450
4. Kamakura WA, Wedel M, De Rosa F, Mazzon JA (2003) Cross-selling through database marketing: a mixed data factor analyzer for data augmentation and prediction. Int J Res Mark 20(1):45–65
5. Stremersch S, Tellis GJ (2002) Strategic bundling of products and prices: a new synthesis for marketing. J Mark 66(1):55–72
6. Harris J, Blair EA (2006) Consumer preference for product bundles: the role of reduced search costs. J Acad Mark Sci 34(4):506–513
7. Chan Y (2006) An analytic hierarchy framework for evaluating balanced scorecards of healthcare organizations. Can J Adm Sci 23(2):85–104
8. Bhagwat R, Sharma MK (2007) Performance measurement of supply chain management using the analytical hierarchy process. Prod Planning Control: Manag Oper 18(8):666–680
9. Varma S, Wadhwa S, Deshmukh SG (2008) Evaluation petroleum supply chain performance: application of analytical hierarchy process to balanced scorecard. Asian Pacific J Marketing Logist 20(3):343–356
10. Jovanovic J, Krivokapic Z (2008) AHP in implementation of balanced scorecard. Int J Qual Res 2(1):59–67
11. Bentes AV, Carneiro J, da Silva JF, Kimura H (2011) Multidimensional assessment of organizational performance: integrating BSC and A
12. Najmi A, Makui A (2012) A conceptual model for measuring supply Chain's performance. Prod Planning Control Manag Oper 23(9):694–706
13. Dey PK, Cheffi W (2012) Green supply chain performance measurement using the analytic hierarchy process: a comparative analysis of manufacturing organisations. Prod Planning Control Manag Oper.https://doi.org/10.1080/09537287.2012.666859
14. Leung L, Lam K, Cao D (2006) Implementing the balanced scorecard using the analytic hierarchy process and the analytic network process. J Oper Res Soc 57(6):682–691
15. Wang G, Huang SH, Dismukes JP (2004) Product driven supply chain selection using integrated multi-criteria decision-making methodology. Int J Prod Econ 91(1):1–15
16. Bhagwat R, Sharma MK (2009) An application of the integrated AHP-PGP model for performance measurement of supply chain management. Prod Planning Control Manag Oper 20(8):678–690
17. Agarwal A, Shankar R, Tiwari MK (2006) Modeling the metrics of lean, agile and Leagile supply chain: an ANP-based approach. Eur J Oper Res 173(1):211–225
18. Unahabhokha C, Platts K, Tan KH (2007) Predictive performance measurement system—a fuzzy expert system approach. Int J Benchmarking 14(1):77–91
19. Tsai WH, Hung SJ (2009) A fuzzy goal programming approach for green supply chain optimisation under activity-based costing and performance evaluation with a value chain structure. Int J Prod Res 47(18):4991–5017
20. Chan FTS, Qi HJ (2003) An innovative performance measurement method for supply chain management. Int J Supply Chain Manag 8(3/4):209–223
21. Seçme NY, Bayrakdaroğlu A, Kahraman C (2009) Fuzzy performance evaluation in Turkish banking sector using analytic hierarchy process and TOPSIS. Expert Syst Appl 36(9):11699–11709
22. Sun CC (2010) A performance evaluation model by integrating fuzzy AHP and fuzzy TOPSIS methods. Expert Syst Appl 37(12):7745–7754
23. Bhatt N, Guru S, Thanki S et al (2021) Analysing the factors affecting the selection of ERP package: a fuzzy AHP approach. Inf Syst E-Bus Manage 19:641–682. https://doi.org/10.1007/s10257-021-00521-8
24. Zhou Y, Liu M, Ma F, Luo N, Yin M (2021) Modelling and solving the supply marketing order allocation problem with time consistency and bundle discounts. J Oper Res Soc.https://doi.org/10.1080/01605682.2021.1932619
25. Wei F, Chen H (2021) Independent sales or bundling? Decisions under different market-dominant powers. J Indus Manag Optim 17(4):1593–1612. https://doi.org/10.3934/jimo.2020036

26. Caiati V, Rasouli S, Timmermans H, Bundling, pricing schemes and extra features preferences for mobility as a service: sequential portfolio choice experiment. https://doi.org/10.1016/j.tra.2019.09.029
27. Zhang X, Yue WT, Hui W (2020) Bundling cloud software to fight piracy: an economic analysis. Internet Res 30(1):191–241. https://doi.org/10.1108/INTR-10-2018-0455
28. Dolla T, Laishram B (2020) Bundling in public–private partnership projects—a conceptual framework. Int J Product Perform Manag 69(6):1177–1203. https://doi.org/10.1108/IJPPM-02-2019-0086
29. Operational role of retail bundling and its implications in a supply chain https://doi.org/10.1111/poms.13017
30. Dominique-Ferreira S, Antunes C (2020) Estimating the price range and the effect of price bundling strategies: an application to the hotel sector. Eur J Manag Bus Econ 29(2):166–181. https://doi.org/10.1108/EJMBE-04-2019-0066
31. Shao L, Li S (2019) Bundling and product strategy in channel competition. Intl Trans Oper Res 26:248–269. https://doi.org/10.1111/itor.12382
32. Liao S-H, Tasi Y-S (2019) Big data analysis on the business process and management for the store layout and bundling sales. Bus Process Manag J 25(7):1783–1801. https://doi.org/10.1108/BPMJ-01-2018-0027
33. Honhon D, Pan XA (2017) Improving profits by bundling vertically differentiated products. Prod Oper Manag 26:1481–1497. https://doi.org/10.1111/poms.12686
34. Xu Q, Xu B, Wang P, He Y (2018) Bundling strategies for complementary products in a horizontal supply chain. Kybernetes 47(6):1158–1177. https://doi.org/10.1108/K-02-2017-0082
35. Wan J, Zhao L, Lu Y, Gupta S (2017) Evaluating app bundling strategy for selling mobile apps: an ambivalent perspective. Inf Technol People 30(1):2–23. https://doi.org/10.1108/ITP-08-2015-0210
36. Chakravarty A, Mild A, Taudes A (2013) Bundling decisions in supply chains. Europ J Oper Res Elsevier 231(3):617–630
37. Setia P, Setia P, Venkatesh V, Joglekar S (2013) Leveraging digital technologies: how information quality leads to localized capabilities and customer service performance
38. Gomes PJ, Dahab S (2010) Bundling resources across supply chain dyads: the role of modularity and coordination capabilities
39. Schoenherr T, Mabert VA, The use of bundling in B2B online reverse auctions, J Oper Manag. https://doi.org/10.1016/j.jom.2007.05.001
40. van der Putten S, Robu V, Poutré HL, Jorritsma A, Gal M (2006) Automating supply chain negotiations using autonomous agents: a case study in transportation logistics
41. Deck CA, Wilson BJ (2006) Tracking customer search to price discriminate. Econ Inq 44:280–295. https://doi.org/10.1093/ei/cbj014
42. Geng X, Stinchcombe MB, Whinston AB (2005) Bundling information goods of decreasing value. Manag Sci 51(4):662–667
43. Personalized pricing and quality differentiation Vidyan and Choudhary, Anindya Ghose, Tridas Mukhopadhyay, and Uday Rajan Management Science 2005 51:7, 1120–1130 markets for product modification information Ganesh Iyer and David Soberman Marketing Science 2000 19:3, 203–225
44. Hitt LM, Chen P-Y (2005) Bundling with customer self-selection: a simple approach to bundling low-marginal-cost goods. Manag Sci 51:10:1481–1493
45. Iyer G, Soberman D (2000) Markets for product modification information. Mark Sci 19:203–225. https://doi.org/10.1287/mksc.19.3.203.11801
46. Saaty TL (1980) What is the analytic hierarchy process? Mathematical models for decision support, pp 109–121. https://doi.org/10.1007/978-3-642-83555-1_5

Chapter 25
Decision Support System Modelling and Analysis for Sustainable Smart Supply Chain Network

C. Sreerag, G. Rajyalakshmi, K. Jayakrishna, and Srinivas Viswanath

Abstract The chapter is more emphatic on the possibility of developing statistical techniques coping with the competition science to create a powerful decision making tool appropriate for sustainable supply chain network analysis. It highlights two independent techniques, to improve the quality of the outcomes in the decision making process. The former one focuses on the deployment of natural risk involved in the network. The concept of conditional probability is assigned to the conventional deterministic adjacency matrix in order to figure the uncertain interactions in the network. This will be useful to identify the crucial nodes in any supply chain and advantageous to classify these nodes based on their natural risk factor. The model concentrates on the delay time deviations from the mean delay time as the main competition. The combination of the two methodologies will find their position in improving the quality of the decision making actively.

Keywords Sustainable supply chain network · Adjacency matrix · Competition science · Orthopedic footwear

1 Introduction

Every established thought and logics behind any decision making technique carries a wide range of statistical and other data science tools. The drastic transformation of transportation techniques stimulated with scientific enhancement provoked the possibilities and capabilities of a good supply chain to a greater extent. Today, the art

C. Sreerag · G. Rajyalakshmi (✉) · K. Jayakrishna
School of Mechanical Engineering, Vellore Institute of Technology, Vellore, India
e-mail: rajyalakshmi@vit.ac.in

K. Jayakrishna
e-mail: jayakrishna.k@vit.ac.in

S. Viswanath
Narayana College of Engineering, Gudur, Andhra Pradesh, India

S. Kautish et al. (eds.), *Computational Intelligence for Modern Business Systems*, Disruptive Technologies and Digital Transformations for Society 5.0,
https://doi.org/10.1007/978-981-99-5354-7_25

of decision making is getting improved day by day to evolve itself into a total solution for uncertainty and risk assessment for complicated supply chain networks. The possibility of happening failures (anything that provides a negative impact to a defined or pre-set goal) cannot be denied to any supply chain network node. However, these failures need not be absolute for every node and, at the same time, be conditional one instead. This brings forth a broader canvas to equip the existing supply chain network with these conditional probabilities. This, therefore, definitely shows the present urge for novel techniques, with the viability to handle both certain and uncertain interactions between nodes in a single frame of reference.

The fundamental limitations of many conventional decision making tool make itself difficult to explore a wide range of practical applications in modern industrial scenarios. The study [1] noticed the limitation of conventional decision-making tools like the Bayesian approach in an economic perspective. The relevant article by Srinivas et al. [2, 3] explored similar findings as well and introduced another probabilistic methodology to address relevant social situations related to e-supply chain management (e-SCM), and [3] extended the technique used in a more generalized way.

The customer fulfilment and economical betterment in decision-making are crucial along with the environmental features of a sustainable supply chain. The blissful positive combination of technology with the supply chain analysis really made impact on many recent literatures with the innovative technical contextual of the modern supply chain; the article by Özceylan et al. [4] conferred the advantage of implementing 3D printing technologies, cooperated with the current traditional supply chain management in health care commerce. The life cycle assessment (LCA) in a sustainable supply chain network is highlighted in the article by Blass et al. [5]. The study focused on the advantages and influence of some LCA software like GaBi and ultimately how the software helped to improve the quality of decision made. This can be very well treated as a work to recognize the significance of LCA in the sustainable SCM and also to realize how well these concepts can be computerized. The technical support also benefited in the framing of proper decision and strategy making, with the help of statistical tools or algorithms that can bring about improvements in specific nodal position of the supply chain.

The article [6] "Metaheuristic optimization for automatic clustering of customer-oriented supply chain data" is yet another relevant case to explain the significance of statistical tools, optimization, and probabilistic techniques in supply chain management. However, there is very little contribution specified to sort out the continuous variations in probability of effecting failures in each node of the supply chain. This failure may come in the way of a transportation delay, supplier delay or on other influential factors thatare capable to bring negative effects on the supply chain network, partly or as a whole. Some research studies while utilizing a probabilistic perspective in supply chain management, focused on risk management, the article titled "Probabilistic Supply Chain Risk Model for Food Safety" by Convertino et al. [7] is a better example to be cited. Even though the study conclusively focused on the specific goals regarding risk management in health and food safety, the model failed in introducing the conditional probabilities that influence the supply chain system as well. Research studies [8–11] have similar issues as mentioned in article [7]. However, [10] is more significant as it is well equipped with the computerized risk management technique and article [11] is identified as a more managerial article, concentrated on trust issues inherent in a supply chain. The article, as desired, pointed out the social commitment to sustainable supply chain management.

Today, industries are trying to function in more dedicated and sustainable ways. Remaining compromised towards environmental pleasantness and ecological commitments is the latest method of developing and performing for the future. The transformation of a raw material to a usable product and the waste material after consumption is termed as the well-known "cradle to grave" movement of any product. The central and core awareness in this direction is well-shared in different articles and books. "Sustainable supply chains: a guide for small- to medium-sized manufacturers" from Jennifer et al. [12] discussed the importance of waste management in smart supply chain networks and development of improved decision making methodologies in view of these issues. The idea of different efforts, including external and internal drivers involved in the sustainable supply chain, discussed by Muhammad Amad Saeed and Wolfgang Kersten [13] found caring documents to identify the purpose of developing another novel methodology in the present scenario. The article [14] is more curious with the social influences of a supply chain and speaks out the social background of supply chain associated with the textile industries. The article calls for the requirements for more managerial and statistical support, vital for sustainable supply chain management. The pros and cons of benchmarking explained by Wai Peng Wong and Kuan Yew Wong [15] in their article "A review on benchmarking of supply chain performance measures" is obviously a fruitful discussion on the same topic. The article [15] is relevant as it covers benchmarking an essential phase among competitive suppliers in the sustainable supply chain. However, the improved methods and methodologies for analyzing the benchmarked competitors are relevant here. Article [16] identified similar ideas regarding benchmarking in the paint industry.

The impact of IoT on the sustainability of the supply chain is also found in some literatures. Instead of stating the direct environmental effects, many of these articles display the possible environmental hazards in the near future due to IoT tools. The research article by Naskar et al. [17], IoT carries a wide range of capabilities to help supply chain management through delivering cost-sparing, accuracy in inventory, and goods tracking. However, the degree of impact of these IoT tools on the individual supply chain network procedures isn't well addressed here. Articles [18, 19], well-established the movements of IoT on total wireless frames. This indicates the possibility of growing electronic gadgets in a few more centuries to come. The development of adequate decision making tools to address this ever-growing IoT tool is not yet materialized. The power of statistical tools can definitely resolve the problem to a greater extent.

The same IoT is accomplished to solve many social and environmental problems relating to the sustainable supply chain network. The best example for the above statement is the intelligent tools that have the potential to identify the amount of hazardous chemicals exhausted from the transporting vehicles tangled in the supply chain network and thereby providing backup to improve the sustainability of the supply chain with environmental considerations. Hence, IoT can be briefed as a positive factor to improve sustainability as well. The research article "An Integrated System for Regional Environmental Monitoring and Management Based on Internet of Things" [20] focused on climate changes, the study made use of IoT cloud computing to forecast the climate bounds mentioned in the article. The IoT-based decision making is a double-edged sword as it causes environmental issues due to the large dumping of e-waste. Correspondingly, IoT tools can build a strong organization, persons and suppliers selection, and so on, while article [21] envisages a similar event and discusses how we can use IoT to create strong governments. This is an interesting section and needs further discussions as a philosophy of science inherent. It is very clear that the possibility of introducing a new statistical methodology along with the existing IoT tools will definitely lead to powerful decision making package and may not be a fallacy for the future. The article demonstrates the decision making capacity of the tool even in the very managerial perspective.

The very affluent uncertainty in every sort of today's decision making can be addressed only through a probabilistic model which can be easily hold up and computerized. The possibility of equipping the supply chain network with conditional probability terms is still not in the arena. Such statistical decision making techniques will definitely be able to full fill the needs of many supply chain network of today with some probabilistic backup. Another important aspect is the integration of competition science into the scenario. Irrespective of the product or supplies, there will be certain competition in between the different suppliers involved in the supply chain network in order to reduce the deviation from their mean delay time. Taking this factor as competition criteria, one model can be developed to perform decision making in a competition perspective. The significance of competition science in a sustainable supply chain network is evident, the output of the model will individually analyze each supplier and their competitive advantage, over other suppliers and above all how the suppliers contribute to the mean delay of the product to the market. The probabilistic results will be very helpful to end up in more proper and practical decisions with reference to delay time in the supply chain; however, currently, very few studies confessed the possibility of probabilistic science in the field of sustainable supply chain network. This is the huge literature gap noticed in the area. On imparting the probabilistic-adjacency metric for network analysis along with the probabilistic competition science model taken together determines the behaviour and trustworthy of individual supplier involved in the chain [22–24]. This will certainly serve better strategy for proper decision making. The feasibility of statistical tools coped up with the existing IoT tools and computerization, in order to attain maximum performance, still remains as a relevant concern of modern industries. This signifies the importance of introducing less complicated statistical approach in the area, with more applicable decisional outcome, based on probabilistic values and current dynamism. The perception of circular economy discloses the importance of intra-supplier network to improve the overall efficiency of the supply chain network as well. This shows the necessity of a model which is suitable to peep into the intra-supplier interactions concurrently [25].

The major focus of the article is on the possibility of developing statistical techniques by considering the dynamic competence to create a powerful decision making tool appropriate for sustainable supply chain network analysis. The objective of the work is to improve the quality of the sustainable supply chain outcomes with a suitable decision support system for decision making process.

2 Mathematical Modelling

The mathematical modelling proposed includes two different sections. The former includes the mathematical model to calculate the natural risk involved at each node of the supply chain network. This probabilistic model helps to identify the deployment of natural risk in terms of probability values using the concept of probabilistic adjacency matrix and conditional probability. The latter part is about including competition science into the system, in order to analyze each supplier individually and their impact on the overall delay of the product to the market. This is done with a view to highlight the mean delay time, corresponding probabilistic values of realization at different nodes and thereby assisting better decision making.

2.1 Supply Chain Network Analysis Using Probabilistic Adjacency Matrix

Adjacency matrix is the matrix representation of all possible interaction between different nodes in any supply chain network. These interactions may be refereed to be of any kind like transportation of data, material manpower correlation and so on. In deterministic perspective the elements of adjacency matric are zeros and ones, because there are only two possibilities that exist, whether there is guaranteed interaction or not. If the interaction is certain then the element in the adjacency matrix will be one *($a_{ij} = 1$)* and if not the corresponding element will be zero *($a_{ij} = 0$)*. The adjacency matrix A can be represented mathematically as follows

$$A = \begin{pmatrix} a_{11} & \dots & a_{1n} \\ \vdots & a_{ij} & \vdots \\ a_{m1} & \cdots & a_{mn} \end{pmatrix} \forall i, j \le m, n \;\&\; a_{ij} = \begin{matrix} 1, & \text{interactions} \\ 0, & \text{non - interactions} \end{matrix} \tag{1}$$

where the numerical values of *m, n* will be the same and are equal and represent the total number of nodes present in the supply chain network. Also, $a_{ij} \neq a_{ji}$ provided there is no double-sided interaction possible, and, $a_{ij} = a_{ji} = 1$. If there is such interactions exist. Consider a four-node network and the corresponding adjacency

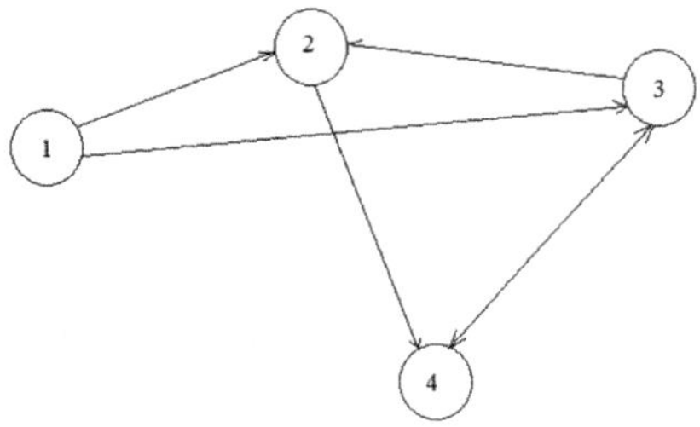

(a)

$$\begin{bmatrix} 0 & 1 & 1 & 0 \\ 0 & 0 & 0 & 1 \\ 0 & 1 & 0 & 1 \\ 0 & 0 & 1 & 0 \end{bmatrix}$$

(b)

Fig. 1 **a** Pictorial representation of a network and **b** Adjacency matrix for the network shown in (**a**)

matrix as shown in Fig. 1a, b meant for a better understanding of the concept of adjacency matrix.

There are certain situations where some interactions may not be completely deterministic in nature and at the same time have some probabilistic constraints. Coming to practical situations most of our interactions are likely to be of probabilistic by nature. Hence, more focus should be given to address the problem in a probabilistic point of view. Let us consider the case Fig. 1a, b for better understanding of the situation described. If all the interactions in Fig. 1a, b are not deterministic ones, then we have to talk about the possibility of nodes getting interacted one another corresponding to each link given. Hence the resulting probabilistic adjacency matrix will be generally different from that of a deterministic case given above and is demonstrated below in Fig. 2a, b.

The idea of conditional probability is more significant as the supply chain network is itself engaged with some product, process and data flow between different nodes. Consider a thought experiment; we are expecting one failure in one of the assembling unit (AU) in our manufacturing process. Suppose this single failure is quite certain to occur, then as per production flow, the probability of happening of this failure ($P(f)$) at the very first unit is least compared to other nodes (assuming the process flow

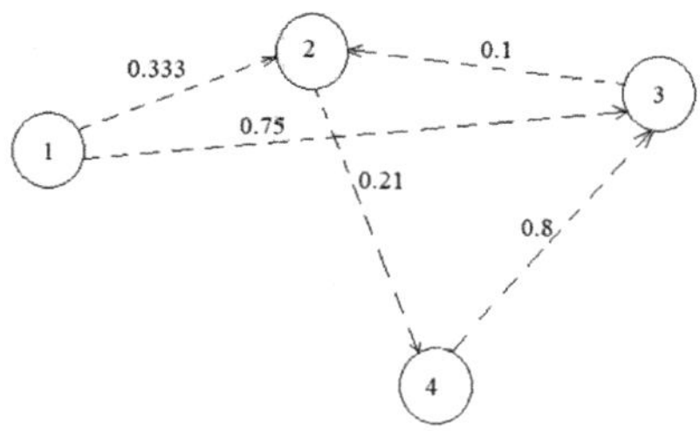

(a)

$$\begin{bmatrix} 0 & 0.33 & 0.75 & 0 \\ 0 & 0 & 0 & 0.21 \\ 0 & 0.10 & 0 & 0 \\ 0 & 0 & 0.80 & 0 \end{bmatrix}$$

(b)

Fig. 2 **a** Pictorial representation of a network with probabilistic interactions at different node **b** Adjacency matrix for the same network (**a**)

to be in series). Moving towards the final unit, the probability of occurrence of this single failure at the node gets increased as well. A similar situation corresponding to a parallel connected AU network is shown in Fig. 3a, b. Here the situation is relatively easy to handle as it is connected in series. However, if the network is more complicated with different branches of interactions, it would be difficult to obtain the distribution of the natural pass through probability associated to the different nodes of the network.

In the case of a supply chain network, the network may not be as simple as judged in the series and parallel connected cases described above. The natural pass through chances and risk probability associated with each node will be a function of the various links tangled with the corresponding node. Let the pass through possibility of a particular *i*th node be expressed as a function, say ϕ_i. Hence the natural risk involved in that particular node will be, $(1-\phi_i)$, Likewise, the risk probability be represented by R_i. The incoming links towards a particular node is expressed as I_{ij}, where i represents the corresponding node to which it is associated with and j represents towards which node respectively. Similarly, the outcoming links towards a particular node can be expressed as O_{ij}, where i represents the corresponding node

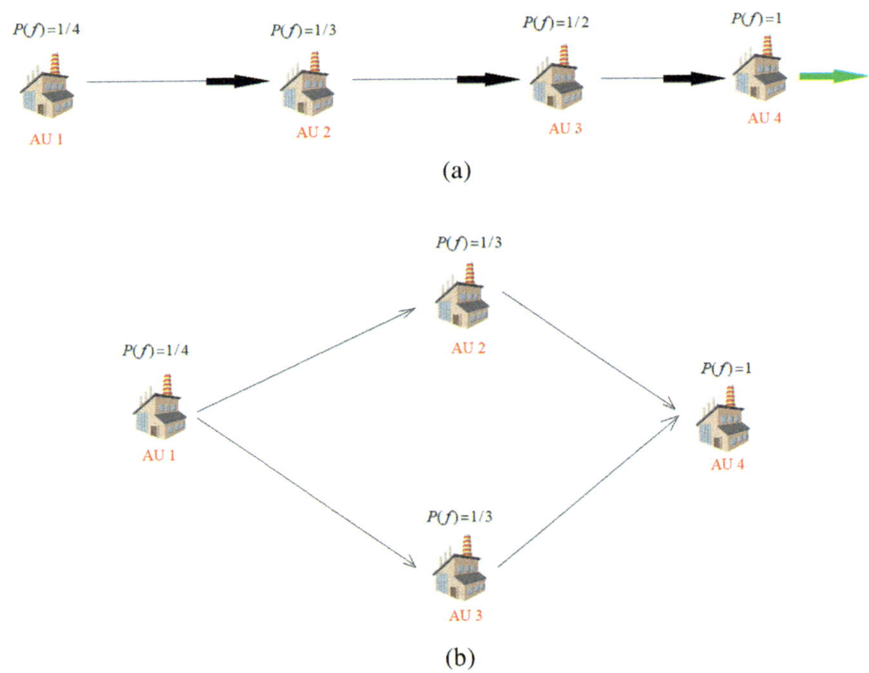

Fig. 3 **a** Pictorial representation of the assembling units in a network, connected in series, of a manufacturing process with corresponding conditional probability of a single failure at one of the nodes and, **b** represents the pictorial representation of the parallel connected network of assembling units of a manufacturing process with corresponding conditional probability of single failure at one of its nodes

to which it is associated with and *j* represents leaving from which node respectively. The total links can be represented into a single set $\{Q_{ij}\}$.

$$\text{Where, } \{Q_{ij}\} = \{I_{ij}, Q_{ij}\} \, \forall i, j, \in N \quad i, j \leq m, n \tag{2}$$

Theorem 1 In order to allocate conditional probability (P_{ij}) for choosing a particular link *(i, j)* (pass through probability), a logical methodology is proposed. That is if the node diverts into different links the probability of the previous link is shared among the newly proceeding links along with a predefined weightage w (O_{iy}) allocated already (based on different factors under consideration). If there is no such weightage allocated to the links, then these probabilities will be uniformly distributed.

If there is no further connectivity available in a node (the node itself is an end of the network), we should assume a hypothetical link with some probabilistic value. If the possibility of further extension of the node is null, we consider this probability as 0 and if quite certain it is taken as 1. However, in most of the cases, it is only probable, not certain and hence, often has to allocate an appropriate probability term subjectively or otherwise better consider 0.5 as this probability, which shows a 50–50 chance. The hypothetical link can be represented as diagonal elements in the conditional probability matrix *(P)*.

$$P_{ij} = \frac{w(O_{iy}) \sum_{r=1}^{m,n} P_{rj}}{\sum_{y=1}^{m,n} w(O_{iy})} \, \forall i, j, r, y \in N, \quad i, j, r, y \leq m, n \tag{3}$$

The elements of probabilistic adjacency matrix (a_{ij}) can be incorporated along with Eq. (3) to estimate the total pass through chances. Hence the natural risk involved in each node of the supply chain network can be expressed as

$$R_i = 1 - \phi(\{a_{ij}.P_{ij}\}) \, \forall i, j \in N, \quad i, j \leq m, n \tag{4}$$

The selection of function ϕ totally depends on the situation where it is used and can be a simple mean or more complicated function. The model is more relevant where the multi-supplier interaction is active. The circular economy is the best example to be incorporated here. Though the model is developed in a more general way, it can be equally implanted well on a circular economic model or environment where complicated network is quite visible in between supplier—supplier, supplier—manufacturer and manufacturer—manufacturer unit. Let's focus on the impact on the corresponding risk probabilities through allocating different known functions of ϕ. This may include functions like those of simple mean, geometric and harmonic mean.

2.1.1 Simple Arithmetic Mean Function

In the simple arithmetic mean function (5), the probabilistic weightage is more towards the higher probabilities and this will give more dominance towards the links getting active earlier than those coming later. Hence this mean function helps to estimate the associated risk probabilities, attaching more weightage towards the initially active links.

$$\phi_i\left(\left\{a_{ij}.P_{ij}\right\}\right) = \frac{\sum_{j=1}^{m,n} a_{ij}.P_{ij}}{n\left(\left\{I_{ij}\right\} \cup \left\{O_{ij}\right\}\right)} \quad \forall i, j \in N \;\; i, j, \leq m, n \tag{5}$$

where $n\left(\{I_{ij}\}\cup\{O_{ij}\}\right)$ represents the total number of links attached with ith node.

Therefore, Eq. (4), in terms of simple arithmetic mean function, will be,

$$R_i = 1 - \frac{\sum_{j=1}^{m,n} a_{ij}.P_{ij}}{n\left(\left\{I_{ij}\right\} \cup \left\{O_{ij}\right\}\right)} \quad \forall i, j \in N \;\; i, j, \leq m, n \tag{6}$$

This arithmetic mean is needed to identify the associated natural risk at different nodes in terms of delay time perspective, as the links with higher delay time are given higher weightage. The same function is also applicable to measure the corresponding data transferring rate on account of risk deployment as well.

2.1.2 Geometric Mean Function

As we are aware, Geometric Mean generally performs better in economic and financial data analysis and is especially used for developing stock indexes as well. Majority of the value line indexes maintained by financial experts make use of GM as their averaging tool. In this type of index, all stocks have equal weights, regardless of their market capitalizations or prices. The index is calculated by taking the geometric average of the percentage change in prices of each stock.

$$\phi_i\left(\left\{a_{ij}.P_{ij}\right\}\right) = \left[\prod_{j=1}^{m,n} a_{ij}.P_{ij}\right]^{1/n\left(\left\{I_{ij}\right\}\cup\left\{O_{ij}\right\}\right)} \quad \forall i, j \in N \tag{7}$$

Therefore Eq. (4), in terms of geometric mean function, will be,

$$R_i = 1 - \left[\prod_{j=1}^{m,n} a_{ij}.P_{ij}\right]^{1/n\left(\left\{I_{ij}\right\}\cup\left\{O_{ij}\right\}\right)} \quad \forall i, j \in N \;\; i, j, \leq m, n \tag{8}$$

2.1.3 Harmonic Mean Function

The harmonic mean is largely used in the risk assessment related to finance. The weighted harmonic mean is used in finance to average the multiples like the price-earnings ratios because it gives equal weight to each data point. Using a weighted arithmetic mean to average these ratios would give greater weight to high data points than low data points because price-earnings ratios aren't price-normalized while the earnings are equalized. This method is more appropriate for rate-related factors like price rate, data transfer rate, rate of delay time, and so forth.

$$\phi_i(\{a_{ij}.P_{ij}\}) = \frac{n(\{I_{ij}\} \cup \{O_{ij}\})}{\sum_{j=1}^{m,n} (a_{ij}.P_{ij})^{-1}} \forall i, j \in N \;\; i, j, \leq m, n \tag{9}$$

Therefore Eq. (4) in terms of harmonic mean function, will be,

$$R_i = 1 - \frac{n(\{I_{ij}\} \cup \{O_{ij}\})}{\sum_{j=1}^{m,n} (a_{ij}.P_{ij})^{-1}} \forall i, j \in N \;\; i, j, \leq m, n \tag{10}$$

The three Eqs. (6), (8), (10) are in fact deduced from the main Eqs. (2), (4) and provide the three different ways to assess the natural risk involved in the supply chain network, in any sector. However, the feasibility of each equation is entirely dependent on the situation involved and the criteria related to the risk factors associated with delay time, data transfer rate, economic and financial constraints, and so on.

2.2 *Intra-node Competition Model in the Supply Chain Network*

Intra-node competition may be referred to as competition between different nodes of the sustainable supply chain network. Despite of the deployment of natural risk along the nodes, the competition between nodes also affect the overall efficiency of the supply chain network. Here, the total number of nodes present in the system is more important than the possible links present in between them. If all the competitors (nodes) are competing to achieve certain goal within a competition criteria, and have defined levels or rank among these nodes, then the outcome probability of each rank or level (here the different clusters of nodes of supply chain network) will be variant in accordance with the number of failure gates assigned. The concept of number of failure gates and the equations of outcome probabilities are well explained by Srinivas in the article [2]. The numerical totality of all adverse effects in the competition, ultimately results in the betterments, in terms of performance, of low standard competitors (nodes of supply chain network), which is termed as the number of failure gates λ.

Theorem 2 If there are different nodes involved in the competition, with some defined competition criterion that permits the competing nodes in the network to be classified themselves into clusters of different ranks (n) or levels, then there exists an equilibrium rank (n_e) or level with some probabilistic value ($P(n_e)$) of its occurrence in the corresponding rank-probability distribution and remains independent of the number of failure gates λ assigned.

From the competition model developed in article [2], the mathematical expression for the outcome probability is,

$$P(n) = \frac{\lambda n_1 R_{M-2} + \left(\frac{M}{2} - \lambda\right) n_2 R_{M-2}}{R_M} \tag{11}$$

where, R_M is the number of all possible competition arrangements for M ranked competition, $R_{M\text{-}2}$ denotes the total number of competition arrangements among M competitors through fixing one of the competitors and, n_1 and n_2 represent the total number of nodes in front of the nth level and below the nth level respectively. M_{C2} is the total number of combinations of M levels and hence, $M = n_1 + n_2 + 1$, $R_{M=}$ M_{C2} $R_{M\text{-}2}$.

The general Eq. (11) together with the Theorem 2 indicates the existence of the equilibrium rank and corresponding equilibrium probability as given below, where n_{e1} and n_{e2} represent the total number of nodes in front of the n_eth level and below the n_eth level, respectively. Thus,

$$P(n_e) = \frac{\lambda n_{e1} + \left(\frac{M}{2} - \lambda\right) n_{e2}}{{}^{M}C_2} \; \forall \lambda \in R \;\; 0 \le \lambda \le M/2 \tag{12}$$

Since we are considering clusters of nodes having the same rank or level, it's reasonable to sum up these individual nodes *($A(n_i)$)* all together (frequency of n_ith rank).

$$\text{Therefore, } \sum_{i=1}^{n_e-1} A(n_i) = n_{e1} \;\text{ and }\; \sum_{i=n_e+1}^{M} A(n_i) = n_{e2} \tag{13}$$

On substituting Eqs. (13) on (12) we have

$$P(n_e) = \frac{\lambda \sum_{i=1}^{n_e-1} A(n_i) + \left(\frac{A}{2} - \lambda\right) \sum_{i=n_e+1}^{M} A(n_i)}{{}^{A}C_2} \; \forall \lambda \in R, \;\; 0 \le \lambda \le A/2 \tag{14}$$

where, $\lambda \sum_{i=1}^{n_e-1} A(n_i)$, the total number of nodes available.

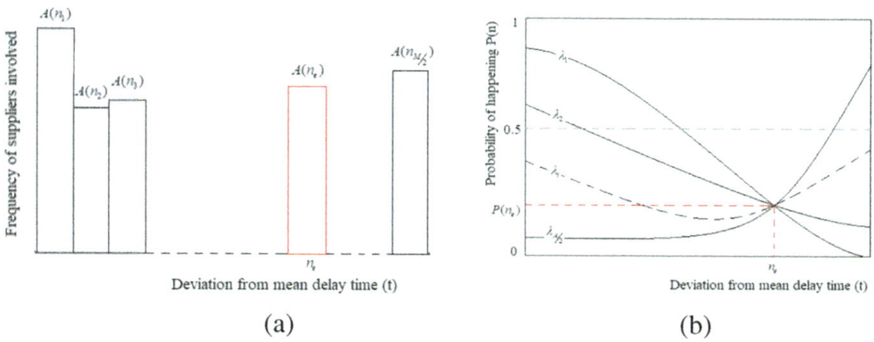

Fig. 4 **a** Frequency distribution of number of suppliers in a cluster having a certain rank or level, and **b** Deviation in delay time from their mean delay time and its corresponding probability of realization or occurrence

The model pictorially represented in Fig. 4a, b is shown below for better understanding of the system mentioned above. The competition criteria considered here is the delay time deviation from the mean delay time. All those suppliers having less deviation from their mean delay time are ranked top and clustered under different level or rank. The frequency distribution of each cluster (represented as strip) is given in Fig. 4a. The corresponding expected plot for mean delay time and its probability of occurrence with reference to the Eq. (14) is given in Fig. 4b. The term equilibrium deviation discussed here is nothing but the expected deviation in delay time from their mean delay time, represented by n_e. The associated probability of the realization of this expected deviation from the mean delay time is obviously related to the distribution of the frequency of each cluster, magnitude of the frequency of suppliers in a particular cluster, and number of clusters formed. Hence, when all the suppliers are in competition to reduce their deviation from the mean delay time, the probability of its happening will be generally different, represented by $P(n_e)$. Moreover, the existence of this equilibrium condition will be independent of the number of failure gates assigned and can be derived through equating Eq. (14), for two different number of failure gates λ_i and λ_k, or simply taking n_e as the nth portion where the better and worse halves get separated, that is, where $\sum_{i=1}^{n-1} A(n_i) = \sum_{i=n+1}^{M} A(n_i)$.

3 Implementation of the Model Developed

The mathematical model developed here helps to manage the risk involved in each node of the supply chain network by measuring the natural risk probability on that particular node using the corresponding probabilistic adjacency matrix and the conditional probability attached. The mathematical philosophy behind this model is already explained in Theorem 1. Whilst the implementation of this competition model paves the path to identify the possibility of imposing a suitable competition criteria like: deviation from mean delay time of the supplier, at different nodes with their corresponding probabilities of occurrence, makes the model itself practically relevant and equally applicable. In addition to this model, one should aware of, other operations like the environment-related LCA analysis that can be incorporated with the existing model for better decision outcomes of the supply chain. This related LCA analysis will very well include the associated environmental impact of the supply chain caused by transportation, industrial smoke, and other factors concerned. However, this article tries to be more specific on the mathematical model, within a probabilistic purview, accounting for the existing dynamism persistent in the modern supply chain network. The flow chart for the methodology for making more accurate decision making is presented in Fig. 5. The problem is how to make the supply chain more sustainable through developing more feasible and powerful decision making tools or revamping the existing decision support system (DSS) itself. While doing so, the two individual objectives of the study have to be well defined. The first one is to improve the quality of the entire decision making process itself and the second one is maintaining the environmental friendliness. The quality of decision making can be improved through empowering the existing DSS by considering the quite inherent supply chain risk management and suitable competitive probabilistic model in account. The environmental factors still remain significant to make a supply chain more sustainable and eco-friendly. However, the environmental friendliness of supply chain and related LCA analysis is out of coverage of this present study. The DSS components are briefly classified into three. The first one is the extraction of the required data; this is nothing but data mining, the extracted data then goes through logical and computerized platform for further analysis and hence leads to productive results. This forms the second component. The final component of the DSS is the user interface system. The results of the analysis after being exported to the user interface system without any muddles, the same user interface system delivers them to the manager or user in a readable and accessible format. This may include graphs, bar charts, and so on. The mathematical chapters in DSS can be therefore generally classified as software and statistical tools.

There are a few different ways to classify the DSS, in the application perspective. DSS may not be fully deployed into a specific category but conveniently a blend of at least two structures is possible. Some authors [22] categorized DSS into the accompanying six structures: database-oriented DSS, text-oriented DSS, spread sheet-oriented DSS, rule-oriented DSS, solver-oriented DSS, and compound DSS. A compound DSS is the most common among these; it is a hybrid framework that incorporates at least two of the five essential structures. The decision support provided by DSS can be divided into three interrelated categories [23], that is, Personal Support, Organizational Support, and Group Support. Finally, the DSS segments can be classified further into four, input, user knowledge and expertise, output, and decisions.

The overall representation of DSS parts in the sustainable supply chain network is represented in the Fig. 6. How the mined data gets transferred in the system and the different steps taken are shown in Fig. 6. The same model is used to analyze the supplier–manufacturer case study in the following section.

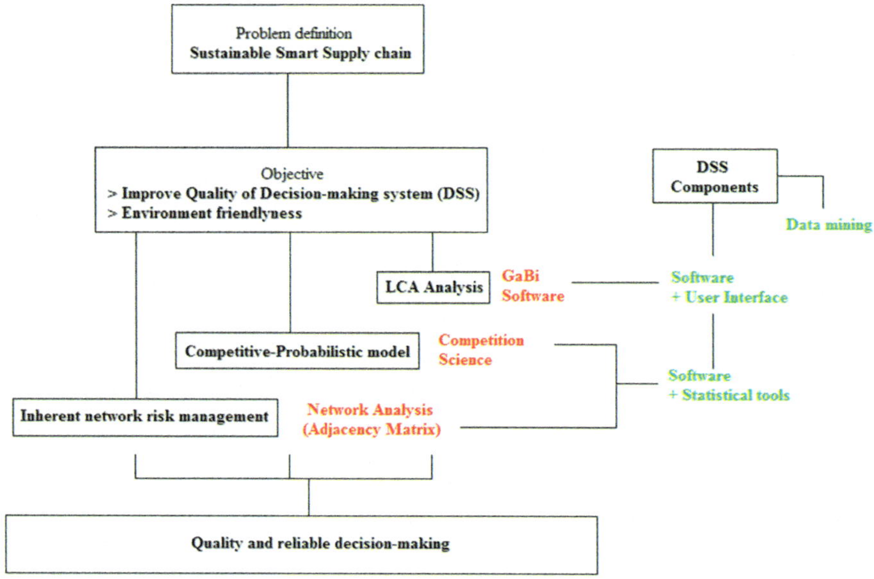

Fig. 5 Flow chart of the objectives and related DSS components to improve the quality of the decision making process

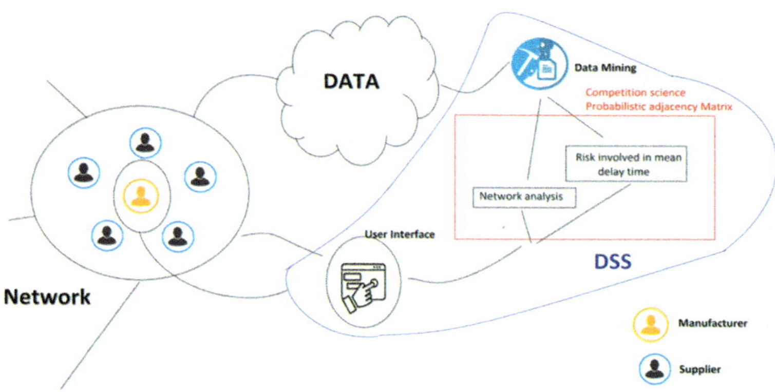

Fig. 6 Schematic model for the DSS equipped with the competitive science deployed in the supply chain network highlighting the flow of data through the DSS

The data from the supplier–manufacturer environment related to the deviation from their mean delay time is extracted using a data mining tool. The extracted data regarding the deviation from the mean delay time for each supplier is further transferred to the decision support system, which is equipped with the inherent risk analysis, using an adjacency matrix and conditional probability and the risk involved in the total delay time for the product to get manufactured in the light of competition science. The objective of the study is mainly focused on these two tools referred to inside the red-colored box shown in Fig. 6 above. The results obtained in the study are then carried to the user interface and from there to the manager. This well-deployed data and results can be transferred to a cloud or to the manufacturer itself, and ultimately be helpful to improve the quality of decision making process.

4 Case Study—In an Orthopedic Footwear Manufacturer Perspective

A contextual study is incorporated along with the mathematical model introduced in the former sections. The key objective of the case study is to understand practical feasibility of the model developed. The case study is carried out in a manufacturer perfective along with the three major influencing suppliers involved. The concept of circular economy is also stitched up on, as the intra-supplier interactions are considered.

4.1 Description of the Problem

The orthopedic footwear manufacturer having their manufacturing unit located in Kunnamangalam, Kerala, India, required three major different supplies for its production. These include fiber sole, rubber, and leather. Three different suppliers are located in different geographical locations, as shown below, to fill the demand of the manufacturer. Fundamentally two networks are assigned here, one is meant to describe the possible data exchange in between different nodes of the supply chain and the second refers to the material transfer network associated with the problem. The two networks along with the geographical deployment of the different suppliers and the manufacturer concerned are given in Fig. 7a, b shown below. The direction of iteration in the same representation is made using the arrow symbol.

(a) (b)

Fig. 7 **a** Network of data transfer in between the suppliers and the manufacturer in the supply chain **b** Material transfer network in between the same suppliers and manufacturer in the supply chain

The possibility of a circular economy is also attached here; focusing on the link (2–3) which represents the material transfer network between suppliers, and where the excess waste chemical and dust powders like dimethyl amine find useful in leather industry to remove the hair parts in the leather. Hence the circular economy exists in between suppliers in order to reduce wastage and other economic benefits, resulting in more sustainability. Likewise, the probabilistic adjacency matrix is provided for both data transfer and material transfer networks discussed above. Since almost all links in the network are almost most likely to occur, they are given a very high probability like 0.9, as done here. However, the probability of occurrence of links like (2–3) in the material transfer network is generally less compared to those that are almost certain to occur, the probability assigned them is 0.5. The probabilistic adjacency matrix for adjacency matrix for data transfer network is represented by (A_{Data}) and the corresponding material transfer network by ($A_{Material}$). Hence the associated A_{Data} and $A_{Material}$ are put in the form given below (considering the manufacturer as the fourth node in the network).

$$A_{Data} = \begin{pmatrix} 1 & 0 & 0 & 0.9 \\ 0 & 1 & 0 & 0.9 \\ 0.9 & 0.9 & 1 & 0.9 \\ 0 & 0 & 0.9 & 1 \end{pmatrix} \tag{15}$$

$$A_{Material} = \begin{pmatrix} 1 & 0 & 0 & 0.9 \\ 0.5 & 1 & 0 & 0.9 \\ 0 & 0 & 1 & 0.9 \\ 0 & 0 & 0 & 1 \end{pmatrix} \tag{16}$$

In addition to this, the deviation of the delay time of different suppliers in the system in the last six months was observed and plotted and treated as the standard deviation of the distribution of the associated delay time data. To achieve complete production, it is required to consider other direct suppliers involved other than the three major suppliers already discussed. Here, in this particular case study, 11 more small suppliers participate in the supply chain. Now the total number of suppliers becomes 14 altogether. The geographical deployment of these suppliers is represented in Fig. 8.

The mean delay time taken by each supplier in the last six month is tabulated and given in Table 1, along with their standard deviation.

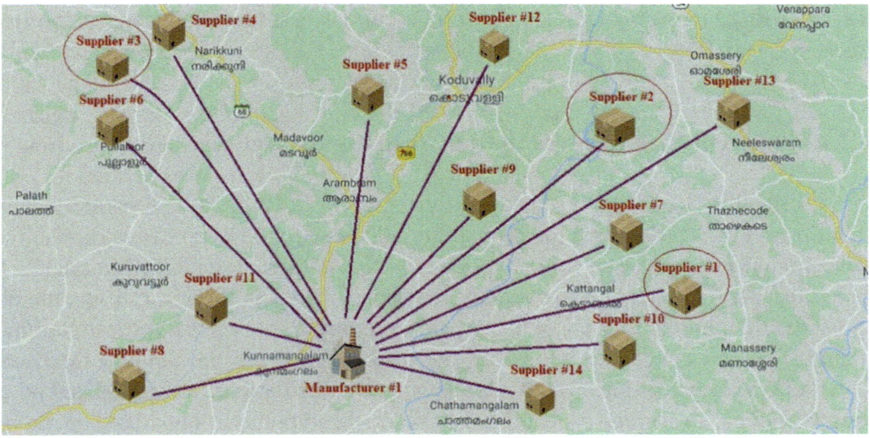

Fig. 8 The geographical deployment of all 14 direct suppliers involved in the supply chain. The three major suppliers are represented inside the red-colored circle

Table 1 Mean delay time and standard deviation of each supplier involved in the supply chain

Supplier	Supply	Mean delay time (days)	Standard deviation
Supplier #1	Leather	1.5	0.2
Supplier #2	Rubber	2	0.6
Supplier #3	Fiber sole	4	0.1
Supplier #4	Supply #4	0.5	0.1
Supplier #5	Supply #5	2	0.2
Supplier #6	Supply #6	3	0.4
Supplier #7	Supply #7	5	0.8
Supplier #8	Supply #8	5	1
Supplier #9	Supply #9	1	0.8
Supplier #10	Supply #10	2.5	0.1
Supplier #11	Supply #11	6	0.2
Supplier #12	Supply #12	5	0.5
Supplier #13	Supply #13	2	0.6
Supplier #14	Supply #14	1	0.3

4.2 *Inherent Risk in the Supply Chain Network*

Inherent risk associated with the supply chain network can be analyzed using the model developed. In this connection, two different networks can be observed. The first one is about data transfer and the second deals with the associated material transfer. Hence the network analysis for inherent risk is studied under two respective Sects. 4.1 and 4.2.

4.2.1 Inherent Risk Involved in Data Transferring Network

As described in Sect. 4.1, the data transferring network (Fig. 7a) is separately identified from Fig. 9 along with the probability allocated to the hypothetical links. The concept and importance of these hypothetical links are well explained in Sect. 2.1. Since the data transferring network is assumed to have null data transfer outside the network, a probability equal to 0.0 can be deployed along the hypothetical links of the suppliers concerned. However, the data output from the manufacturer still remaining as a case of uncertainty, due to possible fluctuations in demand and other factors. Hence, an average probability is given to the hypothetical link corresponding to the manufacturer. Figure 9, therefore, represents the data transferring network and the hypothetical links associated. The conditional probability (using Eq. 3), originating from manufacturer perspective, is allocated, above to each link of the data transferring network, as shown below.

Assuming every link has equal weightage throughout the network. Therefore,

$$w(O_{iy}) = 1 \,\forall i, y \in N, \quad i, y \leq m, n \tag{17}$$

We have the total links available in the network as,

$$\begin{aligned}\{Q_{ij}\} &= \{I_{ij}, O_{ij}\} = \{\{I_{ij}\} \cup \{O_{ij}\}\} \\ &= \{Q_{11}, Q_{14}, Q_{22}, Q_{24}, Q_{31}, Q_{32}, Q_{33}, Q_{44}\} \forall i, j \in N, \quad i, j \leq 4\end{aligned} \tag{18}$$

The conditional probability matrix P obtained from 2.3,

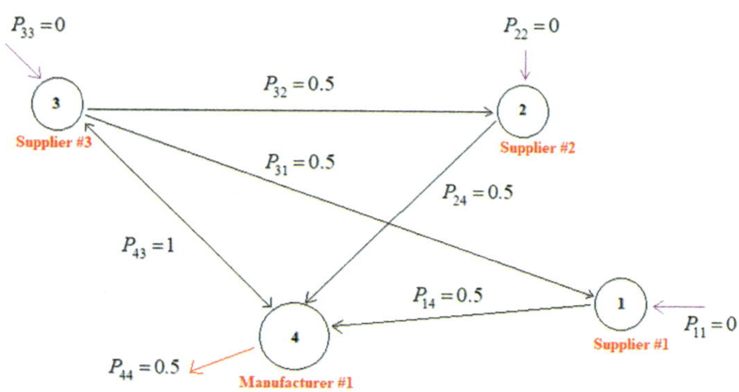

Fig. 9 Data transferring network with the conditional probability allocated above to each link. The plot also comprises the hypothetical link and corresponding probability at each node

$$P = \begin{pmatrix} 0 & 0 & 0 & 0.5 \\ 0 & 0 & 0 & 0.5 \\ 0.5 & 0.5 & 0 & 1 \\ 0 & 0 & 1 & 0.5 \end{pmatrix} \tag{19}$$

Also, the probabilistic adjacency matrix for the data transferring network is given in the problem description (refer to Eq. 15). On account of the Eqs. 15 and 19, the collection of product elements $\{a_{ij}\ P_{ij}\}$ becomes

$$\{a_{ij}.P_{ij}\} = \begin{pmatrix} 0 & 0 & 0 & 0.45 \\ 0 & 0 & 0 & 0.45 \\ 0.45 & 0.45 & 0 & 0.9 \\ 0 & 0 & 0.9 & 0.5 \end{pmatrix} \tag{20}$$

On assuming the data operation as a simple mean function (the later analysis being regarding the delay time, simple mean function is more relevant here), the probability of happening (ϕ) of the product element may be calculated using the Eq. 5 mentioned above in Sect. 2.1.1, as given below.

$$\phi_i(\{a_{ij}.P_{ij}\}) = \frac{\sum_{j=1}^{4} a_{ij}.P_{ij}}{n(\{I_{ij}\} \cup \{O_{ij}\})}\ \forall i, j \in N,\ \ i, j \leq 4 \tag{21}$$

And correspondingly the risk probability attached can be calculated using Eq. 6. Hence, Eq. 6 becomes,

$$R_i = 1 - \frac{\sum_{j=1}^{4} a_{ij}.P_{ij}}{n(\{I_{ij}\} \cup \{O_{ij}\})}\ \forall i, j \in N, i, j \leq 4 \tag{22}$$

The risk probability associated with each node is calculated using the Eq. (22). These results are represented in Table 2.

The maximum risk probability in the data transfer network is with regard to the leather and rubber suppliers and the minimum risk probability is on the side of the

Table 2 Inherent risk probability associated with each node in data transferring network

Node	Supplier	Supply	Probability of happening (ϕ_i)	Risk probability (R_i)
1	Supplier #1	Leather	0.30	0.70
2	Supplier #2	Rubber	0.30	0.70
3	Supplier #3	Fiber sole	0.54	0.46
4	Manufacturer #1	–	0.64	0.36

manufacturer itself. It is clearly visible that data volatility toward the manufacturer is very smooth compared to other nodes since every node itself communicating to the manufacturer will reduce the risk probability of the manufacturer in the data transferring network.

4.2.2 Inherent Risk Involved in Material Transferring Network

The material transferring network (Fig. 7b) can be independently identified from Fig. 10 together with the probability allocated to the hypothetical links created. Since this data-transferring network has an uncertainty over their preceding nodes, it is better to allocate a probability of 0.5 along the supplier's hypothetical links. Similarly, the material output from the manufacturer is still a case of uncertainty, the probability along the manufacturer's hypothetical link can be assumed to be 0.5 (providing 50–50 chances). Figure 10, represents the material transferring network along with the hypothetical links associated. The conditional probability (using Eq. 3), originating from manufacturer perspective, allocated above to each link of the data transferring network is given below.

Assuming every link has equal weightage throughout the network, $W(O_{iy}) = 1\ \forall\ i,y\ \varepsilon\ N,\ i,y \leq m,n$

As in the previous section, we have the total links available in the network as,

$$\{Q_{ij}\} = \{I_{ij},\ O_{ij}\} = \{\{I_{ij}\} \cup \{O_{ij}\}\}$$
$$= \{Q_{11}, Q_{14}, Q_{21}, Q_{22}, Q_{24}, Q_{33}, Q3_4\}\forall i, j\ N,\ i, j \leq 4 \quad (23)$$

The conditional probability matrix P is obtained as

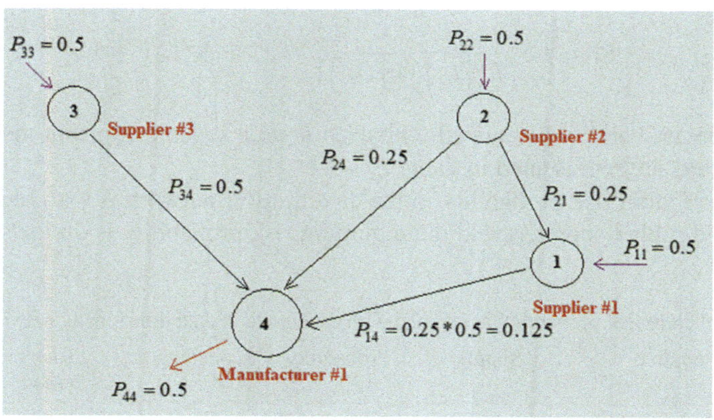

Fig. 10 Material transferring network together with the conditional probability allocated to each link. The plot further comprises the hypothetical link and corresponding probability at each node

$$P = \begin{pmatrix} 0.5 & 0 & 0 & 0.125 \\ 0.25 & 0.5 & 0 & 0.25 \\ 0 & 0 & 0.5 & 0.5 \\ 0 & 0 & 0 & 0.5 \end{pmatrix} \tag{24}$$

Also, the probabilistic adjacency matrix for the material transferring network is given in the problem description (refer Eq. 15). On account of the Eqs. 15 and 24, the collection of product elements $\{a_{ij} P_{ij}\}$ becomes

$$\{a_{ij}.P_{ij}\} = \begin{pmatrix} 0.5 & 0 & 0 & 0.1125 \\ 0.125 & 0.5 & 0 & 0.225 \\ 0 & 0 & 0.5 & 0.45 \\ 0 & 0 & 0 & 0.5 \end{pmatrix} \tag{25}$$

On assuming the data operation as a simple mean function (because the later analysis is regarding the delay time, simple mean function is found more applicable here), the probability of happening (ϕ) may be calculated using the Eq. 5 mentioned in Sect. 2.1.1.

$$\phi_i(\{a_{ij}.P_{ij}\}) = \frac{\sum_{j=1}^{4} a_{ij}.P_{ij}}{n(\{I_{ij}\} \cup \{O_{ij}\})} \forall i, y \in N, \quad i, y, \leq 4 \tag{26}$$

And the corresponding risk probability attached can be calculated using Eq. 6. Hence, Eq. 6 becomes,

$$R_i = 1 - \frac{\sum_{j=1}^{4} a_{ij}.P_{ij}}{n(\{I_{ij}\} \cup \{O_{ij}\})} \forall i, y \in N, \quad i, y \leq 4 \tag{27}$$

The risk probability associated with each node is calculated using the Eq. (27). The results are thus represented in Table 3.

From the table above, the maximum risk probability in the data transfer network is along the leather supplier and the minimum risk probability is with the fiber sole

Table 3 Inherent risk probability associated with each node for material transferring network

Node	Supplier	Supply	Probability of happening (ϕ_i)	Risk probability (R_i)
1	Supplier #1	Leather	0.29	0.71
2	Supplier #2	Rubber	0.30	0.70
3	Supplier #3	Fiber sole	0.50	0.50
4	Manufacturer #1	–	0.34	0.66

supplier. Since the fiber sole supplier has very direct network with the manufacturer, it is reasonable to have minimum risk associated to the same supplier as compared to other suppliers in the material transfer network.

Intra-Supplier competition in terms of deviation from their mean delay time

The mean delay time and corresponding standard deviation are given in the problem description mentioned in Sect. 4.1 (Table 1). The actual plot for the delay time is given in Fig. 11, assuming the delay time normally distributed.

The 14 suppliers can be classified and clustered into different groups based on their corresponding standard deviation of delay time for supplying. Each cluster size is assumed to be 0.2, and the group with less standard deviation is ranked at the top. Table 4, is the classification of the different suppliers based on their standard deviation.

The different ranked clusters can be separately represented using a bar chart where the total number of suppliers involved in the supply chain is nothing but the frequency of the corresponding cluster. According to the competition science, the frequency of the cluster has a significant role in the outcome probability of each rank and thereby the outcome probability of the supplier attached to each cluster. Figure 12 indicates the bar chart representation of the clustered suppliers based on standard deviation as mentioned above.

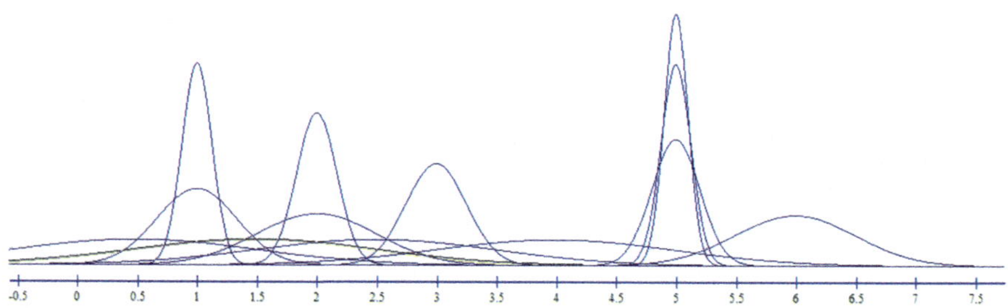

Fig. 11 The graphical representation of delay time for different suppliers with their corresponding standard deviation, assuming the delay time normally distributed

Table 4 Cluster classification of the 14 suppliers based on standard deviation

Cluster size (in terms of standard deviation)	Suppliers	Total number of suppliers
0.0–0.2 (rank #1)	Supplier #1, #3, #4, #5, #10, #11	6
0.2–0.4 (rank #2)	Supplier #6, #14	2
0.4–0.6 (rank #3)	Supplier #2, #12, #13	3
0.6–0.8 (rank #4)	Supplier #7, #9	2
0.8–1.0 (rank #5)	Supplier #8	1

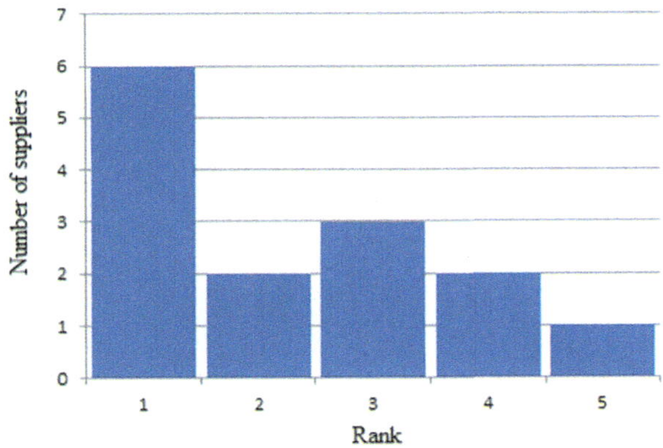

Fig. 12 Bar chart representation of all clustered ranks based on the standard deviation of each supplier

Since there are 14 suppliers available, the maximum number of failure gate is 7, refer [2]. The outcome probability of each rank can be measured using the general equation for (14). That is

$$P(n) = \frac{\lambda \sum_{i=1}^{n-1} A(n_i) + \left(\frac{14}{2} - \lambda\right) \sum_{i=n+1}^{5} A(n_i)}{{}^{14}C_2} \quad \forall \lambda \in R0 \leq \lambda \leq 14/2 \tag{28}$$

where, $A = \sum_{i=1}^{5} A(n_i) = 14$, the total number of suppliers available.

The outcome probability for rank n can be calculated using Eq. 28 and according to Theorem 2, the equilibrium rank and probability can be measured. Let's consider the number of failure gate λ as 0 to 7. Table 5 represents the probability distribution of each rank for different number of failure gates.

The graphical representation for the outcome probabilities is plotted in the graph given below. The number of failure gate is taken from 0 to 7 with an interval of 1, the

Table 5 Probability matrix for different ranked clusters based on number of failure gates

Number of failure gates (λ)	Ranks provided for the clusters				
	1	2	3	4	5
0	0.6153	0.4251	0.2307	0.0769	0.0000
1	0.5274	0.4351	0.2857	0.1868	0.1428
2	0.4395	0.4450	0.3406	0.2967	0.2857
3	0.3516	0.4552	0.3956	0.4065	0.4285
4	0.2637	0.4615	0.4505	0.5164	0.5714
5	0.1758	0.4651	0.5054	0.6263	0.7143
6	0.0879	0.4753	0.5604	0.7362	0.8571
7	0.0000	0.4851	0.6153	0.8461	1.0000

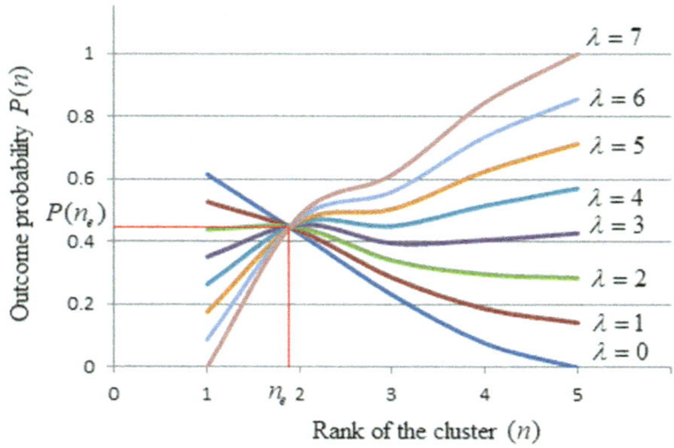

Fig. 13 Graphical representation of outcome probabilities of differently ranked supplier clusters for various number of failure gates, n_e represents the equilibrium cluster rank and $P(n_e)$ represents the corresponding outcome probability

equilibrium rank or position n_e and its corresponding probability $P(n_e)$ is observed and represented in the plot given as Fig. 13.

It is clear from the graph plotted above Fig. 13, the equilibrium rank n_e is in between 1 and 2 and very closer to 2nd position or rank. Therefore, the expected standard deviation of delay time the product gets finished will be near to 2nd rank or clearly near to the standard deviation of delay time ranging 0.2 to 0.4. The outcome probability of realization of this particular deviation from the delay time is observed to be in between 0.4 and 0.5, to be specific $P(n_e) = 0.4615$. The maximum probability it can reach is up to 0.5. Therefore 0.4615 is a descent probabilistic value one can rely on.

5 Results and Discussions

In the context of the outcome from the inherent risk analysis on the supply chain network mentioned in Sect. 4.2, the two independent networks are analyzed. The inherent risk associated with the data transferring network (in terms of percentage) is plotted on a bar chat shown in Fig. 14a. Similarly, network analysis is conducted for the material transferring network and the risk associated with each supplier and the manufacturer itself as represented in Fig. 14b. Since bar graph is a highly effective visual tool for use in presentations especially when comparing two different data sets, it is reasonable to represent the inherent risk probability with the help of a bar chart. These bar charts are highly useful, as they always allow recognizing patterns or trends far easier than grasping through some tabular data sets. Let us concentrate on Fig. 14a, b for further comparison in between the inherent risks in between suppliers and in between the networks itself.

In data transferring network, obviously the inherent risk is very less on the side of the manufacturer position. This shows, in communication perspective, that the current data transferring network is doing its best to reduce the inherent risk probability

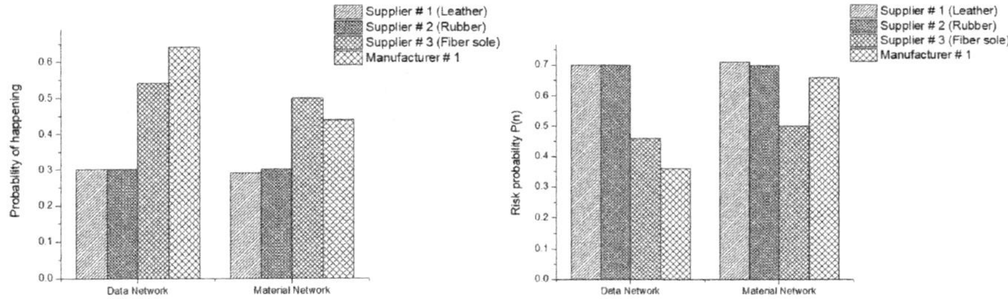

(a) (b)

Fig. 14 **a** Bar chart of the comparison of probability of happening for each major supplier on the data transferring network and material transferring network, **b** Bar chart of the comparison of risk probability for each major supplier on the data transferring network and material transferring network

associated with the manufacturer node to a minimum. The leather supplier and rubber supplier share a risk probability of 0.7 each. This clearly indicates further focusing and intervention on these two suppliers, in particular. The solutions like independent and common data cloud with high reliability together with the data transferring network can equip the system to reduce this risk to a greater extent, associated with the leather and rubber suppliers. Since the fiber sole and design providing supplier has lesser risk compared to the other two suppliers, it may be excluded from the data cloud, these decisions totally depend up on the management and other social indicators.

Coming to the material transferring network, almost all nodes doomed with high risk probability. The maximum risk is yet again shared between leather supplier and rubber supplier respectively with risk probability of 0.71 and 0.70 each. The manufacturer bares a risk probability of 0.66, this is almost to near to the risk associated with the two major suppliers leather and rubber and again quite more than that of the risk involved in the fiber sole supplier.

The impact of hypothetical links and their corresponding probabilities will be very much visible in the risk probability of each node and hence, the relative risk probability of each supplier in the manufacturer frame of reference helps to measure the actual impacts from manufacturer perspective. The risk probability involved in the manufacturer node is standardized into 1 and accordingly the variation in the risk probability associated with the other nodes is measured. This implies,

$$\left(R_i^1\right)_{Manufacturer} \Rightarrow \frac{(R_i)_{manufacturer}}{(R_i)_{manufacturer}} = 1 \tag{29}$$

$$\left(R_i^1\right)_{\text{supplier}} \Rightarrow \frac{(R_i)_{\text{supplier}}}{(R_i)_{manufacturer}} \tag{30}$$

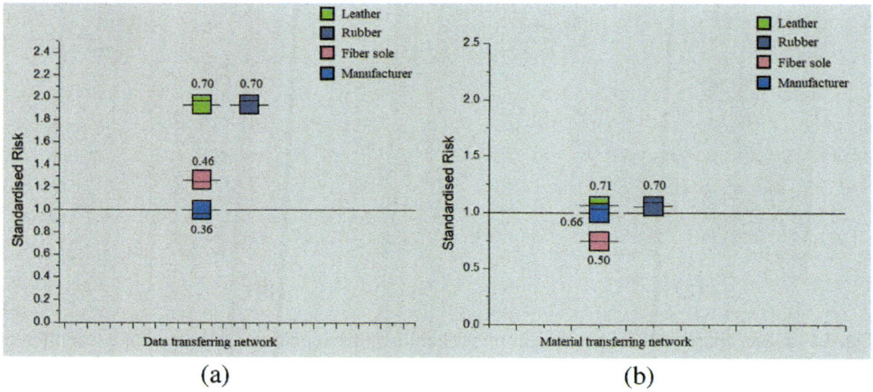

Fig. 15 **a** Graphical deployment of standardized risk number for the data transferring network from a manufacturer perspective. **b** Graphical deployment of standardized risk number for the material transferring network from a manufacturer perspective

where, $R^1{}_i$ represents the standardized risk number. The graphical representation of the standardized risk number $R^1{}_i$ is given in Fig. 15a, b shown below. The figure is drawn from a manufacturer perspective with a standardized risk, $(R^1{}_i)_{manufacturer} = 1$. Figure 15 is associated with the data transferring network whilst Fig. 15b about the corresponding material transferring network.

It is clear from Fig. 15a, b, the standardized risk number is very much closely distributed in the material transferring network compared to the data transferring network. This indicates the risk is more stable and uniformly distributed among the nodes, and can be assumed to be more stable from a manufacturer perspective. Though the inherent risk probability associated with the manufacturer is less in the data transferring network, the network is found to be less stable compared to the material transferring network. Hence, there should be some intervention and action required to minimize the risk involved in the case of leather and rubber suppliers concerned using techniques like reliable data cloud and so on.

The expected deviation in the delay time is observed to be between 0.2 and 0.4 with an outcome probability of 0.4615, the result obtained here is fairly decent. Though some of the suppliers involved can make more deviation from the delay time, the current system tries to manage itself to a lesser deviation in delay time with an excellent probability of occurrence. The maximum equilibrium probability $P(n_e)$ one system can reach is 0.5, here the outcome probability is very much near to 0.5. Hence the current distribution of standard deviation of delay time for major and other suppliers is affordable if the expected standard deviation in product completion is between 0.2 and 0.4.

In the current situation, already the inherent risk involved in leather and rubber suppliers of the material transferring network is fairly high and hence, it is better to reduce the standard deviation of the two suppliers. This may definitely contribute to reducing the overall stress relief in the supply chain network as a whole.

6 Conclusion

The article tried to focus on novel methodologies based on statistical tools that can be very well used for better decision making and strategy formulations. One of the major targets of the study is the introduction of statistical tools for network analysis and the imparting of competition science into the modern supply chain scenario. This includes the mathematical and statistical addressing of networks, how the models and theorems are defined, and implementation of the statistical tools in the current supply chain networks. To emphasize the arguments practically, the article likewise incorporated a case study related to orthopedic footware. The purpose of the case study was to apply the model developed in a real-time practical situation and was clearly found fulfilled in the fourth section of the study. The outcomes from the case study conducted were then analyzed in detail.

- The model developed here consists of two independent sections; one is entirely dedicated to the network analysis, and is used to measure the inherent risk probability involved in each node on a supply chain network. Since the flow direction of any entity in any network will have an influence of conditionality, the model very well incorporated the possible application of conditional probability in a general perspective and thereby used to calculate the inherent risk associated with each node of the supply chain network. As the selection of a particular node matters, the model made use of probabilistic adjacency matrix instead of a deterministic one. This made the model more generalized in character. Though the study introduced a particular method to allot conditional probability over different nodes in a supply chain network; there can be more feasible and powerful option to assign this conditional probability in the same network as, in general, the assignment of conditional probability is more subjective, experience-based, and application-oriented. The second section of the mathematical modelling focused on the clubbing of the competition science in the supply chain. The model discussed the concept of outcome probabilities, number of failure gates that can be used in the supply chain management and the DSS to improve the quality of decision making. The model can be easily coped with the software elements and IoT tool, and hence, leaves a huge scope for attaching the concept of competition science with IoT tools.
- The case study discussed in the article is very much specific from a manufacturer perspective. The mathematical model developed in the former section is applied in the case study to analyze the inherent risk probability involved in the case of each supplier and manufacturer of the supply chain network. Moreover, using the principle of competition science and the corresponding mathematical model developed, the deviation from the mean delay time of the product is calculated with the probability of occurrence of such deviation.
- The impact of the two decision support statistical models already discussed in the result section, highlighted the relevance of clubbing these models together during a decision making process, especially related to the risk involved with regard to the leather and rubber suppliers.

- The above methodology based on less sophisticated statistical tools will definitely find more application in different spheres of strategic formulations and also expected to nurture the already existing decision making tools with additional information and guidance to face the most complicated social dilemmas and associated supply chains of today.

References

1. Gilboa I, Postlewaite AW, Schmeidler D (2008) Probability and uncertainty in economic modeling. J Econ Perspect 22(3):173–188
2. Srinivas MN, Sreerag C, Murty AVSN (2019) Impact of dummy variables in a probabilistic competitive environment. SN Appl Sci 1(9):1115
3. Sreerag C, Srinvas MN, Murty AVSN (2020) Generalized probabilistic ranking competition model with quality inspection technique. Int J Qual Eng Technol
4. Özceylan E, Çetinkaya C, Demirel N, Sabırlıoğlu O (2018) Impacts of additive manufacturing on supply chain flow: a simulation approach in healthcare industry. Logistics 9:2(1), 1
5. Blass V, Corbett CJ (2018) Same supply chain, different models: Integrating perspectives from life cycle assessment and supply chain management. J Ind Ecol 22(1):18–30
6. Mattos CL, Barreto GA, Horstkemper D, Hellingrath B (2017) Metaheuristic optimization for automatic clustering of customer-oriented supply chain data. In: 2017 12th international workshop on self-organizing maps and learning vector quantization, clustering and data visualization (WSOM), pp 1–8. IEEE
7. Convertino M, Liang S (2014) Probabilistic supply chain risk model for food safety. Planet@ Risk 2(3)
8. Abolhasanpour M, Khosrojerdi G, Sadeghi A, Jafari ELAA, Arab S, Hojabri H, Kimiagari A (2009) M, Supply chain planning with scenario based probabilistic demand. Afr J Bus Manage 3(11):605–714
9. Sarkar B (2013) A production-inventory model with probabilistic deterioration in two-echelon supply chain management. Appl Math Model 37(5):3138–3151
10. Klimov RA, Merkuryev YA (2006) Simulation-based risk measurement in supply chains. In: 20th European conference on modelling and simulation (ECMS 2006), Bonn, Germany
11. Haghpanah Y (2011) A probabilistic trust and reputation model for supply chain management. In: Sixteenth AAAI/SIGART doctoral consortium
12. Blackhurst J, Cantor D, O'Donnell M (2012) Sustainable supply chains: a guide for small-to medium-sized manufacturers. CIRAS, Iowa State University
13. Saeed MA, Kersten W (2019) Drivers of sustainable supply chain management: identification and classification. Sustainability 11(4):1137
14. Oelze N (2017) Sustainable supply chain management implementation–enablers and barriers in the textile industry. Sustainability 9(8):1435
15. Peng Wong W, Yew Wong K (2008) A review on benchmarking of supply chain performance measures. Benchmarking Int J 15(1):25–51
16. Mishra P, Sharma RK (2014) Benchmarking SCM performance and empirical analysis: a case from paint industry. Logist Res 7(1):113
17. Naskar S, Basu P, Sen AK (2020) A literature review of the emerging field of IoT using RFID and its applications in supply chain management. In: Securing the internet of things: concepts, methodologies, tools, and applications, pp 1664–1689. IGI Global
18. Jan MA, Khan F, Alam M (Eds) (2019) Recent trends and advances in wireless and IoT-enabled Networks. Springer

19. Mandler B, Marquez-Barja J, Campista MEM, Cagáňová D, Chaouchi H, Zeadally S, Vieriu RL (Eds) (2016) Internet of things. IoT infrastructures: second international summit, IoT 360° 2015, Rome, Italy, October 27–29, 2015. Revised Selected Papers (Vol 169). Springer
20. Fang S, Da Xu L, Zhu Y, Ahati J, Pei H, Yan J, Liu Z (2014) An integrated system for regional environmental monitoring and management based on internet of things. IEEE Trans Indus Inf 10(2):1596–1605
21. Kankanhalli A, Charalabidis Y, Mellouli S (2019) IoT and AI for smart government: a research agenda
22. Holsapple CW, Whinston AB, Benamati JH, Kearns GS (1996) Instructor's manual with test bank to accompany decision support systems: a knowledge-based approach
23. Hackathorn RD, Keen PG (1981) Organizational strategies for personal computing in decision support systems. MIS Quart, 21–27
24. Teniwut W, Hasyim C (2020) Decision support system in supply chain: a systematic literature review. Uncertain Supply Chain Manag 8(1):131–148
25. Paul A, Shukla N, Paul SK, Trianni A (2021) Sustainable supply chain management and multi-criteria decision-making methods: a systematic review. Sustainability 13(13):7104

Chapter 26
Reverse Logistics: An Approach for Sustainable Development

Rashmi Ranjan Swain, Swagatika Mishra, and S. S. Mahapatra

Abstract Any firm's sustainability performance can be enhanced through effective implementation of supply-chain and reverse logistics strategies. Although the reverse logistics concepts have been recognized to be very important in developing countries, however, the companies' progress has been hampered owing to certain prevailing variables responsible for their sustainable developments. Thus, this study focused on identifying the associated variable(s) with reverse logistics practices in the bottle manufacturing companies in Odisha (India), and to rank those variables by using the technique for order of preference by similarity to ideal solution (TOPSIS) method. These research findings will provide a path to the professionals and other decision-makers in planning for suitable strategies in the reverse logistics-based performances in the manufacturing sectors.

Keywords Reverse logistics · Supply-chain management · Sustainable · Development · Bottle manufacturing · Companies · TOPSIS · Ranking · India

1 Introduction

In order to improve any firm's sustainability performance, the role of "Reverse Logistics" has been regarded to be significant in different aspects, such as for cost-savings [7], higher sales' revenue from products that are recovered and/or remanufactured [17], enhanced satisfactions of customers [14] with their loyalty [2], and by considerable reduction in carbon footprints, providing positive impacts on climate-change

R. R. Swain · S. Mishra (✉)
VSSUT, Burla, Sambalpur, Odisha, India
e-mail: swagatika.mechanical@gmail.com

S. S. Mahapatra
NIT Rourkela, Rourkela, India

S. Kautish et al. (eds.), *Computational Intelligence for Modern Business Systems*, Disruptive Technologies and Digital Transformations for Society 5.0,
https://doi.org/10.1007/978-981-99-5354-7_26

as well as global-warming [3], respectively. Moreover, higher competitive advantages are achievable through well-organized and sustainable reverse logistics practices [9]. Thus, for sustainable development, reverse logistics always have positive implications [5, 6, 21].

Although the reverse logistics concepts have been recognized to be very important in developing countries, however, the companies' progress has been hampered owing to certain prevailing variables responsible for their sustainable developments. Thus, this study focused on identifying the associated variables with reverse logistics practices in the bottle manufacturing companies in Odisha (India), and to rank those variables by using suitable "multi-criteria decision-making (MCDM)" method.

1.1 Literature

A wide variety of tools and techniques have been utilized in order to evaluate sustainability in different cases. For instance, on account of weighted sum of product's sustainability aspects, Jaffar et al. [8] represented a model by considering economical, environmental as well as social aspects for product's sustainability assessments. Both environment as well as ecological factors were considered for a "desalination plant" to evaluate sustainability [1]. An "environmental impact assessment" was performed by Vinodh et al. [20] for an automotive sector by means of "eco-indicator." Ticehurst et al. [19] have utilized "Bayesian-network approach" for the evaluation of coastal-lakes' sustainability in New South Wales, Australia. Similarly, different researchers have made use of fuzzy-logic-based techniques for evaluating sustainability in a range of areas, such as in "petroleum corporation" [22], nation's sustainability judgments [11], in chemical-industry [4], in mining as well as mineral segments [10], and in the pandemic situations with the use of Grey-TOPSIS [15], respectively.

2 Methodology

In this work, five bottle manufacturing companies from Odisha (India) were considered to analyze their reverse logistics practices for long-term sustainable development. Different associated variables with reverse logistics practices were identified using the literature and subsequent consultation with industry-based experts for further analysis. Furthermore, when a number of variables/criteria are involved in any problem, then the application of 'multi-criteria decision-making (MCDM)' methods becomes more useful. For instance, the 'Grey-theory' superiority has been reported in situations dealing with uncertain information that help in the study of judgments by human with uncertainties as well as ambiguities [12, 13]. In different areas the "technique for order of preference by similarity to ideal solution (TOPSIS)" method has been successfully utilized [16]. Thus, the ranking of the associated variables with reverse logistics practices in the bottle manufacturing companies/units was made based on the TOPSIS method.

2.1 *TOPSIS Method*

The linguistic variables, i.e., criteria weights as well as the criteria ratings (⊗ GN) in terms of grey numbers by using "1–7 scale" were illustrated in Table 1.

The sequence of steps in the "Grey-TOPSIS" method included the following:

Step 1: With the formation of a decision-makers' committee, the criteria weights were identified for alternatives (variables). By assuming "P" number of persons in the decision group, the criteria weight can be determined as:

$$\otimes w_j = \frac{1}{P}\left(\otimes w_j^1 + \otimes w_j^2 + \cdots + \otimes w_j^P\right) \tag{1}$$

where, $\otimes w_j^P$ (j $= 1, 2, \ldots,$ n) denoted the criteria weight of Pth decision-makers'.

Step 2: By using the linguistic-variables, the rating values were obtained as:

$$\otimes GN_{ij} = \frac{1}{K}\left(\otimes GN_{ij}^1 + \otimes GN_{ij}^2 + \cdots + \otimes GN_{ij}^P\right) \tag{2}$$

where, $\otimes GN_{ij}^P$ (i = 1, 2, ..., m; j = 1, 2, ..., n) denoted 'criteria rating-value of Pth decision-makers'.

Step 3: The grey decision-matrix with $\otimes GN_{ij}$ as linguistic-variables on the basis of grey numbers was established as:

$$GDM = \begin{bmatrix} \otimes GN_{11} & \otimes GN_{12} & \ldots & \otimes GN_{1n} \\ \otimes GN_{21} & \otimes GN_{22} & \cdots & \otimes GN_{2n} \\ \vdots & \vdots & \ddots & \vdots \\ \otimes GN_{m1} & \otimes GN_{m2} & \cdots & \otimes GN_{mn} \end{bmatrix} \tag{3}$$

Table 1 Scale of criteria weights ($\otimes w$) and criteria ratings (⊗GN)

Scale	$\otimes w$	⊗GN
Very-low (VL)	[0.0, 0.1]	[0, 1]
Low (VL)	[0.1, 0.3]	[1, 3]
Medium–low (ML)	[0.3, 0.4]	[3, 4]
Medium (M)	[0.4, 0.5]	[4, 5]
Medium–high (MH)	[0.5, 0.6]	[5, 6]
High (H)	[0.6, 0.9]	[6, 9]
Very-high (VH)	[0.9, 1.0]	[9, 10]

Step 4: Normalizing the grey decision-matrix as:

$$GDM^* = \begin{bmatrix} \otimes GN_{11}^* & \otimes GN_{12}^* & \ldots & \otimes GN_{1n}^* \\ \otimes GN_{21}^* & \otimes GN_{21}^* & \cdots & \otimes GN_{2n}^* \\ \vdots & \vdots & \ddots & \vdots \\ \otimes GN_{m1}^* & \otimes GN_{m2}^* & \cdots & \otimes GN_{mn}^* \end{bmatrix} \tag{4}$$

where, for a benefit-criterion, $\otimes GN_{ij}^*$ was expressed as:

$$GN_{ij}^* = \frac{\underline{GN}_{ij}}{GN_j^{\max}}, \frac{\overline{GN}_{ij}}{GN_j^{\max}}; \ GN_j^{\max} = \max_{1 \le i \le m}\left(\overline{GN}_{ij}\right) \tag{5}$$

where, for a cost-criterion, $\otimes GN_{ij}^*$ was expressed as:

$$GN_{ij}^* = \frac{GN_j^{\min}}{\overline{GN}_{ij}}, \frac{GN_j^{\min}}{\underline{GN}_{ij}}; \ GN_j^{\min} = \min_{1 \le i \le m}\left(\underline{GN}_{ij}\right) \tag{6}$$

Step 5: By considering each criterion with different importance, the "weighted normalized grey decision-matrix" was formulated as:

$$GDM^* = \begin{bmatrix} \otimes VN_{11} & \otimes VN_{12} & \ldots & \otimes VN_{1n} \\ \otimes VN_{21} & \otimes VN_{22} & \cdots & \otimes VN_{2n} \\ \vdots & \vdots & \ddots & \vdots \\ \otimes VN_{m1} & \otimes VN_{m2} & \cdots & \otimes VN_{mn} \end{bmatrix} \tag{7}$$

Where, $\otimes VN_{ij} = \left[\otimes GN_{ij}^* \times \otimes w_j\right]$

Step 6: For "m" possible alternative-sets, i.e. S $= \{S_1, S_2, \ldots, S_m\}$, the ideal referential-alternative was obtained as

$$S_{\max} = \left[\otimes GN_1^{\max}, \otimes GN_2^{\max}, \ldots., \otimes GN_n^{\max}\right] \tag{8}$$

Step 7: The 'grey possibility-degree' was obtained between the compared alternative sets S $= [S_1, S_2, \ldots, S_m]$, and ideal referential-alternative $S^{\max}$ as:

$$P\left[S_i \le S^{\max}\right] = \frac{1}{n}\sum_{j=1}^{n} P\left[\otimes VN_{ij} \le \otimes GN_j^{\max}\right] \tag{9}$$

Step 8: The final ranking of the alternative sets was done by considering the following conditions: (a) For smaller P $[S_i \le S^{\max}]$, the better was the ranking order of S_i. (b) If not, the worse will be the ranking order.

2.2 Selection of Variables for Performance-Based Sustainable Manufacturing

Table 2 illustrates the performance-based sustainable manufacturing for the bottle manufacturing units with effective reverse logistics practices, where the sustainability assessment indices were as suggested by Singh et al. [18], literature, and the suggestions of experts involved in this work.

Table 2 Performance-based sustainable manufacturing

Sustainability-aspects	Associated variable(s) with reverse logistics	Detail(s)
Economic performances (ECP)	Quality (AVRL-1)	Enhancement in reliability of delivery, reductions in scrap-levels as well as rework-levels
	Sensitivity (AVRL-2)	Reductions in "order lead-time," "manufacturing lead-time" and "product-development time"
Environmental performances (EVP)	Stuff handling (AVRL-3)	Reductions in stuff intensity and usages; Enhancement in re-used/re-cycled/re-manufactured materials' usages
	Water and energy utilizations (AVRL-4)	• Reductions in water consumptions; Enhancement in re-cycled water usages • Reductions in total energy consumption Enhancements in usages of renewable energy with energy-saving
	Wastages' and emissions' reductions (AVRL-5)	• Reductions in total wastes' generation, landfills, and unsafe materials utilization • Reductions in CO_2 and BFCs emissions
Social performances (SOP)	Member of staffs, consumers and communities welfare (AVRL-6)	• Number of training hours provided, decrease in ratio of turnovers, augmented job satisfaction, improved working conditions, and active participation of members of staffs in decision-making activities • Enhanced consumers' satisfaction, transparency in products and services information, assessment-level of products' health and safety, and availability of warranty/take-back options • Community-based projects' numbers, reduction in number of non-compliances, availability of policy for child-labour, work-force composition, average provided salary and active participations of the community in decision-making activities

3 Results and Discussion

The associated variable(s) with reverse logistics were assigned with different weight values (W) and the corresponding integrated matrix was obtained (Table 3). This was further followed by formation of the "normalized-matrix" and "weighted normalized-matrix," respectively (Tables 4 and 5).

Table 3 Formation of "integrated-matrix"

Associated variable(s) with reverse logistics	W_{EVP}		W_{SOP}		W_{ECP}	
	0.67	0.95	0.75	0.95	0.65	0.75
AVRL-1	4.60	6.60	4.40	6.00	4.40	6.00
AVRL-2	4.00	5.20	4.40	6.00	4.60	6.60
AVRL-3	3.80	5.00	2.60	3.60	4.20	5.80
AVRL-4	3.60	4.40	1.60	2.80	4.40	6.00
AVRL-5	3.60	4.40	2.00	3.00	3.60	4.40
AVRL-6	4.20	5.40	1.60	2.80	3.40	4.20
MAX	6.60		6.00		6.60	
MIN	3.60		1.60		3.40	

Table 4 Formation of "normalized-matrix"

Associated variable(s) with reverse logistics	W_{EVP}		W_{SOP}		W_{ECP}	
	0.67	0.95	0.75	0.95	0.65	0.75
AVRL-1	0.54	0.78	0.26	0.36	0.56	0.77
AVRL-2	0.69	0.90	0.26	0.36	0.51	0.73
AVRL-3	0.72	0.94	0.44	0.61	0.58	0.81
AVRL-4	0.81	1	0.57	1	0.56	0.77
AVRL-5	0.81	1	0.53	0.80	0.77	0.94
AVRL-6	0.66	0.85	0.57	1	0.81	1

Table 5 Formation of "weighted normalized-matrix"

Associated variable(s) with reverse logistics	EVP		SOP		ECP	
AVRL-1	0.36	0.74	0.20	0.34	0.36	0.58
AVRL-2	0.46	0.85	0.20	0.34	0.33	0.55
AVRL-3	0.48	0.90	0.33	0.58	0.38	0.60
AVRL-4	0.54	0.95	0.42	0.95	0.36	0.58
AVRL-5	0.54	0.95	0.40	0.76	0.50	0.70
AVRL-6	0.44	0.81	0.42	0.95	0.52	0.75
S^+	0.54	0.95	0.42	0.95	0.52	0.75

On the basis of the values of "grey possibility-degree" between the associated variable(s) with reverse logistics, their ranking was done that were based on the minimum values of the "Grey possibility-degree" at first place and so on (Table 6).

It was found from Table 6 that, the associated variable(s) with reverse logistics that required of major attentions included "Member of staffs' Consumers' and Communities welfare" that ranked in the first-level followed by "Wastages' and Emissions' reductions; Water and Energy utilizations; Stuff handling; Sensitivity; and Quality," respectively (Fig. 1).

Table 6 Determination of $P\left[S_i \leq S^{max}\right]$ and the final ranking of associated variable(s) with reverse logistics

Associated variable(s) with reverse logistics	EVP	SOP	ECP	SUM	Value of 'grey possibility-degree'	Ranking
AVRL-1	0.75	1	0.87	2.62	0.65	6th
AVRL-2	0.61	1	0.93	2.55	0.63	5th
AVRL-3	0.57	0.79	0.82	2.18	0.54	4th
AVRL-4	0.50	0.50	0.87	1.87	0.46	3rd
AVRL-5	0.50	0.62	0.57	1.70	0.42	2nd
AVRL-6	0.65	0.50	0.50	1.65	0.41	1st

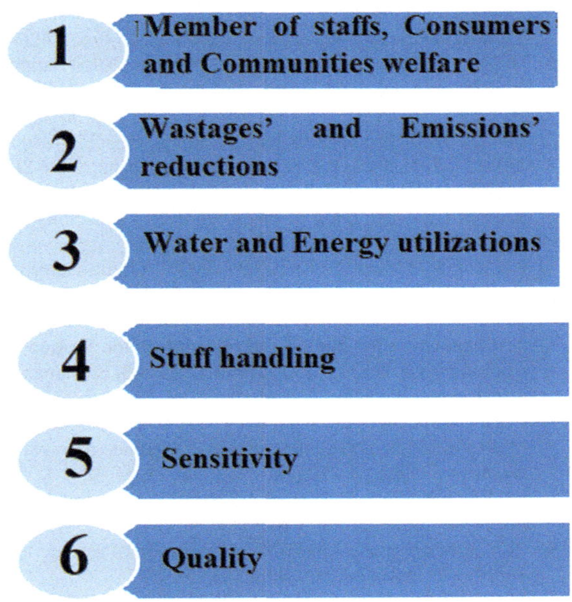

Fig. 1 Rank-order of the associated variable(s) with reverse logistics requiring major attention

4 Conclusion

As the supply-chain management of any company gets more influenced by the reverse logistics practices, which has attracted many professionals and manufacturers to connect all the firm-related activities together. In this study, the associated variable(s) with reverse logistics that required of major attention for sustainable developments of the bottle manufacturing units included "Member of staffs' Consumers' and Communities welfare" which ranked in the first-level followed by "Wastages' and Emissions' reductions; Water and Energy utilizations; Stuff handling; Sensitivity; and Quality," respectively.

References

1. Afghan NH, Darwish M, Carvalho M (1999) Sustainability assessment of desalination plants for water production. Desalination:124
2. Aitken J, Harrison A (2013) Supply governance structures for reverse logistics systems. Int J Oper Prod Manag 33(6):745–764
3. Carter CR, Rogers DS (2008) A framework of sustainable supply chain management: moving toward new theory. Int J Phys Distrib Logist Manag 38(5):360–387
4. Conner J, Phillis Y, Manousiouthakis V (2009) A fuzzy logic global optimization approach to sustainability assessment. In: AIChE annual meeting, environmental division (28d), Nashville
5. Govindan K, Soleimani H (2017) A review of reverse logistics and closed-loop supply chains: a journal of cleaner production focus. J Clean Prod 142(1):371–384
6. Huang YC, Rahman S, Wu YCJ, Huang CJ (2015) Salient task environment, reverse logistics and performance. Int J Phys Distrib Logist Manag 45(9/10):979–1006
7. Jack EP, Powers TL, Skinner L (2010) Reverse logistics capabilities: antecedents and cost savings. Int J Phys Distrib Logist Manag 40(3):228–246
8. Jaffar H, Venkatachalam A, Joshi K, Ungureanu A, De Silva N, Dillon O Jr, Rouch K, Jawahir I (2007) Product design for sustainability: a new assessment methodology and case studies. In: Kutz M (ed) Handbook of environmentally conscious mechanical design. Wiley, New York, pp 25–65
9. Kannan G, Pokharel S, Kumar PS (2009) A hybrid approach using ISM and fuzzy TOPSIS for the selection of reverse logistics provider. Resour Conserv Recycl 54(1):28–36
10. Kommadath B, Sarkar R, Rath B (2012) A fuzzy logic based approach to assess sustainable development of the mining and minerals sector. Sustain Dev 20:386–399
11. Kouloumpis V, Kouikoglou V, Phillis Y (2008) Sustainability assessment of nations and related decision making using fuzzy logic. IEEE Syst J 2:224–236
12. Li GD, Yamaguchi D, Nagai M (2006) A grey-based approach to suppliers selection problem. In: Proceedings of the international conference on parallel, distributed processing techniques and applications, pp 180–186
13. Li GD, Yamaguchi D, Nagai M (2007) A grey-based decision-making approach to the supplier selection problem. Math Comput Model 46:573–581
14. Li X, Olorunniwo F (2008) An exploration of reverse logistics practices in three companies. Supply Chain Manage Int J 13(5):381–386
15. Mishra D, Satapathy S, Jain VK (2022) A Grey-TOPSIS approach to minimize covid-19 transmission for the betterment of public-health in the indian context. Int J Serv Sci Manage Eng Technol 13(1):1–14. https://doi.org/10.4018/IJSSMET.29032
16. Mishra S, Datta S, Mahapatra SS (2013) Grey-based and fuzzy TOPSIS decision-making approach for agility evaluation of mass customization systems. Benchmark Int J 20(4):440–462

17. Mollenkopf DA, Closs DJ (2005) The hidden value in reverse logistics. Supply Chain Manage Rev 9(5):34–43
18. Singh S, Olugu EU, Musa SN, Mahat AB (2014) Proposition of key performance measures for sustainable manufacturing in SMEs. In: MSME Conclave cum conference on sustainable supply chain capabilities of micro, small and medium enterprises: influences, practices, training needs and employment opportunities. Doon University, Dehradun, pp 1–8
19. Ticehurst JL, Newham LTH, Rissik D, Letcher RA, Jakeman AJ (2007) A Bayesian network approach for assessing the sustainability of coastal lakes in New South Wales, Australia. Environ Modell Softw 22:1129–1139
20. Vinodh S, Jayakrishna K, Joy D (2012) Environmental impact assessment of an automotive component using eco-indicator and CML methodologies. Clean Technol Environ 14:333–344
21. Wanjiku E, Mwangangi P (2019) Influence of procurement best practices on the performance of food and beverage manufacturing firms in Kenya. Int J Supply Chain Logist 3(1):26–49
22. Zhang LF (2007) On the assessment of petroleum corporation's sustainability based on linguistic fuzzy method. Lect Notes Comput Sc 4487:562–566

Chapter 27
Applications of Artificial Intelligence in Public Procurement—Case Study of Nigeria

David Edijala, Sandip Rakshit, and Narasimha Rao Vajjhala

Abstract Governments worldwide are exploring the applications of artificial intelligence (AI) technologies to improve public services. Organizations may utilize AI, machine learning, deep learning, and big data together to acquire relevant business insights, increase efficiencies, make better decisions, and advance their objectives. Governments worldwide are rapidly developing and deploying artificial intelligence derived from machine learning to improve operations, public services, compliance, and security activities. Government procurement in Nigeria has had several issues, including corruption. Corruption from both government officials and contractors has been an issue of significant concern. This chapter looks at how several government agencies in developed countries around the world are using artificial intelligence (AI) in their contracting procedures and some of the capabilities that are being considered for future use. This study looks at ways in which the Nigerian government could implement AI in its contracting to maximize efficiency and reduce the rate of corruption. AI has the potential to profoundly change how government agencies handle contracting, from vendor selection through contract closeout and everything in between. However, AI is not error-proof, and issues regarding technology implementation are also reviewed in this chapter.

Keywords Artificial intelligence · Big data · Procurement · Contracting · Machine learning · Public · Nigeria

---

D. Edijala (✉) · S. Rakshit
American University of Nigeria, Yola, Nigeria
e-mail: david.edijala@aun.edu.ng

S. Rakshit
e-mail: sandip.rakshit@aun.edu.ng

N. R. Vajjhala
University of New York Tirana, Tirana, Albania
e-mail: narasimharao@unyt.edu.al

S. Kautish et al. (eds.), *Computational Intelligence for Modern Business Systems*, Disruptive Technologies and Digital Transformations for Society 5.0,
https://doi.org/10.1007/978-981-99-5354-7_27

1 Introduction

The application of AI for public administration services comes under the scope of e-governance under which information and communication technologies are used to improve government services [1]. Digital technologies have been enhancing government processes for decades, and AI is another layer that can completely revolutionize specific government sectors like contracting. Artificial intelligence is being used to provide more sophisticated ways to make decisions in difficult situations, and automated administrative decision-making processes are being broadened. The data generated and processed is also key for governments for future referencing and planning. The science and engineering of creating intelligent devices, particularly intelligent computer programs, is called Artificial Intelligence (AI). It is analogous to the task of utilizing computers to study human intellect, but AI does not have to be limited to physiologically observable ways [2, 3]. Artificial intelligence is a computational method that allows machines to analyze and interpret large amounts of data to answer questions, solve problems, and handle concerns. This is accomplished by an AI-enabled computer doing cognitive processes like humans, but considerably faster and fewer errors [4].

Every year, federal governments all over the world award thousands of contracts individually [5]. Many people depend on government contracts for their livelihoods. The best way to get in on some contracts is to know someone, which sometimes is problematic. The largest consumers on the planet are the governments; thus, they need an efficient procurement system. The federal government usually has the highest purchasing power in many countries, and corporations cannot singlehandedly match them [1]. Contracting can be a terrific way for many to get their feet in the door for their businesses, but it can also be highly time-consuming, especially in Nigeria. Unlike dealing with a conventional firm, interacting with the government necessitates saintly patience; one must patiently go through the government procurement process, which is a long and complicated process [5]. Dealing with the federal government, on the other hand, can be a tremendous and steady source of money for both new and established enterprises, provided they are ready to master the system and be patient and assertive [5]. The use of AI in the Nigerian government contracting process can be key in eradicating corruption in the sector and ensuring quality projects are being done. There is a great demand to understand the drivers, barriers, opportunities, and risks to the adoption of AI in public procurement services [1]. This study will look at ways in which this can be achieved and the limitations that may arise in the implementation of AI in Nigerian government procurement. Finally, we will look at ways to solve the stated issues.

2 Review of Literature

2.1 *Artificial Intelligence*

Generally, computing was initially designed to foster automation in various aspects of life to help increase efficiency. However, although conventional computing can make storage, retrieval, and sorting of large data simple, it still has to be coded with strict instructions and processes to follow. Artificial Intelligence was introduced to interact more flexibly with the human world without the need for hard coding by humans or other software. Although AI is becoming increasingly popular, many still are not familiar with it. A recent study in 2017 shows that out of 1,500 top business leaders in the US, only 17% acknowledged familiarity [6].

Although many are still unfamiliar with AI, it is affecting our day-to-day lives significantly right now. AI is almost everywhere, from the recommendations system on an e-commerce store to mobile personal assistants [4, 7]. It is essential humans deploy even more efficient systems that can reduce workloads and improve the standard of living. Even entire cities and states have begun the implementation of "smart city," which leverages the power of AI to enhance the delivery of urban services [8–10].

Although AI has many advantages, it also comes with new challenges and issues for societies and individuals to overcome. A significant point is that AI needs data to be efficient and grow with time and trends, but access to the needed data calls for policy changes and causes privacy concerns. AI also tends to (in some situations) enable biased or discriminatory practices based on its interpretation of the data it has been fed [11]. The emergence of artificial intelligence needs to be handled properly to reduce the chances of a dystopia forming worldwide. Improved data access, increased government investment in AI, promoting AI workforce development, creating a federal advisory committee, engaging with state and local officials to ensure they enact effective policies are ways governments can promote AI. Other ways of encouraging AI tech are; regulating broad objectives rather than specific algorithms, taking bias seriously as an AI issue, maintaining mechanisms for human control and oversight, penalizing malicious behavior, and promoting cybersecurity [12, 13].

2.2 *AI in Public Governance*

In a classical sense, governance could be regarded as the activity or string of making and enforcing rules and delivering services [14]. In the context of upholding a country's constitutional ideals in the face of changing issues and surroundings, public governance refers to the formal and informal procedures that influence how public choices are made and how public actions are carried out [15]. Individuals, people, organizations, and systems of organizations from the public, private, and nonprofit sectors are the key participants in public governance. To attain the goal of public governance, these actors engage in collective decision-making that is constrained, mandated, and enabled by law, rules, and practice [16]. The e-procurement systems can be divided into two categories, namely, e-sourcing and e-supply chain management systems [17]. The e-sourcing system includes the e-tendering, e-catalog, and e-marketplace systems. The e-supply chain management system includes the transport management system, warehouse management system, and the e-supply chain execution system [17].

With rapid digital technological change, governments worldwide are forced to reinvent their traditional methods to improve citizen engagement, accountability, and interoperability. This can be accomplished by coping with a complex changing environment and becoming resilient through intelligent technologies that facilitate innovation, sustainability, and competitiveness. AI is a powerful tool that can undoubtedly improve public service. AI can assist in freeing up government labor by automating repetitive operations, resulting in faster transactions in the delivery of government services and, more correctly, analyzing the outcomes of policy alternatives [1]. AI has enormous potential in a variety of government sectors, including telecommunications, education, healthcare, physical infrastructure, management, and data security, finance, transportation, research and development, policymaking, and the legal and justice system, among others; and authorities must recognize and implement it to improve citizen quality of life and governance efficiency [18].

2.3 AI in Government Contracting

AI poses to be a handy tool in government contracting as governments need an efficient and seamless way of managing civil servants and contractors to ensure fairness, equality, and quality. Governments can focus on reducing wasteful government spending while giving the project to the most qualified bidder. AI could assist the government in quickly identifying ineligible contractors and debarring or suspending them from engaging further with the bidding process until they are fully compliant [19, 20].

Countries like Nigeria have been trying hard to curb corruption, improve transparency, and make more efficient procurement processes. For example, Nigeria's Public Procurement Act of 2007 (The 2007 Act) is based on the 1994 UNCITRAL Model Law on Procurement and is fashioned after the Bank and other International Bank systems. The Nigerian government has taken further initiatives to combat corruption in addition to the Procurement Act [21, 22]. The Nigerian government, like that of the United States, has adopted debarment legislation. Any agent who is convicted or found guilty of any of the offenses stated in the provision is subject to a fine. The Bureau of Public Procurement (BPP) can assess when sanctionable acts occur and resolve procurement disputes under the 2007 Act. The BPP can also order contractors or agents to disclose "books, records, accounts, or documents" related to the procurement contract.

Even with all the legislation, there is still a significant issue with procurement, and this is because of the inadequate use of technology. Without e-procurement, the government cannot properly list debarred contractors and make them publicly known to other institutions. Additionally, government officials' activities cannot be properly tracked, thus giving way to the misappropriation of funds and resources. With the emergence of artificial intelligence, banks can detect criminal activity or changes in the behavior of their customers and flag it immediately. Similarly, the Nigerian government can leverage the power of AI to profile its civil servants and contractors to detect patterns and change patterns to understand when there might be fraudulent activity or when the commissioned contract was not successfully completed. AI can also be used in audits that are useful for oversight and risk identification. They can be a cost-effective way for remediating control deficiencies if conducted early in a contract or project.

3 Applications of AI in Government Contracting

One primary use of AI in the government procurement process is to increase efficiency through automation. Throughout the lifecycle of a government contract, automation can help speed up manual operations. In developed countries, certain government agencies use AI in their procurements process and only have praises to sing about the new order. For example, the US Department of Health and Human Services (HHS) employs artificial intelligence to streamline contract vehicles [21]. Some routine contract administration processes, such as market research, contract changes, invoicing, and others, could be phased out entirely by automation. AI has the potential to assist agencies in completing procurements more swiftly. In essence, thousands of pages of legislation relating to a particular contract could be fed to the system, and it will analyze it in a few seconds and return recommendations [5].

AI is also used to identify, assess, and allocate risk in the procurement process. Risk allocation in the economic connection between government agencies and contractors is usually managed by choosing a specific contract type, mandating bonds and insurance, and, in some cases, by government-funded contracts. With the introduction of AI, government agencies can estimate demand accurately and easily uncover possible cost reductions, and identify which suppliers are the riskiest. AI can also determine whether a particular supplier can deliver within the stipulated time due to past performance [23].

AI can also boost competition between suppliers because it relegates the need to know someone before getting a contract, thus, limiting human interference and, in extension, limiting corruption. Small and growing firms and first-time contractors may find it much easier to break into the federal contracting market thanks to artificial intelligence. The AI system links the most suitable organization with the current contract on the ground based on the organization's ability to meet the procurement requirements [5, 23]. This means that most suppliers need to be on their top game as they can no longer bribe or ask a family member to get the contracts. Another major use of AI in the government procurement field is monitoring the performance of contractors. Since the system is unbiased, the vendor must submit evidence of meeting objectives and milestones. If the submissions do not meet the requirements, the system will not mark the contract as completed. The contractors will also try their best to do a good job because the system will profile their work and not recommend them for future procurement if they have a poor record.

4 Benefits of Leveraging AI in Public Procurement

E-procurement is the electronic tendering and procurement of goods and services. The first and foremost reason the Nigerian government needs to implement AI in its contracting process is to eradicate corruption. It is widely known that government agencies use their positions to increase nepotism and negatively influence outcomes, which constitutes an abuse of power. At the moment, to get a contract, a vendor would need to know someone within the government agency of a high rank and then pay bribes to that person and other individuals [17]. This is problematic because it means the vendor might not even be qualified to meet the contract requirements, which means re-work, which means the government spends more and the people have to endure further. AI will substitute the officers by analyzing the profiles of all vendors and matching the most suited ones to the contracts thus, eradicating the need to know anyone or make illegal payments.

Another reason would be to minimize the human bias and error from the contracting processes. The truth of the matter is that Nigerian government officials tend to be biased in certain situations based on culture, religion, gender, etc. To minimize the possibility of human bias, AI can be implemented in the procurement process in terms of vendor selection and contract signoff. This would help reduce inequality as well as ensure only quality contracts are carried out within the country.

An enforced and maintained standard for the procurement process and performance quality needs to exist, and AI is a useful tool to help achieve this. The Nigerian government can leverage the power of AI commission projects that have met a certain standard. This will ensure all contracts awarded will be at a certain uniform standard and will only be marked as complete when all the requirements are completed. Essentially, Nigerian citizens will no longer need to endure bad roads due to poor construction because most standard projects would exist [21, 22]. Implementing AI in the country's government procurement processes would mean a lot of data would be generated frequently and adequately stored for easy retrieval and analysis in the future. Things like contractor information and officials relating to certain contracts could easily be retrieved by name, date, type, location, etc. Also, the government could analyze the data, which could help in decision-making and planning for the future.

Cost should always be minimized, and unnecessary costs should be mitigated. With AI, fewer people, resources, and equipment are needed in the procurement process. Many physical meeting points would no longer be necessary as most of the

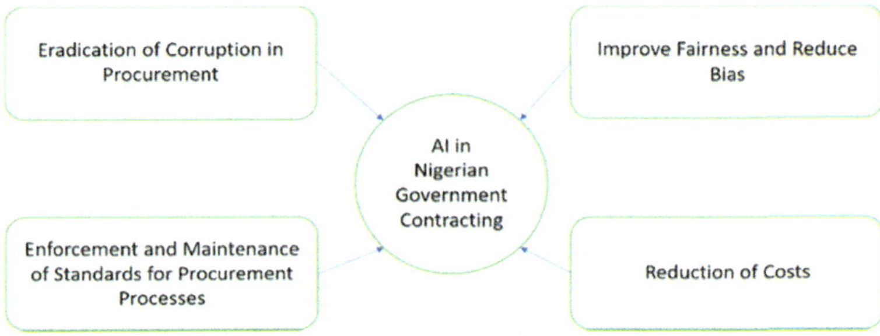

Fig. 1 Benefits of leveraging AI in Nigerian government contracting

activities would be carried out by the system [17]. Additionally, the cost of re-work and re-awarding of contracts will be drastically reduced as contracts done would be more standard and more adequate planning would have been done. Also, the estimation of procurement costs is done more accurately by AI thus, ensuring the government saves more and plans accordingly. The benefits of AI are illustrated in Fig. 1.

5 Limitations and Recommendations for Implementing AI in Nigerian Public Procurement

One of the significant limitations to implementing AI in the Nigerian government's procurement process is the lack of properly organized and kept data from past dealings. Without valuable data, the system cannot be trained to fit the needs of the government. Also, there might be an ineffective use of the data generated by the system because it could come in massive amounts so frequently that the current professionals on the ground might not be able to handle it completely.

Also, in the world, there is a minimal number of data scientists. This means that the Nigerian government may find it difficult or costly to employ people who would manage the system in the short run because local and foreign organizations present very enticing packages that the government might find difficult to compete with to the available AI talent. It may lead to the government converting existing programmers, systems analysts, statisticians, etc., to temporary data scientists, leading to an inadequate system being deployed and used. Adoption of new technologies is a challenge for all businesses. However, due to their established norms and processes, the public sector is less adaptable than its private sector counterparts. Employees are encouraged to innovate in sections of the private sector where a strong culture of experimentation exists and good performance is recognized. Employees in the government may be less encouraged to take risks. This means that it might be challenging to imbibe a new system where some staff might be reassigned, and many processes changed, especially in the current corrupt state of Nigeria's civil service [24–26].

The Nigerian Federal and State Governments need to construct effective data centers all around the country to ensure data is well acquired and managed. To do this, the government needs to shift mainly from the paper methods of generating data to better online ways which could be stored in databases in the preferred formats of the data scientists [21]. The data mined could be further analyzed by data scientists to create a useful model for the AI machine to be better suited to the needs of the Nigerian government. The government could also invest massively in the education of its citizens—especially the younger generations to reduce the level of illiteracy and increase the chances of having meaningful labor in the AI field [22]. The government could even introduce AI and data science courses in the elementary stages of young adults' lives. Without even getting a degree in tech, the citizens are still properly poised to go into the tech and AI-driven public sector.

6 Conclusion

All levels of the Nigerian government should begin an enlightenment program for civil servants and government contractors. This would help mitigate the legacy culture of the public sector and its unwillingness to adopt new processes and technologies. It would also help new contractors understand the new process to spur competition. The use of artificial intelligence in government contracting can revolutionize the Nigerian government's procurement system completely. AI solves many issues and reduces human interference in the procurement process, thereby eradicating corruption. AI is currently being used in developed countries for their various government procurement processes, and it has worked out fine, Nigeria should follow suit in implementing it. There may be some limitations to implementing the AI system in the Nigerian government's AI process. Still, with determination, those limitations can be addressed, and the system could help spur efficiency in the public sector overall.

References

1. Noordt CV, Misuraca G (2020) Evaluating the impact of artificial intelligence technologies in public services: towards an assessment framework. In: Proceedings of the 13th International conference on theory and practice of electronic governance. Association for Computing Machinery, Athens, Greece, pp 8–16
2. Ibrahim DR, Ghnemat R, Hudaib A (2017) Software defect prediction using feature selection and random forest algorithm. In: 2017 International conference on new trends in computing sciences (ICTCS)
3. Min H (2010) Artificial intelligence in supply chain management: theory and applications. Int J Log Res Appl 13(1):13–39

4. Wooldridge M (2020) Artificial intelligence requires more than deep learning—but what, exactly? Artif Intell 289:103386
5. Hebous S, Zimmermann T (2021) Can government demand stimulate private investment? Evidence from U.S. federal procurement. J Monetary Econ 118:178–194
6. Feng M et al (2019) Big data analytics and mining for effective visualization and trends forecasting of crime data. IEEE Access 7:106111–106123
7. Vajjhala NR et al (2021) Novel user preference recommender system based on twitter profile analysis. In: Soft computing techniques and applications. Springer Singapore, Singapore
8. Amgoune H, Mazri T (2018) 5G: interconnection of services and security approaches. In: Proceedings of the 3rd International conference on smart city applications. Association for Computing Machinery, Tetouan, Morocco, p 18
9. Manal R, Fatima R, Tomader M (2019) Authentication for e-health applications in IoT enabled 5G and proposed solution. In: Proceedings of the 4th International conference on smart city applications. Association for Computing Machinery, Casablanca, Morocco, p 75
10. Usman M et al (2018) Integrating smart city applications in 5G networks. In: Proceedings of the 2nd International conference on future networks and distributed systems. Association for Computing Machinery, Amman, Jordan, p 2
11. Le DC, Zincir-Heywood N, Heywood MI (2020) Analyzing data granularity levels for insider threat detection using machine learning. IEEE Trans Netw Serv Manage 17(1):30–44
12. Al-Turkistani HF, Ali H (2021) Enhancing users' wireless network cyber security and privacy concerns during Covid-19. In: 1st International conference on artificial intelligence and data analytics (CAIDA)
13. Thuraisingham B (2020) Cyber security and artificial intelligence for cloud-based internet of transportation systems. In: 2020 7th IEEE International conference on cyber security and cloud computing (CSCloud)/2020 6th IEEE International conference on edge computing and scalable cloud (EdgeCom)
14. Al-Ruithe M, Benkhelifa E (2017) Cloud data governance maturity model. In: Proceedings of the second international conference on internet of things, data and cloud computing. Association for Computing Machinery, Cambridge, United Kingdom, p. 151
15. Awotwi J, Amega-Selorm C (2015) A case Study of an African e-government/e-governance development. In: Proceedings of the 2015 2nd International conference on electronic governance and open society: challenges in Eurasia. Association for Computing Machinery, St. Petersburg, Russian Federation, pp 49–58
16. Gupta R, Muttoo SK, Pal SK (2017) The need of a development assessment index for e-governance in India. In: Proceedings of the 10th International conference on theory and practice of electronic governance. Association for Computing Machinery, New Delhi AA, India, pp 414–422
17. Nicoletti B (2016) Cloud computing and procurement. In: Proceedings of the International conference on internet of things and cloud computing. Association for Computing Machinery, Cambridge, United Kingdom, p 56
18. Fothergill BT et al (2019) Responsible data governance of neuroscience big data. Front Neuroinf 13(28)
19. Seetharaman P (2017) Technology governance challenges in e-government projects: health information systems in India. In: Proceedings of the 10th International conference on theory and practice of electronic governance. Association for Computing Machinery, New Delhi AA, India, pp 504–507
20. Shrivastava S, Pal SN (2017) A big data analytics framework for enterprise service ecosystems in an e-governance scenario. In: Proceedings of the 10th International conference on theory and practice of electronic governance. Association for Computing Machinery, New Delhi AA, India, pp 5–11
21. McAfee RP, McMillan J (1989) Government procurement and international trade. J Int Econ 26(3):291–308
22. Strang KD (2014) Exploring marketing theories to model business web service procurement behavior, In: Goodyear J, Sun Z (eds) Demand-driven web services: theory, technologies and

applications. IGI-Global, Hershey, PA, pp 21–42. https://doi.org/10.4018/978-1-4666-5884-4.ch002
23. Agarwal A, Jayant A (2019) Machine learning and natural language processing in supply chain management: a comprehensive review and future research directions. Int J Bus Insights Transform 13(1):3–19
24. Aigbovo O (2019) Trend and pattern of economic and financial crimes statutes in Nigeria. J Financ Crime 26(4):969–977
25. Bitrus SN, Strang KD, Vajjhala NR (2019) Exploring socio-cultural factors impacting agriculture information system acceptance in rural Nigeria after terrorism. In: Proceedings of the 24th UK Academy for information systems international conference. University of Oxford, London, Academy for Information Systems (AIS)
26. Ogbodo UK, Mieseigha EG (2013) The economic implications of money laundering in Nigeria. Int J Acad Res Account Finance Manage Sci 3(4):170–184

SPRINGER NATURE

GPSR Compliance

The European Union's (EU) General Product Safety Regulation (GPSR) is a set of rules that requires consumer products to be safe and our obligations to ensure this.

If you have any concerns about our products, you can contact us on ProductSafety@springernature.com

In case Publisher is established outside the EU, the EU authorized representative is:

Springer Nature Customer Service Center GmbH
Europaplatz 3
69115 Heidelberg, Germany

The manufacturer's authorised representative in the EU is Springer Nature Customer Service Centre GmbH, Europaplatz 3, 69115 Heidelberg, Germany. If you have any concerns regarding our products, please contact ProductSafety@springernature.com

Printed and bound by CPI Group (UK) Ltd, Croydon, CR0 4YY
01/10/2026
02225867-0016